Bird Families of the World

A series of authoritative, illustrated handbooks of which this is the first volume to be published

Series editors
C. M. PERRINS Chief editor
W. J. BOCK
J. KIKKAWA

THE AUTHOR Alan Kemp is Head Curator, Department of Birds, at the Transvaal Museum in Pretoria. He was born and raised in Zimbabwe and developed early interests in natural history and falconry. He attended Rhodes University, and was awarded a doctorate in 1973 for his research on hornbills. The field work for this thesis was done while he was a research assistant at Cornell University. He then moved to the Transvaal Museum, where he extended his study of hornbills and raptors further north in Africa and into Southeast Asia. He is the author of several other books, a past editor of the journal of the South African Ornithological Society, *The Ostrich*, and a Corresponding Fellow of the American Ornithologists' Union.

THE ARTIST Martin Woodcock has been deeply interested in birds since early childhood and became a full-time ornithological writer and artist in 1974. He has travelled extensively in Europe, Africa, and Asia, and has concentrated on illustrating comprehensive guides covering much of this area, particularly *The Birds of South-East Asia*, *The Birds of Oman*, and the ongoing *Birds of Africa*. He exhibits work regularly in London and elsewhere, and is represented in many private collections. He was chosen 'Bird Illustrator of the Year' by the journal *British Birds* in 1983 and was elected a member of the Society of Wildlife Artists in 1986.

Bird Families of the World

1. The Hornbills
   ALAN KEMP

2. The Penguins
   TONY D. WILLIAMS

3. The Megapodes
   DARRYL N. JONES, RENÉ W. R. J. DEKKER, and CEES S. ROSELAAR

Bird Families of the World

The Hornbills
Bucerotiformes

ALAN KEMP

Illustrated by
MARTIN WOODCOCK

Oxford   New York   Tokyo
OXFORD UNIVERSITY PRESS
1995

Oxford University Press, Walton Street, Oxford OX2 6DP

*Oxford New York
Athens Auckland Bangkok Bombay
Calcutta Cape Town Dar es Salaam Delhi
Florence Hong Kong Istanbul Karachi
Kuala Lumpur Madras Madrid Melbourne
Mexico City Nairobi Paris Singapore
Taipei Tokyo Toronto
and associated companies in
Berlin Ibadan*

Oxford is a trade mark of Oxford University Press

*Published in the United States
by Oxford University Press Inc., New York*

© Text: Alan Kemp, 1995; © Plates 1–2 by Alan Kemp and the individuals listed on p. x;
© Plates 3–14, maps, and drawings Oxford University Press, 1995

All rights reserved. No part of this publication may be
reproduced, stored in a retrieval system, or transmitted, in any
form or by any means, without the prior permission in writing of Oxford
University Press. Within the UK, exceptions are allowed in respect of any
fair dealing for the purpose of research or private study, or criticism or
review, as permitted under the Copyright, Designs and Patents Act, 1988, or
in the case of reprographic reproduction in accordance with the terms of
licences issued by the Copyright Licensing Agency. Enquiries concerning
reproduction outside those terms and in other countries should be sent to
the Rights Department, Oxford University Press, at the address above.

This book is sold subject to the condition that it shall not,
by way of trade or otherwise, be lent, re-sold, hired out, or otherwise
circulated without the publisher's prior consent in any form of binding
or cover other than that in which it is published and without a similar
condition including this condition being imposed
on the subsequent purchaser.

A catalogue record for this book is available from the British Library

Library of Congress Cataloging–in–Publication Data
Kemp, A. C. (Alan C.)
The hornbills: Bucerotiformes / Alan Kemp; illustrated by Martin Woodcock.
(Bird families of the world)
Includes bibliographical references and index.
1. Hornbills.   I. Woodcock, Martin. II. Title. III. Series.
QL696.C729K448   1995   598.8′92—dc20   94-25805
ISBN 0 19 857729 X

Typeset by EXPO Holdings, Malaysia

Printed in Hong Kong

Contents

QL
696
.C729
K448
1995

Acknowledgements		ix
List of colour plates		xi
List of abbreviations		xii
Plan of the book and notes on reading the species accounts		xiii
Topographical diagram of a hornbill		xvi

PART I *General chapters*

1	The world of hornbills	3
2	The design of hornbills	10
3	Non-breeding behaviour and biology	24
4	Feeding ecology	36
5	Breeding biology	46
6	Relationships and evolution of hornbills	61
7	Conservation	75
	Appendix: guidelines for captive management	82

PART II *Species accounts*

Bucerotiformes		89
Bucorvidae		90
Genus *Bucorvus*		91
Northern Ground Hornbill	*Bucorvus abyssinicus*	91
Southern Ground Hornbill	*Bucorvus leadbeateri*	94
Bucerotidae		100
Genus *Anorrhinus*		100
Austen's Brown Hornbill	*Anorrhinus austeni*	101
Tickell's Brown Hornbill	*Anorrhinus tickelli*	104
Bushy-crested Hornbill	*Anorrhinus galeritus*	106
Genus *Tockus*		109
Subgenus *Rhynchaceros*		110
African Crowned Hornbill	*Tockus alboterminatus*	110
Bradfield's Hornbill	*Tockus bradfieldi*	114

African Pied Hornbill	*Tockus fasciatus*	116
Hemprich's Hornbill	*Tockus hemprichii*	119
Subgenus *Lophoceros*		121
Pale-billed Hornbill	*Tockus pallidirostris*	121
African Grey Hornbill	*Tockus nasutus*	123
Subgenus *Tockus*		127
Monteiro's Hornbill	*Tockus monteiri*	128
African Red-billed Hornbill	*Tockus erythrorhynchus*	131
Southern Yellow-billed Hornbill	*Tockus leucomelas*	136
Eastern Yellow-billed Hornbill	*Tockus flavirostris*	140
Von der Decken's Hornbill	*Tockus deckeni*	142
Subgenus *incertae sedis* uncertain		145
Dwarf Black Hornbill	*Tockus hartlaubi*	145
Dwarf Red-billed Hornbill	*Tockus camurus*	147
Long-tailed Hornbill	*Tockus albocristatus*	149
Genus *Ocyceros*		152
Malabar Grey Hornbill	*Ocyceros griseus*	153
Sri Lankan Grey Hornbill	*Ocyceros gingalensis*	155
Indian Grey Hornbill	*Ocyceros birostris*	157
Genus *Anthracoceros*		159
Indian Pied Hornbill	*Anthracoceros coronatus*	161
Oriental Pied Hornbill	*Anthracoceros albirostris*	164
Palawan Hornbill	*Anthracoceros marchei*	170
Malay Black Hornbill	*Anthracoceros malayanus*	172
Sulu Hornbill	*Anthracoceros montani*	175
Genus *Buceros*		177
Great Pied Hornbill	*Buceros bicornis*	178
Great Rhinoceros Hornbill	*Buceros rhinoceros*	184
Great Philippine Hornbill	*Buceros hydrocorax*	189
Great Helmeted Hornbill	*Buceros vigil*	192
Genus *Penelopides*		197
Sulawesi Tarictic Hornbill	*Penelopides exarhatus*	198
Visayan Tarictic Hornbill	*Penelopides panini*	200
Luzon Tarictic Hornbill	*Penelopides manillae*	202
Mindanao Tarictic Hornbill	*Penelopides affinis*	204
Mindoro Tarictic Hornbill	*Penelopides mindorensis*	207
Genus *Aceros*		208
Subgenus *Berenicornis*		209
White-crowned Hornbill	*Aceros comatus*	210

Subgenus *Aceros*		212
Rufous-necked Hornbill	*Aceros nipalensis*	213
Sulawesi Wrinkled Hornbill	*Aceros cassidix*	215
Sunda Wrinkled Hornbill	*Aceros corrugatus*	218
Mindanao Wrinkled Hornbill	*Aceros leucocephalus*	221
Visayan Wrinkled Hornbill	*Aceros waldeni*	223
Subgenus *Rhyticeros*		224
Papuan Wreathed Hornbill	*Aceros plicatus*	224
Narcondam Wreathed Hornbill	*Aceros narcondami*	227
Plain-pouched Wreathed Hornbill	*Aceros subruficollis*	230
Bar-pouched Wreathed Hornbill	*Aceros undulatus*	232
Sumba Wreathed Hornbill	*Aceros everetti*	239
Genus *Ceratogymna*		241
Subgenus *Bycanistes*		242
Piping Hornbill	*Ceratogymna fistulator*	242
Trumpeter Hornbill	*Ceratogymna bucinator*	245
Subgenus *Baryrhynchodes*		248
Brown-cheeked Hornbill	*Ceratogymna cylindricus*	249
Grey-cheeked Hornbill	*Ceratogymna subcylindricus*	252
Silvery-cheeked Hornbill	*Ceratogymna brevis*	256
Subgenus *Ceratogymna*		260
Black-casqued Wattled Hornbill	*Ceratogymna atrata*	260
Yellow-casqued Wattled Hornbill	*Ceratogymna elata*	263
Glossary		266
References		270
Index		291

Acknowledgements

Many people have assisted me over the past 25 years with my addiction to hornbills and I am deeply grateful to them all. During this time I have been privileged to meet 31 species in the wild and a further eight in captivity, and to visit eight African and five Asian countries. Foremost in her support has been Meg, latterly with the help of Lucy and Justin. Others who greatly influenced my life are, in the following sequence, my parents Michael, Valeska and step-mother Jean, Bob Brain, Gordon Maclean, Brian Allanson, Tom Cade, Johan Kloppers, Petiros Mundhlovu, Banggan Empulu, S. A. Hussain, Elizabeth Vrba, Hugh Paterson, Charles Sibley, and Pilai Poonswad.

Assistance in ways great or small, but all essential and much appreciated, came from: Alan Abrey, the late Salim Ali, Hans Bartels, Professor S. Basalingappa, Phyll Beaumont, Keith Begg, Linda Birch, David Bishop, Stephen Botha, Laura Brain, Koen Brouwer, the late Leslie Brown, Richard Brooke, Chris Brown, Murray Bruce, Phillip Burton, Tamar Cassidy, Hugh Chittenden, Phillip Clancey, the late Charles Clinning, Peter Colston, Tim Crowe, James Culverwell, Leobert de Boer, Jared Diamond, Moussa Diop, Velvet Douglas, Françoise Dowsett-Lemaire, Morné du Plessis, Heinz Dullart, Bob Elbel, Celia Falzone, Alex Forbes-Watson, Joe Forshaw, Laura Freysen, Cliff and Dawn Frith, Nick and Jane Gerard-Pearce, Llewellyn Grimes, James Gulledge, Clem Haagner, Tony Harris, the late George Henry, Derek Holmes, David Houston, Michael Irwin, Des Jackson, Mary Jensen, Rolf Jensen, Andrew Johns, David Johnson, Salomon Joubert, Jan Kalina, Rangupathy Kannan, Tim Kemp, Robert Kennedy, Ben King, Lim Chong Keat, Margaret Kinnaird, Joris Komen, Ron Krupa, Lawrence Kuah, Peter Lack, Mark Leighton, Dede Leighton, Michel and Kris Louette, Charles Luthin, Eugene Marais, Joe Marshall, Elliot McClure, John Mendelsohn, Bernd-Ulrich and Christina Meyburg, Modse Sanjeevareddy, Gérard Morel, Chris and Mary Perrins, Major W. W. A. Phillips, Attila and Karen Port, Elizabeth and Michel Prager, Gray Ranger, Anne Rasa, Paul Reddish, Rudolf Reinhard, Ben Riekert, Michael Riffel, Art Risser, Frank Rozendaal, John Rushworth, Kurt Sanft, Günter Schleussner, Karl Schuchmann, Ken Scriven, James Shade, Peter Shannon, Christine Sheppard, Roy Siegfried, Siti Hawa Yatim, John and Mary-Dean Snelling, Walter and Sally Spofford, Mark Stanback, Peter Steyn, Ernst Sutter, Bernard Treca, Atsuo Tsuji, Dr C. A. Van Ee, Renate van den Elsen, Anthony van Zyl, Carl Vernon, Oscar Wambuguh, Timothy Watt, David Wells, Fred White, Mark Witmer, Martin and Barbara Woodcock, and Wendy Worth.

Logistical and financial support has come from the Transvaal Museum, South African Council for Scientific and Industrial Research, Foundation for Research Development, United States Public Health Service (Environmental Health ES 00008 and ES 00261), Cornell University, American Museum of Natural History, Frank Chapman Memorial Fund, Forestry Department of the Sarawak

Government (Malaysia), World Wildlife Fund (Malaysia), British Museum (Natural History) at Tring, Delaware Museum of Natural History, Sarawak Museum, Basel Museum, Museum Alexander Koenig, Humbolt Museum, Edward Grey Institute of Field Ornithology, Bombay Natural History Society, Alexander Library, Percy FitzPatrick Institute of African Ornithology, South African National Parks Board, Vögelpark Walsrode, Jurong Birdpark, Zimbabwe Department of Fisheries and Wildlife Management, Peregrine Fund, Finch-Davies Research Fund, Library of Natural Sounds (Cornell University), the IUCN Captive Breeding Specialist Group, and The Asian Hornbill Network.

In the production of the book, I received special support from librarians Ronel Goode and Linda Birch, from artist Martin Woodcock, from series editor Christopher Perrins, and from the staff of Oxford University Press.

Any omissions from the above lists are entirely my own fault and intend no lack of appreciation.

Photographs

I am grateful to the following people for kindly providing photographs which appear in Plates 1 and 2:

Philip Burton (Plate 2, part (f))
Tony Harris (Plate 2, part (f))
Atsuo Tsuji (Plate 2, parts (b) and (c))
Adisak Vidhidharm (Plate 1, part (b))
T. S. U. de Zylva (Plate 2, part (a))

Colour plates

Colour plates fall between pages 96 and 97.

Plate 1   Extremes of hornbill habitats and morphology

Plate 2   Nesting habits, social organization, and human use of various hornbill species

Plate 3   Indomalayan crested hornbills, genus *Anorrhinus*, and African ground hornbills, genus *Bucorvus*

Plate 4   Small African arboreal hornbills, genus *Tockus* and subgenera *Rhynchaceros* and *Lophoceros*

Plate 5   Small African terrestrial hornbills, genus *Tockus* and subgenus *Tockus*

Plate 6   Indian grey hornbills, genus *Ocyceros*, and small African forest hornbills, genus *Tockus* but subgenus uncertain

Plate 7   Indomalayan pied hornbills, genus *Anthracoceros*

Plate 8   Indomalayan great hornbills, genus *Buceros*

Plate 9   Indonesian and Philippine tarictic hornbills, genus *Penelopides*

Plate 10   Indomalayan white-crowned and wrinkled hornbills, genus *Aceros* and subgenera *Berenicornis* and *Aceros*

Plate 11   Indomalayan wreathed hornbills, genus *Aceros* and subgenus *Rhyticeros*

Plate 12   African white-rumped and wattled hornbills, genus *Ceratogymna* and subgenera *Bycanistes*, *Baryrhynchodes*, and *Ceratogymna*

Plate 13   Appearance in flight of African hornbills, genera *Bucorvus*, *Tockus*, and *Ceratogymna*

Plate 14   Appearance in flight of Indomalayan hornbills, genera *Anorrhinus*, *Ocyceros*, *Anthracoceros*, *Buceros*, *Penelopides*, and *Aceros*

Abbreviations

♂/♂♂	male/males	ml	millilitre(s)
♀/♀♀	female/females	mm	millimetre(s)
ad(s)	adult(s)	Mt/Mts	Mount/Mountain(s)
allosp./spp.	allospecies (singular/plural)	n	number in sample
		$2n$	diploid number
c/number	clutch size	P.	primary
DBH	tree diameter at breast height	pers. comm.	unpublished information received verbally
DNH	tree diameter at nest height	pers. obs.	personal observation(s)
		R.	River
g	gram(s)	s	second(s)
ha	hectare(s)	St.	Strait
hr	hour(s) (time period)	sp./spp.	species (singular/plural)
hrs	hours (time of day)	subad(s)	subadult(s)
imm(s)	immature(s)	subsp./spp.	subspecies (singular/plural)
in litt.	unpublished information received in writing	supersp./spp.	superspecies (singular/plural)
I./Is	Island(s)		
juv(s)	juvenile(s)	T.	tail feather / tail-feather pair
km	kilometre(s)		
m	metre(s)	wt	weight
min	minute(s)		

Plan of the book and notes on reading the species accounts

This book attempts to summarize the present state of our knowledge about hornbills and to provide a springboard for further research and conservation. The seven general chapters in Part I introduce and describe the entire group (an order containing two families), with comprehensive chapters on their general biology, feeding ecology, breeding biology, evolutionary relationships, and conservation.

This first half of the book is written in colloquial style, with relatively few scientific terms or references to other sources of literature. It offers an overview of hornbill biology and ecology, extracts general characteristics common to many species, and highlights features of particular interest. Each chapter includes a framework of the basic scientific theory applicable to that subject, such as foraging strategies or evolutionary relationships, within which to discuss the biology of hornbills. In each chapter, scientific names are given after the first mention of a species, but not thereafter.

The text is interspersed throughout with information on what humans are doing to hornbills or have discovered about them. The aim of this book is to improve human–hornbill relationships and to introduce hornbills to those who do not yet wonder at and admire their diversity. The relationship between humans and hornbills may serve as a barometer of the health of both parties, and it will certainly decide the fate of these fascinating birds.

Part II comprises detailed accounts for each of the 54 hornbill species, presented in a more abbreviated style and complete with references to important sources of literature. These individual accounts attempt to summarize all that is known, and to highlight what is unknown, about each species of hornbill. The species are divided into groups according to the families, genera, and, in some cases, subgenera to which they have been assigned and which indicate their relationships. Diagnostic features and traits common to members of each family, genus, and subgenus are presented. These summarize the principal features of each taxon and so reduce the amount of repetition necessary for features common to each species within any such group.

At the start of each account, the individual species is identified by its English name, its scientific name and the numbers of the plates and figures in which it is illustrated. This is followed by the original scientific names assigned to the species, the generic name indicating the species group in which it was placed plus the unique specific name, and these are qualified by details of the author who described the first specimen to science, the date on which this description was published, the literature reference to the publication and the locality from which the original type specimen was collected. Other English names are then listed when relevant.

All species and subspecies are illustrated in colour by paintings of an adult of each sex and an immature, drawn to scale on each plate. Related species are grouped together on Plates 3–12 which show most of the relevant details of form and coloration for each species. Plates 13 and 14 show an adult male of each species in flight, but not to scale, with Afrotropical and Indomalayan species respectively grouped together on separate plates to assist comparisons for field identification. Plates 1 and 2 illustrate aspects of hornbill habitat, biology, and behaviour.

In the species accounts, the information is arranged under separate headings. It is based on personal research, a thorough survey of the scientific literature located up to the end of 1993 and unpublished information from colleagues past and present. It is as complete as possible and particular sources of data are identified wherever practicable, with full references listed alphabetically in the bibliography. The information is arranged in the following sections:

Description This presents details of colour and external form of the plumage, bill, legs, and soft parts, first for adult males, then for adult females, and finally for immatures. Where subspecies have been recognized within the species, the initial description is of the nominate form. This is followed by abbreviated descriptions of other subspecies, with special attention to differences which distinguish them from one another and sometimes with mensural differences between different geographical populations. The authorship and date for the original description of each subspecies is indicated, and any other taxonomic discussion is also included. The section ends with any information on ontogeny of plumage, anatomy, and coloration, or details of moult. Readers may wish to refer to the labelled anatomical diagram on p. xvi.

Measurements and weights Data are presented on standard measurements of adult wing length, tail length, bill length (chord taken from its junction with the skull), tarsus length (all in millimetres), and body weight (in grams). Samples are separated for each sex and presented in the form: (n = sample size) range (mean). Data are presented separately for each subspecies where appropriate. The majority of data come directly from the extensive monograph of Sanft (1960), the sources of any additional measurements being acknowledged where not my own.

Field characters Basic indications of size (estimated length when perched), colour, form, and behaviour, by which the species can be recognized in the field, are presented in relation to other hornbill species that may occur in the same area. Attention is drawn to distinctive calls where these are especially useful in identification, but details of vocal communication are provided in the following section.

Voice All vocal and non-vocal noises reported for each species are described in detail, together with an indication of the context in which they are produced. Whether any of the noises have been tape-recorded is mentioned, and details of where they are available, plus a selection of sonograms, can be obtained from the FitzPatrick Bird Communication Library (Transvaal Museum, Box 413, Pretoria, 0001 South Africa).

Range, habitat, and status The range of each species is described country by country, divided into subspecies where appropriate, and illustrated on a map. The basis for the distributions is the detailed records of specimen localities plotted by Sanft (1960) and, for African species, the atlas of Snow (1978). This has been augmented with additional localities gleaned from the literature or other sources. The basic habitat requirements of each species are then described, and finally any observations on the density and status of the species in different areas are presented. Any human uses of and associations with the species are also mentioned.

The variety of terms used to describe different types of forest is confusing, but the terminology used by Collins *et al.* (1991) is followed here. Forests are divided into moist and dry forests, most of which are situated in the tropics. Moist evergreen forests, or rainforests, enjoy precipitation throughout the year, while deciduous or monsoon forests experience only seasonal rainfall. These moist forests include mangrove and freshwater swamp (or peat) forests, and can be separated altitudinally into lowland and montane forests. Dry forests, forming a closed canopy over less than 40% of their area, include deciduous woodland and thorn forests. They grade into various densities of wooded savanna and finally into grassland, shrubby steppe or desert.

Forests can also be classified according to whether they are primary and pristine, or whether they have been cut to some degree to produce selectively logged, secondary, or otherwise degraded structure. For many hornbill species, especially those of tropical rainforest, this account represents their historical range, which has been much reduced by habitat destruction. Exact information on the current range is not available for most species. Comparison of comments under habitat requirements and status, together with maps of the remaining forest types in Africa (Sayer *et al.* 1992) and Southeast Asia (Collins *et al.* 1991), do, however, allow estimation of the present situation. The historical range remains useful for studies of biogeography and systematics, but it should not be confused with the present range when planning conservation and management of the species.

A historical range map, using the best available information, is given for each genus as it is introduced. The species accounts also contain historical range maps for the individual species, although this information is usually approximate, as explained above.

Feeding and general habits Social organization, habitat use, foraging behaviour, and diet are described in as much detail as possible. Information on seasonal and diurnal activity patterns, roosting behaviour, maintenance activities, and parasites is also included.

Displays and breeding behaviour Social organization during breeding, any ritualized displays described, and any form of reproductive behaviour prior to egg-laying are recorded.

Breeding and life cycle An initial summary is presented of the duration of phases of the nesting cycle, estimated laying dates for each country, and clutch and egg sizes. A more detailed account follows of the nest site and nest cavity, preparation of the nest, egg-laying, and the incubation, nestling and post-fledging periods. It includes the roles of each sex or of helpers at the nest and details of chick development. The account ends with a summary of any information related to the population dynamics of the species. Where only few data on breeding are available, this is indicated before such information is summarized in one or two paragraphs.

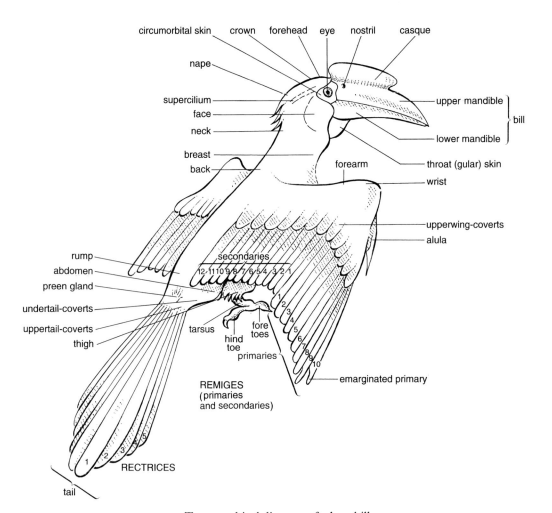

Topographical diagram of a hornbill.

PART I

General chapters

1

The world of hornbills

Imagine being on the tropical island of Borneo and drifting quietly down a stream in a dug-out canoe. The giant trees of the rainforest rise on each bank like cathedral spires, and the creepers which festoon them form cloisters that conceal the dark damp interior. Raindrops pattering on the foliage and the distant rumble of a retreating thunderstorm form a backdrop of sound, through which penetrates a single mournful hoot. More hoots follow at intervals, accelerating in tempo until they break suddenly into peals of maniacal laughter. Two huge birds then burst across the dome of the sky, their naked red heads extended and metre-long tail feathers trailing behind. Cackling loudly, they ram into one another like mountain sheep. Their yellow casques, solid as ivory, clash with a loud crack before they return from their jousting to perch at pavilions in the forest canopy on each side of the stream. Male Great Helmeted Hornbills *Buceros vigil* are busy in defence of their territorial boundaries.

By contrast, the thundershower which fell during the night on the mountains bordering the Namib Desert may be the total precipitation for the year. The freshly laundered dawn reveals a flash flood in the previously dry riverbed, but no obvious effect, as yet, on the leafless thorn bushes scattered over the rocky hillsides. A deep clucking call emanates from the valley and provides rhythm to the chorus of singing birds, all excited by the promise of plenty which water brings to their arid land. The clucking call comes from a pair of hornbills, bobbing up and down with each note, bowing their heads and fanning their white outer tail feathers. Their territorial display over, they fly to the riverbed, gather mud in their long red bills, carry it to a cliff, and apply it to the rim of a cavity among the rocks. A pair of Monteiro's Hornbills *Tockus monteiri* is wasting no time in preparing to breed.

These two examples indicate the extremes of habitat occupied by hornbills, from moist evergreen forests that measure their rainfall in metres (Plate 1a) to arid steppes where every millimetre of rain is precious (Plate 1c). They also illustrate that hornbills are conspicuous birds throughout their Old World distribution, be it in the savannas and forests of Africa or in the forests of Southeast Asia. Even if they are not always easily visible within some of these habitats, each species has a loud and distinctive call that is among the most notable sounds of the region. Most are easily detected by their pied, boldly marked plumage and direct, often noisy, flight. At close quarters, their large bill is a striking feature, as is the prominent casque that arises from the top of the bill on many species (Fig. 1.1). These structures, together with areas of bare facial skin, are also often brightly coloured in reds, yellows, blues, and greens.

Such conspicuousness, together with special elements of hornbill biology such as the female sealing herself into the nest cavity, have long attracted the attention of humans sharing their environment. It is not surprising, therefore, that hornbills feature prominently among the cultures, emblems, lores, and ceremonies of many human societies (Figs 1.2, 1.3; Plate 2f: Camman 1951; Harrison 1951; Smythies 1960; Chin 1971).

4 The Hornbills

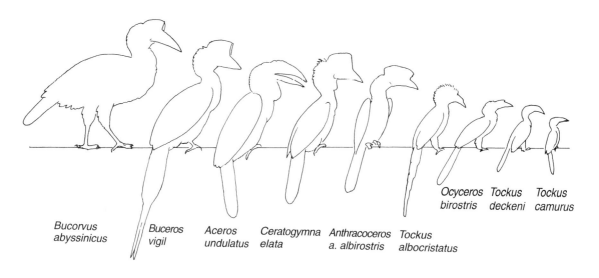

1.1 The avian order Bucerotiformes: illustrations of adult males of selected species of hornbills to indicate the range of shapes and sizes within the order.

1.2 The skull of a Northern Ground Hornbill *Bucorvus abyssinicus*, attached to a carved wooden neck and with a strip of elastic inner tube on the end. The contraption is worn on the forehead of Cameroonian tribesmen as camouflage while stalking game mammals. (Courtesy of Michel Louette.)

1.3 A ceremonial head-dress from the Philippines, including a skull of a Great Philippine Hornbill *Buceros hydrocorax* in the design. (Courtesy of Lim Chong Keat.)

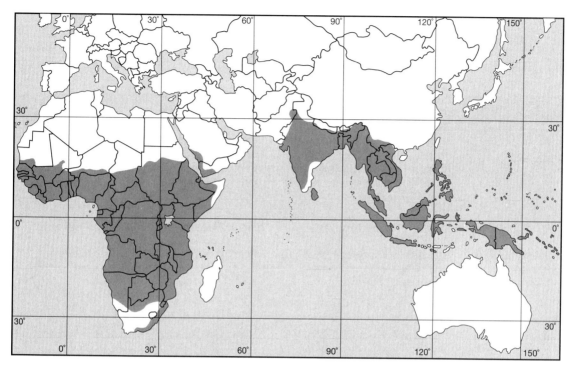

1.4 The total world distribution of hornbills within the avian order Bucerotiformes.

Anatomy and biology

The decurved bill and projecting casque, from which hornbills derive their common name, is one of the most obvious features which assigns them to the avian order Bucerotiformes (Sibley and Monroe 1990; Sibley and Ahlquist 1991); previously considered a family, the Bucerotidae. The world distribution of hornbills extends from sub-Saharan Africa, through India and the islands of the Malaysian archipelago, to as far east as New Guinea and the Solomon Islands (Fig. 1.4). There are no hornbills in the New World, but their role there is filled, to some extent, by toucans; large-billed and distant relatives classified in the barbet family Ramphastidae within the woodpecker order Piciformes.

Most of the 54 species currently recognized as comprising the Bucerotiformes occupy forest of some description, with only about one quarter in savanna or steppe and all but one of these in Africa (Table 1.1). This wide range of habitats and extensive geographical range also mean that different hornbill species exist among quite different communities of plants and share their world with quite different suites of animal neighbours, predators, and parasites.

Features of the environment which are most important to hornbills are those related to their feeding and breeding requirements. Hornbills come in a variety of shapes and sizes (Fig. 1.1), and these two aspects will determine how each species locates, captures, and consumes its preferred foods and where it will site its nest. Species range in size from the 4-kg, turkey-sized *Buceros* ground hornbills of the African savannas to the 100-g, dove-size dwarf *Tockus* hornbills of the African evergreen forests (Table 1.1). All are omnivorous, but some, such as the *Bucorvus* ground hornbills, are almost entirely carnivorous while others, such as the wattled *Ceratogymna* hornbills of African forests or the wreathed *Aceros* hornbills of Asia, are largely frugivorous. This

Table 1.1 A list of the scientific and common names of all nine genera and 54 species of hornbills described in this study, with indications of their body size, distribution and habitat preference

Scientific name	Common name	Mean mass ($\male$, g)	Distribution and habitat preference
Bucorvus			
B. abyssinicus	Northern Ground Hornbill	4000	African savanna N of Equator
B. leadbeateri	Southern Ground Hornbill	4191	African savanna S of Equator
Anorrhinus			
A. austeni	Austen's Brown Hornbill	933*	SE Asian mainland forests
A. tickelli	Tickell's Brown Hornbill	869	Myanmar and Thailand hill forests
A. galeritus	Bushy-crested Hornbill	1172	Sunda Shelf forests
Tockus			
T. alboterminatus	African Crowned Hornbill	244	North to South African montane and coastal forests
T. bradfieldi	Bradfield's Hornbill	464*	Central African teak woodlands
T. fasciatus	African Pied Hornbill	275	W and central African forests
T. hemprichii	Hemprich's Hornbill	639*	Abyssinian Highland forests
T. pallidirostris	Pale-billed Hornbill	256	Central African miombo woodlands
T. nasutus	African Grey Hornbill	200	Sub-Saharan African and S Arabian savanna
T. monteiri	Monteiro's Hornbill	268*	SW African steppe and arid savanna
T. erythrorhynchus	African Red-billed Hornbill	177	Sub-Saharan African savanna
T. leucomelas	Southern Yellow-billed Hornbill	211	S African savanna
T. flavirostris	Eastern Yellow-billed Hornbill	258	NE African savanna
T. deckeni	Von der Decken's Hornbill	194	NE African savanna
T. hartlaubi	Dwarf Black Hornbill	112	W and central African forests
T. camurus	Dwarf Red-billed Hornbill	111	W and central African forests
T. albocristatus	Long-tailed Hornbill	297	W and central African forests
Ocyceros			
O. griseus	Malabar Grey Hornbill	310	W Indian forests
O. gingalensis	Sri Lankan Grey Hornbill	238	Sri Lankan forests
O. birostris	Indian Grey Hornbill	375	Indian subcontinent savanna
Anthracoceros			
A. coronatus	Indian Pied Hornbill	1054*	Indian and Sri Lankan forests
A. albirostris	Oriental Pied Hornbill	810	SE Asian mainland and Sunda Shelf forests
A. marchei	Palawan Hornbill	713*	Palawan Archipelago forests
A. malayanus	Malay Black Hornbill	1050	Sunda Shelf forests
A. montani	Sulu Hornbill	741*	Sulu Archipelago forests
Buceros			
B. bicornis	Great Pied Hornbill	3007	Indian, SE Asian and W Sunda Shelf forests
B. rhinoceros	Great Rhinoceros Hornbill	2580	Sunda Shelf forests
B. hydrocorax	Great Philippine Hornbill	1600	Philippine forests
B. vigil	Great Helmeted Hornbill	3060	Sunda Shelf forests
Penelopides			
P. exarhatus	Sulawesi Tarictic Hornbill	350*	Sulawesi forests
P. panini	Visayan Tarictic Hornbill	486*	Philippine forests (Panay I.)
P. manillae	Luzon Tarictic Hornbill	450	Philippine forests (Luzon I.)
P. affinis	Mindanao Tarictic Hornbill	456	Philippine forests (Mindanao I.)
P. mindorensis	Mindoro Tarictic Hornbill	367*	Philippine forests (Mindoro I.)

Table 1.1 (cont.)

Scientific name	Common name	Mean mass (δ, g)	Distribution and habitat preference
Aceros			
A. comatus	White-crowned Hornbill	1476*	Sunda Shelf forests
A. nipalensis	Rufous-necked Hornbill	2500	SE Asian mainland forests
A. cassidix	Sulawesi Wrinkled Hornbill	2360	Sulawesi forests
A. corrugatus	Sunda Wrinkled Hornbill	1590	Sunda Shelf forests
A. leucocephalus	Mindanao Wrinkled Hornbill	1086	Philippine forests (Mindanao I.)
A. waldeni	Visayan Wrinkled Hornbill	1146*	Philippine forests (Panay I.)
A. plicatus	Papuan Wreathed Hornbill	1725	Moluccas, New Guinea, Solomon Is forests
A. narcondami	Narcondam Wreathed Hornbill	725	Narcondam I. forests
A. subruficollis	Plain-pouched Wreathed Hornbill	2042	SE Asian mainland forests
A. undulatus	Bar-pouched Wreathed Hornbill	2515	SE Asian mainland and Sunda Shelf forests
A. everetti	Sumba Wreathed Hornbill	1090*	Sumba I. forests
Ceratogymna			
C. fistulator	Piping Hornbill	540	W and central African forests
C. bucinator	Trumpeter Hornbill	721	East to South African montane and riverine forests
C. cylindricus	Brown-cheeked Hornbill	1330	W and central African forests
C. subcylindricus	Grey-cheeked Hornbill	1311	W and central African forests
C. brevis	Silvery-cheeked Hornbill	1308	North to East African montane and coastal forests
C. atrata	Black-casqued Wattled Hornbill	1348	W and central African forests
C. elata	Yellow-casqued Wattled Hornbill	2100	W African forests

* Body mass (averaged across subspecies, if any) is estimated as an index of body size for those species for which it has not been recorded, based on the correlation with wing length derived from a sample of well-documented species (see Chapter 2, Table 2.2, Fig. 2.1)

variety of hornbill species and the dynamic nature of their resources, such as availability of fruiting trees or nest cavities, result in a complex web of interactions which determines their movements, social structure, breeding season, productivity, and moult.

All hornbills are hole-nesters, like their close relatives the hoopoes and wood-hoopoes (order Upupiformes) and their more distant relatives the trogons (order Trogoniformes) and the rollers, bee-eaters, and kingfishers (order Coraciiformes). All prefer natural cavities for their nests, be they holes in trees (Plate 1b) or crevices in rock faces (Plate 1d), and they make little or no attempt to modify the interior. Hornbills, however, are unique among birds in that the female seals the entrance to leave a narrow vertical slit, through which she, and later the chicks, receive food from the male and out of which food remains and droppings are voided (Fig. 1.5). In most species the female also simultaneously moults all her flight feathers, growing them back while she incubates the eggs and broods the chicks.

The main exceptions are the two species of *Bucorvus* ground hornbills, placed in their own family Bucorvidae. These do not seal their nest and, while in the majority of cases they nest in natural holes, they occasionally breed in an old stick nest of another bird or excavate their own chamber in an earth bank.

Within these basic patterns of breeding behaviour there exists a whole range of detailed differences, such as the social organization

1.5 Cross-section of an active nest of African Red-billed Hornbills *Tockus erythrorhynchus*, showing the female squirting her droppings out through the entrance for nest sanitation. When the chicks are large enough, and their legs sufficiently well-developed for them to perform similar sanitation, the female emerges and they reseal the nest. By this time the female's flight feathers on wings and tail are almost completely regrown. Note the high ceiling to the nest, which can serve as an escape-hole or refuge for the inmates should danger threaten.

during breeding, whether the male takes part in the nest-sealing process, how the parents moult during breeding, when the female emerges from the nest, how nest sanitation is maintained, and how the nest inmates are provisioned (Plate 2a–d).

The details of behaviour and biology exhibited by each species of hornbill also prove to be similar to or different from those of other species. These patterns of shared or unique characters form the basis for assessing relationships between species and for determining the historical pathways along which they evolved. These pathways can be confirmed by examining the distribution of other characters, be they morphological, anatomical, behavioural, biochemical, or parasitological. The final evolutionary pathways that emerge then form the basis for understanding and comparing the geological, environmental, and biological history by which the present diversity of hornbills arose.

Research and conservation

Humankind has affected the world of many hornbill species for centuries, something of which these birds are not always, of course, aware but which is now of vital importance to them, given the burgeoning human population. Hornbills and their evolutionary history provide an important example for understanding the wonderful biological diversity that currently occupies the planet Earth.

This variety, however, will itself become history if humans cannot limit their effects on the environment. So many hornbill species occupy forest, the habitat most under pressure from humankind at the moment (Collins 1990; Collins *et al.* 1991; Lewis, D., 1992), and so many are large birds, which require extensive areas of habitat to support viable populations, that the outlook for many of them is bleak. On the positive side, some human beings have gone to considerable lengths to study and conserve hornbills. It is to them that this book is dedicated, for it is on their work that any salvation of hornbills will be based.

Hornbills were already known to early Classical writers, such as Pliny, and their skulls accurately described as those of those mythical birds the 'tragopans' (Newton 1893). The medieval author Aldrovandus described and figured the *'Rhinoceros Avis'* in 1599, the same species being subsequently described to science as the Great Rhinoceros Hornbill and named *Buceros rhinoceros* by Carl Linnaeus in 1758. Most appropriately, this is also one of the largest and most spectacular hornbill species, with a fine example of the casque that typifies the order. The generic name *Buceros* was, however, first given by Linnaeus to its close relative the Great Pied Hornbill *Buceros bicornis*, the word being derived from the Latin *bucerus* meaning 'having ox's horns'. It followed, by convention, that the same word was used to name the hornbill order Bucerotiformes and the principal family Bucerotidae.

The first comprehensive account of the order was the magnificent work by Daniel Giraud Elliot, *A monograph of the Bucerotidae, or family of the hornbills*, issued in 10 parts between 1877 and 1882 and lavishly illustrated with 60 hand-coloured lithographs by John Gerrard Keulemans. It was almost fifty years later before studies of hornbills in the field were pioneered: by Reg and Winnie Moreau, with their work on the Silvery-cheeked Hornbill *Ceratogymna brevis* at the East African Agricultural Research Station at Amani in Tanzania (1934–41); by Hans Bartels and his brother, with their detailed observations and amazing black-and-white photographs of several forest species around their tea plantations in Java and Sumatra (1937–56); by Gordon Ranger, with his studies of the African Crowned Hornbill *Tockus alboterminatus* and other hornbills on his farm (Gleniffer) in the Cape Province of South Africa (1941–52; unpublished notes at Transvaal Museum, Pretoria); and by Lawrence Kilham, with his work on the Grey-cheeked Hornbill *C. subcylindricus* in the botanical gardens of Kampala, Uganda (1956).

The first truly scientific survey of the whole order was the comprehensive and meticulous treatise compiled in 1960 by Dr Kurt Sanft of Berlin. The study was based on extensive examination of museum specimens and the scientific literature, bringing together all that was then known of the taxonomy, morphology, measurements, distribution, and biology for the 45 hornbill species that were recognized. It was an enormous and accurate contribution, and remains the foundation of the present book.

Interest in hornbills since that time has continued to expand. Among more prominent studies are those co-ordinated by the author in southern Africa, by Cliff and Dawn Frith in Southeast Asia, by S. A. Hussain on Narcondam Island, by Pilai Poonswad in Thailand, by Jan Kalina in Uganda, by Mark Leighton in Indonesia, and by Modse Sanjeevareddy in India. These people, and many others, have contributed to the activities of an IUCN/ICBP Hornbill Specialist Group formed in 1983. This has led subsequently to formation of subgroups among aviculturists in Europe and North America and among conservationists in Southeast Asia, all of which bodes well for the future study and conservation of hornbills.

2

The design of hornbills

Each species of organism has evolved a body form that enables it to survive and reproduce within its chosen habitat. Part of this design is a historical consequence of the evolutionary pathway by which the particular species came into being, and this will be considered later (Chapter 6). Many aspects of the body form, especially overall size, are, however, governed by inanimate physical laws that set many of the limits and abilities of the organism. These limits are so pervading that they correlate closely with numerous features of the organism, including its metabolism, growth patterns, and details of its life history. Understanding the range of sizes, shapes, and proportions among hornbills and their effects on hornbill biology is fundamental to describing the variety of feeding and breeding systems which exist among the Bucerotiformes.

Body size

The size of an animal can be expressed as a length, an area, or a volume. Body mass, or weight, is the most commonly used equivalent of volume, but it is inaccurate, in that it assumes a uniform density for the animal's tissues, and impractical, in that it varies diurnally and seasonally with body and behavioural condition; moreover, it is often unknown for rarer species. Nevertheless, body mass often correlates closely with some linear dimension, such as wing length, and such correlations are of particular accuracy in the case of birds, since they grow early in life to virtually their maximum size.

In the individual species accounts (Part II), the mean wing length is given for both sexes of all species and subspecies of hornbill (Table 2.1). When plotted against mean body mass, for those taxa where the latter parameter is known (Fig. 2.1), it provides a strong correlation, and this fact therefore allows body mass

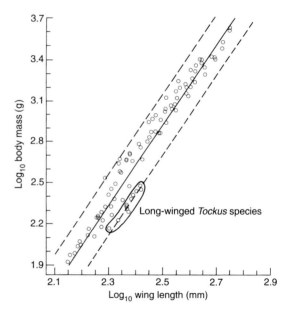

2.1 A graph showing the relationship between body mass and wing length for both sexes of some species and subspecies of hornbill ($n = 94$, from Table 2.1). Both are plotted on a $\log_{10}$ scale for the best correlation (R = 0.9708, S.E. = 0.1111, R^2 = 94.24%). The line of best fit calculated for the regression can be used to predict body mass for those species for which it is not yet recorded (or the equation: $\log_{10}$ of mass in g = $\log_{10}$ of -4.373 + $2.917.\log_{10}$ of wing length in mm).

Table 2.1 Mean wing length and body mass for each sex and all species and subspecies of hornbill (data from individual species accounts in Part II. Asterisk (*) denotes data estimated as an index of body size for those species/subspecies for which it has not been recorded, based on the correlation with wing length derived from a sample of well-documented species (Fig. 2.1)).

Hornbill species and subspecies		Mean wing length (mm) ♂	♀	Mean body mass (g) ♂	♀
Bucorvus					
B. abyssinicus	Northern Ground Hornbill	560	512	4000	3390*
B. leadbeateri	Southern Ground Hornbill	560	528	4191	3344
Anorrhinus					
A. austeni	Austen's Brown Hornbill	329	306	933*	755*
A. tickelli	Tickell's Brown Hornbill	326	308	869	723
A. galeritus	Bushy-crested Hornbill	351	329	1172	933*
Tockus					
T. alboterminatus	African Crowned Hornbill	255	234	244	205
T. bradfieldi	Bradfield's Hornbill	259	237	464*	193
T. f. fasciatus	Lower Guinea Pied Hornbill	265	243	278	241
T. f. semifasciatus	Upper Guinea Pied Hornbill	254	234	271	215
T. hemprichii	Hemprich's Hornbill	289	262	639*	297*
T. p. pallidirostris	Western Pale-billed Hornbill	247	233	256	212
T. p. neumanni	Eastern Pale-billed Hornbill	227	205	316*	235*
T. n. nasutus	North African Grey Hornbill	225	206	233	183
T. n. epirhinus	South African Grey Hornbill	217	201	167	146
T. monteiri	Monteiro's Hornbill	215	194	268*	340
T. e. erythrorhynchus	North African Red-billed Hornbill	182	167	177	131
T. e. rufirostris	South African Red-billed Hornbill	188	177	150	128
T. e. damarensis	Damaraland Red-billed Hornbill	195	183	210	175
T. leucomelas	Southern Yellow-billed Hornbill	205	191	211	168
T. flavirostris	Eastern Yellow-billed Hornbill	200	183	258	182
T. d. deckeni	Von der Decken's Hornbill	188	169	194	145
T. d. jacksoni	Jackson's Hornbill	183	167	163*	116*
T. h. hartlaubi	Upper Guinea Dwarf Black Hornbill	154	141	109	83
T. h. granti	Lower Guinea Dwarf Black Hornbill	156	144	118	94
T. camurus	Dwarf Red-billed Hornbill	157	150	111	97
T. a. albocristatus	Western Guinea Long-tailed Hornbill	239	205	297	282
T. a. macrourus	Eastern Guinea Long-tailed Hornbill	240	213	371*	263*
T. a. cassini	Lower Guinea Long-tailed Hornbill	253	209	310	248*
Ocyceros					
O. griseus	Malabar Grey Hornbill	212	195	310	203*
O. gingalensis	Sri Lankan Grey Hornbill	210	197	238	209*
O. birostris	Indian Grey Hornbill	220	203	375	228*
Anthracoceros					
A. coronatus	Indian Pied Hornbill	343	320	1054*	861*
A. a. albirostris	Asian Pied Hornbill	299	279	738	624
A. a. convexus	Sunda Pied Hornbill	312	281	907	879
A. marchei	Palawan Hornbill	300	283	713*	601*
A. malayanus	Malay Black Hornbill	324	288	1050	633*
A. montani	Sulu Hornbill	304	283	741*	601*
Buceros					
B. bicornis	Great Pied Hornbill	527	480	3007	2211
B. r. rhinoceros	Malayan Great Rhinoceros Hornbill	493	453	2580	2180
B. r. borneoensis	Bornean Great Rhinoceros Hornbill	475	435	2724*	2108*

Table 2.1 (*cont.*)

Hornbill species and subspecies		Mean wing length (mm)		Mean body mass (g)	
		♂	♀	♂	♀
Buceros (*cont.*)					
B. r. sylvestris	Javan Great Rhinoceros Hornbill	508	458	3314*	2449*
B. h. hydrocorax	Great Luzon Hornbill	409	380	1824	1421*
B. h. mindanensis	Great Mindanao Hornbill	389	366	1574	1557
B. h. semigaleatus	Great Samar Hornbill	386	360	1424	1209
B. vigil	Great Helmeted Hornbill	494	443	3060	2500
Penelopides					
P. e. exarhatus	North Sulawesi Tarictic Hornbill	235	223	350*	370
P. e. sanfordi	South Sulawesi Tarictic Hornbill	237	222	358*	296*
P. p. panini	Visayan Tarictic Hornbill	263	249	486*	414*
P. p. ticaensis	Ticao Tarictic Hornbill	292	274	659*	547*
P. m. manillae	Luzon Tarictic Hornbill	233	220	450	473
P. m. subnigra	Polillo Tarictic Hornbill	261	259	475*	464*
P. a. affinis	Mindanao Tarictic Hornbill	233	218	456	281*
P. a. samarensis	Samar Tarictic Hornbill	240	218	513	435
P. a. basilanica	Basilan Tarictic Hornbill	236	227	354*	316*
P. mindorensis	Mindoro Tarictic Hornbill	239	230	367*	328*
Aceros					
A. comatus	White-crowned Hornbill	385	355	1476*	1470
A. nipalensis	Rufous-necked Hornbill	434	420	2500	2270
A. cassidix	Sulawesi Wrinkled Hornbill	441	396	2360	1602*
A. corrugatus	Sunda Wrinkled Hornbill	402	366	1590	1273*
A. leucocephalus	Mindanao Wrinkled Hornbill	344	305	1086	814*
A. waldeni	Visayan Wrinkled Hornbill	353	314	1146*	748*
A. plicatus	Papuan Wreathed Hornbill	417	385	1725	1667
A. narcondami	Narcondam Wreathed Hornbill	307	283	725	675
A. subruficollis	Plain-pouched Wreathed Hornbill	419	381	2042	1431*
A. undulatus	Bar-pouched Wreathed Hornbill	497	454	2515	1950
A. everetti	Sumba Wreathed Hornbill	347	320	1090*	861*
Ceratogymna					
C. f. fistulator	Upper Guinea Piping Hornbill	241	226	376*	312*
C. f. sharpii	Sharpe's Piping Hornbill	259	240	600	500
C. f. duboisi	Dubois' Piping Hornbill	267	246	463	449
C. bucinator	Trumpeter Hornbill	288	263	721	567
C. c. cylindricus	Upper Guinea Brown-cheeked Hornbill	322	291	908*	921
C. c. albotibialis	White-thighed Hornbill	327	298	1330	978
C. s. subcylindricus	Upper Guinea Grey-cheeked Hornbill	326	298	876*	699*
C. s. subquadratus	Lower Guinea Grey-cheeked Hornbill	358	325	1311	1090
C. brevis	Silvery-cheeked Hornbill	369	340	1308	1162
C. atrata	Black-casqued Wattled Hornbill	395	354	1348	1059
C. elata	Yellow-casqued Wattled Hornbill	402	362	2100	1750

to be estimated for those species for which it is unknown. These measures of body mass, whether empirical or estimated, can then be used in subsequent comparisons between body size and other parameters.

Before going further, some caution is necessary in assessing this and subsequent relationships between size and other factors. The estimate of body mass is only an approximation based on a general, but not perfect, relationship

Table 2.2 Comparison of the body size of the largest species in the hornbill order Bucerotiformes with the largest species in all other avian orders (classification after Sibley and Monroe 1990)

Avian order	Largest body size (approx. mass in g)	Species and sex
Struthioniformes	130 000	Ostrich *Struthio camelus*, ♂
Tinamiformes	<1 200	Solitary Tinamou *Tinamus solitarius*, ♀
Galliformes	10 000	Turkey *Meleagris gallopavo*, ♂
Anseriformes	12 500	Trumpeter Swan *Cygnus cygnus*, ♂
Turniciformes	<100	Spotted Button-quail *Turnix ocellata*, ♀
Piciformes	<550	Imperial Woodpecker *Campephilus imperialis*, ♂
Galbuliformes	<250	Black-fronted Nunbird *Monasa nigrifrons*, ♂
Bucerotiformes	**4 191**	**Southern Ground Hornbill *Bucorvus leadbeateri*, ♂**
Upupiformes	96	Violet Wood-hoopoe *Phoeniculus damarensis*, ♂
Trogoniformes	189	Resplendent Quetzal *Pharomachrus mocinno*, ♀
Coraciiformes	350	Giant Kingfisher *Megaceryle maxima*, ♂
Coliiformes	62	Red-backed Mousebird *Colius castanotus*, ♂
Cuculiformes	<400	Common Crow-pheasant *Centropus sinensis*, ♀
Psittaciformes	2 000	Kakapo *Strigops habroptilus*, ♂
Apodiformes	180	Purple Needletail *Hirundapus celebensis*, ♂
Trochiliformes	21	Giant Hummingbird *Patagona gigas*, ♂
Musophagiformes	981	Great Blue Turaco *Corythaeola cristata*, ♀
Strigiformes	3 164	Eurasian Eagle Owl *Bubo bubo*, ♀
Columbiformes (flighted)	<1 500	New Guinea Crowned Pigeon *Goura cristata*, ♂
(flightless)	23 000	Dodo *Raphus cucullatus*, ♂
Gruiformes	17 000	Great Bustard *Otis tarda*, ♂
Ciconiiformes (flighted)	14 000	California Condor *Gymnogyps californianus*, ♂
(flightless)	42 500	Emperor Penguin *Aptenodytes forsteri*, ♂
Passeriformes	1 700	Eurasian Raven *Corvus corax*, ♂

between wing length and body mass for a subset of hornbill species. Even for those species for which body mass and wing length are recorded, the data are often inadequate owing to small sample sizes and inadvertent inclusion of records for aberrant individuals (young, captive, or starving). An example of this is the lower-than-predicted mass of Black-casqued Wattled Hornbills *Ceratogymna atrata*, which suggests them to be quite different from the similarly sized Yellow-casqued Wattled Hornbill *C. elata* in all subsequent comparisons.

Most individual measurements do not fall exactly on the line describing the relationship between body mass and wing length (Fig. 2.1), and for some, such as the long-winged but lightweight aerial-foraging species of *Tockus* hornbills, may even fall so far below the line that their mass is likely to be overestimated (Fig. 2.1). Such limitations of the method will compromise the accuracy of further comparisons. It does, however, allow, for the first time, a general comparison of relationships to body size across all hornbill species — something which has been attempted for very few orders of birds.

The 54 living species of hornbill exhibit a wide range of body sizes. The largest species weighs fifty times more than the smallest, with average body mass ranging from 4191 g for a male Southern Ground Hornbill *Bucorvus leadbeateri* down to 83 g for a female Dwarf Black Hornbill *Tockus hartlaubi*. Hornbills differ from most other orders of non-passerine birds, and especially from the orders to which they are considered most closely related, in the range of sizes represented (Table 2.2). The distribution of body size across all bird families is skewed in favour of smaller species and hence there are relatively few larger species. This suggests that there may be a limited number of ecological niches available to large birds, at least in comparison with what is available to smaller species, and that paucity or lack of larger species may be the result of competition for or exclusion from such opportunities.

A search for patterns in the number and diversity of niches occupied by large birds in different parts of the world, of the families that are represented in them, and of the relevant ecological positions occupied by hornbills would be revealing. One possible example concerns the Great Helmeted Hornbill *Buceros vigil*, which is suspected to be a major predator of small vertebrates in the forests of Malaysia, Borneo and Sumatra. Hornbills share most other large landmasses on which they exist with a large forest eagle (Crowned Eagle *Stephanoaetus coronatus* in Africa, Changeable Hawk-eagle *Spizaetus nipalensis* on mainland Asia, Monkey-eating Eagle *Pithecophaga jefferyi* in the Philippines, and Papuan Harpy Eagle *Harpyopsis novaeguineae* and Gurney's Eagle *Aquila gurneyi* in New Guinea). Maybe the lack of such an eagle on the landmasses of the Sunda Shelf is due to the presence of a large predatory hornbill?

Size, shape, and proportions

The products of several laws of physics are also considerably influenced by size, so that the exact size which an animal attains will inevitably affect many aspects of its design, development, and performance, such as the cross-section of limbs, wing area, metabolic rate, and growth trajectory (Calder 1984). These physical limitations based on size affect in turn, and therefore are correlated with, many other areas of hornbill biology, such as food requirements, energy requirements of locomotion, rate of travel, home-range size, sex roles, incubation and brooding requirements, fecundity, age at maturity, and longevity.

Hornbills occupy a diversity of habitats, within which the particular shape and proportions of each species, as illustrated in Fig. 1.1, also have to function appropriately. The shape of any animal is also, like its size, a compromise between its evolutionary or genetic history and its immediate patterns of growth and development. Size and shape combine to determine the proportions of an animal, which, in turn, set the limits to what it can do and how well it can persist in its chosen environment.

The size and shape of hornbills also correlate with many other aspects of their breeding biology, as will be shown in Chapter 5. Unfortunately, these data are not complete for all species, yet the correlations that do emerge allow preliminary predictions of values to be made for poorly known species. These values can always be tested empirically, but in the meantime they offer an important service for the study, management, and conservation of rare or inaccessible forms.

Correlations with size

The relationships between body size and various biological parameters have been worked out in detail for non-passerine birds, such as hornbills (Calder 1984). Estimation of metabolic rate is especially valuable (Table 2.3), since this determines the food requirements of a species and the costs of locomotion. These in turn influence such aspects of its life style as foraging strategies, activity patterns, and home-range requirements.

Flight is an especially energetic and expensive form of locomotion. Body mass, wing area, and wingspan have been shown to be the basic parameters necessary to estimate energetic costs of flight for birds (Maasman and Klaasen 1987). It has also been demonstrated, this time for raptors which have similar long broad wings to hornbills, that wing area and wingspan can be accurately predicted from measurements of the ulna, outer secondary, and longest primary, dimensions which are easily obtainable from most study skins in a museum (Mendelsohn *et al.* 1989). Assuming that the same relationships apply to hornbills, and using samples of

Table 2.3 Estimated basal metabolic rate, flight metabolism, wing loading (mass and linear), and aspect ratio for males of hornbill species (calculated from formulae in Aschoff and Pohl 1970; Maasman and Klaasen 1985; Mendelsohn *et al.* 1989; derived from data in Tables 2.1 and 2.4). The species follow the classification of Sanft (1960), and species since divided are placed within single quotation marks. See text for further details

Hornbill species	Basal metabolic rate (BMR, in watts)	Flight metabolism (FM, in watts)	FM/BMR	Wing loading		Aspect ratio
				Mass	Linear	
Bucorvus						
B. abyssinicus	1.52	4.73	3.1	0.655	0.202	5.04
B. leadbeateri	1.58	4.86	3.1	0.685	0.206	5.08
Anorrhinus						
A. 'tickelli'*	0.51	2.00	3.9	0.479	0.221	4.20
A. galeritus	0.61	2.18	3.6	0.608	0.240	4.48
Tockus						
T. alboterminatus	0.16	0.58	3.6	0.195	0.176	4.09
T. bradfieldi*	0.17	0.53	3.2	0.192	0.174	4.31
T. fasciatus	0.18	0.68	3.7	0.212	0.180	4.04
T. hemprichii*	0.20	0.73	3.7	0.202	0.174	4.00
T. pallidirostris	0.17	1.05	6.2	0.265	0.214	3.25
T. nasutus	0.16	0.89	5.7	0.265	0.207	3.38
T. monteiri*	0.18	0.87	4.9	0.275	0.206	3.62
T. erythrorhynchus	0.10	0.53	5.0	0.192	0.190	3.56
T. leucomelas	0.14	0.86	6.0	0.264	0.210	3.32
T. deckeni	0.13	0.90	6.7	0.274	0.217	3.16
T. albocristatus	0.19	0.53	2.7	0.204	0.175	4.65
T. hartlaubi	0.08	0.70	9.0	0.290	0.246	2.85
T. camurus	0.08	0.88	11.2	0.287	0.244	2.58
Ocyceros						
O. 'griseus'	0.20	1.21	6.0	0.334	0.222	3.34
O. birostris	0.24	1.08	4.5	0.452	0.250	3.91
Anthracoceros						
A. 'coronatus'	0.42	1.20	2.9	0.351	0.197	4.72
A. marchei*	0.41	1.60	3.9	0.431	0.219	4.14
A. malayanus	0.56	2.54	4.6	0.522	0.226	3.94
A. montani*	0.42	4.21	10.0	0.490	0.323	2.70
Buceros						
B. bicornis	1.24	3.30	2.7	0.629	0.208	5.33
B. rhinoceros	1.11	3.84	3.5	0.626	0.213	4.67
B. hydrocorax	0.86	2.66	3.1	0.595	0.220	4.83
B. vigil	1.26	3.70	2.9	0.689	0.217	5.12
Penelopides						
P. exarhatus*	0.22	0.74	3.3	0.276	0.197	4.35
P. 'panini'	0.28	1.31	4.7	0.404	0.229	3.79
Aceros						
A. comatus*	0.73	2.61	3.6	0.494	0.208	4.41
A. nipalensis	1.08	3.58	3.3	0.674	0.222	4.80
A. cassidix	1.04	3.67	3.5	0.551	0.203	4.55
A. corrugatus	0.77	2.38	3.1	0.614	0.229	4.84
A. 'leucocephalus'	0.57	1.37	2.4	0.446	0.208	5.26
A. plicatus	0.82	1.89	2.3	0.530	0.210	5.50
A. narcondami	0.41	1.80	4.3	0.497	0.235	4.00
A. 'undulatus'	1.09	2.86	2.6	0.622	0.213	5.30
A. everetti*	0.57	3.44	6.0	0.586	0.238	3.50

Table 2.3 (*cont.*)

Hornbill species	Basal metabolic rate (BMR, in watts)	Flight metabolism (FM, in watts)	FM/BMR	Wing loading		Aspect ratio
				Mass	Linear	
Ceratogymna						
C. fistulator	0.29	0.97	3.4	0.390	0.224	4.39
C. bucinator	0.42	2.18	5.2	0.547	0.245	3.71
C. cylindricus	0.67	2.44	3.6	0.624	0.238	4.47
C. subcylindricus	0.66	2.08	3.1	0.564	0.227	4.75
C. brevis	0.66	1.97	3.0	0.556	0.225	4.86
C. atrata	0.68	1.48	2.2	0.404	0.191	5.47
C. elata	0.95	3.54	3.7	0.709	0.235	4.53

* Body mass (averaged across subspecies, if any) is estimated for those species for which it has not been recorded, based on the correlation with wing length derived from a sample of well-documented species (Table 2.1, Fig. 2.1)

these measurements taken at the British and Transvaal Museums (Table 2.4), it is possible to offer the first estimates of basal metabolic rate and the energetic costs of flight for all hornbill species (Table 2.3).

The ratio between these two estimates offers an index of the flight efficiency of each species and genus, with a wide range of values from under 3.0 for the most efficient to over 10.0 for the least efficient. Flight ability is also related to the wing loading and aspect ratio of a bird (Table 2.3). Wing loading, the mass borne per unit area of the wing, can be expressed in two ways. Mass loading is expressed as the number of grams of body mass supported by each square centimetre of wing area, and indicates whether low loading is likely to result in light and buoyant flight (under 0.2 g/cm^2 for some small *Tockus* species) or heavy loading in more laboured flight (over 0.7 g/cm^2 for Yellow-casqued Wattled Hornbills). Values for linear loading (cubed root of body mass over square root of wing area) cover a smaller range but reduce the effects of body size and allow direct comparisons between species of different sizes, from broad-winged small *Tockus* species under 0.18 to various larger but relatively smaller-winged species over 0.23.

The estimated aspect ratio (wingspan^2/wing area, Table 2.3) is related mainly to flight efficiency, long narrow wings with a high aspect ratio offering the least drag and permitting the highest speeds. Among hornbills, aspect ratios are both low and rather conservative, being within the range of 2.58 to 5.50. This indicates an order with rather short and broad wings, but capable of slow and controlled flight. There are many other variables influencing flight efficiency and speed, although the basic estimates in Table 2.3 offer the best means of comparing species. For example, efficiency is affected by flight behaviour, such as the energy-saving breaks to glide or swoop between bouts of flapping that are shown by most hornbill species. Flight speed is also related to size, so that larger species with their higher mass wing loading have to fly faster to keep aloft.

Other factors, such as brain size, are also related to flight co-ordination, as well as to longevity (Western and Ssemakula 1982), and the positive and significant relationship of this parameter to body size has already been analysed for 14 hornbill species (Mlikovsky 1989; brain size = $0.102 \times$ mass$^{0.669}$). Further correlations between body size and aspects of breeding ecology and developmental biology are indicated in Chapter 5.

Table 2.4 Mean body dimensions** of males of selected hornbill species, expressed as a proportion of the cubed root of the body mass (from Table 2.1). The species follow the classification of Sanft (1960), and species since divided are placed within single quotation marks

Hornbill species	n	Bill length (mm)	Bill volume (mm³)	Casque volume (mm³)	Secondary length (mm)	Ulna length (mm)	Primary area (cm²)	Secondary area (cm²)	Wing length (mm)	Tail length (mm)	Tarsus length (mm)	Footspan (mm)
Bucorvus												
B. abyssinicus	8	16.7	17.0	3.0	29.1	14.6	5.15	4.25	35.3	22.3	9.54	8.46
B. leadbeateri	10	13.2	12.8	1.0	28.5	15.1	5.36	4.35	35.0	21.4	9.20	9.41
Anorrhinus												
*A. 'tickelli'**	10	12.9	20.8	3.0	26.8	12.1	4.07	3.28	33.4	29.4	4.61	10.16
A. galeritus	10	13.6	27.5	2.8	24.0	13.4	4.84	3.90	33.1	24.8	4.56	7.52
Tockus												
T. alboterminatus	10	14.8	21.5	3.2	32.5	14.4	5.29	4.37	39.5	34.7	5.45	9.79
*T. bradfieldi**	1	17.8	26.6	3.2	32.3	14.5	5.63	4.40	41.4	35.7	6.21	10.35
T. fasciatus	10	15.4	27.7	3.1	32.2	14.5	5.65	4.54	40.1	36.3	5.07	8.93
*T. hemprichii**	10	14.9	31.1	3.0	34.0	15.1	6.35	5.12	42.1	37.5	5.83	10.33
T. pallidirostris	2	12.6	24.9	1.8	29.8	13.2	5.31	4.26	37.4	34.2	5.20	9.14
T. nasutus	10	13.7	20.0	2.9	28.5	13.2	5.05	3.95	36.3	32.4	5.86	9.27
*T. monteiri**	5	18.1	30.2	3.1	28.3	13.2	4.20	3.60	33.1	36.2	7.16	11.66
T. erythrorhynchus	10	15.2	16.5	1.1	29.9	12.9	4.12	3.52	35.3	37.5	8.10	9.64
T. leucomelas	10	14.5	27.6	3.9	27.8	12.9	4.52	3.65	34.4	36.0	7.40	9.75
T. deckeni	10	14.5	22.3	4.0	26.9	12.1	3.87	3.23	32.2	36.0	6.75	10.21
T. hartlaubi	10	10.5	21.4	1.9	21.5	10.1	3.40	3.57	32.7	34.6	5.25	9.64
T. camurus	10	10.9	23.2	2.9	22.6	9.7	3.21	3.60	32.1	30.9	6.26	9.59
T. albocristatus	10	14.8	16.7	4.8	31.6	13.9	4.34	3.71	37.1	66.1	5.40	10.34
Ocyceros												
O. 'griseus'	10	14.9	28.7	1.0	26.9	12.6	4.00	3.41	31.4	33.6	6.07	10.07
O. birostris	10	13.2	20.5	3.1	22.1	11.1	3.19	1.91	29.6	37.3	5.95	9.31
Anthracoceros												
A. 'coronatus'	10	20.4	37.7	22.2	28.8	14.0	4.25	3.65	33.7	31.4	6.21	10.09
*A. marchei**	2	17.0	33.3	13.5	26.8	12.6	3.78	3.24	31.2	25.1	5.73	9.87
A. malayanus	10	16.3	31.2	11.9	27.0	12.2	3.84	3.32	31.2	31.2	5.03	8.77
*A. montani**	1	16.6	29.8	11.3	29.9	—	6.20	1.80	35.4	26.5	5.20	11.62
Buceros												
B. bicornis	10	20.4	37.1	23.0	27.3	14.8	5.26	4.06	35.7	30.9	4.93	9.73
B. rhinoceros	10	20.4	38.2	23.1	28.0	14.3	4.58	3.98	32.3	27.4	4.93	9.79
B. hydrocorax	10	15.6	22.4	6.4	26.3	12.3	3.62	3.18	31.5	25.5	4.74	10.58
B. vigil	9	15.2	24.2	5.0	26.5	14.0	4.63	3.57	33.3	27.6**	4.76	8.02

Table 2.4 (cont.)

Hornbill species	n	Bill length (mm)	Bill volume (mm³)	Casque volume (mm³)	Secondary length (mm)	Ulna length (mm)	Primary area (cm²)	Secondary area (cm²)	Wing length (mm)	Tail length (mm)	Tarsus length (mm)	Footspan (mm)
Penelopides												
*P. exarhatus**	3	14.2	22.6	4.5	28.3	12.9	3.87	3.45	34.0	27.7	4.28	9.94
P. 'panini'	10	12.1	17.6	2.1	25.4	11.7	3.41	2.89	30.2	25.7	5.97	9.28
Aceros												
*A. comatus**	6	14.5	21.0	3.2	28.9	12.4	3.88	3.47	32.9	37.7	5.36	9.77
A. nipalensis	10	16.8	32.1	2.0	26.4	13.3	4.28	3.50	32.2	31.0	5.02	8.64
A. cassidix	4	16.9	28.2	10.1	29.8	15.2	4.96	4.50	32.9	21.2	4.73	8.81
A. corrugatus	8	14.7	26.1	4.0	24.9	12.2	4.00	2.95	33.0	22.1	4.74	8.25
A. 'leucocephalus'	9	12.1	19.4	3.4	26.3	12.5	3.64	2.90	33.1	23.3	4.29	8.87
A. plicatus	10	18.9	33.1	5.2	26.1	15.3	5.12	3.91	34.7	20.1	4.76	9.27
A. narcondami	1	13.4	28.3	4.4	25.1	13.9	5.34	3.94	34.0	20.9	4.24	8.69
A. 'undulatus'	10	15.3	26.0	3.0	26.4	14.7	5.12	3.88	34.8	21.1	4.27	8.71
*A. everetti**	1	13.7	37.0	6.6	26.8	13.6	5.58	4.44	33.6	25.8	4.68	8.77
Ceratogymna												
C. fistulator	10	13.0	18.5	4.1	24.7	13.3	4.19	3.20	32.5	24.4	4.79	8.80
C. bucinator	10	15.2	30.5	14.8	24.5	12.4	3.51	3.13	27.9	30.1	3.08	9.30
C. cylindricus	10	14.6	20.9	20.9	24.5	12.9	3.91	3.18	30.0	22.4	4.65	8.75
C. subcylindricus	10	15.1	25.0	20.0	25.2	13.1	4.28	3.30	32.7	24.1	4.77	8.98
C. brevis	10	15.6	26.0	38.0	25.1	13.0	4.36	3.28	33.5	24.9	4.86	9.07
C. atrata	10	16.5	24.4	22.1	28.8	13.8	4.16	3.42	35.0	27.9	4.26	9.89
C. elata	8	16.3	27.1	22.1	25.3	12.3	3.71	3.11	30.2	23.8	4.07	8.92

* Body mass is estimated for those species for which it has not been recorded, based on the correlation with wing length derived from a sample of well-documented species (Table 2.1, Fig. 2.1)
** Bill and bill-volume index = chord × breadth at base × width at base; secondary length = length from point of wrist to end of outermost secondary; primary-area index = secondary length × wing length; secondary-area index = ulna length × secondary length

Correlations with shape and proportions

Differences in bodily proportions, such as relative length of leg or breadth of wing, can be expressed in two main ways. First, ratios between different elements can be calculated. This is most appropriate for comparing species of similar size and the same basic proportions. Elements can then be expected to vary more or less in tandem with one another, in response to basic physical properties such as body size. It is least useful when one or more elements vary independently, for instance in response to demand for long tail feathers for display.

Each linear dimension can also be expressed as a proportion of the cubed root of the body mass. Body mass (or volume) changes as the cubed power, and flight surfaces (or areas) as the squared power, of changes in linear dimensions. For instance, imagine the changes in the length of the sides, in the area of the faces, and in the volume of a cube with each side first of 1 cm and then of 2 cm in length. Such comparison reduces biases that arise when one linear dimension changes independently of another. It also allows comparison between species that differ widely in body size, where relative changes in proportions will depend on the extent to which each element changes, or scales itself allometrically, according to body size.

Comparison of dimensions adjusted for body size shows some consistent differences between different genera and species of hornbill (Table 2.4). For example, differences in wing-area indices for different *Tockus* species or the relatively short wing dimensions (ulna, wing, secondary) of the Indian Grey Hornbill *Ocyceros birostris* can be related to differences in flight ability and behaviour for these species. The relative size of the casque (such as the voluminous casques of most *Buceros* and *Ceratogymna* species) can also be assessed, as can the long legs of *Bucorvus* ground hornbills. The relative sizes, proportions, and flight efficiency of each species should also be borne in mind when reading the individual accounts in Part II.

Sexual differences in size and shape

Another aspect of size and shape is the extent of differences between the sexes, or sexual dimorphism, within a species. Ratios of male to female mass and wing length provide indices of sexual dimorphism, and these are compared with indices of sexual differences in casque volume and coloration (Table 2.5). Differences in size, shape, and colour between different age classes may also be important to functioning and survival, but are not studied further.

Males are equal to or up to 17 per cent heavier than females (variation among some *Penelopides* species). Wing length of males varies from 1 per cent to 21 per cent greater than that of females and bill length from 8 per cent to 30 per cent. This variation in sexual differences among species, and even among subspecies, is not consistent for each set of indices, including subjective assessments of sexual dimorphism in casque volume and general coloration. Why the extent and pattern of sexual differences vary for each parameter, and how these differences are related to one another and to finer details of differences in coloration, voice, and social behaviour, is a puzzle still awaiting analysis.

Casque form and function

One example of how sexual differences in design and proportions may be effective is in the development of the casque, a structure the possible function of which has drawn numerous proposals. The casque is formed from the horny layer of keratin that covers the entire bill and, in most species, contains little or no bone in its structure. In a number of species, the casque is no more than a raised ridge along the base of the upper mandible. It may have arisen to provide support to the base of the long decurved upper bill, which would be subject to considerable stress given the strong biting forces which the jaw muscles can transfer to the tip. Such reinforcement is evident in the triangular cross-section of the bill of other long-billed relatives, such as kingfishers.

Table 2.5 Indices of the differences in size between the sexes (sexual size dimorphism) for all species and subspecies of hornbill. The indices are expressed as the ratio of male:female for the cubed root of mean body mass, mean wing length and mean bill length, each × 100 (Table 2.1). These indices are compared with subjective assessments of the extent of sexual differences in casque volume and colour of plumage or soft parts

Hornbill species/subspecies	Sexual size dimorphism			Extent of sexual differences	
	Body mass	Wing length	Bill length	Casque volume	Coloration
Bucorvus					
B. abyssinicus	106*	109	118	+	+++
B. leadbeateri	108	106	108	+	++
Anorrhinus					
A. austeni	107*	108	114	+	++
A. tickelli	106	106	109	+	++
A. galeritus	108*	107	107	+	+++
Tockus					
T. alboterminatus	106	109	116	++	+
T. bradfieldi	134*	109	115	+	+
T. f. fasciatus	105	109	115	+	+
T. f. semifasciatus	108	108	115	+	+
T. hemprichii	129*	110	119	+	+
T. p. pallidirostris	106	106	112	+	+
T. p. neumanni	110*	111	111	+	+
T. n. nasutus	108	109	127	+	+++
T. n. epirhinus	105	108	114	++	+++
T. monteiri	92*	111	121	+	+
T. e. erythrorhynchus	111	109	124	+	++
T. e. rufirostris	105	106	122	+	++
T. e. damarensis	106	106	120	+	++
T. leucomelas	108	107	122	++	+
T. flavirostris	112	109	120	++	+
T. d. deckeni	110	111	128	+	+++
T. d. jacksoni	111*	112	121	+	+++
T. h. hartlaubi	109	109	126	+	++
T. h. granti	108	108	123	+	++
T. camurus	105	105	112	+	+
T. a. albocristatus	102	117	137	+	+
T. a. macrourus	112*	113	125	+	+
T. a. cassini	120*	121	139	+	+
Ocyceros					
O. griseus	115*	109	128	+	+++
O. gingalensis	104*	106	123	+	++
O. birostris	118*	108	116	++	+
Anthracoceros					
A. coronatus	107*	107	113	+	++
A. a. albirostris	106	107	119	+	++
A. a. convexus	101	111	127	+	++
A. marchei	106*	106	111	+	+
A. malayanus	118*	113	138	++	+++
A. montani	107*	107	–	+	++
Buceros					
B. bicornis	105	110	119	+	++

Table 2.5 (cont.)

Hornbill species/subspecies	Sexual size dimorphism			Extent of sexual differences	
	Body mass	Wing length	Bill length	Casque volume	Coloration
B. r. rhinoceros	106	109	115	+	++
B. r. borneoensis	109*	109	115	+	++
B. r. sylvestris	111*	111	113	+	++
B. h. hydrocorax	109*	108	110	+	++
B. h. mindanensis	100	106	110	+	+
B. h. semigaleatus	106	107	108	+	+
B. vigil	107	112	114	+	+++
Penelopides					
P. e. exarhatus	98*	105	114	+	+++
P. e. sanfordi	107*	107	106	+	+++
P. p. panini	105*	106	115	+	+++
P. p. ticaensis	106*	107	-	+	+++
P. m. manillae	98	106	125	+	+++
P. m. subnigra	101*	101	118	+	+++
P. a. affinis	117*	107	115	+	+++
P. a. samarensis	106	110	125	+	+++
P. a. basilanica	104*	104	-	+	+++
P. mindorensis	104*	104	111	+	+
Aceros					
A. comatus	100*	108	111	+	+++
A. nipalensis	103	103	120	+	+++
A. cassidix	113*	111	134	+	+++
A. corrugatus	108*	110	123	+	+++
A. leucocephalus	113*	113	128	+	+++
A. waldeni	112*	112	121	+	+++
A. plicatus	101	108	124	+	+++
A. narcondami	102	108	118	+	+++
A. subruficollis	112*	110	124	+	+++
A. undulatus	108	109	125	+	+++
A. everetti	108*	108	104	+	+++
Ceratogymna					
C. f. fistulator	106*	107	118	+	++
C. f. sharpii	106	108	122	+	++
C. f. duboisi	101	109	130	+	++
C. bucinator	109	110	121	+++	+
C. c. cylindricus	98*	111	122	+++	++
C. c. albotibialis	110	110	117	+++	++
C. s. subcylindricus	109*	109	125	+++	++
C. s. subquadratus	106	110	120	+++	++
C. brevis	104	109	115	+++	++
C. atrata	108	112	128	+++	+++
C. elata	106	111	128	+++	+++

* Body mass is estimated for one or both sexes of those species/subspecies for which it has not been recorded, based on the correlation with wing length derived from a sample of well-documented species (Table 2.1, Fig. 2.1)

When the casque is well developed, it consists of a keratin skin covering an uninterrupted airspace, with only a small portion in the rear supported by bony reinforcing struts similar to those in the rest of the upper mandible. The casque is virtually non-existent on young birds, other than being a paler area, well supplied with blood vessels, from which subsequent growth will originate. The final shape and volume of the casque vary from one species to another, and in many, but not all, the adult male has a larger (sometimes much larger) casque than the adult female.

The casque and bill probably have a role in communicating the age, sex, and status of an individual, in conjunction with parallel changes in eye, skin, and plumage colours. The bill changes in size, form, and colour and, in some species, also in the number of ridges, bars, or wreaths that have accumulated. The exact role of these changes in signalling age or sex to conspecifics remains to be tested experimentally. Such knowledge would be of considerable practical value in the study and management of a number of species, but it is already a useful means of separating ages, sexes, and individuals in the field.

The large unimpeded air cavity within the casque, together with the loud and nasal quality of the calls of many hornbills, also suggests an acoustic function (Alexander 1991; Alexander et al. 1994). This would be most useful if it amplified calls of certain frequencies, especially those appropriate to certain 'sound-windows' that might exist in a given environment, whereby their transmission would be enhanced. The most likely design is for the air volume within the casque to expand and contract, the frequency of such Helmholtz resonance being dependent primarily on the volume of air and the size of aperture leading into the casque.

Experiments, comparing the frequencies produced in calls with casque form and function as measured on fibreglass casts, showed a correlation between the structures of call and casque. The results were close to those predicted from Helmholtz theory, which is based on a spherical air volume. There was no obvious difference between forest and savanna species in the dominant frequency or its transmission, which was often a harmonic or multiple of the fundamental or basic frequency of the call. Casque resonance frequencies were, however, correlated with the fundamental frequency.

The casque may function as a feedback resonator to the syrinx, where the calls originate, or as an amplifier in its own right, but these options and the details of nasal apertures leading into the casque (Technau 1936) await further study. The long trachea and inflatable throat-pouch in the genera *Bucorvus* and *Aceros*, species without voluminous casques, may also add resonance to their calls, although details of these structures and their function are undescribed.

In all hornbill species, males have larger bills and casques than females, this possibly being related to their dominant role in territory-defence, mate-attraction, and gathering food for the breeding female and chicks. The casque is almost the same size in both sexes of some species, and its primary function can then be expected to be more practical, as in foraging or defence. In these species, secondary functions of the casque and bill, showing gender or status, can be added to the casque merely with colours or details of form, as for species in the genera *Anthracoceros* and *Buceros*. Among these, the Great Helmeted Hornbill has an exceptional casque, in that the front end is a block of solid 'ivory', such that the whole skull forms some 11% of the total body weight (Manger Cats-Kuenen 1961; see species account for details); the structure is similar in both sexes and is probably used primarily for feeding, although it is also used by males in fighting.

In species in which the casque differs notably between the sexes, the differences may reflect a primarily sexual or epigamic function. The details and extent of differences probably reflect different selection forces imposed on each sex, in part by the opposite sex, and may also be associated with differences in sex ratio and social organization.

Age differences

The same theory holds for differences related to age, where juveniles that differ very much from adults may enjoy a special status. This is apparently the case for many of the species known to breed co-operatively, in which immatures remain within the parental territory, are accepted within the social group, and often appear to delay adoption of mature colours compared with immatures of non-cooperative relatives. In other species, juveniles may resemble adults of either sex, and this dichotomy is especially striking in hornbills. For example, in the wreathed and wrinkled hornbills of the genus *Aceros* juveniles of both sexes resemble adult males, yet in wattled hornbills of the subgenus *Ceratogymna* they resemble adult females, with almost identical differences in plumage colours in each case. Genetic and demographic factors that correlate with these differences in juvenile mimicry would be most interesting to determine, given the relatively consistent roles of the adult male and female hornbill in courtship behaviour and breeding biology.

3

Non-breeding behaviour and biology

The normal day-to-day activities which hornbills perform are the subject of this chapter. There is obviously variation in these daily activities, depending on such variables as the weather, availability of food, and stage in the annual cycle, so that what follows must be taken only as a generalized account. The feeding ecology of hornbills, which applies daily but varies seasonally, and their breeding biology, which occupies only part of the annual cycle, are treated in subsequent chapters.

Emergence from the roost

Hornbills usually emerge from their roost at first light, often flying to a nearby and prominent perch on which they preen and stretch. This period of waking may be extended, and often leads to bouts of territorial calling and displaying in those species which exhibit such behaviour. Some species, many of which are frugivorous, roost communally, and for most of these the waking period is brief, since the first birds to leave the area often stimulate the rest to follow suit. Such roosts may also serve as information or social centres (see p. 43), so departure of the first birds may leave the others no choice if they wish to follow experienced birds to fruiting trees.

Most hornbills begin to feed once they leave the roost area, or at least start out on flights which will take them to feeding areas. Such flights may be undertaken singly, but in most species either a pair, a family group, a co-operative group, or a communal flock will move off together. However, even at the largest communal roosts, containing several hundred birds, most hornbills move in pairs or families which remain discrete and closely associated within the larger group. These initial flights of the day may cover anything from a few metres to tens of kilometres.

Location and collection of food

The exact location of food depends on the dispersion and density of those items that comprise the preferred diet of each species. Species with more evenly dispersed food sources, such as *Bucorvus* ground hornbills which feed on small animals, may begin searching for food from close to the roost, and continue feeding for long periods. Others with patchy food sources, such as *Aceros* wreathed hornbills which feed at fruiting trees, may have to travel far to reach their food, but on arrival can quickly obtain their fill. Typically, those that feed off patchy resources show behaviour that probably helps to communicate the location of patches, such as sharing communal roosts, moving in flocks, being audible in flight, and calling to one another when on the move.

Capture of food by any species can take place at any time of day and depends to a large extent on the habitat occupied and foraging

Non-breeding behaviour and biology 25

3.1 An adult female Great Pied Hornbill *Buceros bicornis* picking a small fruit from the ground and tossing it back into the throat. The species only rarely forages on the ground, but the position illustrates the short legs typical of most hornbills and the fused bases to the anterior toes.

abilities expressed by each species or individual. There are a number of basic feeding methods, the simplest of which is (1) merely **picking** up the food item wherever it is found (Fig. 3.1). More complex and energetic feeding methods include (2) **levering** over objects with the bill in search of items underneath; (3) **digging** into the ground or rotten wood in search of prey; (4) **snatching** food items taken on foot or in flight; (5) **swooping** down from a perch and picking an item off the ground; (6) **plucking** an item from the ground or foliage while still in flight; or (7) **hawking** airborne prey while on the wing. Each feeding method varies in the time and energy devoted to each capture, the postures adopted, and the success with which it is executed.

Manipulation of the food, once it has been captured, also varies. Many items are simply swallowed whole, which, in hornbills, owing to their long bill and short tongue, consists simply of tossing the item back from the bill tip into the throat. Some items are divested of extraneous parts before being swallowed, such as fruits of their skins or insects of their wings and legs. Others are softened before swallowing, the hornbill passing them through and crushing them in the bill, which is serrated in many species, or are cleaned of unwanted coverings by being wiped back and forth over a perch or along the ground, as in the case of hairy caterpillars, slimy toads, or juicy fruits. Some food is not swallowed immediately but carried in the bill tip for long periods, to be used in social interactions or, most frequently, for courtship-feeding or provisioning to nest-occupants. Feeding usually ends with cleaning the bill, either by scratching with the foot or, more commonly, by wiping the inside and outside of the bill on a perch.

There is also a succession of more physiological processes that relate to feeding and are discussed in more detail under feeding ecology (Chapter 4). These include how much hornbills eat and the effects of such feeding behaviour on their environment, including the rate and extent of digestion; the delay in, frequency of, and proportion of indigestible components, such as seeds, produced during regurgitation; the volume, content, and control of defecation; and the necessity and frequency of drinking.

Progression and maintenance behaviour

Hornbills progress mainly by flying, typically with an undulating flightpath in which bursts of flapping flight are interspersed with short periods of gliding. The lack of underwing-coverts, unique to the Bucerotiformes, means that some air passes through the base of the primaries, and in the larger species this produces a loud rushing noise, audible at over a kilometre. This sound may be augmented by that of air passing through the two short outer primaries, especially in those genera in which these primaries are emarginated and have stiff narrow tips. All hornbills also progress by hopping when on perches or on the ground, bounding along with both feet together. Only a few species spend much time on the ground and are capable of walking or running, the two *Bucorvus* ground hornbills and the five species of ground-dwelling *Tockus* hornbills. These species may also walk when moving along their perches.

3.2 An adult Southern Ground Hornbill *Bucorvus leadbeateri* rubbing its head and neck over the preen gland at the base of its tail. Members of this genus have an especially well-developed preen tuft, which they share with *Buceros* great hornbills. Note also the long legs for a terrestrial existence, the habit of standing only on the tips of the toes, and the fused bases to the anterior toes.

At intervals throughout the day, and sometimes at night, hornbills perform a variety of maintenance behaviours. Preening is the most obvious of these. This involves nibbling at individual feathers or drawing them through the bill, oiling the plumage by nibbling at the preen gland and smearing preen-oil over the feathers or by rubbing the head and neck over the gland (Fig. 3.2). The long bill and long flexible neck, together with the habit of completely opening one wing or fanning the tail during preening, lead to a variety of complex and characteristic postures during feather-maintenance.

Various stretching behaviours are also undertaken, either separately or during bouts of preening, including extending down one wing while stretching the leg on the same side beneath it, leaning forward and hitching up both unopened wings, or yawning with the jaw wide open and neck arched. In most hornbill species, the head is scratched indirectly, by bringing the leg over the top of the wing, but those species that walk on the ground scratch directly, with the leg raised in front of the wing.

Feather-maintenance and stretching are especially common after one of the popular bathing routines. Sunbathing is most widespread, a variety of postures being shown by different species, from just drooping the wings with the back to the sun through to lying prostrate with both wings and the tail fully spread (Figs 3.3, 3.4). Many species have also been recorded dustbathing, lying on the ground with neck outstretched and using the feet to stir up dust and the wings to distribute it over the plumage. No hornbill species has been recorded bathing in open water, even though some have been seen to enter standing water, probably to obtain food. Many species, however, bathe in wet foliage, and this, like the other bathing behaviours, can often be elicited from captive birds by providing them with an appropriate stimulus. The drying posture adopted after foliage-bathing, of raising the body plumage, fanning the tail, and drooping the wings, appears to be common to all species.

Hornbills occur in a variety of climates, from hot lowlands to cold mountaintops, and many show characteristic postures to gain or lose body heat. Heat-gaining postures, recorded only among smaller species, include turning the body side-on or back-on to the sun, drooping the wings (often only the one on the

3.3 Prostrate sunbathing posture by an immature African Grey Hornbill *Tockus nasutus*, usually performed only when completely at ease.

Non-breeding behaviour and biology 27

3.4 Sunbathing postures of three species of Asian hornbills (after Frith and Douglas 1978): (a) an adult Oriental Pied Hornbill *Anthracoceros albirostris* lying spread-eagled on the ground; (b) a juvenile Great Rhinoceros Hornbill *Buceros rhinoceros* with its back to the sun and wings drooped to expose the back and (c) an immature Great Pied Hornbill *B. bicornis* perched with its back to the sun and its tail and one wing spread.

3.5 An adult African Red-billed Hornbill *Tockus erythrorhynchus* warming up with its back to the sun, the wing nearest the sun lowered and the black flanks exposed.

3.6 Heat-loss posture of an immature Great Pied Hornbill *Buceros bicornis*, with bill agape, wrists apart to expose the bare underwings, and during which gular fluttering or panting may be performed. (After Frith and Douglas 1978.)

appropriate side), and raising the back feathers, which in some species are coloured differently from the rest of the plumage (Fig. 3.5). Heat-loss behaviours include gaping with the bill open, fluttering the gular region of the throat, and, in a few species at very high temperatures, even panting (Fig. 3.6). They also include moving into the shade, facing into the wind, often with the wrists apart to allow passage of air over the bare underwing surfaces devoid of coverts and, in the *Bucorvus* ground hornbills, elevation of the upperwing-coverts (Fig. 3.7).

Communication

Communication during the course of the day between members of the same and also of different species is obviously important. It is achieved by use of visual signals, vocalizations, or a combination of both. All hornbills possess various patterns of markings and coloration to their plumages and soft parts which serve as visual signals. This makes it relatively simple to distinguish the sexes in almost all species and also to separate various age classes in many of them. Most such markings remain relatively constant throughout the life of an individual, but in a few *Buceros* and *Aceros* species the coloration is applied cosmetically, as coloured oils from the preen gland, and this leads to considerable variation in the intensity and extent of these red, orange, or yellow pigments on the bill and white areas of plumage. The hornbills themselves undoubtedly use these visual signals to recognize the sex, age, and condition of a conspecific, and hopefully further study of the details and development of such markings will allow us to attain the levels of discrimination available to the hornbills themselves.

Visual signals may be passive but are often enhanced by movement and may be either stereotyped or variable across species within the order.

Basic aggressive, or agonistic, behaviour appears to be uniform throughout the order. It consists of raising the bill to expose the throat, which is often bare, brightly coloured, and indicative of age and sex in many species. Less overt, possibly redirected aggression is shown by banging the bill on a perch when close to an opponent, sometimes with sudden flicking movements of the wings. Only in a few species are the wings or tail regularly fanned in special threat postures, such as the crouch of the Great Helmeted Hornbill *Buceros vigil* or the heraldic pose of the White-crowned Hornbill *Aceros comatus* (Fig. 3.8). Aggression only rarely leads to fighting, which involves grappling with the bill or butting with the casque, either on the ground or in the air, and usually directed at the head and neck but sometimes at the wings or legs.

The most ritualized form of aggression is the territorial display, lacking in many forest and fruit-eating species but elaborate and characteristic of others, especially the more carnivorous species of open savanna habitats. Such display is usually performed by both members of a pair, often joined by juveniles or helpers within family or co-operative groups. It is often performed by young birds even before they leave the nest, suggesting primarily an aggressive rather than a sexual motivation. The postures adopted may be similar among related species, but they differ between groups of species in whether the head is held raised or lowered and the extent to which the wings and tail are expanded during display (Figs 3.9–3.14). Various territorial calls may be uttered either alone or in conjunction with the special displays.

The calls of larger species tend to be deeper and louder than those of small species, but a number of calls are similar and probably homologous for all hornbill species, while others differ and are specific to only one species even though they may be used in the same context. Furthermore, calls of selected savanna- and forest-dwelling species do not show significant differences in call frequency as might be expected from their range of body sizes and habitats (Alexander 1991). This indicates that other factors besides environmental ones may also be influencing optimal call structures and communication frequencies.

Non-breeding behaviour and biology

3.9 An adult Southern Ground Hornbill *Bucorvus leadbeateri* giving the loud, deep territorial call with the head lowered and while walking on the ground.

3.7 Heat-loss posture of an adult Southern Ground Hornbill *Bucorvus leadbeateri*, with bill open, wrists apart, and wing-covert feathers raised.

Calls which are similar among species include the fright call, a harsh squawk uttered when surprised or under duress, as when mobbing or being caught by a predator, being accosted while in the nest, or being handled during research. Another harsh squawking call, the acceptance call, is uttered by the occupants of a nest when taking food delivered to them. It is usually given as a single squawk by the female but is often extended by the chicks into a continuous series, the begging

3.8 An adult female White-crowned Hornbill *Aceros comatus* in aggressive posture, with bill open, wings extended, and tail fanned.

call, from which the louder acceptance call may be derived. The female and chicks also utter this call when out of the nest, the female when being courtship-fed by the male before breeding and the chicks when being fed after leaving the nest. The begging call of small chicks is a much quieter cheeping note, although no less insistent. The only other call common to most species is the displeasure call, a low growl uttered when approached too closely by another bird of the same or another species.

The loudest calls, probably with territorial and mate-attraction connotations, are the most ritualized, most aggressive, and most specific to each species. They are also the ones most often combined with elaborate and ritualized display movements. They may be derived from simple warning, anxiety, or alarm calls, extending the single whistle, cluck, or grunt given under such circumstances into loud and multiple versions of the same notes with variations in volume, tempo, and frequency. Simpler versions of the loud calls also serve to maintain contact between mates and their young, building up from a series of simple notes into the full loud call at the highest possible intensity.

The loud calls of many of the bigger hornbills are notable for their high volume and nasal tone, which indicate the existence of amplification and of several harmonic frequen-

3.10 An adult Hemprich's Hornbill *Tockus hemprichii* at the end of its territorial display, given with the head raised, while uttering high whistling calls, and terminated when the tail is fanned over the back. A display with similar movements and whistling calls, but lacking the tail-raising finale, is performed by three other related species, the African Crowned Hornbill *T. alboterminatus*, Bradfield's Hornbill *T. bradfieldi* and the African Pied Hornbill *T. fasciatus*.

3.11 An African Grey Hornbill *Tockus nasutus* in territorial display, uttering high piping calls with the bill raised and the wings flicked in and out with each phrase. The related Pale-billed Hornbill *T. pallidirostris* performs a similar display.

cies in the call. Many of these species also have large casques which are hollow internally and connected to the mouth by narrow openings. It has been suggested that the casque may amplify the calls, and it is possible that the inflated throat-sac and large trachea of such species as the *Bucorvus* ground hornbills may also enhance production of their low-frequency calls (see p. 22).

Some aspects of behaviour have no immediately obvious function, are most often performed by juvenile birds, and appear to fulfil components of play or training. These include bouts of manipulating objects such as sticks and mammal dung with the bill or levering over bark, which are typical foraging behaviours. They also involve high-speed chases of other birds, whether conspecific or not, undertaken on foot or in flight and very similar to aggressive encounters, or forms of bodily contact such as bill-wrestling or jumping on conspecifics, which are similar to fighting behaviour shown by adults. Such play is most often observed when departure from a roost is delayed, as by territorial disputes, or when the day ends with some free time before going to sleep. Adults may break up these bouts of play if they become too robust, or if movements to feeding or roosting areas are being disrupted.

Lastly, there is one form of communication that is not easily interpreted. Chicks that have left the nest but are still dependent on their

Non-breeding behaviour and biology 31

3.12 An adult male Southern Red-billed Hornbill *Tockus erythrorhynchus rufirostris* performing its territorial display, with head bowed, wings slightly raised, and uttering high clucking calls. The display is similar to that of the parapatric Monteiro's Hornbill *T. monteiri*.

3.14 An adult Southern Yellow-billed Hornbill *Tockus leucomelas* in territorial display, with head bowed, wings fanned wide open, and giving a series of clucking calls. A similar display is performed by two other species, the Eastern Yellow-billed Hornbill *T. flavirostris* and Von der Decken's Hornbill *T. deckeni*.

3.13 An adult male Northern Red-billed Hornbill *Tockus e. erythrorhynchus* performing its territorial display, with head raised, wings partly open, and uttering high clucking calls. The display is different from that of the southern subspecies (Fig. 3.12), suggesting possible specific separation.

parents often approach them with their head and neck feathers raised, uttering a growling call similar to the displeasure call and stopping only when they reach the parent. Such behaviour may reflect motivational conflict and indecision between soliciting food, greeting an adult, and parent-offspring conflict.

Roosting behaviour

Roost sites appear to be rather specific in form and regularly used by those species for which such behaviour has been observed in detail. Species that form communal roosts, notably *Aceros* wreathed hornbills in Southeast Asia and *Ceratogymna* species in Africa, may use the same roost for months on end and return to it over long distances at the end of each day. Such roosts are often high up in stands of particularly large trees, with perches either protruding as dead snags or concealed within dense foliage. Most territorial species also have regular roost sites within their home-range, situated at the tips of slender branches, deep within foliage, or tucked under large boughs, depending on the species. Only the large *Bucorvus* ground hornbills, with unusually extensive territories, do not use regular roosts, probably since the energetic cost of returning each night to the same roost is prohibitive.

Breeding females are obviously restrained to roosting in their nests, but breeding males, or other group-members in co-operative species, may either roost near the nest or travel some distance away to a more regular site, even to a communal roost. In all instances, the choice of roost site is probably determined primarily by how safe it is from predators, and secondarily by how conveniently situated it is in relation to feeding areas or nests. Any changes in the status of a roost with regard to these factors will induce movement to another site, the result of which is that the use of and number of birds at available roosts are rarely constant.

Once at their roost, hornbills adopt a typical sleeping posture, crouching down with the belly against the feet and drawing the head back between the shoulders so that the bill points upwards at a characteristic angle (Fig. 3.15). The exact time at which each species enters the roost varies during the year, but some tend to be earlier and others later in relation to sunset. Some territorial species, such as the *Bucorvus* ground hornbills or African Crowned Hornbill *Tockus alboterminatus*, are suspected of mock-roosting, entering an apparent roost site at dusk and squatting down as if to sleep, only to fly off in the last rays of light to a final roost site several hundred metres away. Such behaviour may serve to confuse any predators that may be watching the hornbills going to roost.

Interactions with other animals

Some forms of behaviour are most often seen during interactions with other animals. The characteristic inquisitive posture is adopted where a hornbill watches, and is uncertain of, a creature below its perch. It results from the body being kept horizontal, with the neck hung below the perch on one side, the tail on the other, and the head turned slowly from side to side to view the object with alternate eyes.

Predators also elicit different responses from hornbills. When they are distant or, through their small size or less rapacious abilities only mildly threatening, predators are greeted with the warning call and kept under observation. If they come closer or appear suddenly, they may elicit the fright call and either a retreat to a safe distance or an emergence to an open perch. Dangerous predators, such as large raptors, man, or the big cats, usually produce a precipitous flight to cover, followed by a notable silence and stillness. If the predator, judging by its species or lack of hunting behaviour, does not pose an immediate risk, hornbills may start mobbing behaviour. They then perch by or flutter over the

3.15 An adult Bar-pouched Wreathed Hornbill *Aceros undulatus* crouched down on its perch to roost, with head drawn back and bill raised in the posture typical of sleeping hornbills.

3.16 An adult male Southern Ground Hornbill *Bucorvus leadbeateri* preening the female prior to copulation, the grooming being performed so heavily that the female is forced into a crouching position.

predator, utter the warning call, or, at higher intensities, repeat the loud and penetrating fright call. Certain animals regularly attract intense mobbing, especially during the breeding season, such as leopards and lions by *Bucorvus* ground hornbills or African Gymnogenes *Polyboroides typus* by small *Tockus* hornbills, suggesting that they are recognized as important nest-predators.

Most social interactions with other hornbills involve members of the same species. They take place between mated pairs, parents and offspring, or members of co-operative groups, most of which are likely to be genetically related to one another. Mutual grooming, or allopreening, is the most common activity and takes place between birds of any age or either sex, although in some species it is also a prelude to copulation (Fig. 3.16). Mutual exchange of food, or allofeeding, is also common, especially among members of co-operative groups, where it may serve to establish and reinforce hierarchies of dominance between members. Other co-ordinated activities, such as cohesive movements of flocks, territorial defence, or bouts of calling, may further facilitate and reinforce interaction between individuals.

Pre-breeding behaviour

Behaviour associated with reproduction may be observed at any time of the year but, obviously, is most prevalent just prior to the breeding season. This includes courtship, nest-selection, nest-preparation, copulation and nest-sealing. Courtship is often signalled by closely co-ordinated activities between members of a pair, when they fly, perch, and feed close together and react swiftly to each other's movements. Allopreening and allofeeding increase and become directed mainly by the male at the female. This leads later to courtship-feeding, until eventually the female no longer needs to collect any of her own food.

Inspection of nest-holes often begins or increases during this period and is usually led by the female, with the male following close behind. The female perches at any possible nest-holes, pecks at the entrance rim, inserts her head, and later enters the cavity to remain inside for longer and longer periods. The male continues to deliver food to the female, although the attraction of the nest site also becomes so strong that the male often offers food into the empty nest, even while the female is perched in the open nearby. The female crouches quietly within the nest for much of the time, her behaviour interspersed with bouts of chipping away at the walls of the interior with her bill and rearranging the lining which the male now starts to deliver between visits with food. When the female emerges after each visit, the pair may remain near the nest for some time, depart from the area for the rest of the day, or return only at intervals, sometimes of several months, before nesting finally gets under way.

A commitment to breed is usually heralded by the start of regular copulation and by preliminary sealing of the nest. Copulation often takes place near the nest after the female emerges from a bout of nest-preparation, and it may or may not be preceded by courtship-feeding. With most hornbill species it is silent and takes place without any preliminaries, although in some it is preceded by heavy allo-preening that forces the female into a crouching position, as with *Bucorvus* ground hornbills (Fig. 3.16), or by the male leaping back and forth over the female's back, as with arboreal *Ceratogymna* and *Buceros* species. Copulation is also frequently observed away from the nest in savanna species of *Bucorvus* and *Tockus*, but this may be a bias resulting from the difficulty of observing forest species for long periods once they leave the nest area.

Preliminary sealing of the nest may be conducted either from outside, sealing any cracks that might lead into the cavity, or, more often, by the female working from inside. The extent to which the male helps depends on the species, and most assistance is in the form of ferrying soil, mud, and sticky food items, such as millipedes and fruits, to the female within. Only rarely does the male himself apply some sealing material to the outside, this done in a desultory fashion.

Moulting

There is one final set of activities which occurs more or less on an annual basis, in both sexes and all ages classes, but which is not necessarily connected with either feeding or breeding. This is the regular replacement of feathers by moulting, which is an integral part of basic hornbill biology, essential to their survival, and may place some metabolic load on the individual.

Juveniles of some species appear to undergo a partial moult of their body feathers, at least of their first plumage of rather fluffy head and neck feathers. This takes place in the period between when they leave the nest and before they start their first full moult into adult plumage at about one year of age. These changes are often initiated in those areas of the plumage that undergo the most obvious alteration in colour when becoming adult, even among species in which the body feathers are not replaced until the first complete moult. For example, in juvenile female *Aceros* or juvenile male *Ceratogymna* species, the head and neck areas that change from brown to black are the first areas to moult, and such signs of maturity are often synchronized with maturation of soft-part colours in the bill, casque, eyes, or bare facial skin. These changes usually take place before their first breeding season, which for smaller species with fairly regular breeding cycles is the one following that in which they were raised.

Smaller species of hornbill most often exhibit a post-juvenile body moult, and thereafter their moults are also the most regular, most frequent, and shortest in duration. In larger species, mature plumage may be developed only after two or three years, and moult, especially that of the larger flight feathers, may be suspended between seasons. This means that a complete replacement of all feathers may be spread over several years, and that successive waves of moult may be initiated and be evident among the flight feathers.

Moult often overlaps with breeding, an unusual attribute of hornbills in that it imposes increased nutritional requirements during what is already a demanding period. Breeding females of many species undergo a virtually simultaneous moult of all the major flight feathers, the remiges and rectrices, but with a normal sequential moult of the body feathers (see Chapter 5 for further details). However, moult is often coincident with breeding even among those species where moult of all feather tracts proceeds by a normal sequential progression of feather loss. Males may also undergo a gradual sequential moult while breeding, although in some species or individuals the moult may be suspended until late in the breeding cycle, once foraging demands on the male are reduced.

Non-breeding adults and juveniles usually moult during the period when most other birds in the population are breeding, but there is a need for much more accurate information on details of moulting and its integration into the annual cycle. The exact sequence of feather moult is also not well established, either within or between species and genera. Examination of study skins in museums suggests that the 10 primaries, 12 secondaries and 10 tail feathers are usually not moulted symmetrically in pairs, one from each side of the body (Stresemann and Stresemann 1966; Kemp 1976a). This has been considered exceptional for the elongated central pair of tail feathers of the Great Helmeted Hornbill, where apparently only one feather is shed at a time, possibly on an annual basis (Wetmore 1915; Friedmann 1930). Such asymmetry in feather loss extends to the other pairs of tail feathers, and is a trait shared with hoopoes among other groups of birds (Stresemann and Stresemann 1966).

There appears to be some consistency in moulting patterns of the major wing and tail feathers within each of the genera that has been studied (Stresemann and Stresemann 1966; Kemp 1976a), but there is a need for much additional research. In all species studied, moult of the tail feathers usually started with the central pair (T1), then the outermost pair (T5), before moving across between these pairs (sequence T1 or T5, T2 or T4, T3). Moult of the wing feathers was less systematic, usually starting with the outermost primary (P10), followed by a second centre further in (P3 or P4), after which moult spread from both centres, especially the inner one. No clear sequence existed among the secondary feathers, other than that at least two moult centres were initiated, and once again the patterns were much obscured in larger species by suspended moult and successive waves of moult. Moult of the body feathers is virtually unstudied, but in smaller *Tockus* species was initiated in the sequence abdomen–upper back–breast–lower back–head–neck, although several areas are in moult together, which confuses the pattern (Kemp 1976a).

4

Feeding ecology

Feeding ecology and life cycle

The types of food required by each hornbill species are the primary determinant of when, where, and how they feed during their daily activities. These requirements may be divided into general ones, such as those fulfilling daily subsistence requirements of materials, energy, and moisture, and more specific ones, such as minerals needed for eggshell-formation or special proteins for feather growth. The general requirements have to be available on a day-to-day basis throughout the year, while the specific needs may be required only at certain times and not necessarily in large quantities, even though they are just as essential for completion of the life cycle.

Availability of different foods within the environment has further implications for how hornbills will respond on a given day and when they will undertake such elements of their annual cycle as partial migrations, moult, and breeding. Immediate or proximate information about food, such as its density (high or low), dispersion (uniform or clumped), geographical availability (widespread or local), temporal availability (seasonal or perennial), nutritional content (water, energy, and trace substances), and attractiveness (ripeness, proportion of waste, and harvesting costs), all bear on what daily returns a hornbill can anticipate from its preferred food sources.

Other factors which operate over longer time periods are less obvious in action but do ultimately affect which feeding ecology will evolve. These include the influence of food on timing and success of reproduction, the effect of food on timing and extent of mortality, the extent of social behaviour that will develop within a species, and the role of hornbills within their community as dispersers of fruits or predators of seeds and small animals.

Feeding habits

The feeding habits of only 15 species of hornbill (28 per cent) have been studied in any detail, but they represent a sufficient diversity of diets and feeding behaviours to permit some wider interpretations within the order (Table 4.1). The species for which aspects of feeding behaviour, diet, and food dispersion are known include a *Bucorvus* ground hornbill, the largest and most carnivorous species; four species of *Tockus* hornbill, among the smallest and most insectivorous species; and a total of nine medium-sized to large forest-dwelling hornbills in the genera *Ceratogymna* (one), *Anthracoceros* (two), *Anorrhinus* (two), *Aceros* (two), and *Buceros* (two), which are mainly frugivorous but include various proportions of animal food within their diets. These species also represent a range of social organization, from solitary foragers, through others that live in co-operative groups, to ones which form into large communal feeding-flocks. Sufficient information is also available on at least the diet of most other hornbill species, or of their close relatives, to assign them to a basic food guild

Feeding ecology

Table 4.1 Subjective comparison of the diet and feeding habits of 15 well-studied hornbill species (see individual species accounts for details and references)

Species	Diet		Feeding behaviour			
	Frugivorous	Carnivorous	Aerial	Perched	Terrestrial	Special abilities
Bucorvus leadbeateri		+++++		+	+++++	Digging, pursuit, grouping
Anorrhinus austeni	+++++	+++	+	++++	+	Grouping, peeling husks
Tockus alboterminatus	++	++++	++	++++	++	Flight buoyancy
T. nasutus	+	+++++	+++	+++	++	Flight buoyancy
T. monteiri		+++++		+	+++++	Digging
T. erythrorhynchus		+++++	+		+++++	Digging
T. leucomelas	+	+++++	+	++	++++	Versatility
Anthracoceros coronatus	+++++	++	++	++++	++	Twisting/hammering fruit
A. albirostris	++++	+	++	++++	++	Versatility
Buceros bicornis	+++++	++		++++	++	Peeling off husks/bark
B. rhinoceros	++++	++		++++	+	Peeling off husks/bark
Aceros narcondami	+++++			+++++		Frugivory, social
A. undulatus	+++++	+		++++	++	Wide-ranging, social
Ceratogymna subcylindricus	+++++	+	+	++++	+	Frugivory, group hunting
C. brevis	+++++	+	+	++++	+	Frugivory, group hunting

and hence to allow some predictions about their feeding ecology (Table 4.2). A comparison with feeding habits and ecology of toucans in the New World would be interesting, since none attain the large size of hornbills but many have long bills and all have long tongues to employ in food acquisition.

The basic motor patterns by which hornbills capture and manipulate their food, and which reflect the overall size and design of each species, have been described together with other aspects of their basic biology in Chapter 3. All hornbills have relatively long and heavy bills, the mandibles of which meet exactly at the tips but do not make contact further back, where they are often heavily serrated. The bill operates principally as a pair of forceps, grasping objects held in the tip with great dexterity and force and then tossing them back into the throat (Burton 1984), sometimes after crushing and softening them between the open serrated areas. The long bill, combined with the long, flexible neck and stout, grasping feet, enables food to be reached over a wide area and usually without having to adopt any special body positions. It also allows dangerous food items, such as poisonous animals or plants, to be held well away from the fleshy areas of the face, body, and short tongue while being rendered harmless in the horny mandibles.

Patterns of food capture include picking, levering, digging, chasing, swooping, plucking, and hawking, while forms of preparation include crushing, softening, wiping, carrying, and swallowing (see p. 25). Some of these categories may be either combined or subdivided further, depending on the level of investigation and differences in feeding behaviour shown by different species. The costs of food capture and preparation are estimated to be of relatively little importance in the overall energy and time budgets of hornbills, compared with the costs of locating widely dispersed food sources and of digesting food once secured. The relative amounts of time and energy involved do, however, differ for each method of food capture and preparation, and may be important in understanding differences in feeding ecology between closely related species or different sex and age classes.

Table 4.2 Assignment of lesser-known hornbill species to feeding guilds, based on what is known of their diet and by comparison with well-studied species (marked with an asterisk, from Table 4.1)

Species	Arboreal	Terrestrial	Frugivorous	Carnivorous	Special attributes
Bucorvus					
B. abyssinicus		+++++		+++++	
*B. leadbeateri**	+	+++++	+	+++++	Digging, grouping
Anorrhinus					
*A. austeni**	+++++	++	+++++	+++	Grouping
A. tickelli	+++++		+++++		Grouping
A. galeritus	+++++	+	++++	++	Grouping, peeling husks/bark
Tockus					
*T. alboterminatus**	+++++	+	++	++++	Buoyant flight
T. bradfieldi	+++	+++	+	+++++	Digging, buoyant flight
T. fasciatus	++++	++	+++	++++	Buoyant flight
T. hemprichii	+++	+++	+	+++++	Buoyant flight
T. pallidirostris	++++	++	+	+++++	Buoyant flight
*T. nasutus**	++++	++	+	+++++	Buoyant flight
*T. monteiri**		+++++		+++++	Digging
*T. erythrorhynchus**		+++++		+++++	Digging, running
*T. leucomelas**	+	+++++	+	+++++	Versatility, running
T. flavirostris	+	+++++	++	++++	Running
T. deckeni	+	+++++	++	++++	Running
T. hartlaubi	+++++	+	+	+++++	Ant-following
T. camurus	++++	++	+	+++++	Grouping, ant-following
T. albocristatus	+++++	+	+	+++++	Agile flight
Ocyceros					
O. griseus	+++++	+	++++	++	
O. gingalensis	+++++	+	++++	++	
O. birostris	++++	++	+++	+++	
Anthracoceros					
*A. coronatus**	+++++	++	++++	++	Versatility
*A. albirostris**	+++++	++	++++	+	Versatility
A. marchei	+++++	++	++++	+	
A. malayanus	+++++	+	++++	+	
A. montani	+++++	+	+++++	+	
Buceros					
*B. bicornis**	+++++	+	+++++	++	Peeling off husks/bark
*B. rhinoceros**	+++++	++	+++++	++	Peeling off husks/bark
B. hydrocorax	+++++		+++++	++	Grouping
B. vigil	+++++		++++	++	Axe-like bill
Penelopides					
P. exarhatus	+++++		++++	++	Grouping
P. panini	+++++		++++	++	
P. manillae	+++++		++++	++	
P. affinis	+++++		++++	++	
P. mindorensis	+++++		++++	++	
Aceros					
A. comatus	++++	++	+++	+++	Grouping

Table 4.2 (cont.)

Species	Feeding guild				Special attributes
	Arboreal	Terrestrial	Frugivorous	Carnivorous	
A. nipalensis	+++++	+	+++++	+	
A. cassidix	+++++		+++++		Social, mobile
A. corrugatus	+++++		++++	++	Mobile
A. leucocephalus	+++++		+++++	+	
A. waldeni	+++++		+++++	+	
A. narcondami★	+++++		+++++		Social
A. plicatus	++++	+	+++++	+	Social, mobile
A. subruficollis	++++	+	+++++	+	Social, mobile
A. undulatus★	++++	++	+++++	+	Versatile, social, mobile
A. everetti	++++	+	++++	+	
Ceratogymna					
C. fistulator	+++++	+	+++++		Social, mobile
C. bucinator	+++++	+	+++++	+	Social, mobile
C. cylindricus	+++++	+	+++++	++	Mobile
C. subcylindricus★	+++++	+	+++++	++	Social, mobile
C. brevis★	+++++	+	+++++	++	Social, mobile
C. atrata	+++++	+	+++++	+	Grouping
C. elata	+++++	+	+++++	+	Grouping

Drinking

Drinking of open water is probably not essential to any hornbill species, regardless of its diet. Indeed, drinking behaviour has been recorded for only four species, and even then only as a remarkable event. It has been noted for Black-casqued Wattled Hornbill *Ceratogymna atrata*, both for a captive and from a sequence apparently filmed in nature, for captive African Red-billed Hornbill *Tockus erythrorhynchus* and Malay Black Hornbill *Anthracoceros malayanus*, and for wild Eastern Yellow-billed Hornbill *Tockus flavirostris*. This infrequency of drinking means that hornbills must obtain all moisture requirements from their food, a factor to be borne in mind when considering whether some food items, such as figs, are being selected primarily for their nutritional or for their moisture content.

Some species, especially in the genera *Anthracoceros* and *Aceros*, are known to enter water, but apparently only to feed on crabs or small fish. Most species that have been studied in detail show no interest in open water, either for bathing, feeding, or drinking. This is despite the fact that water is available to many species, in both forest and savanna, as streams, seasonal pools, or perennial rivers. Furthermore, many species use mud taken from beside water for sealing their nests.

Lack of drinking and control of salt-levels in the bloodstream are of special interest in relation to carnivorous species whose diet contains excessive quantities of such salts, such as the *Bucorvus* ground hornbills. In other predatory and marine species, such as raptors and seabirds, the kidneys are assisted in excretion of excess salts by special glands situated above the eyes in depressions visible on the skull. Ground hornbills show no evidence on their skull or under their facial skin of possessing such glands, yet they are carnivorous and at times have a trickle of fluid running from their nostrils along a groove in the bill, much as in birds of prey when they secrete salts after a meal. Possibly another structure or mechanism is involved, and the special bi-lobed rather than tri-lobed kidney unique to the Bucerotiformes may also have unusual properties (see Chapter 6).

Diet-selection

All hornbill species are omnivorous, eating animal and plant foods in proportions that differ between species (Tables 4.1 and 4.2), and probably also within species between seasons and different sex and age classes. Certain properties of each food category, be it animal (mainly arthropods and small vertebrates) or plant (mainly fruits), have different effects on hornbill feeding ecology and need to be considered separately. Hornbills, in turn, will have had some effects on the food sources alongside which they have evolved. The range of hornbills extends over four great floral assemblages, the African, Indian, Indo-Chinese, and Malesian, together with all the other animal communities which they support.

The geographical area occupied by each hornbill species obviously determines the diversity of available foods. For example, the savannas of Africa are especially rich in small animals, while the forests of Southeast Asia, such as those in Borneo, are renowned for their variety of edible fruits, especially of the family Lauraceae with their nutritious pericarps (Crome 1975). Regional timing of availability is also relevant, such as annual peaks in fruiting on the Malay Peninsula and in Thailand at the end of the dry season and start of the monsoon (Medway 1972), or wide variation in fruit abundance between years in the less seasonal climate of Kalimantan, where a few weeks without rain at any time of the year may trigger a bout of flowering and fruiting (Leighton 1982; Leighton and Leighton 1983).

Animal prey tends to be more evenly and thinly distributed than fruits, both spatially and seasonally. It is usually also more elusive and difficult to harvest, all of which means that hornbills have to spend longer periods foraging for animals to reap equivalent returns (Beehler 1983; Leighton 1986). This has the effect that hornbills with a primarily animal diet tend, like their food source, to occur rather evenly and at low densities within suitable habitat. They also tend to be sedentary and territorial within such habitat, defending a fixed foraging area in which members of a pair often hunt independently and where they are familiar with seasonal and spatial variations in prey availability.

The total area which they defend is also related to their foraging behaviour, so that arboreal-foraging species in the open tree savanna, such as the African Grey Hornbill *Tockus nasutus*, require about three to four times the area of the sympatric ground-foraging African Red-billed Hornbill or Southern Yellow-billed Hornbill *T. leucomelas* (Kemp 1976a). Similarly, species reliant on forest may not be able to exist in patches which are too small for their requirements, as in the case of the African Crowned Hornbill *T. alboterminatus*, or in which insufficient dense tangled habitat is available, as for the White-crowned Hornbill *Aceros comatus*. It is notable that most species defend territories only against their own species, which would compete directly for resources, with the result that up to eight species of hornbill with different feeding habits often coexist in a complex mosaic of overlapping territories.

Obviously, animal food items are sometimes available abundantly and locally, as with rodent, caterpillar, or locust outbreaks, but such feasts are usually short-lived and often occur at times when other food is also relatively abundant. The general sparseness of animal sources of food also means that the more carnivorous hornbills frequently have to adopt specialized feeding techniques, such as digging or plucking. This is especially evident during times of low food availability, such as the dry season in savanna or the wet season in monsoon forests.

Some carnivorous hornbill species even associate with other animals for the prey they disturb, such as the insects flushed by armies of ants, parties of birds, bands of squirrels, or troops of monkeys (Moynihan 1978; Willis 1983; Brosset and Erard 1986; Terborgh 1990). There may be additional benefits from these associations, such as increased vigilance for predators, which could be of benefit to both hornbills and other vertebrates attracted to such beaters. At least one of these associations, the hunting of grasshoppers by some *Tockus* hornbills in collaboration with Dwarf Mongooses, has reached the level of a mutualism, where both partners benefit and both have behaviours by which to initiate and enhance the association (Rasa 1983).

In a few instances, the hornbills themselves may serve as beaters, but with no apparent benefit to themselves, as when chanting goshawks *Melierax* follow *Bucorvus* ground hornbills for the quail they flush. To date, almost all these associations have been described with reference to African hornbill species. Nevertheless, a *Spizaetus* hawk-eagle has been seen hunting with a group of Bushy-crested Hornbills *Anorrhinus galeritus* for over 3.5 hours, chasing prey flushed by the hornbills, and several other Asian species are recorded interacting with this genus of eagles. It is likely that more such associations will be discovered by observant fieldworkers in forests, especially in Southeast Asia, and that observation of undisturbed animals will reveal even greater complexity among the associations already reported.

Fruits eaten by hornbills fall into three groups: those rich in carbohydrates and water, especially figs; a variety of drupes and capsules that are rich in fats or lipids; and some watery fruits in thick husks (Leighton 1982). Figs are particularly important as a basic food item, since they occur widely in most forest types, exist as a variety of species, are often common, fruit frequently, produce heavy crops, the individual trees often fruit asynchronously, their fruits have little waste with a high ratio of flesh to seed, and some are relatively rich in protein, including the pollinating wasps and their parasites (Jansen 1979; Lambert 1989; Lambert and Marshall 1991). Figs provide, therefore, a plentiful and easily harvested source of short-term energy (sugars) and moisture, on condition that the location of trees currently in fruit can be ascertained. Some figs, and maybe also fruits of *Diospyros* species and the families Burseraceae and Lauraceae, may also be important during breeding for their relatively high protein content, which in some species is almost sufficient for rearing chicks (Kalina 1988).

Drupaceous, capsular and husked fruits are produced by a greater variety of plant species, but are usually available in lower numbers than figs on any particular plant. They also tend to have a higher proportion of waste than figs, either as a hard covering that must be removed or by the comparatively large seeds they contain. These factors make them relatively expensive to harvest, inefficient to store in the stomach, or demanding to digest. Furthermore, many of these plants tend to fruit seasonally or to show considerable variation in fruit crops between years, resulting in temporary shortages during the intervening months (Leighton 1982; Leighton and Leighton 1983). However, the high lipid content of many of these fruits, such as *Trichilia* species in Africa or *Connarus*, *Aglaia*, *Ocotea*, or *Virola* in Asia, provides an efficient source of energy that can readily be converted into fat reserves by the hornbills. These fruits, therefore, although relatively expensive to acquire and assimilate, and usually only seasonally abundant, are important to many species in forming reserves for lean periods or for use during reproduction (Leighton 1982; Kalina 1988).

Animal foods tend to have a higher protein content than fruits and to come in larger packages with less waste, especially when the prey is small vertebrates. Sources of protein are most important for tissue-formation and body growth, and hence are in high demand during the time that chicks are growing up in the nest. This is indicated by the increase in the animal component of the diet delivered to nests during the nestling period, especially evident in those species which are most frugivorous during the rest of the year. Fruit still predominates in the diet of the female of these species during incubation and is also included in the diet of chicks, especially later in the nestling period and in the form of lipid-rich drupes and protein-rich figs and other fruits.

It is probable that vegetable food is harvested when harmful substances, functioning as defences for the plants, are at their lowest and nutritional components at their highest. Fruits are usually taken only when perfectly ripe, being then low in phenol compounds but high in sugars, and foliage and flowers are taken as buds, low in tannins but high in sugars. The colours of the various fruit species eaten by hornbills when ripe are mainly red, black, or purple, a phenomenon typical of most avian frugivores. Species with dehiscent capsules (those which split open), red and black colour, and fleshy, oily arils around the seed are especially attractive to hornbills (Gautier-Hion *et al.* 1985). The diversity of fruiting species and fruit structures in evergreen forest suggests that considerable learning and experience in selecting fruits is necessary on the part of the birds.

Particular foods may also be sought for special substances they contain, such as selection of empty snail shells prior to egg-laying, possibly as a source of calcium salts for eggshell-formation. This may also apply to other items eaten only rarely, such as flowers with pollen rich in certain proteins, buds rich in certain sugars, or fungi rich in vitamins. Other foods may be taken regularly but eaten only rarely, such as millipedes or some sticky fruits which, at least during breeding, serve mainly as a source of sealing material for the female or chicks in the nest. These rather specific items, which might appear negligible in the overall diet and foraging behaviour, could be essential during particular parts of the annual cycle, and deserve special attention during research and management.

Location and collection of food

Much of the feeding ecology of hornbills has to do with how each species locates and harvests its preferred foods. The better-studied and most omnivorous hornbill species are known to harvest just over 100 different taxa of small animals within their home-range on the African savanna, and certainly many more if every taxon could be identified to its constituent species and if all those missed by the survey procedures were included (Southern Yellow-billed Hornbill: Kemp 1976a). They also include in their diet at least 45 taxa of small fruits and seeds.

Several Asian forest hornbills include, in their breeding-season diet alone, 30–35 species of fruit (Poonswad *et al.* 1983, 1987), which is only a subset of all fruiting plants available. Where seven species of hornbill coexisted within 3 km^2 in Kalimantan, among the richest moist evergreen forests in the world, at least 900 species of fruiting trees and lianas were present, but only 128 were definitely used by hornbills and a further 144 probably were (Leighton 1982). Even in the less rich African forests, 67 species of fruit in 42 genera were recorded in the diet of the Grey-cheeked Hornbill *Ceratogymna subcylindricus* in Uganda (Kalina 1988), and there were at least 120 species of fruit available to frugivores in 2 km^2 of forest in Gabon (Gautier-Hion *et al.* 1985).

Finding and collecting such a wide variety of food items, some of which, such as certain liana fruit or insects, are quite rare within the habitat, indicates the intimate relationship between hornbills and their environment. Hornbills tend to favour larger fruits taken from the relatively few larger trees within the forest. This behaviour, together with their ability to open various capsules, raises them above competition from many smaller frugivores, but demands of them efficient location and harvesting behaviour (Leighton 1982). Taken in conjunction with the manner in which each food type is processed by each hornbill species, this must lead to complex and dynamic interactions between the birds and their sources of food. It will also have important long-term effects on the population dynamics of their food-base, through their predation on animals and fruits or their distribution of seeds.

Those hornbill species which are primarily carnivorous or insectivorous, and which reside within a territory for all or most of their lives, tend to harvest their food by methodical searching. They spend whatever time is necessary in proclamation and vigorous defence of their territory, by calling, displays, or combat, and their steady movements through the area, at least in the case of *Bucorvus* ground hornbills, probably serve both to locate prey and to patrol their extensive and exclusive hunting reserve.

Prey is usually encountered at a relatively low rate during much of the year, and special foraging techniques, such as digging or excavating (Kemp 1976a; Kemp and Kemp 1978), or use of alternative food sources, such as fruit (Lack 1985), are employed by most species during times of food shortage. Considerable time and energy may be spent, therefore, in both locating and capturing prey. However, the exact diurnal and seasonal patterns by which any hornbill species harvests prey within its territory, and the effect this has on the abundance and dynamics of prey populations, remain to be studied.

Fruiting trees, on the other hand, tend to provide more patchy sources of food, erratic in both timing and dispersion (Hallé *et al.* 1978). For example, in Malaysia there may be two to three fig trees per hectare of forest, with up to 38 species in 2 km^2, making some fig fruits available throughout the year provided those in fruit can be located. However, fig species with the larger fruits (over 12 mm in diameter) that are favoured by hornbills presented only 16 crops per km^2 of forest per annum (Lambert and Marshall 1991). This makes them demanding to locate, but the food, once found, is usually abundant and relatively easy to harvest. Furthermore, fruit crops are present for only a limited period, ranging from 25 to 83 days (mean 38, $n = 75$) for most non-fig species in Borneo (Leighton 1982), but averaging only two weeks for most bird-dispersed fig species in Malaysia (Lambert and Marshall 1991). Even so, it appears that some figs, and certainly many other fruits, would always be available to an average large forest hornbill with a home-range of about 300 km^2.

Frugivorous hornbills tend to congregate in numbers at fruiting trees, especially figs, even though most still move around as pairs or family groups within the feeding flocks. This communal behaviour frequently extends to roosting, when, depending on the time of year and species concerned, tens or even hundreds of birds will gather at favourite and often traditional sites. Such behaviour often results in mixed-species flocks and roosts, such as those of *Ceratogymna* species in Africa or *Aceros* and *Buceros* species in Asia. Even in more sedentary and territorial species, immatures and non-breeding adults will often form into flocks that range widely, roost communally, and probably are more frugivorous than residents.

It is possible that these communal roosts act as information centres, allowing newcomers to an area to follow residents to trees currently in fruit and enabling inexperienced foragers to follow successful birds in search of food. When foraging during the day, birds appear to return to flocks rather than set off alone, suggesting that flocking is important to efficient feeding. While at a fruiting tree, individuals of several *Ceratogymna* and *Aceros* species will even call to others, passing overhead and audible by their wingbeats, and so attract them to join the feast. The large aggregations could attract predators, although there is some safety in numbers and the risk may not be too critical for the many larger hornbills involved. Roosts are often situated in swamps, on hillsides, and in tall trees, out of reach of and away from concentrations of most predators, and flocks at fruiting trees are notably alert and wary.

There is usually little aggression between members of the same species at fruiting trees, although it varies between species, and males, larger species, or ones operating as co-operative groups can establish dominance when necessary (Leighton 1986). However, fruiting trees also attract many other species of frugivore, both avian and mammalian, so that competition, both direct and indirect, for the resources may sometimes be intense. This depends initially on timing, numbers, and consumption by each species of the fruits available. It is also affected ultimately by their roles in seed-dispersion, seed-predation (Howe 1977; Breitwisch 1983; Jordano 1983; Coates-Estrada and Estrada 1986), and subsequent placement of seeds at sites suitable for germination. This might involve removal of seeds from the parent tree, below which seed mortality is usually highest, or transport to well-lit gaps produced by falling trees (Becker and Wong 1985; Schupp 1988; Raich and Christensen 1989) or to within tangles safe from antelope predators (pers. obs.). The habit of hornbills of quickly gathering fruit and then retiring to rest in dense cover contributes to their effectiveness in fruit-transport (Gautier-Hion et al. 1985), besides reducing their exposure to predators.

Hornbills that harvest widely dispersed fruits tend to travel long distances in search of food. They often fly high above the forest and may pass from horizon to horizon without stopping. This makes them conspicuous in an area, even though their overall density may be lower than that of more sedentary species. Their return at night to communal roosts also entails flights over ten of kilometres, a measure of the value of such roosting behaviour to these species. Their wide dispersion makes assessment of numbers and densities difficult, even at roosts, where there may be marked fluctuations in numbers both within and between sites. However, the precision with which they locate isolated trees, just as the fruit crop is ripening, indicates that much of the population is nomadic, constantly sampling resources over a wide area and with a well-developed information system (Leighton 1986). The details by which they locate fruiting trees, move between communal roosts, or swap one area of fruiting trees for another remain undescribed.

Communal use and sharing of fruit applies mainly to figs, which are widespread in the habitat, generally available throughout the year, and occur in large numbers on the tree, even if over a restricted fruiting and ripening period. Hornbills that feed more on drupes, which are fewer and more restricted in time and space, or which include a reasonable component of animal food in their diet tend not to show social behaviour and are less likely to broadcast their whereabouts. Adults tend to reside as pairs within territories, at least during the breeding season, and within these territories they use their residency to be early at fruiting trees, including figs, and to spend the rest of the day in the more laborious and time-consuming harvesting of drupes and animals.

It is notable, however, that in most of these species the juveniles and non-breeding adults form into flocks and occupy communal roosts, and that these flocks reside for only short periods in an area and are not driven out by territorial adults. Even the more communal and frugivorous species, in which breeding pairs do not defend a territory and which often travel far from the nest for food, spend time after feeding at fruit trees in searching alone for drupes and animals. Obviously, the balance between the amounts of animal and fruit foods in the diet affects the social organization and movements of each species, and acts differently on each age and sex class.

Processing of food

Digestion, absorption, and excretion are the processes that follow consumption of food. The rate and extent of digestion determine how much of a given food is available for absorption, how quickly it is available, how much will be excreted, and how long is required before subsequent items can be consumed. Capacity increases with body size, from about 100 ml for medium-sized Bushy-crested Hornbills to 300 ml for large Bar-pouched Wreathed Hornbills *Aceros undulatus* (Leighton 1982). Such a volume can be filled by as few as four of the larger fruits or by more than 200 small figs. Hence there is considerable variation in the size of loads and surface areas exposed to digestion depending on the food source selected.

Little is known of the extent and rate of digestion by hornbills, and most evidence comes from the interval between feeding on and voiding of indigestible seeds and insect cuticle, the latter action by regurgitation of larger items or by defecation of finer remains. Digestion rates of most fruits appear to be highly variable, with regurgitation of seeds after 19–110 min (mean 67 min). Digestion often appears to be incomplete, since the hornbill feels the fruit in its bill and may then swallow it again (Leighton 1982).

Smaller seeds, those under 5 mm in diameter for large hornbill species, are defecated after a longer period (mean 83 min for fig seeds: Leighton 1982) than those which are regurgitated. Digestion efficiency, faecal content, and extent of defecation may alter through the annual cycle. This is especially likely given the role of the female's faeces in nest-sealing, and later their expulsion through the entrance slit in nest sanitation. The high incidence of germination below hornbill nests and its assistance in determining the stage of the breeding cycle are well known to several local cultures.

Of 35 tree species in Uganda, 16 would have had their seeds dispersed by hornbill regurgitation, 11 by defecation and eight by both methods (Kalina 1988). Further information on movements after feeding and on voiding behaviour is especially important in order to determine how far a hornbill might travel from a fruiting tree before dispersing a seed, besides what proportion of seeds is dispersed and how digestion might affect their germination (Dowsett-Lemaire 1988).

Effectiveness in seed dispersal

Hornbills are among the principal frugivores in many of the forests which they occupy, and their role in dispersal, germination, and predation of seeds warrants further study. Their ability compared with other avian forest frugivores, to break up and swallow large fruits (50–200 mm in diameter), and their regurgitation of the seeds undamaged, make them ideal dispersers (Leighton and Leighton 1983). Their method of selecting fruit also aids dispersion, such as opening fresh capsules before other seed-predators can enter, selecting ripe and insect-free fruit, dropping many fruits, and not damaging most seeds during ingestion. They are also capable of dispersing seed over several kilometres, resulting in a rather even spread of seedlings. It has been estimated for several Bornean species that 80 per cent of seeds are regurgitated after 80 min, within which time the hornbill can be expected to have moved up to a kilometre, given ranging velocities of 200–1200 m/hr for different species (Leighton 1982).

Their effectiveness as seed-dispersers has been demonstrated fortuitously by the role of Silvery-cheeked Hornbills *Ceratogymna brevis* in the rather even dispersion of seedlings of an exotic timber tree *Maesopsis eminii* into adjacent indigenous African forests (Binggeli 1989). Daily movements of Trumpeter Hornbills *C. bucinator* and other congeneric species also indicate complex patterns of visitation to fruiting trees and important daytime resting behaviour. This results in considerable success in setting seed of such food-plants as *Trichilia* species, which 'reward' the hornbills with their fatty arils (Kalina 1988; pers. obs.). The Malay Black Hornbill has also been shown to be a principal seed-disperser for several large-seeded *Aglaia* species (Meliaceae) with capsular fruits, again taking the oily seed coat or aril for food (Becker and Wong 1985) and bird-dispersed seeds generally have a higher oil or lipid content (Pannell and Koziot 1987). Further studies may well reveal hornbills as 'keystone' species in the maintenance of the very forests on which they, and many humans, depend for their livelihood.

5
Breeding biology

It is the nesting habits of hornbills that have most attracted the attention of naturalists, and in particular the unique habit of the female of sealing herself into the nest, moulting all her flight feathers, and being fed throughout the nesting cycle by her mate. All species make their nest within a natural cavity (except for a few smaller species in old barbet or woodpecker holes), but there is considerable variation among genera in details of the breeding biology (Moreau 1937; Kemp 1979; Table 5.1). Many of the differences between species in the duration of stages of breeding and of development are a consequence of differences in body size (Table 5.2; see Chapter 2).

Social and spatial organization

All hornbills appear to nest as monogamous pairs, in which each partner has an essential and demanding role to play in the rearing of the offspring. This remains to be confirmed by molecular or other measures of paternity, but in many species pairs are territorial, some

Table 5.1 Variation in details of the nesting habits recorded for various hornbill genera

Genus	No. of species	Breeding feature						
		Nest sealed	Nest sanitation	Food-delivery	♀ moult	Chick skin	♀ emergence	Co-operative breeding
Bucorvus	2	No	No	Bolus in bill	No	Black	Before chick	1 species
Anorrhinus	3	Yes	Yes	Regurgitate	Yes	Pink	With chick	All species
Tockus	14	Yes	Yes	Items in bill	Yes	Pink	Before chick	None*
Ocyceros	3	Yes	Yes	Regurgitate	Yes	Pink	With chick	None*
Anthracoceros	5	Yes	Yes	Regurgitate	Yes/no**	Pink	Before/with	None*
Buceros	4	Yes	Yes	Regurgitate	Yes	Pink/black	Before chick	1 species*
Penelopides	5	Yes	Yes	Regurgitate	Yes	Pink	With chick	1 species*
Aceros	11	Yes	Yes	Regurgitate	Yes	Black	With chick	1 species
Ceratogymna	7	Yes	Yes	Regurgitate	Yes/no**	Black	With chick	1 species*

* Additional species within the genus are thought to breed co-operatively, although the behaviour is not yet confirmed (see Table 5.3).
** Breeding females, especially those of larger species, may or may not undergo a simultaneous moult of their flight feathers when breeding.

Table 5.2 Some parameters of breeding and developmental biology recorded for various hornbill species (see individual species accounts for more details). All these parameters show a basic relationship to body size, which is indicated by arranging the species in descending order of female body mass (from Table 2.1). Missing values, including those for unstudied species, can be estimated by comparison with species of similar size, ideally within the same genus

Species	♀ mass (g)	Highest density (ha)	Mean egg volume (ml)**	Largest clutch size (eggs)	Incubation period (days)	Nestling period (days)	♀ in nest (days)***	Age at maturity (years)
Bucorvus abyssinicus	3390*		89	2	37–41	80–90		2–3
Bucorvus leadbeateri	3344	10 000	100	3	38	86	70+	4–6
Buceros vigil	2500	180	no further details available for this species					
Aceros nipalensis	2270		56	2			130	
Buceros bicornis	2211	180	71	3	38–40	72–96	112–134+	4–5
Buceros rhinoceros	2180	120	65	2	37–46	78–80	86–97+	4–5
Aceros undulatus	1950	260	60	2	40	90	113–137	3
Ceratogymna elata	1750		no further details available for this species					
Aceros plicatus	1667		48					
Aceros cassidix	1602*		no further details available for this species					
Buceros hydrocorax	1557		46					
Aceros comatus	1470	160		2				
Aceros subruficollis	1431*		51	2				
Aceros corrugatus	1273*	130	43		29	63–73		
Ceratogymna brevis	1162		37	2	40	77–80	107–120	
Ceratogymna subcylindricus	1090		35	2	42	70–79	112–132	
Ceratogymna atrata	1059		42	2				
Ceratogymna cylindricus	978			2				
Anorrhinus galeritus	933*	120		2				2
Anthracoceros coronatus	861*		39	3	29	49		
Aceros everetti	861*		no further details available for this species					
Aceros leucocephalus	814*			2		92		
Anorrhinus austeni	755*		29	5	30	62		
Aceros waldeni	748*		no further details available for this species					
Anorrhinus tickelli	723		27	5			92	
Aceros narcondami	675	3	25	2				
Anthracoceros malayanus	633*	50	25	2				2
Anthracoceros albirostris	624	100	29	4	25–29	42–55	48+	2
Anthracoceros montani	601*		no further details available for this species					
Anthracoceros marchei	601*		no further details available for this species					
Ceratogymna bucinator	567	50	29	4	28	50		
Penelopides manillae	473			5	28–31	50–65	80–102	
Penelopides affinis	435		no further details available for this species					
Penelopides panini	414*		26	2			95	
Penelopides exarhatus	370		29	2				
Tockus monteiri	340	10	15	8	24–27	43–46	48–63+	1
Penelopides mindorensis	328*		no further details available for this species					
Ceratogymna fistulator	312*		31	2				
Tockus hemprichii	297		15	3				
Tockus albocristatus	282			2				
Tockus fasciatus	241		17	4				
Ocyceros birostris	228*		19	5				
Tockus pallidirostris	212		17	5				
Ocyceros gingalensis	209*		17	3	29			
Tockus alboterminatus	205	300	15	5	25–27	46–55	61–69+	1

Table 5.2 (cont.)

Species	♀ mass (g)	Highest density (ha)	Mean egg volume (ml)**	Largest clutch size (eggs)	Incubation period (days)	Nestling period (days)	♀ in nest (days)***	Age at maturity (years)
Ocyceros griseus	203*		20	4				
Tockus bradfieldi	193		14					
Tockus flavirostris	182		13	6				
Tockus leucomelas	168	14	13	6	24	42–47	45–60+	1
Tockus nasutus	146	22	13	5	24–26	43–49	46–65+	2
Tockus deckeni	145		9	3		48	56+	
Tockus erythrorhynchus	131	5	10	7	23–25	39–50	42–55+	1
Tockus camurus	97	20		2				
Tockus hartlaubi	83	25		4				

* Body mass is estimated for those species for which it has not been recorded, based on the correlation with wing length derived from a sample of well-documented species (Table 2.1, Fig. 2.1).
** Egg volume calculated from Hoyt's (1979) formula: eggshell length × breadth^2 × 0.51.
*** Female emerges when chicks only half grown in species marked +.

throughout the year, so that mate-attraction and mate-guarding are not an obvious precursor to breeding. Even in non-territorial species, which form communal or immature flocks, birds consort in pairs for much of the year so that, once again, pair-formation long precedes breeding. Any form of polygamy is unlikely, given the special role of each sex in the breeding system of hornbills and the prolonged period of mate-interdependence.

Co-operative breeding by small groups containing one nesting pair has been recorded for at least eight and is suspected, at least facultatively, for a further 10 species of Bucerotiformes (Table 5.3). This comprises 15 per cent or possibly even 33 per cent of all hornbill species, a higher proportion than for any other order of birds and well above the overall avian average of 2.5 per cent. Such a high incidence of co-operative breeding is possibly a consequence of the special status of the female and her demands for provisioning during the breeding cycle. The *alpha* or breeding pair that dominates within each co-operative breeding group is assisted in most aspects of territory-defence and chick-raising by other members of the group. It is thought likely that these group-members are closely related, many of them offspring from previous breeding attempts, but some are themselves sexually mature and usually these are males.

Nest sites are fairly evenly spaced in territorial species, but in non-territorial species nests may be clumped within suitable habitat and only short distances apart. All species tend to defend the area immediately around the nest from intruders of their own and other species, but a larger feeding territory is usually defended only against members of the same species. Since as many as eight species of hornbill may coexist within an area, however, nests of different species may end up within a few metres of one another. Indeed, two small *Tockus* species may even have their nests in different holes in the same tree, alongside other hole-nesting animals such as starlings, owls, and squirrels. Interchange of nest sites between hornbill species is also common, but what constitutes an ideal cavity for each species, availability of cavities, and competition for sites between hornbill and other animal species is complex and little studied.

Timing of breeding

Most hornbill species that have been studied exhibit a particular breeding season or rhythm.

Table 5.3 Hornbill species known or suspected to breed as co-operative groups at least some of the time

Species	Co-operative breeding	
	Confirmed	Suspected
Bucorvus		
B. leadbeateri	×	
Anorrhinus		
A. austeni	×	
A. tickelli	×	
A. galeritus	×	
Tockus		
T. camurus		×
Ocyceros		
O. gingalensis		×
Anthracoceros		
A. malayanus		×
Buceros		
B. rhinoceros		×
B. hydrocorax	×	
Penelopides		
P. exarhatus	×	
P. panini		×
P. manillae		×
P. affinis		×
P. mindorensis		×
Aceros		
A. comatus	×	
Ceratogymna		
C. bucinator	×	
C. atrata		×
C. elata		×

Either this season is common to most members of the population, governed mainly by extraneous factors such as the effects of rainfall on food supply (Kemp 1973, 1976b; Poonswad et al. 1987), or each pair adopts its own breeding rhythm in areas with limited seasonality, apparently governed by the rate at which individual birds complete one breeding cycle and regain condition for the next (Leighton 1982). The timing of breeding is governed ultimately by the success with which each pair provides recruits to the next generation.

In seasonal habitats, success may depend mainly on getting an early start, such as Southern Ground Hornbills *Bucorvus leadbeateri* taking advantage of the effects of good rainfall (Kemp and Kemp 1991). It may also depend on having the young emerge into a hospitable environment as suggested by four forest species of different sizes adjusting their cycles so that all their chicks leave the nest at about the same time (Poonswad et al. 1983, 1987). In less seasonal habitats, the ability to take advantage of periods of abundance, which are unpredictable in magnitude or duration, may be more important. This requires breeding as productively and as often as possible during times of plenty, and surviving as well as possible during intervening periods of shortage.

Activities which lead up to fertilization and egg-laying include attraction and courtship of a mate, establishment and maintenance of a territory, selection and preparation of a nest-cavity, and accumulation of adequate body reserves (Chapter 3). Such pre-laying behaviours usually increase in frequency only once the annual period of moult is more or less complete.

Calls and displays

The first sign of breeding behaviour is usually an increase in the frequency, duration, and intensity of calls and displays. These may be given initially by single birds, but where pairs are already in existence the male and female co-ordinate their movements more closely and begin to roost in close proximity to one another. Some species also exhibit an increase in intensity of soft-part coloration, especially in males, which seems to result from an increased blood flow to areas of bare skin and which is probably under hormonal control.

Territorial calling and display may serve several functions, often simultaneously, including proclamation and defence of territory, mate-attraction, and pair-synchronization or pair-bonding. Once pairs have formed, other signs of preparation for breeding follow, including presentation of food to the female by the male in courtship-feeding and visitation of potential nest sites by both members of the pair. Nests are usually sited within a defended area, the permanent home-range for territorial species or the focus of a small nest zone for non-territorial species. Territory size is related to body size (Table 5.2), ranging from under 10 ha in small *Tockus* species to 100 km^2 in the large *Bucorvus* ground hornbills.

Courtship-feeding

Courtship-feeding is simply the feeding of a female by her mate, or by other members of co-operative groups, but theoretically it has several components. First, accepting food from another bird includes aspects of juvenile behaviour, last exhibited when the female was fed by its parents, and this is evident in the way that adult females solicit or beg for food with quivering wings, fluffed-out head feathers, and utterance of the acceptance-call on receipt. Second, in several species, especially co-operative breeders, proffering of food is an expression of social position, and passing of food back and forth between mates has been reported for many species. Third, courtship-feeding precedes production of eggs and the female's total dependence on her mate or group for food during incubation; such feeding augments reserves of fat and protein needed for egg-laying, incubation, or brooding, and supplies special nutrients such as for eggshell-formation. Fourth, the ability of the male or group to supply food may indicate to the female what she can expect before committing herself to breeding and sealing herself into the nest.

Preparation of nest sites

Selection and preparation of nest sites also has a simple goal, the provision of a suitable breeding environment, but it, too, has a number of components. First, the availability of desirable cavities is likely to be limited, depending on the habitat, and on whether the species is territorial or not so constrained in where it can look for nests. Second, the site must be within reasonable range of food sources that can be expected to endure throughout the long nesting cycle. Third, the site has to be defendable against other hornbills, of the same and other species, and against other hole-living animals. It should also be as free of predators and parasites as possible, through either its inaccessibility or its defendability. Fourth, the cavity must have a suitable microhabitat in which to incubate the eggs and rear the young. Aspects of a suitable microhabitat include an entrance with a rim to which sealing material can adhere; an entrance-hole small enough to be sealed up, but large enough to admit the female, to compensate for tree growth, and to allow the female and chicks to make their exit after breeding; an even floor not too far below the entrance; a chimney or funk-hole above the nest into which the inmates can withdraw if threatened; sufficiently thick and robust walls to insulate and protect the contents; and lack of protrusions to interfere with sealing, so that a vertical slit results through which, with the warm occupants below, proper ventilation can flow.

The practical expression of these demands can be seen in the thoroughness with which pairs check any apparently suitable cavities within their range. Both members of the pair, but especially the female, go around poking their heads into holes and pecking around within them. Nest-selection occurs at any time during the year in territorial pairs, but usually only at the beginning of the breeding season in non-territorial ones. A pair soon focuses its attention on one or two sites, especially if one used in previous years is still suitable. The female then spends more and more time looking into the hole, chipping away at or applying sealing to irregularities on the exterior, sometimes with the help of her mate, and later entering and remaining within the cavity for extended periods.

Lining and preparatory sealing

Once the female begins to spend time in the nest, the male or group-members continue to feed her in courtship and begin to bring her lining and sealing materials. The exact materials chosen and how they are delivered depend on the species concerned. Males often develop an attraction to the nest-hole independently of the female, presenting material to the hole when the female is not inside or alternating presentations between the hole and the female if she is nearby. This is not a case of the male attracting the female to the hole, since she undertakes most of the nest-selection, and the male continues to present to the hole even if she is removed. Copulation also begins to take place, but usually it is not frequently recorded for hornbills.

Sealing of the nest-entrance, and of any additional holes leading into the cavity, commences well before laying. Sealing is applied with the broad, flattened sides of the bill (Burton 1984), using rapid and stereotyped side-to-side vibrations of the skull. Sealing material is held in the bill tip and is squeezed out of the sides during application to form a series of thin layers. The effect of this technique is to produce a vertical slit at the entrance, unless protrusions within the cavity force the sealer to bend its head to one side and so curve or otherwise alter the shape of the slit.

The role of the sexes in sealing varies among species. In the majority of cases, only the female seals, using mud or sticky foodstuffs when still outside and mainly her own droppings once inside. No studies have been done to assess any changes in faeces compositon or production by the breeding female. In most other hornbill species, the male assists to varying extents. In some species, sealing materials are carried in the bill and passed to the female while she is either inside or outside the nest. Rarely, the male makes some sealing movements on the outside with some of the material which he has brought. In its most advanced form in *Ceratogymna* species, the male swallows soil, forms it into pellets in the gullet, regurgitates these at the nest, and either passes them to the female or uses them himself.

Most nest-cavities are well insulated, with thick walls of wood or rock and few or no openings to the exterior save the vertical slit of the sealed entrance. This could pose problems of temperature-regulation and ventilation for the warm-blooded inhabitants of the cavity, which can include a female and up to six large chicks. Preliminary studies, however, suggest that this is not a problem (White *et al.* 1975): when the sun rises, the nest-walls delay rises in internal temperature until later in the day; when the sun sets, heat absorbed during the day delays the nocturnal drop in temperature. This lag effect provides the nest-cavity with a more even temperature regime than the exterior. The warmth of the occupants situated below the entrance also acts as a heat pump which, coupled with the vertical conformation of the slit, draws in a plentiful and well-separated flow of cool air below while forcing out the warmer air above. Such ventilation is assisted by any other small openings that may lead into the cavity, especially if they lead off the high funk-hole that exists above most nests.

Males also bring lining to the nest, both before and after it is finally sealed. Again, there is variation among species in how much lining is used, what materials are brought, and how the lining is delivered. Some species prefer a moist lining, such as green leaves and grass, while others prefer a dry lining, such as bark flakes and dry leaves. Some deliver single items of material, in the same way as they deliver food, and others form bundles with multiple items gathered in the bill (Fig. 5.1). The lining contributes towards levelling out the floor of the nest, which in cavities used in successive seasons is augmented by food remains such as fruit stones, by moulted feathers, and by other debris chipped off from the walls during nest-preparation.

The lining may also absorb some food and excretory matter, especially when it consists of dry material. Soiled lining is thrown from the nest by many species, and dry lining may also favour ground hornbills, which do not practise nest sanitation but do apply copious amounts of dry leaves and grass as lining. The materials may also possess other hygienic properties, which encourage insect commensals to keep the nest clean or reduce parasites and pathogens, but this awaits further testing.

5.1 An adult male Southern Ground Hornbill *Bucorvus leadbeateri* carrying a bundle of lining material for the nest, mainly dry leaves and grass, within which is secreted a small item of food. Food is later ferried to the nest in the same fashion, as a collection of small animals caught individually but carried as a bundle in the bill.

Finally, the lining may also serve as a food substitute in social interactions. For example, in co-operative groups of Southern Ground Hornbills, a food item is always included in the lining and presented only by adult males within the group, even though, later, all ages and both sexes will bring food to the female and chick. The two species of *Bucorvus* ground hornbill are also the only members of the order that do not seal the nest-entrance.

The pre-laying period

Once the female has finally entered and sealed herself into the nest, which often takes less than a day to achieve, she then waits for several days or even weeks before starting to lay. The period is usually 4–6 days, the maximum recorded being 24 days, during which time she is liable to desert the nest if disturbed or if something happens to the male. This sensitivity supports the contention that the pre-laying period serves as a final monitor of nest security, and of the ability of the male to provide food before the female commits herself to breeding. The irrevocability of this commitment depends on whether the breeding female, as in the majority of hornbill species, moults her flight feathers within a few days of starting to lay (Table 5.1, p. 46). It also provides the female with a period of minimal energetic demands, during which she can form the eggs and add to her body reserves.

Egg-laying and female moult

The number of eggs in a clutch varies from one or two for the larger species to up to eight for the smaller species (Table 5.2, p. 47). All hornbills lay oval, rather elongated eggs with white shells, the larger the hornbill the larger the egg (Table 5.2). The shells are usually pitted and become stained as incubation proceeds, the degree of staining offering some insight into their stage of incubation. Eggs are laid at intervals of one to several days, depending partly on the size of the species and also on their sequence within the clutch. Larger species tend to lay their eggs 3–5 days apart, while smaller species lay them on consecutive or alternate days. As laying of the clutch proceeds, however, so the interval between eggs often extends to as long as six days. The total period for egg-laying by small species with large clutches may therefore span as much as 20 days. Breeding female hornbills appear not to moult their belly feathers to develop a bare area well supplied with blood vessels (the brood-patch), but the abdomen is normally unfeathered and this area of bare skin is placed directly against the eggs.

The moult and energy demands of breeding hornbills deserve further study, having been examined, superficially, for only a few species. In smaller species moult is usually confined to a specific season, but in larger ones, or those resident in habitats without obvious seasons, both breeding and non-breeding individuals can be found with some moult at any time of the year. In some species, such as ground

Breeding biology

5.2 A female Southern Yellow-billed Hornbill *Tockus leucomelas*, removed from the nest during incubation to show how all the main flight feathers of the wing and tail have been moulted almost simultaneously. The feathers grow back while the female is incubating the eggs and brooding the chicks, and growth will be almost complete by the time the chicks are about half grown.

hornbills and larger *Ceratogymna* species, the female either suspends moult during breeding or moults in normal fashion, dropping a few feathers sequentially and regrowing them before dropping more. In most other species, the breeding female moults her flight feathers coincidently with or soon after the start of egg-laying, resulting in an almost simultaneous shedding of all primary, secondary, and tail feathers (Fig. 5.2). In a few cases, females have been found in nests with either complete flight-feather moult or no moult, suggesting that the exact nutritional condition of each female may determine the exact timing, pattern, and rate of moult.

The tail feathers are usually dropped first, and in small species all feathers may be out on the day of laying of the first egg, with the wing feathers being dropped over the next few days. Where breeding has been delayed, as by late arrival of the rainy season, a feather that had been moulted normally before the onset of breeding is not dropped during the simultaneous moult of the breeding female. This can result in a rather irregular appearance of the wing and tail, besides considerable wear to the new feathers so exposed. In captivity, an African Grey Hornbill *Tockus nasutus* even moulted the flight feathers twice when laying successive clutches (Wilkinson and McLeod 1990), which adds further to the mystery of the exact hormonal and energetic control of moult.

The main period of moult for non-breeders, breeding males, and other members of co-operative groups usually coincides with the breeding season and the higher supply of food with which it is associated. This means that the breeding male may himself be moulting at the same time as he is feeding the moulting female and their chicks in the nest. This differs from the normal avian pattern, in which the female and young are not so dependent on the male and where moult is usually postponed by both sexes until later in the breeding season. In some cases, however, hornbill moult may be suspended or delayed until after breeding, or at least until later in the nesting cycle when demands for food are reduced, again stressing the need to consider the nutritional status of each individual. This unusual allocation by hornbills of time, energy, and nutrients within the annual cycle, and the extra demands made on the breeding male, remain to be described in detail.

Incubation

Once the breeding female has sealed herself into the nest, she has no choice but to begin incubation with the first egg, sitting on the clutch on the nest-floor, often in cramped circumstances with her tail held vertically over her back. Usually the female does not have to leave the eggs to take food from the male at the nest-entrance; she just reaches out her long neck. Apart from the usual shuffling about on and turning of the eggs, she normally leaves the eggs only to hide away up the funk-hole or to defecate. The size of the eggs, the clutch size, and the duration of incubation are all closely correlated with the body size of each species (Table 5.2, p. 47). The incubation period extends from 23 days in the smaller species to 42 days in the largest hornbills.

The female in the nest spends long periods with one eye pressed to the entrance slit and is capable of discerning, by sight and sound, many activities around the nest. Any foreign object or animal appearing suddenly at the nest-entrance is usually attacked with the bill, to the accompaniment of loud fear-calls. If the intruder's attention at the slit is determined, however, or if a dangerous animal is spotted, then the female will bolt up the funk-hole that exists above the majority of nests. In this way, the female, and later the chicks, often disappear completely from sight and are beyond the reach of most predators that might break away the nest sealing.

The male, and other members of the group in co-operative breeders, ferry food to the incubating female. Depending on the species, and to some extent on the type of food, items are carried singly in the bill tip (Fig. 5.3), as a bundle gathered into the bill (cf. Fig. 5.1, p. 52), or as a load within the gullet which is then regurgitated as individual items into the bill and passed to the female (Plate 2a, b and c). The feeding bird sits by the nest if a convenient perch is available, or else hangs at the entrance using the tail as a brace (Fig. 5.3). The number of visits to the nest depends on whether the food is carried singly or collectively, on how many birds are provisioning the nest, and on the stage of the nesting cycle, with maximum demand coming later, when growing chicks have to be satisfied.

Some male hornbills have been reported to shed the lining of their stomachs (Bartlett 1869; Curl 1911), as has also been recorded for a penguin species (Beintema 1991). The exact physiology, histology, and biology of this observation, however, remain to be discovered. Reasons proposed for this habit range from providing a sack in which to cast indigestible food remains, to its being a mechanism for voiding dangerous chemicals built up in the gut-wall during digestion.

The female is probably heavier than usual on entering the nest, her ovaries, oviduct, and egg glands swollen in preparation for egg-laying. She may also accumulate some body reserves, evident as yellow fat through the skin of the abdomen, but her activity is now minimal and the male is providing food. In Monteiro's Hornbill *Tockus monteiri*, body weight continues to increase during incubation, suggesting deposition of additional reserves and that the regrowth of flight feathers is not too taxing. It is highest by the end of incubation, and lowest by the time the female emerges half-way through the nestling period, with losses of 9–27 per cent ($n = 8$) from the peak weight (J. Mendelsohn *in litt.* 1992). This suggests that, after the hatch, females may use up body reserves, rather than draw too heavily on food provided by the male and intended for the chicks. Further investigation, including of other species, would establish how food and reserves are allocated among the members of hornbill families during breeding.

5.3 A male African Red-billed Hornbill *Tockus erythrorhynchus*, hanging at the nest-entrance, bracing itself with the long tail and about to pass to the female within a single item of food which it has carried to the nest in the tip of its bill.

Nest sanitation

Food remains and excreta are dealt with in a most hygienic fashion by most hornbill species. Only ground hornbills show no special behaviour, but, since their nest is not sealed, the female often defecates, maybe exclusively so, away from the nest during short breaks in incubation. Their nest is also deeply lined with dry leaves and grass, which help absorb droppings of the single chick, and the nest only sometimes become odorous when the chick is large or the cavity poorly drained.

In the majority of species, the excreta are squirted out through the nest-entrance by an elaborate and stereotyped behaviour (Fig. 1.5, p. 8). This involves turning away from the entrance and the only source of light, reversing up to the slit, positioning the anus, and then defecating with considerable force. It results in a faecal 'shadow' on the ground and vegetation below the nest, which is often useful in revealing the latter's position and the nesting diet. The extent of the splash and the growth of seeds so ejected may also indicate the stage of the nesting cycle.

Other food remains within the nest, such as regurgitations or broken fragments, are tossed from the entrance with the bill. The same behaviour is shown to other debris within the nest, such as soiled lining from small chicks, or moulted feathers. Some items, such as fruit stones, may be left to accumulate, adding to the lining, raising and levelling the floor, and providing a useful record of the diet. Some objects, especially moulted feathers, are poked into crevices around the nest, and these indicate the continual fiddling in which the enclosed birds indulge.

The nest may also support a diverse insect fauna (Britton 1940), some of which is eaten, but the extent to which this fauna is attracted to the nest debris or plays a role in the sanitation and health of the inmates remains unstudied. Of course, when necessary, excreta can be voided gently, to be mixed with food remains and debris and used as sealing material. This occurs either during the initial closure, or when repairs are needed, or when chicks reseal the nest in those species in which the female emerges before they are fully grown.

Hatching and chick development

The eggs hatch at intervals, since incubation begins with laying of the first egg, and the sequence of and interval between hatchings usually correspond closely to when the eggs were laid. The empty shells are either left in the nest, to become broken into the lining, or thrown out through the slit, providing a sure sign that hatching has occurred. The newly hatched chick is altricial, pink-skinned, naked, with the eyes closed, and with the upper mandible, on which the egg-tooth is prominent, noticeably shorter than the lower (Fig. 5.4a). Such a chick is typical of all allied families of birds in the orders Upupiformes and Coraciiformes.

The chick responds immediately to auditory and tactile stimuli. It begs with a weak cheep-

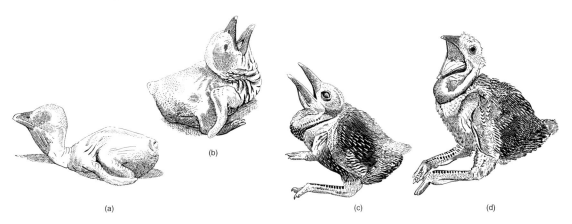

5.4 Stages in the development of a chick of the Southern Ground Hornbill *Bucorvus leadbeateri*: (a) newly hatched; (b) 10 days old; (c) 20 days old; (d) 30 days old and about one third grown. See text for details.

ing sound for food, which is placed whole in its bill by the female after being selected from items delivered to the nest by the male. The chick has a strong neck and well-developed swallowing reflex, balancing on the rotund belly with little help from the small legs while ingesting food. The nestling period is correlated with adult body size, and extends from just under five weeks in the smallest species to over three months in the largest (Table 5.2, p. 47). The stages of nestling development appear to be consistent for all species studied (Fig. 5.4), although the duration of each stage depends on the final body size achieved, being longer the larger the species.

An air sac develops under the skin of the shoulder region within a day of hatching. It starts as two separate pockets, but spreads within days to cover the entire dorsal surface of the neck, forearms, back, and thighs, with extensions down the sides of the breast (Fig. 5.4b). This extensive system of air sacs has not been carefully described anatomically; it is apparently unique to hornbills, and could function as thermal insulation, as a mechanical protection from other inmates in the confined nest, or in some other role. It is most obvious in hornbill chicks before their feathers emerge, giving them the feel and appearance of an inflated plastic bag!

Feather growth is evident within a few days of hatching, as dark marks under the pink skin, although this soon becomes less obvious in some larger species where skin colour blackens within a week of hatching. The feather-quills emerge from the skin at about the time that the eyes begin to open (Fig. 5.4c). They emerge first on the breast, then on the back and wings, next on the head and tail, and finally on the neck, the growth differential between feather tracts being retained throughout their development. The sheath enclosing each feather-quill is not shed immediately, giving the chick a prickly appearance as the quills lengthen and aptly termed the 'porcupine' or 'pin-cushion' stage (Prozesky 1965; Fig. 5.4d). Such feather development is again shared with the related families of hoopoes, rollers, kingfishers, and bee-eaters. Once the feather-sheaths begin to be shed then their loss is quite rapid, and the chick becomes extensively feathered over the body and wings, well before the larger flight feathers are fully developed. Thereafter, most feather growth is evident in elongation of the flight feathers and covering of the neck.

Development of limbs and co-ordination also follow a basic pattern. Begging movements of the young chick become stronger and more directed, especially once the eyes have opened and visual cues can be used to locate food in the mother's bill. The hind limbs develop especially rapidly and are almost fully grown by the time that the chick comes to the end of the 'porcupine' stage. It is then able to move easily around the nest, take its own food from the male, and defecate through the nest slit. At this stage, the females of some species may emerge from the nest, and the chick's well-developed legs and strong neck are essential for it to be able to reseal the entrance unaided. Most growth thereafter takes place in the bill, wings, and flight feathers.

The chick also shows steady development in behaviour and vocalizations. The begging-call develops from a weak cheeping into strident cries, which may be uttered for long periods, are audible at some distance, and often reveal the presence of a nest. Larger chicks, like the female, utter a loud acceptance-call on taking food from the male, and this can often be detected as a crescendo between the insistent sounds of begging. The chick may also utter loud fear- and threat-calls if attacked while in the nest, or make soft contact-calls to the male during its frequent monocular scanning of the outside world. Nearer to fledging, the chick may begin to utter the full loud call of the species, which also serves as the highest-intensity contact-call. Some even attempt a full territorial display during such calling, although the performance is usually poorly co-ordinated and somewhat cramped within the confines of the nest.

The male, or all members in co-operative groups, continues to ferry food to the nest throughout the nestling period, and also brings some lining material. The female, when she emerges before the chicks, usually assists the male with provisioning to the nest, but even where her help is minimal she is at least finding her own food. The rate of food-delivery probably depends on many factors, including how it is carried by each species, the types of food available, the location of food sources in relation to the nest, and the size of the brood. Details of feeding rates, where these are available, are presented in the individual species accounts. The general trend is for rates to be considerably higher during the nestling period than during incubation, to rise to a peak when the chicks are growing at their highest rate, and to be positively correlated with the size of the brood.

Emergence and the post-fledging period

Breaking-down of the nest sealing is achieved by persistent hard pecking with the bill, undertaken by the female and chick together, or by the chick alone if the female has already left. It may take from a few hours to several days to complete, depending on the motivation, the extent of sealing, and the size of the hornbill involved. At nests which initially required extensive sealing, only as much is removed as is necessary to allow emergence of the inmates. The female and chick fly well on emergence, although the female sometimes appears a little stiff and the chick is never as agile as its parents. The chick now appears almost as large as its parents, but its bill is noticeably smaller, it lacks any form of casque, and the wing and tail feathers are shorter since their growth is not quite complete.

The chick remains in the vicinity of the nest for a few days after emergence, secreting itself amongst dense cover, but neither it nor the female ever re-enters the nest-cavity until the following breeding season. The chick continues to be fed by its parents and to develop its flying skills. It often shows exaggerated behaviour, apparently in greeting or solicitation, in which it rushes silently up to its parents with crown feathers raised, head withdrawn, and wings drooped, only to stop on reaching them and to return to a normal posture. The chick becomes confident within a short time,

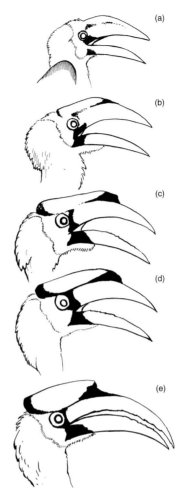

5.5 Stages in the post-fledging development of the head, bill, and casque of an immature male Great Pied Hornbill *Buceros bicornis*: (a) 4 months old; (b) 6 months old; (c) 13 months old; (d) 28 months old; (e) almost mature at about 48 months old. (After Frith and Douglas 1978.)

coming out into the open, flying to its parents, and later following them on foraging forays. The chick also starts to show play or learning behaviour, flying at great speed between the vegetation with much twisting and turning, manipulating objects such as sticks and leaves in the bill, trying to catch passing insects, or bill-grappling with its siblings or parents.

The final stages of development, between fledging and maturity, are least studied. The exact pattern of growth, maturation of coloration, and development of structures such as casques and long tail feathers varies from species to species (Fig. 5.5), and is also related to body size (Table 5.2, p. 47). Juveniles may remain with their parents for months or even years, especially in the case of co-operative breeding species, where they may remain even long after attaining sexual maturity. The interval between fledging and starting to breed varies from less than one year in the smallest hornbills to at least 4–6 years in the larger species, assuming that attainment of adult coloration and form is coincident with onset of sexual maturity. The existence among some species of wandering flocks of subadult birds indicates that they roam widely before settling down to breed. Details and patterns of juvenile dispersion and pair-formation are, however, largely unknown, despite their importance for understanding population structure and management.

Population biology

The basic breeding biology of hornbills, as described above, must also be related to the more dynamic but even less well-studied determinants of breeding success. Logistical problems, especially of finding and examining nests of forest species or catching and marking adults, have resulted in few data on clutch size, hatching success, nestling survival, juvenile survival, age at first breeding, adult turnover, lifetime reproductive success, or other major determinants of the population dynamics and life histories of hornbills.

Factors affecting onset and success of breeding are also poorly known, except for some savanna species where the link between rainfall and food supply is apparent (Kemp 1973, 1976*b*; Kemp and Kemp 1991). There is evidence among smaller savanna species that larger clutches are laid during years of higher rainfall, presumably of higher food availability (Kemp 1973, 1976*b*; Riekert 1988). There is also evidence that the pre-laying interval after sealing into the nest and the inter-egg interval within clutches are shorter when food is abundant. It also appears, however, that the larger the clutch, and therefore the longer the interval between sealing in and laying of the last egg, the higher the chance of failure of later eggs. This may be due to limitations in body reserves for forming eggs, or to infertility arising from the impossibility of copulation after the female has sealed herself into the nest. In the large ground hornbills, there is also evidence of eggs being larger, more similar in size within a clutch, and laid at closer intervals during years of abundant early rainfall and good food supply.

Chicks within a brood are always staggered in size (Fig. 5.6), since incubation begins with the first egg and the eggs hatch at approximately the same intervals as they were laid. Any shortage of food then results in starvation of the smallest chick, owing to domination by larger chicks of positions closest to the nest-entrance where food is presented. Such reduction serves to tailor the brood to the available food supply, without detriment to the larger chicks, even in extreme cases where only a single survivor fledges from a clutch of five eggs (Kemp and Kemp 1972). Compared with other birds, hornbills, along with many other hole-nesting species, have a longer incubation and nestling period than might be predicted from body size alone. This makes them more prone to alterations in food supply while nesting, but plentiful food has the effect of shortening each stage of the nesting cycle and so reducing the period of exposure to risk (Kemp 1976*a*).

5.6 Four chicks of the African Grey Hornbill *Tockus nasutus*, 10 days after hatching of the eldest and on the day of hatching of the youngest, to show the staggered size of the chicks within the brood.

A special case involves death of the younger sibling in the largest species, where usually two or three eggs are laid but only one chick is raised. The second chick to hatch dies of starvation with a few days, as has been studied in detail for Southern Ground Hornbills and as is suspected for several other large forest species (Kemp 1988*a*). The two eggs in a clutch are often markedly different in size, both between females and between seasons, and variation also occurs in the inter-egg laying interval and in the egg-size and incidence of single-egg clutches. This suggests that under ideal conditions two eggs of similar size may be laid close together and both chicks raised, although in all but two cases ($n = 30$) the second-laid egg has been smaller and the chick has perished almost immediately. There is also the suggestion that egg and clutch sizes may correlate with overall population density, through the influence of long-term prospects for larger or smaller chicks when they eventually try to enter the breeding population (Simmons 1988).

The prolonged nesting period of hornbills also extends their exposure to all the normal risks associated with breeding at a fixed site, such as predators and disease, or bad weather which may reduce food supplies, produce flooding, or cause overheating. Loss of the male parent is fatal to the nest inmates of most species, given the total reliance of the flightless female and their offspring on his provisioning. A less obvious risk of the long nesting period is growth of the tree bark, so reducing the entrance hole that the female and chicks are unable to emerge at the end of the cycle.

Despite these limitations, the breeding success of most hornbills is high. Among smaller savanna species, about 90% of all nesting attempts resulted in some chicks being fledged. This was notably higher than for roller and starling species which nested in tree holes in the same habitat but did not seal their nests, suggesting that the sealed nest offers some protection from predation (Kemp 1970). In some larger forest species, the sealed nest is also important in deterring attacks from conspecifics, apparently in competition for the scarce nest-holes (Kalina 1989). It would be useful to try to assess experimentally what are the primary and secondary functions of nest-sealing. Factors that affect availability, suitability, and durability of nest-cavities, such as patterns of breakage, rotting, and mortality in different tree species, or their vulnerability to elephant or forestry damage, may also affect hornbill numbers and breeding success in different habitats.

Survival of different age classes and the turnover that results among breeding adults are barely known for any hornbill species. Various predators of hornbills have been identified, ranging from leopards and eagles to puff-adders, mongooses, and falcons, with avian predators the most commonly reported. Records of individuals without tails, for example African Crowned Hornbill *Tockus alboterminatus* or Oriental Pied Hornbill *Anthracoceros albirostris*, probably indicate narrow escapes from predation. Accidental deaths due to drowning or becoming hooked in thorns are also reported. Diseases and parasites of hornbills have been little studied, although a survey of African species reveals a variety of haematozoans in the blood (Bennett *et al.* 1992), and all species carry mallophagan feather-lice (Elbel 1967, 1969*a*, 1969*b*, 1976, 1977*a*, 1977*b*).

In four small *Tockus* species, anything from 39 per cent to 100 per cent of eggs hatched in a given year resulted in fledged chicks, survival depending mainly on food availability during the nesting cycle (Kemp and Kemp 1972; Kemp 1976*a*). Taking into account those nests which failed altogether, these translated into annual replacement rates of 0.63–1.70 chicks per breeding adult. Assuming that the population is stable, the replacement rate will balance adult mortality which would have been of the order of 38–61 per cent per annum. This is similar to that for many other bird species of similar size, although on the high side for some tropical species. The apparently high turnover of adults may have resulted from their frequent choice as prey by the many large eagles nesting in the area (Snelling 1969, 1971). Artificial predation, in the form of shooting specimens for analysis of feeding and breeding condition, indicated that similar predation levels could affect the population size (Kemp 1976a).

In the largest species, the Southern Ground Hornbill, turnover among breeding adults may be as low as 2 per cent per annum. The co-operatively breeding groups, however, fledge on average only one chick every 9–10 years, and there is an estimated 75 per cent mortality between fledging and attaining mature colours at 4–6 years old (Kemp 1988a). The average age at commencement of breeding is unknown, but is probably well after maturity given that there is only one breeding pair per group of 2–11 birds. Average life expectancy is also unknown, but is likely to be of the order of 35–40 years, with the oldest birds reaching ages of 70 or more.

Even this sparse evidence suggests that the larger hornbills may be among the slowest-breeding, longest-lived, and most vulnerable of terrestrial birds, maybe of all birds. It serves only to enhance the fascinating breeding biology and interesting population dynamics shown by the order Bucerotiformes, further study of which is essential for their effective conservation.

6

Relationships and evolution of hornbills

Study of the relationships between species is an attempt to recover the historical and evolutionary pathways by which they arose. Only two assumptions are necessary in deciding whether such pathways exist. The first is that all organisms have a finite lifespan, reproduce, and therefore are connected to their ancestors by descent, going back indefinitely in time. The second is that separation of populations has led to independent pathways. These pathways then comprise the natural history of the organisms involved. Because these historical connections are already completed, they are a pattern or truth waiting to be recovered, with only one solution, rather than a theory that can be tested repeatedly and empirically.

The most scientifically rigorous approach currently available to try and recover these ancestral and demographic links from the past is that of phylogenetic systematics, using the method of cladistic analysis (Wiley 1981). The method assumes, first, that organisms within the group under study have been accurately assigned to their respective species. Second, it requires evidence that all organisms in the group are descended from a common ancestor, that is they are monophyletic. Third, it requires identification of close relatives for use as outgroups when making comparisons between species within the study group. The cladistic method then compares the distribution of as many characters as possible, across all species in the monophyletic group under study and their nearest relatives. Finally, only those characters judged to have been derived for the first time within the study group are then used to assess relationships between species.

Confidence in relationships indicated by the cladistic method can be enhanced by comparing patterns of relationship as shown by different sets of characters. For example, relationships based on details of behaviour and calls can be compared with those recovered from morphological, molecular, physiological, or developmental characters. They can also be tested for congruence with more independent factors, such as relationships between species of parasites that evolved in concert with the group of organisms under study. Lastly, the evolutionary history can be examined for consistency with geological and climatic histories of the areas in which the study species are thought to have evolved.

The value of recovering the evolutionary pathways between species is often not appreciated, although this historical perspective is of fundamental importance to all comparative biologists. A well-resolved tree of relationships, and the classification by which this is represented, offers an accurate prediction of which features of organisms are shared through a common history (homologies) and which are responses to common environmental pressures (analogies, convergences). Without separating characters on this basis, it is impossible for biologists to predict the basis and response of any character or trait that they may choose to study.

Recognizing species of hornbill

Some phylogenetic and cladistic studies have already been performed on hornbills (Kemp 1979, 1988b; Kemp and Crowe 1985), and the results are summarized below. The best characters to use when deciding whether a hornbill belongs to one species or to another are probably those used by the hornbills themselves when recognizing a mate. Characters employed in communication between individuals are thought to include colours of soft parts, plumage, and bill, development of the casque, and the loud calls used for territorial defence and mate-attraction. In many cases, these characters were also more consistent across the range of a species than were details of size, shades of plumage, or details of casque structure.

Particular attention was paid to consistent differences between different ages and sexes, but it was obviously not possible, in most cases, to confirm that these characters are in fact used in communication with mates. They were, however, known to be used by a few species, such as Southern Ground Hornbill *Bucorvus leadbeateri*, where use of paint on models to alter colours indicating age or sex status, or playback of calls, produced appropriate responses. Most of the characters used in mate-recognition are amenable to testing in the field, should any doubts arise, and further clarity will emerge when better descriptions of behaviours, colours, and calls are available for each species.

This approach indicated at least 54 species of hornbill, some ten more than had been recognized previously by Sanft in 1960. There even remain a few possibilities for recognition of additional species, where information was inadequate at the time of analysis, and these problems are described in more detail in the individual species accounts. This approach also reduced the number of subspecies recognized in the order, partly by raising some of them to the status of species and partly by rejecting distinctions between populations based only on slight, often graded differences in size or in colour of the plumage.

Hornbills as a monophyletic group

Having decided on the limits of each species, and thereby defined the 'leaves' on the evolutionary tree, it was necessary to ascertain that all the 'leaves' belonged on the same 'branch'; in other words, that they were likely to have descended from a common ancestor and therefore to be monophyletic.

All species of hornbill share a number of characters that set them apart from all other birds and which are expected to be evidence of their relatedness and common ancestry. Most obvious of these is the development of the casque on top of the upper mandible of the bill, a structure found in no other birds and which is more or less elaborately developed depending on the species of hornbill examined. Even more telling are features of the internal anatomy of hornbills. They are the only birds in which the first two neck vertebrae that support the skull, the axis and atlas, are fused together (Starck 1940; Bock and Andors 1992), and they also possess an accessory supraoccipital condyle, in addition to the normal basioccipital condyle, which stops the head from being jerked back too far (Bock and Andors 1992), although the neck as a whole remains long and flexible (Desselberger 1930). They are also the only birds with the kidney separated into two lobes and lacking the third, middle, lobe that is present in all other species (Johnson 1979). Other features that characterize hornbills, although not quite so conclusively, include the long, flattened 'eyelashes' and the habit (in all but the two *Bucorvus* ground hornbills) of sealing the nest-entrance during breeding.

It is on the basis of these characters, which so consistently link all hornbill species, that they are assumed to be monophyletic. Indeed, few people, primitive or modern, have had any doubt about what is and what is not a hornbill. Additional characters uniting hornbills may also be found with further study. For example, in the few hornbills that have been examined, the structure of the chromosomes bearing the genetic material within the nuclei of the cells, the karyotype, is consistent in having the Z chromosome as large as or larger than any of the others. This is one of the pair that determines the sex of the individual along with either another Z chromosome (male) or a W chromosome (female). In other respects the karyotype is rather variable, with diploid numbers (2*n*) of 40–88, and with one species, the Trumpeter Hornbill *Ceratogymna bucinator*, even showing the lowest chromosome number (2*n* = 40) of all bird species studied to date (Belterman and De Boer 1984, 1990).

The relatives of hornbills

Trying to decide what other groups of birds are most closely related to hornbills is not quite so easy. The current consensus is that hoopoes, woodhoopoes, and scimitarbills (in the avian families Upupidae, Phoeniculidae, and Rhinopomastidae) are the nearest relatives to hornbills, forming a sister group with and principal outgroup to them as the order Upupiformes. This assessment was based originally on a variety of anatomical and behavioural features (Sibley and Ahlquist 1972; Maurer and Raikow 1981; Maurer 1984; Kemp and Crowe 1985) and has since been supported by molecular studies of DNA (Sibley and Ahlquist 1991). The various families in the order Upupiformes share with hornbills such features as lack of bright plumage colours, and elongate oval eggs with pitted shells. Like some hornbills, they also call with the head bowed, show terrestrial and arboreal radiations of species, and include co-operative breeding among their social organizations.

Similar analyses have also indicated that rollers, kingfishers, bee-eaters, motmots, and todies (families Coraciidae, Alcedinidae, Meropidae, Momotidae, and Todidae, respectively) are closely related to hornbills, some of which are represented partially or exclusively in the New World. All these groups were united traditionally with hornbills in the inclusive order Coraciiformes. This grouping was based mainly on the structure of the foot, in which the bases of the front three toes are partly fused (syndactyl), but the families also share features of their chicks, such as a protruding lower mandible and retention of feathers in quill during a 'porcupine' or 'pin-cushion' stage of development. Recent molecular studies suggest that these families are best united with others without syndactyl feet, such as ground-rollers, cuckoo-rollers, puffbirds, jacamars, and trogons (families Brachypteraciidae, Leptosomatidae, Bucconidae, Galbulidae, and Trogonidae), most of which have New World or Madagascan distributions. They differ from hornbills in most species by having bright plumage pigments and laying round shiny eggs. In combination with hornbills, however, they all seem to form a rather loosely related and primitive radiation of orders, families, and species within the class Aves, within which hornbills are immediately distinguished by the large size attained by many species (Table 2.2, p. 13).

Relationships between hornbill families, genera, and species

Relationships within the hornbill order Bucerotiformes indicate two distinct families, based on the distribution of some 26 characters of anatomy, behaviour, and development, both within the order and across the various related outgroups (Fig. 6.1). The two species of ground hornbill in the genus *Bucorvus* emerge as the earliest surviving offshoot within the order, sufficiently distinct and long separated to be placed in their own family, the Bucorvidae. They were already evident as a mid-Miocene fossil from Morocco some 15 million years ago (Olson 1985). The remainder of the hornbill species, which lack a credible fossil record, then fall within the family Bucerotidae.

The genera within the Bucerotidae can mostly be clearly defined and easily separated, but the relationships between them are still not well resolved (Fig. 6.1; Kemp and Crowe 1985). Traditionally, 14 or more genera were recognized (Peters 1945; Sanft 1960), based either on superficial assessments of overall similarity between species, such as possession

64 The hornbills

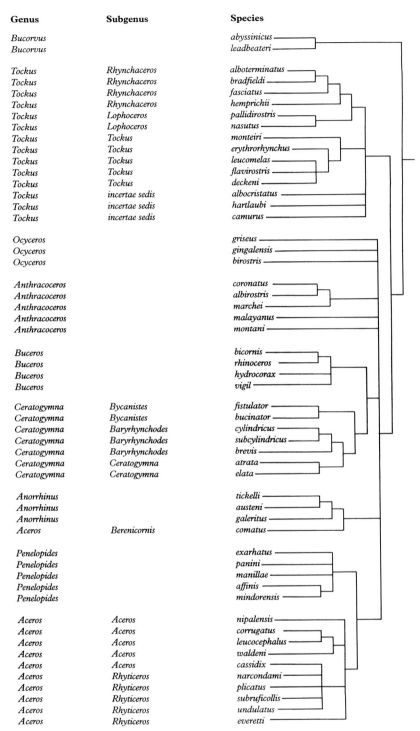

Genus	Subgenus	Species
Bucorvus		*abyssinicus*
Bucorvus		*leadbeateri*
Tockus	*Rhynchaceros*	*alboterminatus*
Tockus	*Rhynchaceros*	*bradfieldi*
Tockus	*Rhynchaceros*	*fasciatus*
Tockus	*Rhynchaceros*	*hemprichii*
Tockus	*Lophoceros*	*pallidirostris*
Tockus	*Lophoceros*	*nasutus*
Tockus	*Tockus*	*monteiri*
Tockus	*Tockus*	*erythrorhynchus*
Tockus	*Tockus*	*leucomelas*
Tockus	*Tockus*	*flavirostris*
Tockus	*Tockus*	*deckeni*
Tockus	*incertae sedis*	*albocristatus*
Tockus	*incertae sedis*	*hartlaubi*
Tockus	*incertae sedis*	*camurus*
Ocyceros		*griseus*
Ocyceros		*gingalensis*
Ocyceros		*birostris*
Anthracoceros		*coronatus*
Anthracoceros		*albirostris*
Anthracoceros		*marchei*
Anthracoceros		*malayanus*
Anthracoceros		*montani*
Buceros		*bicornis*
Buceros		*rhinoceros*
Buceros		*hydrocorax*
Buceros		*vigil*
Ceratogymna	*Bycanistes*	*fistulator*
Ceratogymna	*Bycanistes*	*bucinator*
Ceratogymna	*Baryrhynchodes*	*cylindricus*
Ceratogymna	*Baryrhynchodes*	*subcylindricus*
Ceratogymna	*Baryrhynchodes*	*brevis*
Ceratogymna	*Ceratogymna*	*atrata*
Ceratogymna	*Ceratogymna*	*elata*
Anorrhinus		*tickelli*
Anorrhinus		*austeni*
Anorrhinus		*galeritus*
Aceros	*Berenicornis*	*comatus*
Penelopides		*exarhatus*
Penelopides		*panini*
Penelopides		*manillae*
Penelopides		*affinis*
Penelopides		*mindorensis*
Aceros	*Aceros*	*nipalensis*
Aceros	*Aceros*	*corrugatus*
Aceros	*Aceros*	*leucocephalus*
Aceros	*Aceros*	*waldeni*
Aceros	*Aceros*	*cassidix*
Aceros	*Rhyticeros*	*narcondami*
Aceros	*Rhyticeros*	*plicatus*
Aceros	*Rhyticeros*	*subruficollis*
Aceros	*Rhyticeros*	*undulatus*
Aceros	*Rhyticeros*	*everetti*

6.1 Relationships between the two families, nine genera, and 54 species of hornbill, based on a cladistic analysis of 26 characters of anatomy, behaviour, and ontogeny (from Kemp and Crowe 1985, Kemp 1988*b*, and reanalysed using the computer programme HENNIG86 with the *mhennig* option rooted on an underived ancestor and the *branch-and-bound* method and *nelsen* consensus applied to the two resulting trees).

of a white crest, or on undue emphasis of a few special characters in a single species. The cladistic analysis of relationships, however, suggests that several traditional genera either require rearrangement or belong within more inclusive 'twigs' of the hornbill branching pattern. Only nine genera are now proposed (Kemp and Crowe 1985), but with a few of these further divided into subgenera to reflect the complexities of the evolutionary tree.

The earliest offshoot within the Bucerotidae, based on cladistic analysis (Fig. 6.1), appears to be the radiation of 14 small African species in the genus *Tockus* (including one previously in its own genus, *Tropicranus*). Not surprisingly, they are also most similar to hoopoes and woodhoopoes in many aspects of their biology. The relationships between the remaining genera are not well defined with current analyses, although some affinities between genera do emerge.

Closely allied to *Tockus*, and previously placed with them in the same genus, are three other small species of the Indian subregion that are now separated in the genus *Ocyceros*. Some *Ocyceros* species show close affinities to five species of the Indomalayan region in the genus *Anthracoceros*, but the exact limits of both these genera require further analysis. Some of the larger *Anthracoceros* species share distinctive features with the four species of very large Indomalayan hornbills in the genus *Buceros* (including one rather aberrant species previously in its own genus, *Rhinoplax*). The African equivalent and apparently the closest relatives of this genus of large forest species are the seven medium-sized to large hornbills in the genus *Ceratogymna* (including five species previously placed in the genus *Bycanistes*).

The remaining three genera are all Indomalayan in distribution and not obviously closely allied to the previous genera. Least derived are the three species in the genus *Anorrhinus* (including two species previously considered a single species in their own genus *Ptilolaemus*). Some *Anorrhinus* species show their closest affinities to the two other Indomalayan genera, sharing with them the distinctive plumage of the adult female. These latter are the five relatively small species of *Penelopides* and 11 larger and more widespread species of *Aceros* (including one species previously in *Berenicornis* and nine species previously in *Rhyticeros*). The scientific names chosen for genera are based on strict rules of precedence in date of description, rather than on number of species involved.

The relationships between species within the various generic groupings have also been analysed to some extent (Fig. 6.1), and these indicate the relevant use of subgenera (some of them previously generic names) to enhance the accuracy of the classification. They also indicate why some single species which had previously been placed in their own monotypic genera have now been united with other species in more inclusive genera.

Corroboration of hornbill relationships

Evidence in support of the evolutionary pathways proposed for hornbills comes at this time from three main sources: first, the distribution and relationships of the parasitic feather-lice, or Mallophaga, that are found on hornbills (Elbel 1967, 1969a, 1969b, 1976, 1977a, 1977b; R. Elbel *in litt.* 1988–90); second, hybridization of the protein DNA, which is extracted from chromosomes within the cell nuclei of red blood corpuscles collected from different hornbill species (Sibley and Ahlquist 1985, 1991); third, comparison of the number and structure of the chromosomes, the paired chains of genetic material found within the nucleus of every living cell (Belterman and De Boer 1984, 1990).

Most feather-lice live out their entire lives on the one hornbill which they selected as a host while still in their nymphal stage. Most species of lice are very specialized in the genus and species of hornbill on which they can survive, and it is only on such species that they occur regularly and are able to reproduce. Hence, the evolution and speciation of the lice can be expected to parallel closely that of their avian hosts (Clay 1957). Knowing the distribution of lice species on hornbill species (Table 6.1), and

Table 6.1 The genera and species of feather-lice (Mallophaga) found on hornbills and their distribution across hornbill species and genera (after Elbel 1967, 1969a, 1969b, 1976, 1977a, 1977b, *in litt.* 1988–90)

Hornbill genera, subgenera, and species	Louse families and genera (columns) and species (names in columns)							
	Family: Menoponidae (Amblycera)			Family: Philopteridae (Ischnocera)				
	Chapinia	*Bucerocpocephalum*	*Bucerophagus*	*Paroncophorus*	*Buceronirmus*	*Buceroemersonia*	*Bucerocophorus*	*Bucorvellus*
Bucorvus								
B. abyssinicus			productus/ africanus					docophorus
B. leadbeateri			productus/ africanus					docophorus
Anorrhinus								
A. austeni		emersoni			new sp. 4			
A. tickelli — no lice collected								
A. galeritus	vaniti	deignani			new sp. 5			
Tockus								
Rhynchaceros								
T. alboterminatus	fasciati				albipes	clarkei/ brelihi	latifrons	
T. bradfieldi								
T. fasciatus	fasciati				albipes	clarkei	latifrons	
T. hemprichii					albipes	clarkei		
Lophoceros								
T. pallidirostris						clarkei	latifrons	
T. nasutus	lophocerus					clarkei/ brelihi		
Tockus								
T. monteiri						brelihi		
T. erythrorhynchus	lophocerus					brelihi		
T. leucomelas	lophocerus					brelihi		
T. flavirostris	lophocerus					brelihi		
T. deckeni	lophocerus					brelihi		
incertae sedis								
T. hartlaubi	fasciati				arcellus			
T. camurus	camuri				albipes			
T. albocristatus					arcellus			

Table 6.1 *(cont.)*

Hornbill genera, subgenera, and species	Louse families and genera (columns) and species (names in columns)							
	Family: Menoponidae (Amblycera)			Family: Philopteridae (Ischnocera)				
	Chapinia	*Bucerocolpocephalum*	*Bucerophagus*	*Paroncophorus*	*Bucero-nirmus*	*Bucero-emersonia*	*Bucero-cophorus*	*Bucorvellus*
Ocyceros								
O. griseus	clayae				moucheti			
O. gingalensis					moucheti			
O. birostris	clayae							
Anthracoceros								
A. coronatus	acutovulvata							
A. albirostris	acutovulvata			new sp. 1	new sp. 2 albescens/ new sp. 3			
A. marchei	acutovulvata				new sp. 3			
A. malayanus	malayensis				new sp. 3			
A. montani	hoplai				new sp. 3			
Buceros								
B. bicornis			forcipatus		cephalotes			
B. rhinoceros			forcipatus		cephalotes			
B. hydrocorax	traylori				new sp. 8			
B. vigil			forcipatus		new sp. 9			
Penelopides								
P. exarhatus	muesebecki				new sp. 3			
P. panini					grandiceps			
P. manillae	wenzeli				grandiceps			
P. affinis	wenzeli				grandiceps			
P. mindorensis	wenzeli				grandiceps			
Aceros								
Berenicornis								
A. comatus—no lice collected								
Aceros								
A. nipalensis				major	zonatus			
A. cassidix	lydae			javanicus	new sp. 7			

Table 6.1 (cont.)

Hornbill genera, subgenera, and species	Louse families and genera (columns) and species (names in columns)						
	Family: Menoponidae (Amblycera)			Family: Philopteridae (Ischnocera)			
	Chapinia	*Bucerocol-pocephalum*	*Bucero-phagus*	*Paronco-phorus*	*Bucero-nirmus*	*Bucero-emersonia*	*Bucero-cophorus*	*Bucorvellus*
A. corrugatus	*blakei*			new sp. 2	new sp. 7			
A. leucocephalus	*blakei*				new sp. 6			
A. waldeni					new sp. 6			
Rhyticeros								
A. narcondami					*trabeculus*			
A. plicatus	*hirta*			*javinicus*	*thompsoni*			
A. subruficollis	*hirta*			*javinicus*	*deignani*			
A. undulatus	*boonsongi*			*javinicus*	*deignani*			
A. everetti	*hirta*				new sp. 3			
Ceratogymna								
Bycanistes								
C. fistulator	*bucerotis*				*taurus*		*pachycnemis*	
C. bucinator	*bucerotis*				*longicuneatus*		*pachycnemis*	
Baryrhynchodes								
C. cylindricus	*bucerotis*				*taurus*		*watsoni*	
C. subcylindricus	*bucerotis*				*longicuneatus*		*watsoni*	
C. brevis	*bucerotis*				*longicuneatus*		*watsoni*	
Cerotogymna								
C. atrata	*robusta*				new sp. 10		*watsoni*	
C. elata	*robusta*				new sp. 10		*watsoni*	

Relationships and evolution of hornbills 69

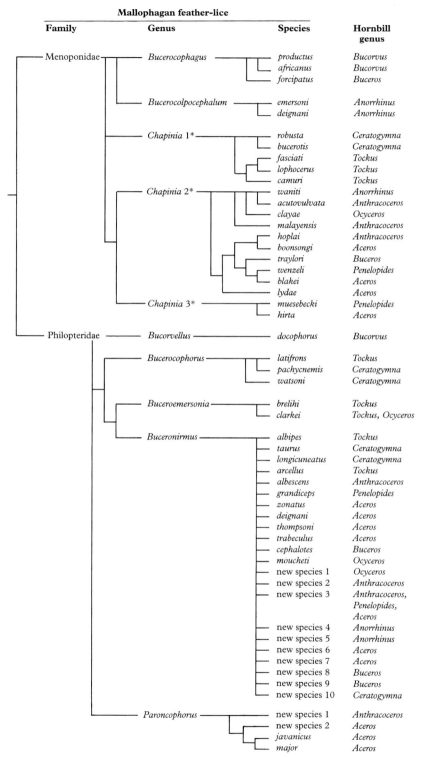

6.2 A possible pattern of relationships between Mallophaga feather-lice that are parasitic on the hornbills (from Table 6.1), compared with their distribution on different genera of hornbills (from Fig. 6.1).
Chapinia lice are proposed to occur in three distinct groups, named 'lophocerus', 'acutovulvata', and 'hirta' after representative species.

assessing the relationships between species of lice (Fig. 6.2), allow some comparison between the evolutionary trees of parasite and host.

The hornbill genera *Bucorvus* and *Anorrhinus* are the only ones each to possess its own unique genus of feather-lice. That the lice have diverged so far along their own evolutionary pathways to have differentiated into new genera also suggests an early separation of these hornbill genera from the rest of the order. In the case of ground hornbills *Bucorvus*, this is supported by their separation as a distinct family. In the case of *Anorrhinus*, it suggests that the genus may, like *Tockus*, be an early offshoot within the Bucerotidae.

Other branching patterns between hornbill genera, subgenera, and species, as identified by cladistic analysis, are also supported to a large extent by the distribution of the remaining lice genera and species (Table 6.1, Fig. 6.2). An exact match is not to be expected, since there is some opportunity for transfer of lice between different host species. This might occur either when hornbills or other birds of different species are feeding together at the same tree or, especially in the case of hornbills, when the same nest-holes are used sequentially by different species. Such transfer between genera may account for the frequent restriction of a group of lice species to either Afrotropical or Indomalayan genera, such as *Chapinia* '1' or *Bucerocophorus* to *Tockus* and *Ceratogymna* species. It may also account for the restricted distribution on hornbills of lice genera widespread on other birds, as in the case of *Paroncophorus*.

Evidence from the nuclear DNA is also largely supportive of the branching pattern suggested by cladistic and parasite analyses. It again shows the *Bucorvus* ground hornbills to be well separated, and *Anorrhinus* and *Tockus* species as the first offshoots within the Bucerotidae (Fig. 6.3). Unfortunately, not all genera and species have yet been analysed by this method, but the results are encouraging since the other genera studied are also arranged more or less as predicted, with the *Buceros*–*Ceratogymna* relationship supported and separated from the other Indomalayan genera *Anthracoceros* and *Aceros*. Even more encouragingly, those species studied within the genus *Tockus* are ordered in exactly the same branching pattern as that derived from the earlier cladistic analysis.

Only seven species of hornbill and one species of hoopoe have been studied for the number and form of their chromosomes (Belterman and De Boer 1984, 1990). Assum-

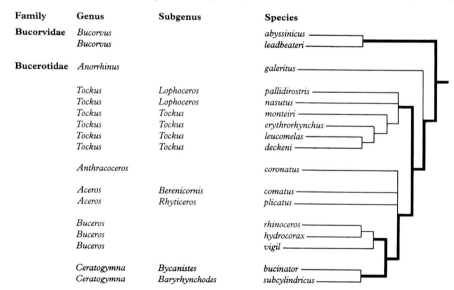

Family	Genus	Subgenus	Species
Bucorvidae	*Bucorvus*		*abyssinicus*
	Bucorvus		*leadbeateri*
Bucerotidae	*Anorrhinus*		*galeritus*
	Tockus	*Lophoceros*	*pallidirostris*
	Tockus	*Lophoceros*	*nasutus*
	Tockus	*Tockus*	*monteiri*
	Tockus	*Tockus*	*erythrorhynchus*
	Tockus	*Tockus*	*leucomelas*
	Tockus	*Tockus*	*deckeni*
	Anthracoceros		*coronatus*
	Aceros	*Berenicornis*	*comatus*
	Aceros	*Rhyticeros*	*plicatus*
	Buceros		*rhinoceros*
	Buceros		*hydrocorax*
	Buceros		*vigil*
	Ceratogymna	*Bycanistes*	*bucinator*
	Ceratogymna	*Baryrhynchodes*	*subcylindricus*

6.3 Relationships between some genera and species of hornbills based on hybridization of their nuclear DNA (after Sibley and Ahlquist 1985, 1991). Relationships also supported by Fitch tree analysis of the data are connected with heavy lines.

ing that a decrease in chromosome number is a derived condition, the hornbill species fall into three groups based on number of pairs of chromosomes (diploid number, or $2n$) and on the relative size and structure of each pair (Fig. 6.4). The variation shown by this small sample, and its lack of obvious conflict with relationships derived from cladistic or parasite analysis, suggests that studies of additional species would be valuable and rewarding.

A major constraint on further studies of hornbill relationships is lack of comparative information of equivalent detail for all species and genera involved. A good example is the placement of the White-crowned Hornbill *Aceros (Berenicornis) comatus* from Malaysia. Originally it was joined in the same genus *Berenicornis* as the Long-tailed Hornbill *Tockus albocristatus* of the African forests, on the basis of their white crests and long, graduated tails (Peters 1945), even though the two species differed considerably in their size, ecology, and behaviour. Later, each species was separated into its own special or monotypic genus, *Berenicornis* and *Tropicranus* respectively (Sanft 1960), which no longer confused them but still gave no indication of their relationships. *Tropicranus* was then shown to share many features with *Tockus* and was united with the genus (Kemp and Crowe 1985), but *Berenicornis* remained a problem.

Cladistic analysis (Fig. 6.1; Kemp and Crowe 1985) suggested a link with other bushy-crested and co-operative-breeding species in the relatively underived genus *Anorrhinus*, while DNA-hybridization studies indicated a relationship with *Aceros* species (Fig. 6.3). No lice have yet been collected from the White-crowned Hornbill, which could be decisive given the restriction of *Bucerocolpocephalum* lice to the hornbill genus *Anorrhinus*. In fact, the genera *Anorrhinus* and *Aceros* may not be too far removed from one another, given the special adult-female plumage, so as a compromise in this study the White-crowned Hornbill is placed within the genus *Aceros* on the basis of its large size, but is still distinguished by its own subgenus *Berenicornis*.

Further details about the anatomy, biology, parasites, or genetic molecules will be able to confirm if the placement of this and other species is correct. Much detail could come from careful field studies (Bennett *et al.* 1993), museum collecting, and recording of behaviour. Other insights, from molecular or karyological studies, might come through analysis of tissues collected from live specimens, such as blood or feather pulp. The most exciting and practical development, applicable in theory to the whole order, might come from analysis and sequencing of the DNA sealed as dried pulp within the shaft of feathers. Such material can be collected non-invasively, either when feathers are shed during the normal moult or from museum specimens (Reddish 1992*a*).

Family, genus, and species	Diploid number ($2n$)	Chromosome changes		
		Microsomes lost	Macrosomes lost	Fusion
Family: Upupidae				
Upupa epops	120			
Family: Bucorvidae				
Bucorvus abyssinicus	88	x		
Bucorvus leadbeateri	88	x		
Family: Bucerotidae				
Tockus fasciatus	68	x x		
Buceros bicornis	70	x x	x	
Aceros undulatus	70	x x	x	
Ceratogymna subcylindricus	44	x x x	x	
Ceratogymna bucinator	40	x x x	x	x

6.4 Possible relationships between some genera and species of hornbills based on examination of their chromosomes or karyology (after Belterman and de Boer 1984, 1990). Extent of loss of microsomes, resulting in reduced diploid number, is scored from x to xxx. Loss of macrosomes or indication of fusion of microsomes is also indicated.

Comparison of evolutionary pathways

If it is assumed that the cladistic or other analysis of relationships between hornbill species is well supported, then it is possible to compare this branching pattern with the distribution across species of various other features. The pattern can be compared with those characters not considered when analysing the evolution of the group. In particular, it offers a relatively independent means of identifying those characters which appear similar, but which arose through either convergent or parallel evolution in response to similar selection pressures.

Examples of such comparisons for hornbills are abundant. The evolution of dark skin colour in small chicks soon after hatching has apparently arisen independently in the genera *Bucorvus*, *Aceros* and *Ceratogymna*. Certain plumage patterns, such as white underparts or all-white tail feathers, are evident in several unrelated genera. The development of the casque and the resemblance of immatures to adults of either sex also follow no obvious evolutionary pathway, and the behaviour of breeding co-operatively is scattered across different genera with no obvious pattern. In each case, lack of congruence between the evolutionary pathway and development of a certain trait becomes a challenge to discover what are the underlying selection pressures which produced that trait.

Numerous other analyses will, it is hoped, become possible by refinement of our understanding of hornbill evolution and by comparison of details available for each species in the second section of this book. The evolutionary patterns can also be compared with those of other animal groups that share a similar geographical distribution. This has been attempted for the hornbills and francolins of Africa, in search of congruent evolutionary pathways that are expected to result when both groups have faced the same historical changes in climate and vegetation (Crowe and Kemp 1988).

Comparisons might also be extended to ecological considerations, such as the correlation of hornbill evolution with the distribution of plant families, such as the Lauraceae, Burseraceae, and Palmae, the fruits of which they eat and the seeds of which they disperse. These plants produce large seeds with high nutrient quality and are favoured by specialist frugivores, yet they are poorly represented in Africa (Snow 1981). This is possibly due to greater reductions of African forest habitats in the past compared with other tropical regions, but it is also correlated with fewer species of large forest hornbills. Other aspects of fruit availability, such as differences in abundance of strangler figs between the Afrotropical and Indomalayan realms (Gautier-Hion and Michaloud 1989; Lambert and Marshall 1991), may also relate to hornbill diversity and abundance.

Evolution and biogeography

The pattern of evolutionary relationships of hornbills can also be overlaid on that of the current geographical distribution of each genus and species (for details, see Table 1.1, p. 6, or individual species accounts). Such an approach assumes that each species has always had similar habitat preferences to those which it exhibits today, and that its present range includes at least part of the historic range in which it evolved. This geographical pattern, or area cladogram, can then be compared with past palaeontological changes in climate, distribution of vegetation, and exposure of landmasses.

One of the best examples of such changes, thought to be important on the large and geologically stable African continent, is the expansion and contraction of lowland and montane evergreen forests (Moreau 1966). Separation of the forest blocks is indicated by separation of subspecies among forest-dwelling *Ceratogymna* and *Tockus* species, most obviously between the Upper Guinea and Lower Guinea forests, but also within the Upper Guinea forests at the Dahomey Gap. These changes were probably responsible, at their most extreme, for driving so many *Bucorvus* and *Tockus* species out into the African savanna and, when less intense, for dividing the species and subspecies of *Ceratogymna* and *Tockus* forest hornbills.

Another example is the extent of the belt of miombo or *Brachystegia* woodland across central Africa (Dean 1988), which may originally have separated the two species of *Bucorvus* ground hornbill. More recently, it probably separated the two species of yellow-billed hornbill *Tockus flavirostris* and *T. leucomelas* and the subspecies of African Grey Hornbill *T. nasutus* and African Red-billed Hornbill *T. erythrorhynchus*. When this woodland was itself bisected, by the arid Luangwa valley, it probably divided the subspecies of Pale-billed Hornbill *T. pallidirostris*.

Wooded mountains isolated within arid savanna probably also produced their crop of new species, such as Hemprich's Hornbill *T. hemprichii* on the highlands of Ethiopia and Bradfield's Hornbill *T. bradfieldi* or Monteiro's Hornbill *T. monteiri* within Namibia. The latter isolation probably led to the transition from an arboreal to a terrestrial life style and the subsequent radiation of terrestrial species in the subgenus *Tockus*. The end result is that 13 of 23 African species (57 per cent) occupy savanna, against only one of 31 species in Southeast Asia. Such a proliferation of savanna species is typical of many elements of the African fauna, including bovids, primates, and even hominids, and equally typical is the radiation of forest species within Southeast Asia.

Similar patterns of changes in vegetation distribution are proposed for the forest–savanna mozaics of the Indian subcontinent (Ripley and Beehler 1990), even though the geological history of the area remains uncertain (Briggs 1989). Much of the evolution in the rest of the Indomalayan Region, however, is best related to geological movements and climatic exposures of land, since most of the region is thought to have remained largely forested since at least the Miocene, some 20 million years ago.

Availability of land can be predicted from two sources. The geology of plate tectonics, by which landmasses are moved about, raised, or lowered on drifting continental plates, determines what land is available and its proximity to other pieces. Superimposed on this, past climates indicate how much land was exposed, as sea-level fluctuated through water being either trapped in or released from the polar ice-caps. For example, sea-levels have varied from 6 m above to 120 m below the present level during the last 5 million years of the Pleistocene (Dingle and Rogers 1972; Siesser and Dingle 1981; Dingle *et al.* 1983; Miller and Fairbanks 1985), with the most recent drop only some 18 000 years ago (Collins *et al.* 1991), and with drops of up to 180 m during earlier Pliocene and Miocene periods up to 20 million years ago. Examination of the 100-fathom marine contour, shown in many atlases, reveals which landmasses, separated at present, would have become interconnected when the sea was at its lowest levels.

Such evidence is especially important for understanding what brought about connection and separation of the islands of Southeast Asia: in particular, the islands of Borneo, Sumatra, and Java and the Malay Peninsula on the Sunda Shelf; the islands of the Philippines on a complex of shelves; and the islands of Indonesia and New Guinea on the Sahul Shelf (Heaney 1984; Hutchison 1989). Distributions of such species as Bushy-crested Hornbill *Anorrhinus galeritus*, Great Rhinoceros Hornbill *Buceros rhinoceros*, and White-crowned Hornbill *Aceros comatus* show the connections across the islands of the Sunda Shelf, as do the distributions of *Penelopides* species across the islands of the Philippines.

Within Africa, and those sections of Southeast Asia that are united on the same tectonic plate or shelf, movement between different areas can be a gradual response to changes in vegetation or land distribution. Major movements over long distances to colonize new habitats, although they may have occurred, do not have to be invoked to explain the present distribution of hornbills. However, the islands of Malesia, or the Malay Archipelago, which stretch from the Asian mainland to Australia, are divided between different tectonic plates which have apparently always been separated by deep ocean channels. Hornbills could have crossed these barriers only by direct dispersal. Only by flying across open water, such as the channel between Bali and Lombok, the Makassar Straight, or the Sulu Sea, would the first colonists have reached such islands as the Philippines, Palawan, Sulu, Narcondam, Sulawesi, New Guinea, and Sumba. In each instance, new hornbill species arose, most of which bear the respective island's name, and in one case even a whole new genus, *Penelopides*, came into being as new species developed on adjacent islands.

Species in the genus *Aceros* appear to be the most intrepid travellers and successful colonists. This ability might be predicted from the daily habit of several species of flying long distances in search of food. They also travel habitually in flocks, which on arrival would provide a good population nucleus for establishment. Their members extend throughout Southeast Asia and have colonized such far-flung islands as Sumba in the Lesser Sundas and Narcondam in the Andaman chain. It is notable that the faunas of Andaman and Nicobar, including Narcondam, show much stronger affinities with Malesia than with India, which supports only one *Aceros* species on its eastern periphery (Ripley and Beehler 1989) and which seems an unlikely source of the original colonists of Narcondam. It also comes as no surprise that an *Aceros* species is the only hornbill to reach the Australasian continental plate, where the range of the Papuan Wreathed Hornbill *A. plicatus* extends across islands from the Moluccas, through New Guinea to the Solomons. Problems of colonization from Malesia might also explain the lack of hornbill species in Australia itself, despite extensive moist evergreen forests in that area during much of the Tertiary (Kemp, E. 1978).

Evolutionary relationships and conservation

Study of the evolutionary and phylogenetic relationships among hornbill species would be an esoteric pursuit if it did not have important implications for the conservation and management of their populations.

The first contribution is in defining the genetic and geographical limits of each species. If the species-units are not accurately delimited, then considerable confusion and wastage of resources can result when planning conservation in the field and management in captivity. The likely confusion of at least four species within the tarictic hornbill *Penelopides 'panini'* complex of the Philippines (Hauge *et al.* 1986), and the recognition of at least five distinct populations within the African Red-billed Hornbill *Tockus erythrorhynchus* are cases in point. A special example is the two very distinct subspecies of the Brown-cheeked Hornbill *Ceratogymna cylindricus*. If they prove to be separate species, then the nominate form in the Upper Guinea forests of West Africa becomes the rarest hornbill on the continent, even more so than the vulnerable Yellow-casqued Wattled Hornbill *C. elata* with which it shares its range and rapidly disappearing habitat.

The second contribution is the ability to extrapolate biological information from well-known species to rare or inaccessible species. Such extrapolation has already been shown for several biological features based on their relationship to body size (Chapters 2, 5). This information and other comparisons are most accurate when extended between closely related species, and only by phylogenetic reconstruction can such relationships be accurately defined. Much of this information is essential for planning conservation, but it is often needed urgently, is unattainable, or is expensive to collect. Such phylogenetic analyses also highlight any poorly known species that are markedly different from their relatives and which might warrant the focus of special conservation efforts.

A third contribution is somewhat more controversial. The limited resources available for conservation, the growing human pressure on the environment, and the accelerating rate of loss of biodiversity suggest that it will not be possible to save all living species. Some form of triage will have to be considered, and the choice made of whether to save many species of the same basic life form or a few species each representing different life forms. Only a well-resolved evolutionary tree can provide a sound biological basis for making such decisions; it may also raise conservation efforts to new heights by highlighting the wonderful complexity and diversity which is at risk.

7
Conservation

Some principles for conservation

Three basic approaches are relevant to the conservation of hornbills. First, one can try to save as much of the habitat of each species as possible. This type of conservation would benefit not only the hornbill concerned but all other organisms in the area. It would be especially important in tropical moist evergreen forest, where some 80 per cent of resident bird species are entirely dependent on this habitat (Collins *et al.* 1991). This may include management of the habitat to maintain it in the best possible condition. Second, one can apply management practices within the habitat aimed directly at the hornbill species of concern. These may include such activities as supplementation of food or nest sites. Third, one can take threatened hornbills into captivity and attempt to breed additional stock. This can then be used either to bolster small natural populations or for reintroduction into restored habitat.

These three options are presented in order of increasing focus on individuals, rather than whole populations, and also in increasing order of cost. They are also considered to be in decreasing order of preference, although conservation ideals and ethics may have little place in deciding which approach is the most pragmatic in a given situation.

Before considering each of these approaches in more detail, some general perspectives are presented on how the demographic functioning of a population relates to its conservation. It is assumed that the goal of all conservation is maintenance of a stable population of the species under consideration, allowing for the relatively small fluctuations in numbers that can be expected as responses to natural cycles of climate or food. Large populations are preferred to small ones because, on average, small populations are more vulnerable to and less likely to recover from random accidents of climate, such as cyclones, of genetics, such as inbreeding, or of demography, such as skewed age or sex ratios. This makes small populations more prone to extinction than large ones.

Increase in population size is generally to be welcomed, so long as it does not have a prolonged negative impact on other animal or plant species within the habitat. Decrease in population size is regarded, justifiably, with concern. However, it is not the individuals lost during a period of decline that are important so much as the rate at which the decline occurs. Any steady rate of decline leads inevitably, like a bank balance losing compound interest, to an exponential fall in total numbers. This occurs no matter how large the initial population, how low the rate of decline, or over how many generations it is spread. Since the probability of extinction increases as population size decreases, even a large population can, within a relatively short period of time, become exposed to a high risk of extinction. For these reasons, even a small population that is stable is preferred to a larger one that is declining.

Whether a population increases, decreases, or remains stable is simply the sum of how many individuals are introduced into versus how many are lost from that population. More particularly, the vital rate of change is the ratio of how many breeding adults are lost compared with how many new recruits are drafted into the breeding segment of the population. In reality, only those breeders whose offspring survive, to become successful breeders in their turn, are really making a contribution to maintenance of the population.

This population-based perspective of conservation deliberately makes no mention of the fates of individuals within that population. This is because much of the conflict that arises around conservation issues results from confusion of the fate of individuals with the fate of populations. The considerable proportion of individuals within a population that fail to contribute to its stability is generally not appreciated, especially by people in developed nations who enjoy abnormally high survival of their offspring. The smaller the population, the greater the role of each individual in population changes, but most such small populations are usually already vulnerable for other reasons.

Many studies of bird populations have shown, firstly, that only a small percentage of eggs laid can be expected to result in successful breeders, and, secondly, that only a few breeders contribute a disproportionate number of recruits to the next generation. The majority of individuals can be expected to die before reaching breeding age, to fail in their breeding attempts, or to produce offspring that fail to become breeders. To focus conservation attention and resources on just any area where mortality or failure is encountered, especially at the individual level, has, therefore, a high probability of failure. Unnatural factors that cause abnormally high levels of mortality or failure, such as electrocution or poisoning, are certainly legitimate targets for conservation action, especially if they affect the breeding segment of the population. Only as a last resort, however, is it useful to try to ameliorate natural causes and levels of attrition.

Some populations may even be able to sustain high levels of failure or harvesting, as for the millions of domestic chickens killed annually but whose 'habitat' and breeding are managed so efficiently. This does not mean that ethical concerns for the fates of individuals must not be carefully integrated into conservation policy, but they can never override factors that apply at the level of the population. The saying 'for the good of the species' must predominate in conservation deliberations, even if it is an incorrect paraphrase of Darwinian natural selection and repugnant to proponents of individual animal rights.

Conservation of habitat

The acquisition of habitat for conservation of hornbills is outside the scope of this book, resting as it does primarily on financial, political, and moral considerations, probably in that order (Pearce 1990). It is still the most important and cost-effective method of conserving hornbills, or any other organism, since if sufficiently large tracts of pristine habitat can be conserved then no further action is necessary. However, it is also the method with the most limited future, given that forests, particularly moist evergreen forests, are the habitat preferred by many hornbill species and also the habitat being reduced most rapidly by logging and agriculture.

Within the range of the hornbill order, the forests of West Africa and particularly those of Southeast Asia are now the most reduced; indeed, they are among the most damaged tropical evergreen forests in the world for both swamp and dryland forests at all elevations (Bennett 1988; Barnes 1990; Collins 1990; Bodmer *et al.* 1991; Collins *et al.* 1991; Lewis, D. 1992; Sayer *et al.* 1992). These forests have been altered through the millennia, first by global climatic fluctuations and latterly by the hand of early man. Alteration of forests has been traced to at least 500 000 years ago on Java (Collins *et al.* 1991), and probably longer in Africa, where the controlled use of fire is known since at least one million years ago (Brain and Sillen 1988). Indeed, none of the deciduous monsoon forest of Southeast Asia may exist in its natural form. Therefore, hornbills must have had to adjust to such changes for a long time, although obviously we have no evidence of any change in distribution or loss of species as a result.

Since the beginning of the twentieth century, the tempo of forest loss has reached unprecedented heights, with all forms suffering extensive cutting following the establishment of plantation crops and industrial logging. Latterly, the forests have also been cut increasingly for slash-and-burn farming to support expanding human populations and their remnants have become prone to massive fires, especially during droughts induced by the marine temperature changes of the El Niño–Southern Ocean Oscillations of 1982/83 and 1990/91 (Anonymous 1984; Helton 1991; Tickell 1992a, b). It is quite possible that all tropical evergreen forests will have been degraded to some extent by the end of the century, with only small pockets reserved for conservation.

The only factor that might moderate the effects of this decline of forests for hornbills is the ability of some species to recolonize, after anything from 5 to 13 years, areas of forest that were logged previously (Wilson and Wilson 1975; Wilson and Johns 1982; Johns 1987, 1988a, b, c, 1989; Blankespoor 1991). This interval approximates the 12–15-year cycle adopted in most traditional and sustainable forms of shifting or swidden agriculture, although logging is more damaging to the habitat and will require longer cycles to be sustainable. The problem of conservation of habitat for hornbills, however, is compounded because so many species are large birds, have extensive spatial requirements, and hence demand huge areas for support of viable populations, such as an estimated minimum of 2000 km^2 of primary forest in Malaysia for large species such as Great Rhinoceros *Buceros rhinoceros* and Bar-pouched Wreathed Hornbill *Aceros undulatus* (Medway and Wells 1971). Few, if any, of the current forest reserves in Africa or Asia seem capable, in isolation, of supporting a viable population of larger hornbill species over the next century.

All these factors place the future of a number of hornbill species in jeopardy. They also indicate that two more manipulative approaches to hornbill conservation, habitat management and captive propagation, may be the only possible alternatives in many instances.

Management of habitat

Management of habitat especially for hornbills has been limited in scope. Identification of important food-plants, such as fig trees, has led to recommendations that such trees be protected from injury during logging operations (Lambert 1991). Logging practices have a particular effect on availability of figs, one of the 'keystone' groups of plants at least in Asian forests and a staple food of many hornbills and other fruit-eating birds (Lambert and Marshall 1991, but see Gautier-Hion and Michaloud 1989 for the relative paucity of figs in Africa). Epiphytic and strangling figs, whose large fruit is especially attractive to hornbills, are particularly prone to reduction since the figs prefer larger size classes of host trees, many of which are important timber species (Lambert 1991). Other immediate forms of management might be cheap and relatively simple. For example, in western India, large fig trees in fruit occur at densities of about 2.5/km^2, but their availability to Great Pied Hornbills *Buceros bicornis* is much lower. This is due to human disturbance at many of the trees (R. Kannan *in litt.* 1992), which might be reduced by education alone.

Management of such essential food resources has important implications, not only directly for the birds, but also indirectly for re-establishment of cut or burnt forests. As pristine forests are continually reduced, regeneration of damaged forest must become an increasingly important conservation activity, in which fruit-eating and seed-dispersing birds will play an important role (Howe 1984).

No attempts at supplementing food supplies for hornbills have been applied intentionally. The possible success of such management is indicated, however, by the readiness with which hornbills avail themselves of new food supplies, such as plantations of exotic food-trees (e.g. oil palms in Southeast Asia; Binggeli 1989) or scraps at tourist and logging camps. Nor have attempts been made to reduce the impact of predators on hornbill populations, other than that of humans. Hornbills are hunted for food and artefacts in many regions, and this increased mortality is thought to have locally eliminated several species, such as *Ceratogymna* wattled hornbills in West Africa and *Buceros* species in Thailand, Malaysia, and Indonesia. Such hunting pressure may also inhibit recolonization of areas of regenerating forest.

Provision of nest sites has also been tried only on a limited scale and confined mainly to small savanna species of African *Tockus* hornbills (Büttiker 1960; Riekert and Clinning 1985). It has been successful in most instances, especially in the drier areas of Namibia where large trees with natural nest sites are limited and where several species resort frequently to nesting in holes in rock faces. Artificial nestboxes have been readily accepted, as they often are by a wide range of species in captivity, but no assessment has been made of what constitutes the most attractive or productive design, nor of their effect on overall population dynamics. Re-erection of a natural nest taken from a forest tree that had fallen was difficult but successful (Poonswad *et al.* 1987); development of nests using lightweight artificial materials seems indicated.

Indeed, the provision of artificial nest sites may be the most important avenue to follow in future conservation efforts in moist evergreen forests. Studies in Malaysia have shown that hornbills are among the least affected species in areas subjected to selective logging, at least in the sense that their numbers return to normal within about a decade of logging (Johns 1987, 1988*a*, *c*, 1989). This suggests that their food supply is not seriously affected, but their breeding may be impaired since the total number of large trees in such an area is considerably reduced by the logging activities. The number of natural nest sites, however, may not be reduced in quite the same proportion, since many nests are in partially hollow or rotten trees that are uneconomical for felling. If availability of nest sites is shown to be limiting in such a situation, then provision of artificial sites may easily redress this shortage.

Management in captivity

Active manipulation of individual hornbills also has a role to play in their conservation. The range of at least two hornbill species is so small and their numbers probably so low that it is considered only prudent to establish a captive population as insurance against extinction of the natural population. Both species occupy the forest habitat of a small island, a typical scenario for threatened species (Johnson and Stattersfield 1990).

The Narcondam Wreathed Hornbill *Aceros narcondami* is found only on the one forested island from which it takes its name, a volcanic peak in the Andaman Sea with only 6.82 km^2 protruding above the surface. The island belongs to India and was uninhabited, but its strategic position, the existence of a small bay for landing, and a spring of fresh water have made it possible for a police presence to be maintained there intermittently since 1968 (Abdulali 1971). Education of new recruits about the uniqueness of the hornbills has been recommended, but this still does not guarantee the birds protection from exploitation, degradation of their habitat, or disasters such as hurricanes.

The Indonesian island of Sumba in the Timor Sea is considerably larger, at 13 720 km^2, but it is well populated by humans and much of its limited monsoon-deciduous-forest habitat has been eliminated. Sumba Wreathed Hornbills *A. everetti* still exist on the island, as do most of six other endemic bird species, but the exact number of hornbills has not been determined, the forests are still being cut to some

Table 7.1 Hornbill genera and species/subspecies known to have raised young in captivity, or where female sealed into the nest and laid eggs (in parentheses). See individual species accounts for details and references (also *International Zoo Yearbook* to 1991; Brouwer 1990, 1991, *in litt.* 1992; Wilkinson 1992)

Genus	Species/subspecies
Bucorvus	*abyssinicus, leadbeateri*
Anorrhinus	none
Tockus	*nasutus, e. erythrorhynchus, e. rufirostris, leucomelas, flavirostris, d. deckeni, d. jacksoni*
Ocyceros	none
Anthracoceros	*coronatus, albirostris, malayanus*
Buceros	*bicornis, rhinoceros,* (*hydrocorax, vigil*)
Penelopides	*panini, manillae*
Aceros	(*comatus*), *cassidix, corrugatus, leucocephalus, plicatus, undulatus*
Ceratogymna	*bucinator, subcylindricus, brevis,* (*atrata*)

extent, and hornbills still appear for sale or as items on the menu of local eating-houses. Education of the Sumba people about the uniqueness and potential value of their avian compatriots seems necessary, as well as finding ways to reduce deforestation and persecution of the hornbills. However, recent surveys and conservation plans are most hopeful.

The world populations of most other hornbills do not appear threatened, although there is concern about the vulnerability of some other island endemics, such as the Palawan Hornbill *Anthracoceros marchei*, several *Penelopides* species and subspecies on islands in the Philippines, and, in particular, the Visayan Wrinkled Hornbill *Aceros waldeni* and the Sulu Hornbill *Anthracoceros montani*. There is also concern about mainland species where logging is rapidly reducing the core areas of their total range, as with Rufous-necked Hornbill *Aceros nipalensis* and Plain-pouched Wreathed Hornbill *A. subruficollis* in Thailand and Myanmar, and the Yellow-casqued Wattled Hornbill *Ceratogymna elata* and the nominate form of the Brown-cheeked Hornbill *C. cylindricus* in West Africa. In all these areas, information about population sizes and whether the populations are stable within the remaining habitats is just not available, taxonomy is still unresolved, and some species are at risk of disappearing even before they are properly studied. There are also a number of populations, covering an even wider range of species, that are threatened locally, and this is mentioned in the texts for individual species; but in no instance is this threat inimical to the species as a whole, or even to recognized subspecies.

Hornbills make spectacular cagebirds, are amusing pets, and are popular in the collections of zoos and aviculturists (Brouwer 1990, 1991). Most are potentially long-lived species, yet much of the stock currently in collections consists of wild-taken birds, trapped by hunters or, more often, taken as chicks from nests (Bell and Seibels 1990). Wild-caught adults of several species do not easily become tame in captivity (Coupe 1967; Benson 1968), and they also—especially females brooding their chicks—represent a large source of proteinaceous food for the hunter. The chicks are quite easy to raise, and they grow up tame and trusting, although often with a psychological imprinting on their human foster parents. Full-grown hornbills fetch good prices within the aviculture trade, although, with middlemen, transport costs, and losses in transit, not much of this currency reaches the local hunter who coexists with the birds.

A number of species have been bred in captivity since at least 1927 (von Wieschke 1928; Wilkinson 1992) (Table 7.1) but few so successfully or regularly that their offspring can meet even the demand of the aviculture trade, let alone provide adequate stocks for reintroduction. Furthermore, none of the rarer species has been bred in captivity, despite concern for their populations in the wild, nor have any offspring from common species been used in trial reintroductions into the wild. There remains a great deal of work and experimentation to be done to improve the success of captive management and breeding, and to develop techniques for re-establishment of wild populations. Regional studbooks have been started for some larger species (*Bucorvus abyssinicus*, *B. leadbeateri*, *Buceros bicornis*; Brouwer 1993). A world centre for propagation, study, and management of hornbills would greatly facilitate future developments.

What future for hornbills?

My personal perspective is that total protection of hornbills seems unlikely in all but a few small and scattered reserves, none of which is likely to contain viable populations of all species present. A broader perspective of conservation seems essential, one that accommodates some integration of the many and rapidly increasing humans that share Africa and Southeast Asia with hornbills. Previous interest in hornbills came mainly from colonial and expatriate workers, but there is now a growing interest and expertise among local people. Their skill in educating their own governments and people about the needs of hornbills holds the best prospects for the future. In more tolerant or wealthier societies, protection of hornbills simply for their ecological and aesthetic roles may be successful, but in many areas some form of utilization already exists, and its increase seems inevitable.

Any legitimate form of utilization should be sustainable, which means that in practice it be done in a way that minimizes effects on wild populations. Exploitation with minimal effect is especially feasible for larger species of hornbill, where the second chick to hatch can immediately be removed since it normally dies of starvation within a few days. That this technique has no obvious effects on the population and that the second chick is quite viable have been demonstrated for Southern Ground Hornbills *Bucorvus leadbeateri* (Kemp 1988a; Marais 1993).

Practical considerations may favour methods of exploitation with greater effects on the population. These are, in increasing order of impact: (1) removal of one or more eggs for artificial incubation and rearing, leaving the pair to raise the remaining egg or to re-lay in the same or subsequent seasons; (2) removal of one or more chicks; and (3) harvesting of full-grown birds by trapping for export or by shooting for food, medicinal, and religious requirements. Ideally, any form of removal should be confined strictly to juveniles, so that the essential adult breeding sector of the population is not affected. Juveniles of most species are easily recognizable by their undeveloped casques and different soft-part colours.

An alternative approach, with minimal impact but with greater requirements of time, funds, and organization, would be to harvest a few juveniles initially, use them for captive breeding, and then offer their offspring for sale. This approach has the potential of integrating conservation, management, and utilization, and would also accommodate those birds confiscated by conservation authorities and which often pose problems of placement.

Any trade in hornbills, together with other forest products and wildlife, must be regulated in such a way that a greater proportion of the profits is returned to people living alongside the resource. This will allow them to appreciate the resource in economic terms, in addition to the aesthetic, religious, and material appreciation which already exists in many instances. Trade in natural products either has been very selective, with considerable wastage of the total resource (as when logging for timber), or has been controlled on protective principles which limit normal trade, such as that in wild animals and plants. Efforts that have been made to place embargoes on trade in natural products, while usually unsuccessful and open to circumvention, have highlighted the strong demand and potentially strong markets which exist.

An open and holistic approach to utilization of wildlife is urgently required, ranging from innovative ideas on the range of products available through to their recognition as assets in national budgets. Natural habitats must not be exploited just for a small temporary portion of their resources. Any harvesting must consider marketing all resources affected in such operations: for example, not just timber, but also rattan for furniture, orchids for collectors, collections of dead animals for museums and plants for herbaria, live animals for zoos, or plants for nurseries, botanical gardens, and pharmaceuticals. Normal market forces and practices should operate, including free trade, advertising, substitution of commoner species, development of new markets, and enhancement of products. Much of present conservation and trading legislation also draws undue attention to rarer species, subjecting them to short-term exploitation and failing to encourage use of alternatives.

Exclusivity and enhancement at the point of origin, to provide additional jobs and revenue and to raise the value of the product, seem especially important. In the case of hornbills, this might involve employment of labour to harvest, house, feed, and maintain a captive stock; use of labour-intensive techniques to eliminate imprinting; purchase of specially collected or specially grown food supplies; manufacture of cages for holding, transport, and display; and provision of transport, care, and couriers to point of sale. Exported birds might even be neutered, to prevent competition from birds bred in captivity outside the range of the species.

Current conservation efforts to avert habitat destruction in the tropics and eliminate trade in wildlife have obviously been a failure. Nations with the highest extra-territorial consumption of natural resources have failed to include within their economic planning sufficient cover for conservation. The minority conservation ethics of these same nations predominate and so prevent normal trade in renewable natural resources, all of which has devalued resource utilization. More pragmatic approaches to conservation must be considered, ones which appreciate the right of local human populations to the economic, material, religious, and aesthetic benefits of their own resources.

Hornbills are among the most conspicuous birds in their environment and can serve as 'flagships' for the conservation of their habitats (Kemp and Kemp 1974; Poonswad and Kemp 1994). Fortunately for them, they also appear rather resilient to habitat alteration, have aspects of their biology that facilitate utilization, are attractive to humans, and have a good market value. It may be these attributes, rather than any higher sentiments, which will eventually determine their survival.

'Hornbill Prayer'

Not long, but evermore may there be hornbills; and trees; and forest for their nests and food.

May the human spirit of good will and regard prevail over ignorance, indifference.

(Gordon Ranger, unpublished notes, 1947)

Appendix: guidelines for captive management

These comments are not based on long personal experience of keeping hornbills in captivity, but are a synthesis of experience with several species in the field, visits to several captive collections, and a survey of the literature. They are not exhaustive, are intended merely as guidelines, and are expected to be easily surpassed by the experience of serious students and keepers. However, they may assist conservation officials in search of basic management principles and isolated students without access to specialist literature. Further details and specialist references are available under the individual species accounts.

Captive maintenance

The main factor in maintenance of healthy hornbills is that they be relaxed in their surroundings. This results from tameness from being handreared, good cage design to exclude disturbing influences, careful introduction to their facilities, or a combination of these factors. Cage size and seclusion do not appear too critical if birds are properly acclimatized to their surroundings. Even large species have been kept in excellent condition, and have bred in cages no larger than five times their own body length and in busy public display areas. Larger cages and seclusion may facilitate breeding by birds less at ease with either observers or their mates. Positioning of perches is important, especially that they are firm, that they are not too close to the roof to cause abrasion, and that some are in the most secluded portions of the aviary. Since hornbills are mainly tropical species, many require access to sheltered and heated areas during subtropical winters. Routine care of beak and claw growth seems rarely to be necessary.

Diet depends on the hornbill species and on what ingredients are available locally. In the wild, frugivorous hornbills often eat from only one or two fruiting species each day, but change species between days and take some animal food in between. With other hornbill species and at other times, several fruit species will be taken during one day, and all hornbills apparently relish any small animals they encounter. Hence, a diverse fruit diet is indicated, including sugar- and water-rich fruits such as fresh figs or grapes and lipid-rich nuts and seeds, with some protein in the form of arthropods, such as insects, or small vertebrates such as baby mice or day-old chicks.

More carnivorous species require higher levels of protein in their diet, and for all species such a diet is indicated when chicks are being reared. More protein, as delivered in courtship-feeding, and calcium, such as snail shells, might also be indicated prior to egg-laying. All food for the breeding female and the chicks should be in chunks that allow for handling, swallowing, regurgitation, and transfer by the male to occupants of the nest. Excessive protein, such as a purely animal diet, is to be avoided, since excess urea has been recorded in chicks of ground hornbills under such a regime, whereas natural food often contains considerable roughage in the form of stomach contents of herbivorous prey.

Hornbills in transit benefit by having the casque taped and padded with cardboard; otherwise, damage to the casque may be repaired with fibreglass (Anonymous 1990). Taping of the long tail with gummed paper may also prevent feather abrasion and breakage.

Artificial nests

The exact dimensions and details required for an artificial nest will depend on the size and species of hornbill concerned. Dimensions of natural nests and other details are given wherever possible under the species accounts; where not available, they can be estimated by taking those of a congeneric species of similar size (see Table 1.1, p. 7).

The common denominator in the design of most hornbill nests (Fig. 7.1) is an entrance hole in a vertical surface, just wide enough to admit the species (which can be judged as 0.5–1.0 cm wider than the width of the shoulders), preferably with a vertical oval shape to allow the wings to be drawn in when entering. The walls of the cavity around the entrance

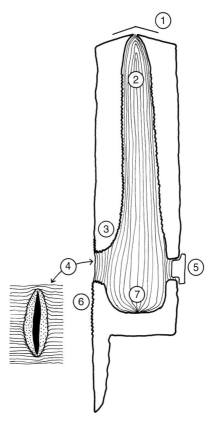

7.1 Cross-section of a basic design for any artificial hornbill nest, the exact dimensions to be estimated from the size of each species and any measurements of nests recorded in the individual species accounts. Rough dimensions for a very large *Buceros* hornbill would be a chamber of 40 cm diameter, entrance 10 × 50 cm, floor depth 10 cm, and funk-hole 200 cm.

1: ventilation opening at top of nest-cavity, with weather shield suspended above.
2: escape- or funk-hole above nest-cavity, with rough walls to provide grip when being climbed.
3: well-insulated walls, with loose flakes or softwood panels on the interior to allow the female some chipping behaviour during nest-preparation.
4: nest-entrance, with walls wide and rough enough for adherence of sealing material, plus anterior view to show oval form and approximate extent of sealing (stippled) to leave a narrow vertical feeding slit (solid).
5: inspection hatch, tightly closed with block of equivalent insulation to nest-walls.
6: rough exterior below nest-entrance for birds to grip on when feeding or sealing, with extension below to allow for bracing with the long tail eathers; alternatively, a perch can be provided just below the entrance
7: nest-chamber, with the floor sufficiently far below the entrance to allow lining material to be added, but not so far below that an incubating female must stand up to take food or the female and chicks cannot reach up to defecate; small drainage holes may be included if there is a risk of water flooding the nest.

should be thick and rough, to supply the surface area and texture necessary for easy adherence of sealing material. The nest-cavity should be of reasonably soft wood, to allow minor alterations by the female from inside, and the exterior should be rough enough or have an appropriate perch for the birds to hold on to when visiting the nest. In some situations it may be more convenient to construct a nest-hole in a wall, with the same attention to basic details and dimensions, since several species have been recorded nesting in holes in rock faces. Such a nest would take up less space within the cage, allow for easier cleaning, fumigation, and insertion of heating aids, and facilitate construction of an inspection hatch or viewing facilities at the back.

The floor of the nest should be below the entrance hole, but not so far below that the incubating female has to leave the eggs to reach food presented at the entrance or is prevented from placing her anus at the entrance for defecating. The floor can be uneven, but should allow for drainage. Plenty of appropriate materials for lining should be supplied outside, for delivery to the nest in courtship, levelling of the nest-floor, and formation of a bed for the eggs. The walls of the nest-chamber should be rough or have perches to which the hornbill can cling, and the roof should extend far above the nest, preferably tapering as it goes, to provide an escape-chamber into which the inmates can crawl.

The walls of the nest should be fairly thick and of a non-conductive material, to provide insulation against fluctuations in nest temperature. The entrance hole should be in such a position that the head of the female is not obstructed while applying sealing, in order that a vertical slit can be formed for ventilation. A vent in the roof of the nest, which can be opened and closed, will allow increase of ventilation if the nest becomes too hot or damp. Preferred sealing materials differ with each genus, but provision of ample food (to produce plentiful droppings), some sticky foods (such as millipedes, sticky fruits, or maize porridge), and lumps of damp clay soil (but not too wet) will anticipate the needs of all species.

Ground hornbills do not seal their nests, nor do they have such specific requirements of cavity form. Their preferred nest conformation, a vertical shaft in a hollow trunk with the floor up to 2 m down, may not be accessible to most captives, which are pinioned and incapable of proper flight. Many natural nests, however, have large lateral openings, with the floor not far below. These could easily be duplicated, and captives have even been known to excavate their own cavity into a vertical earth bank.

Captive breeding

The basic requirement is a compatible pair, but the criteria by which individual hornbills make their mate choice and the behaviours which indicate compatibility are poorly understood. Immatures and non-breeding adults of the larger forest frugivores (genera *Buceros*, *Aceros*, *Ceratogymna*) usually form flocks in which much courtship and pair-formation is observed. This suggests that housing young birds in groups may be indicated, but several species have shown that, once pairs are formed or maturity is attained, they become very aggressive to conspecifics. No attempt appears to have been made to breed any of the co-operatively breeding species in the groups in which they normally occur, which may account for why so few of such species, notably in the genus *Anorrhinus*, have been bred in captivity.

Provision of an artificial nest appears to stimulate potential breeders, and even its removal between seasons to renew stimulation on its replacement has been suggested. Provision of alternative nest sites, which elicit active nest-inspection in the wild, might also provide stimulation. Factors resulting from siting of the nest are poorly known, but seclusion from observers and cleaners seems important, as may selection of a warm area to minimize temperature fluctuations and thereby to reduce energy requirements of birds in the nest.

Great care is normally given to avoid interference at the nest once it is active, but this may be unwarranted, at least in those species in which the female becomes flightless after a simultaneous moult of its flight feathers. It would be advisable to shut the male away during nest-inspections, to minimize the stress of nest-defence, but the flightless female and the nest contents can normally be handled in the wild without fear of desertion. Installation of a sliding false bottom to the nest, on to which the female and chicks can climb during nest-inspection, can make this operation even simpler. Viewing the nest contents and listening to sounds in the nest, by using an infrared-sensitive video camera, infrared light, and microphone, are the ideal solution to the problem. Accurate knowledge of what is going on in the nest is important if maximum productivity is to be achieved, such as by taking weak or late-hatching chicks for handrearing, or by removing the second chick to hatch in those larger species where it is usually doomed to starvation.

Considerable advances in nutrition, management, and breeding of hornbills in captivity would be possible with well-planned and carefully executed research. This will be enhanced by incorporation of information based on the biology of each species in the wild, and together the results will make an essential contribution to hornbill conservation.

Handrearing of chicks

Hornbill chicks are naked, altricial, and totally reliant on the female for food and warmth. No detailed feeding, brooding, or temperature regimes for chicks have been recorded under natural conditions. Observations suggests that whole food, be it fruits or insects, is fed from hatching, and that the frequency of food-delivery is higher in small *Tockus* species (4–6 items/hour) than in large ground hornbill species (4–8 billfuls/day). Chicks of several species have been reared from hatching using standard brooding temperatures and feeding regimes. Newly hatched ground hornbill chicks have proved especially sensitive to intestinal blockages caused by fine roughage in their diet, such as down from day-old chicks or fur from mice.

PART II

Species accounts

Bucerotiformes

Unique among birds in having fused axis and atlas neck vertebrae, a special casque structure on the base of the upper mandible, kidneys lacking the third central lobe, and long flattened eyelashes on the upper lid. Medium-sized to very large birds, most spp. immediately recognizable by large bill with casque on top. Casque varies from simple ridge at base of bill to huge and elaborate crest, cylinder, or other structure: area of tissue for growth of casque obvious at base of bill. In most spp., ♂ about 10 per cent larger than ♀, and bill 15–20 per cent longer.

Plumage black, grey, or brown from melanistic pigments, with contrasting patches of white, but not with any other bright colours. Some spp. have long tail feathers or voluminous crests. Ages and sexes of all spp. separable, by combination of casque structure, soft-part colours, and plumage. Variable amount of bare skin around eyes and on throat, the latter inflatable in some spp. Most closely related to hoopoes, woodhoopoes, and scimitarbills (order Upupiformes) on many characters, including oval eggs with pitted shells, and recently joined with them in Superorder Bucerotimorphae (Maurer and Raikow 1981; Kemp and Crowe 1985; Sibley and Monroe 1990; Sibley and Ahlquist 1991).

Possess 10 primaries and 10 rectrices, but details of moult (Stresemann and Stresemann 1966) and feather tracts (Shelford 1899) known only for some spp. Underwing-coverts lacking, hence large spp. often noisy in flight as air rushes through bases of remiges and two outermost emarginated primaries. Tarsi stout and short, toes 2 and 4 fused at base to central toe 3, in formation known as syndactyly and shared with sister orders Upupiformes and Coraciiformes. Wings broad and tail long, flight often with flapping followed by gliding; skeleton highly pneumatized to produce low wing loading. Neck and feet extended in flight.

All omnivorous, but vary from mainly carnivorous to largely frugivorous, from social to solitary, with considerable variation in basic biology from genus to genus (Kemp 1979). Breeding biology extraordinary (Kemp 1970, 1979) in all but two *Bucorvus* spp. Eggshell white and pitted, rough with tubercles in larger spp. Chick hatches naked and blind, with upper mandible shorter than lower; skin pale pink, but in some spp. turns black after few days. Extensive air sacs develop over back and shoulders. Emerging feathers remain enquilled for a while to produce a prickly, 'porcupine' appearance, as with Upupiformes and Coraciiformes.

Fifty-four spp. in order, with Afrotropical (23 spp., 3 genera), Indomalayan (30 spp., 6 genera), and Australasian (1 sp.) distribution (Fig. 1.4, p. 5). Separated into two families, Bucorvidae (ground hornbills, 2 spp.) and Bucerotidae (true hornbills, 52 spp.). Virtually confined to tropical forest, but for 13 spp. in savanna of which all but one African. Recent systematic reviews of the order (Kemp and Crowe 1985; Kemp 1988*b*) are augmented with measurements and descriptions from Sanft (1960).

Additional formal anatomical characters include: holorhinal, impervious nares; short tongue barbed at base; basipterygoid process present but not reaching pterygoid; 14–15 cervical vertebrae, 8 complete ribs; syrinx tracheo-bronchial with one pair of syringeal muscles; subcutaneous pneumaticity highly developed, appendicular skeleton pneumatic to terminal phalanges but not dorsally (to vertebrae, ribs, sternum, scapula, furcula); oil gland bilobed and tufted; no caecae; no aftershaft; no down; eutaxic; only left carotid; brood-patch feathered.

Bucorvidae

Very large birds, the largest hornbills, with all-black plumage except for the white primaries exposed in flight. Two spp. in single genus *Bucorvus*, living on African savannas and leading a mainly terrestrial existence. Separable from true hornbills by having 15 instead of 14 neck vertebrae, by the elongated tarsi (with more than 8 anterior scutes and posterior scales reticulated), by a special tendon between pelvis and femur, and by not sealing the entrance to the nest-cavity (Maurer and Raikow 1981; Kemp and Crowe 1985; Sibley and Ahlquist 1991).

Unique among birds in having no carotid arteries, these evident only as fibrous cords and their function assumed by other non-homologous blood vessels (Garrod 1876). Support their own unique genus of mallophagan feather-lice *Bucorvellus*. Share with true hornbills *Buceros* spp. an especially well-feathered tuft on the oil gland and also feather-lice in the genus *Bucerophagus* (Kemp 1979). Also the only predominantly carnivorous hornbills. Walk rather than hop along the ground, with a long stride enhanced by the long legs and walking on the tips of the toes. Tail relatively short, nares covered with tuft of bristly feathers, eyelashes well developed and flattened to form screen above eye, and outer two primaries (9 and 10) emarginated (only in *B. abyssinicus*).

Breed in rock and tree cavities, not sealing the nest, showing no nest sanitation, but with the ♀ being fed at the nest while incubating and brooding. Food is carried to the nest as a bundle of several items held in the bill tip. ♀ does not moult flight feathers simultaneously while breeding. Chick's skin turns from pink to black a few days after hatching, and chick left alone in nest well before fledging. *B. leadbeateri* is the largest bird sp. known to breed co-operatively, and one of only five African hornbills thought to exhibit this social organization.

Fossil remains assigned to this family are known from mid-Miocene deposits in Morocco (Olson 1985).

Genus *Bucorvus* Lesson, 1830

Only genus in monotypic family Bucorvidae, hence shares same features as that family.

Two spp., endemic to Africa, differ in form of the casque, colour of the bill, colours of the extensive areas of bare skin around the eyes and on the inflatable throat and foreneck, and in their social organization. Their ranges are parapatric over a small area in Kenya and Uganda, but each sp. differs sufficiently for the two not to be considered allospp. forming a supersp. (but see Short *et al.* 1990).

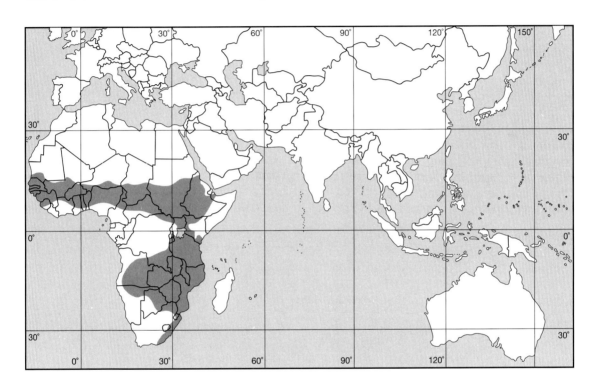

Northern Ground Hornbill *Bucorvus abyssinicus*

Buceros abyssinicus Boddaert, 1783. *Table des Planches Enluminées d'Histoire Naturelle, de M. D'Aubenton*, p. 48 (after Daubenton *Planches Enluminées*, plate 779). Abyssinia, restricted to Gondar, Ethiopia.

PLATES 3, 13
FIGURES 1.1, 1.2

Other English name: Abyssinian Ground Hornbill

Description

ADULT ♂: plumage black except for white primaries. Bill black with triangular yellow to orange patch at base of upper mandible, possibly coloured cosmetically. Casque arises above skull at base of bill in short, high, cowl-like curve, with two ridges along each side and ending abruptly with the anterior end open. Circumorbital skin blue; extensive inflatable bare area on throat and inflatable foreneck red, with blue area at front of throat. Eyes dark brown, legs and feet black.

Northern Ground Hornbill *Bucorvus abyssinicus*

ADULT ♀: like ♂ in overall coloration but smaller, bare throat and neck entirely dark blue.

IMMATURE: like ad but plumage browner, primaries with some small irregular black marks, bill greyer with only small pale yellow spot at base (Browning 1992), casque only a slightly raised area on base of upper mandible, facial skin pale grey (C. Falzone *in litt.* 1992).

SUBADULT: casque becomes a small bulb by 4–5 months, but yellow patch remains small and insignificant. Develops recognizable but paler ad facial skin colours within a year of fledging, and therefore can be sexed from an early age (Anonymous 1974; Penny 1975). Yellow spot on bill begins to enlarge by 6 months, like ad by 2 years, by which time the casque is a high ridge, but not yet so large, curved, or open at the front as on ad (Browning 1992; C. Falzone *in litt.* 1992).

MEASUREMENTS AND WEIGHTS
SIZE: wing, ♂ ($n = 19$) 520–600 (560), ♀ ($n = 17$) 470–545 (512); tail, ♂ ($n = 16$) 320–388 (348), ♀ ($n = 15$) 280–345 (318); bill, ♂ ($n = 14$) 245–290 (264), ♀ ($n = 13$) 195–245 (222); tarsus, ♂ ($n = 18$) 140–158 (151), ♀ ($n = 15$) 125–145 (135). WEIGHT: ♂ ($n = 1$) 4000.

Field characters
Length 90–100 cm. Large, turkey-sized bird with black plumage, very long bill, high casque, and terrestrial habits. Utters deep booming call notes, especially at dawn. Easily distinguished from all except Southern Ground Hornbill *B. leadbeateri*, which has all-black bill without yellow spot or high casque, red skin on face and throat, and slightly different voice. Both spp. show white primaries in flight.

Voice
Series of deep booming notes, *uu-h uh-uh* or *uu-h uh-uh-uh*. Differs from Southern Ground

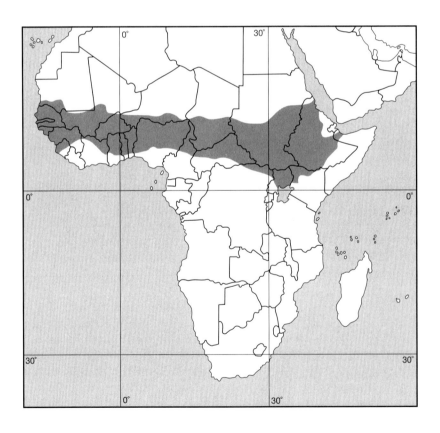

Hornbill in that all notes except the initial one are run together, the pitch slightly higher and tempo faster. May call for lengthy periods. Tape-recordings available.

Range, habitat, and status
S Mauritania; Senegal; The Gambia; Guinea; Sierra Leone; upper Niger R. in Mali; Burkina Faso; S Niger; N Côte d'Ivoire; Ghana; Togo; Benin; Nigeria; N Cameroon; Chad N to 16°N; Central African Republic; S Sudan; Ethiopia; NW Somalia; NE Zaïre; N and E Uganda; NE Kenya.

Sub-Saharan African savannas north of the Equator, inhabiting areas with poor to good grass cover, including among rocky outcrops, at up to 3257 m on Semien Mts of Ethiopia. Generally occupies drier savanna and steppe than its southern congener (Short *et al.* 1990). Overlap as wanderers but not as breeders in W Kenya (Lewis and Pomeroy 1989). Home-range estimated at 260 km^2 (Friedmann and Loveridge 1937). Resident and common, but more local in W Africa.

Enjoys a totemic function in many areas, and generally unmolested. Stuffed heads are worn as a disguise by Sudanese, Cameroon, and Hausa hunters when stalking game, strapped to their heads on a long wooden neck as they crouch to imitate ground hornbills walking through savanna (Fig. 1.2, p. 4). Local people reportedly chase these birds in The Gambia, since considered bad luck if not seen in flight (Cawkell and Moreau 1963).

Feeding and general habits
Usually found as ad pairs, or as trios or quartets with young birds. Mean group size in Ethiopia 2.2 ($n = 32$, range 2–5; L. Brown *in litt.* 1975), in Chad 2.3 ($n = 6$; Blancou 1939); groups of up to six recorded rarely, usually when concentrated at food sources. Flies only if disturbed or when crossing areas of tall grass or dense bush, preferring to walk or run away. One recorded roosting 3 m up in tree on open plains (Duckworth *et al.* 1992).

Small animals form bulk of diet, including tortoises, lizards, amphibians, mammals, birds, spiders, and insects, including beetles and caterpillars. Also eats carrion and some fruits and seeds, including ground nuts, suggesting it is more omnivorous than *B. leadbeateri*. A captive bird regularly cached excess food (Golding 1935). Almost all food taken from the ground, although occasionally jumps up to snatch prey from a bush (Butler 1905). The longer, more slender bill and neck suggest it may forage differently from *B. leadbeateri*, possibly with more reaching for agile prey and less digging.

Displays and breeding behaviour
Breeds in pairs; no record of co-operative breeding, although occasional large groups, with juvs and/or an ad ♂ and 2 ♀♀ (Gore 1981*b*; Schifter 1986), suggest that it may sometimes occur. Most breeding information from captivity (Anonymous 1974; Penny 1975; Falzone 1989), and even here juv shows no interest in breeding activities (Falzone 1989).

Breeding initiated by nest-site inspection, courtship-feeding, and beak-slapping between mates. A captive pair excavated a hole in an earth bank with their bills, scraping out soil with their feet (Anonymous 1974; Penny 1975; Toffic 1985; Falzone 1989). Both sexes, but mainly the ♂, brought lining of dry leaves and debris which the ♀ placed in the nest. Copulation occurred on the ground after the ♂ walked around the ♀ with feathers raised, making booming calls, and repeatedly striking the bill on the ground. Also taps perch and sits with wings hanging and half open.

Breeding and life cycle
NESTING CYCLE: 118–131 days in captivity: 37–41 days incubation, 80–90 days nestling period (Anonymous 1974). ESTIMATED LAYING DATES: Senegal, July; The Gambia, Sept; Mali, June; Nigeria, Feb–Apr; Ghana, Sept; Sudan, Apr, June, Nov; Ethiopia, Mar, May, July, Sept; NE Zaïre, Aug, Nov; E Africa (rainfall region A: see Brown and Britton 1980), Jan–May, July, Nov; Kenya, June; Uganda, Jan–Feb. Not so clearly seasonal as southern congener (Kemp 1976*b*). Captives:

California, Texas, Jan, Mar–Apr, July–Aug, Dec; bred repeatedly without any seasonality, re-laying within 14 days of chick fledging and being removed from cage (Fairfield 1973; Anonymous 1974; Penny 1975; Falzone 1989). c/1–2, but only one chick ever recorded fledged. EGG SIZES: ($n = 6$) 61.0–72.5 × 47.0–51.1 (67.3 × 49.2; Fry *et al.* 1988) or 61.0–85.0 × 47.0– 56.0 (71.0 × 49.6; Schönwetter 1967). Eggs are white with rough and pitted shells, rather pointed at one end. Captive pair laid eggs of mean 71 × 51 mm about 5 days apart (Falzone 1989).

Nests in natural holes in large tree trunks or palm stumps, such as *Adansonia, Acacia albida, Borassus aethiopium*, especially baobabs and even if in agricultural land. Also uses hollow logs and baskets erected as beehives, and may also nest in holes on rock faces (Koenig 1937; Guichard 1950). Holes in earth banks were twice excavated in captivity, about 0.75–1.2 m deep with an entrance 25 × 46 or 35 cm in diameter (Fairfield 1973; Anonymous 1974; Penny 1974), and once an old stick nest of another bird was used (Chapin 1939). Deep lining of dry leaves provided by ♂ (Falzone 1989), which may obscure eggs, and same site may be used year after year.

Incubation by ♀ alone, beginning with first egg. Eggs apparently laid at intervals of several days, judging from embryo development (Lipscomb 1938) and captive chicks hatching a week apart when the first-hatched may weigh five times its 65 g sibling (Rau 1988). Captive ♀ remained in the nest throughout incubation and was fed by ♂ (Anonymous 1974). Hatching normally takes less than 48 hrs (Toffic 1985), and mean weight of newly hatched chick is 69.7 g. When the second egg hatched in captivity the younger chick disappeared within 4 days ($n = 2$), otherwise removal of the first chick for hand-rearing enabled both chicks to be raised (Falzone 1989). Chick's skin colour starts to change from pink to black immediately after hatching, completed by 11 days (C. Falzone *in litt.* 1992) and chick development well documented (Toffic 1985). Darkening of skin and development of chick appears slower than in congeneric *B. leadbeateri*. ♀ remained in nest until 10 days after hatching, then started coming out more and more frequently, until by 15–20 days after hatching the chick was left alone for long periods.

A Martial Eagle *Polemaetus bellicosus* has been found feeding on freshly killed ♂ (Gérard Morel *in litt.* 1984). A pair, received as ads, was still alive in captivity after 40 years (K. Brouwer *in litt.* 1992).

Southern Ground Hornbill *Bucorvus leadbeateri*

Buceros leadbeateri Vigors, 1825. *Transactions of the Linnean Society, London*, **14**, 460. Africa interiori Septentrionali, restricted to Lower Bushman River, Cape Province, South Africa.

PLATES 1e, 1f, 2d, 2e, 3, 13
FIGURES 3.2, 3.7 3.9, 3.16, 5.1, 5.4

Other English names: Ground Hornbill, African Ground Hornbill

TAXONOMY: the correct specific name to apply to this sp. depends on interpretation of a written description by Vigors (1825) of a specimen which is now lost. Browning (1992) presents important evidence that the specimen was a female of this sp., in which case the name *leadbeateri* rather than *cafer* should be applied.

Description

ADULT ♂: black plumage except for white primaries, but becomes sooty-brown when worn. Bill black, with casque only a low ridge at base. Bare skin around eyes and on throat and inflatable upper neck red. Eyes yellow, legs and feet black.

ADULT ♀: like ♂ in overall coloration but smaller, with patch of violet-blue on throat, sometimes extending down sides of neck and as small spots on facial skin.

IMMATURE: sooty-brown with flecks of black in white primaries; bill smaller than ad and more grey distally, eyes grey, facial skin pale grey-brown. Noticeably smaller and slighter than ad of either sex.

SUBADULT: begins to assume ad colours by third year (van Someren 1922), but may not be fully ad until 4–6 years, with possible individual and sexual variation.

MOULT
Main period Nov–June (Brooke and Kemp 1973), complete primary moult possibly extending over 3 years.

MEASUREMENTS AND WEIGHTS
SIZE: wing, ♂ (n = 33) 469–618 (560), ♀ (n = 10) 495–550 (528); tail, ♂ (n = 22) 300–360 (345), ♀ (n = 9) 290–340 (324); bill, ♂ (n = 30) 190–221, (207) ♀ (n = 10) 168–215 (192); tarsus, ♂ (n = 21) 130–155 (143), ♀ (n = 7) 130–140 (135). WEIGHT: ♂ (n = 8) 3459–6180 (4191); ♀ (n = 8) 2230–4580 (3344); juv ♀ 2211 (D. Jackson *in litt.* 1973).

Field characters
Length 90–100 cm. Large size, heavy bill, black plumage, and red skin of face and throat render it unmistakable over most of sub-equatorial Africa. White primaries conspicuous in flight, and utters deep booming call, especially at dawn. Confusable only with *B. abyssinicus*, whose range it approaches in SE Uganda and NW Kenya: the latter has blue facial skin, cowl-like casque, yellow patch at base of upper mandible, and slightly different call.

Voice
Main loud call is a deep, resonant, 4-note booming or *hoo hoo hoo-hoo*, accompanied by three body contractions the last of which produces the double note. Given by all ad group-members, usually just before first light from roost, but also while walking or even while flying. Usually heard at two pitches, ♂ lower than ♀, but an individual may call at either pitch, regardless of sex, when necessary to maintain duet effect. Calls sound deep and mellow from afar, rather soft and tinny close by; may carry up to 5 km to human ears on a still morning (Verheyen 1953). Low-intensity contact-call is 2–3 soft hooting notes, *u-hu-hu*, sometimes used as duet (Seibt and Wickler 1977), often repeated, and may lead into the main booming call, which also functions as a high-intensity contact-call. Single deep *hu*, almost a grunt, uttered in alarm. A harsh growl of threat (Ranger 1931). Harsh nasal bray made by ♀ when accepting food and by begging imm, varying in volume and harshness and often continuing for long periods. Small chick utters soft peeping notes. Tape recordings available.

Range, habitat, and status
Rwanda; Burundi; S Kenya, north to Eldoret and Turkwell R.; SE Zaïre; Tanzania (accidental to Zanzibar); Angola; Zambia; Malawi; N Namibia; N and E Botswana; Zimbabwe; Mozambique; N and E South Africa in Transvaal, Natal, and E Cape Provinces.

African savannas south of the Equator at up to 3000 m. Common in woodland, savanna, and grassland (Plate 1e), but avoiding forest or more continuous woodland and hence uncommon in central African woodlands. Extends into short scrub with few large trees, such as agricultural areas or montane grassland, but only where there are adjacent forests and gorges to roost and breed. Occurs at density of about 100 km^2 per group in Transvaal lowveld, South Africa (Kemp and Kemp 1980), but with overall variations in density (Kemp *et al.* 1989). Nest-spacing averages 8.9 km in less dense areas of Transvaal lowveld (n = 14; Kemp and Kemp 1980) to 4.0 km in dense areas (n = 10), and nests averaged about 10 km apart in Natal reserves and farmland (n = 4) (Jones 1969; Knight 1990).

Declining before dense human populations and habitat destruction, but revered and pro-

Southern Ground Hornbill *Bucorvus leadbeateri*

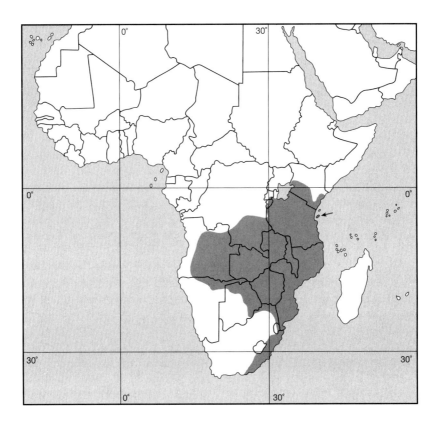

tected by many local tribes, except in times of drought (Stark and Sclater 1903; Godfrey 1941; Vernon 1984). Persecuted in developed areas, where its aggressive territoriality leads to attacks on and shattering of its reflection in window panes. Disappeared from much of its former range in South Africa (70%) and Zimbabwe, probably as a result of direct persecution and inadvertent poisoning during campaigns against livestock-predators and rabies-carriers. May be dissuaded from breaking windows by smearing them with mud (van Someren 1958), painting them (Vernon 1982), or screening them off with wire mesh. Persecuted in some areas of S Africa since at least 1940s (G Ranger unpublished notes; pers. obs.), but can still exist alongside wide range of agricultural practices if protected (Knight 1990). Regular digging for food in newly turned soil renders them vulnerable to land-mines (Masterson 1979).

Feeding and general habits

Resident, group-territorial, and lives in co-operatively breeding groups of 2–11 birds. All members of a group co-ordinate their activities and remain close together throughout the day (Plate 1f). Mean group size 3.5 ($n = 1167$ group-seasons), 98% of population in groups, 72% in groups of 3–5, and only 2% solitary (Transvaal: Kemp *et al.* 1989). Group composition variable, but always only one dominant pair of ads that undertakes breeding. Usually only one ad ♀, often more than one ad ♂, and usually one or more imms of different ages. Most lone birds are ad ♀♀ (Kemp 1988a; G. Ranger unpublished notes). Ad sex ratio 1.4 ♂♂ per ♀, with imms comprising about 20% of the population. Social organization maintained by allopreening and complex interactions involving giving and withholding food.

Calls before leaving roost, usually just before first light, and then descends to ground

Plates

Plate 1
Extremes of hornbill habitats and morphology

(a) Moist evergreen forest along the banks of the Sungai Melinau, Gunung Mulu National Park, Sarawak, Malaysia, typical habitat for seven species of hornbills, including the highly specialized Great Helmeted Hornbill *Buceros vigil*.

(c) The Hakos Mountains of central Namibia, showing the arid steppes and scattered thorn scrub that comprises the preferred habitat of Monteiro's Hornbill *Tockus monteiri* and peripheral habitat for two other smaller *Tockus* species. The canvas hide is placed for observation of a nest among the rocks.

(e) A trio of Southern Ground Hornbills *Bucorvus leadbeateri* foraging near their nest in the wooded savanna of the Kruger National Park, South Africa, a habitat which they share with five other smaller hornbill species.

(b) A male Great Helmeted Hornbill *Buceros vigil*, perched above its nest which is placed within a cavity high in the trunk of a large forest tree. The female has sealed herself in behind a wall of mud and excreta and is fed by the male through the narrow vertical slit which remains open to the exterior. (*Photo Adisak Vidhidharm.*)

(d) A male Monteiro's Hornbill *Tockus monteiri* clings by a rocky cavity within a cliff face. This comprises its nest, the entrance of which has been walled in with extensive sealing material of mud and excreta, applied earlier by both members of the pair.

(f) A group of Southern Ground Hornbills *Bucorvus leadbeateri* walking in search of their diet of small animals, led by a year-old juvenile, with the adult female in the rear and two adult males in the centre. All members cooperate in territorial defence and bring food to the nest but only the dominant adult pair undertakes breeding.

Plate 2
Nesting habits, social organization, and human use of various hornbill species

(a) A male Sri Lankan Grey Hornbill *Ocyceros gingalensis* clings by the nest entrance. He passes food to the female inside, after regurgitation of each item from within his gullet. The female has sealed herself into the nest, undergoes a complete and simultaneous moult of the flight feathers of her wings and tail, and then emerges later, refeathered, when the chicks are about half-grown. (*Photo T. S. U. de Zylva.*)

(c) The female of Austen's Brown Hornbill *Anorrhinus austeni* performs all sealing of the nest but is fed not only by her mate but also by other adult male helpers and by immature birds from previous breeding attempts. The female emerges from the nest only when the chicks fledge, having undergone a complete and simultaneous moult of all her flight feathers while nesting. (*Photo Atsuo Tsuji.*)

(e) A pink newly-hatched chick of the Southern Ground Hornbill *Bucorvus leadbeateri* contrasted with its much larger four-day old sibling whose skin has already turned black. Similar development and hatching interval is found in other genera of large hornbills. Usually the smaller chick is unable to compete for food and dies of starvation within a few days.

(b) A pair of Great Pied Hornbills *Buceros bicornis* at their nest. The pair share sealing of the nest-entrance and later, when the chicks are a few weeks old, the female emerges to assist the male with their provisioning. The female usually undergoes a complete moult of her flight feathers while nesting. (*Photo Atsuo Tsuji.*)

(d) The Southern Ground Hornbill *Bucorvus leadbeateri* nests in an unsealed cavity, usually within the stump of a large savanna tree, and the female emerges daily to defecate, preen, dustbathe, and obtain some of her own food. Most food is delivered to her in the nest, not only by her mate but also by adult and immature helpers of both sexes, and she later leaves the nest to help the group in foraging for the single chick.

(f) The skull of a male Great Helmeted Hornbill *Buceros vigil* with the keratin covering of the bill and casque still intact. It was probably collected by a Dayak tribesman in Malaysia and the ivory-like section at the anterior end of the casque was later carved by a Chinese craftsman, as has been done since the Ming dynasty of the fourteenth Century. (*Photo Tony Harris, courtesy of Dr Phillip Burton.*)

Plate 3
Indomalayan crested hornbills, genus *Anorrhinus*, and African ground hornbills, genus *Bucorvus*

Indomalayan crested hornbills, genus *Anorrhinus*. Medium-sized brown hornbills with white tips to the tail feathers and large loose nape feathers. Usually found in noisy social groups.

1. Bushy-crested Hornbill
Anorrhinus galeritus p. 106
Forests of the Sunda Shelf landmasses. Dark brown, with russet base to tail. Ad ♂ with black bill, ad ♀ with yellow bill marked with black as base, juv with uniform pale olive bill. Spend most time in forest canopy and dense tangles, usually in social groups of 6–8, and often noisy with high squawking group-chorus.

2. Austen's Brown Hornbill
Anorrhinus austeni p. 101
SE Asian mainland forests. Bill pale yellow in both sexes, ad ♂ and juv with white throat and blue facial skin, ad ♀ with pale brown throat and pink circumorbital skin. Spend most time in forest canopy, usually in social groups of 4–10 birds, and call with high cackling notes.

3. Tickell's Brown Hornbill
Anorrhinus tickelli p. 104
Myanmar and Thailand hill forests. Bill yellow in ad ♂, dark brown to black in ad ♀. Ad ♂ and juv darker and more rufous than ad ♀. Spend most time in forest canopy, usually in social groups, and call with high cackling notes.

African ground hornbills, genus *Bucorvus*. Very large, with all-black plumage except for pure white primary feathers. Long legs, short tail, and extensive bare facial skin notable.

4. Northern Ground Hornbill
Bucorvus abyssinicus p. 91
African savanna N of Equator. High casque, yellow patch on bill, and blue circumorbital skin distinctive. Red on throat in ♂, throat all blue in ♀, juv with duller facial colours. Spend much time on the ground, usually in pairs or family trios, and call with deep booming notes.

5. Southern Ground Hornbill
Bucorvus leadbeateri p. 94
African savanna S of Equator. Low casque, all-black bill, and red circumorbital skin distinctive. ♂ and subads all red on throat, ad ♀ with blue patch under throat. Juv with dull facial skin colours, increasing in redness until mature. Spend much time on ground, usually in social groups of 4–8 birds, and call with deep booming notes.

Plate 4
Small African arboreal hornbills, genus *Tockus* and subgenera *Rhynchaceros* and *Lophoceros*

Small African arboreal hornbills, genus *Tockus*. Black, brown, or grey plumage and white underparts. Flight buoyant on large broad wings. Display with whistling calls and the bill pointing skywards.

African pied hornbills, subgenus *Rhynchaceros*, keep the wings closed during display. Ad ♂ with dark facial skin, ad ♀ with orange throat patches.

1. African Pied Hornbill
Tockus fasciatus p. 116
Moist lowland forests of W and central Africa, with 2 subspp. A black hornbill with white underparts and yellow bill. **1a**. Lower Guinea Pied Hornbill *T. f. fasciatus* with all-white outer tail feathers and some deep orange on bill. **1b**. Upper Guinea Pied Hornbill *T. f. semifasciatus* with white tips to outer tail feathers, no orange on bill. Usually in family groups, but congregate at food sources and form flocks at times.

2. Hemprich's Hornbill
Tockus hemprichii p. 119
Forests of Abyssinian highlands and N Kenya. Dark grey with white underparts and all-white outer tail feathers. Bill dull orange with stripes at base, larger in ad ♂. Wings closed but tail raised in display. Usually in small family groups, but may form flocks in dry seasons.

3. Bradfield's Hornbill
Tockus bradfieldi p. 114
Restricted range in teak woodlands of central Africa. Orange bill, with yellow line at base, and white underparts distinctive. Most like African Crowned Hornbill, but bill and plumage paler and orange eye more prominent. Ad ♂ with larger bill than ad ♀, juv with smaller and darker bill. Separate into family groups in summer, but form large flocks that often feed on the ground in the dry season.

4. African Crowned Hornbill
Tockus alboterminatus p. 110
Montane and coastal forests from NE to S Africa. White underparts, bright red bill with yellow line across base, and white streaks above eye distinctive. Ad ♂ with prominent casque ridge along top of bill, juv with small pale bill without yellow line. Usually in small family groups, but may form flocks in dry seasons.

African grey hornbills, subgenus *Lophoceros*. Flick the wings and tail open and closed during display. Ad ♂ with grey facial skin, ad ♀ with pale green throat patches.

5. Pale-billed Hornbill
Tockus pallidirostris p. 121
Miombo woodlands of central Africa, with 2 subspp. Pale grey plumage with white stripe above eye, ad ♂ with more prominent casque than ♀. **5a**. Western Pale-billed Hornbill *T p. pallidirostris* with dusky tip to bill. **5b**. Eastern Pale-billed Hornbill *T. p. neumanni* with dull red tip to bill. Most like African Grey Hornbill, but bill plain yellow in both sexes. Usually in pairs or small family groups.

6. African Grey Hornbill
Tockus nasutus p. 123
Savanna of sub-Saharan Africa and S Arabia, with 2 subspp. Dark grey plumage, with white stripe above eye and down centre of back. Bill black with white stripe at base in ad ♂, yellow with maroon tip and dark bands at base in ad ♀. Ad ♂ of **6a**. North African Grey Hornbill *T. n. nasutus* lacks the prominent casque of **6b**. Southern African Grey Hornbill *T. n. epirhinus*. Usually in pairs or small parties.

Plate 5
Small African terrestrial hornbills, genus *Tockus* and subgenus *Tockus*

Small African terrestrial hornbills, genus *Tockus* and subgenus *Tockus*. White-spotted wing-coverts, white secondaries, and white tail pattern. Direct flap-and-glide flight, clucking calls, and the head bowed during display.

1. **Monteiro's Hornbill**
Tockus monteiri p. 128
Arid savanna and steppe of SW Africa. Dark head and neck, large orange bill, and white outer tail feathers distinctive. Ad ♂ with very heavy bill and dark facial skin, ad ♀ with smaller bill and green throat patches, and juv with pale orange bill. Bob up and down while uttering deep clucking calls, keeping the wings closed.

2. **Southern Yellow-billed Hornbill**
Tockus leucomelas p. 136
Savannas of S Africa. The heavy yellow bill is notably larger in the ad ♂ and smaller and duller in ad ♀ and juvs. Facial skin pink. Display with the wings wide open and fanned above the back.

3. **Eastern Yellow-billed Hornbill**
Tockus flavirostris p. 140
Savannas of NE Africa. Deep yellow bill edged with orange, heavier in ad ♂ than ad ♀ and blotched with dull brown in juv. Facial skin black but for pink throat patches in ad ♂. Utter deep clucking calls in display, with the wings wide open and fanned above the back.

4. **African Red-billed Hornbill**
Tockus erythrorhynchus p. 131
Savannas of sub-Saharan Africa, with at least 3 subspp. Thin red bill and pale head distinctive. **4a.** North African Red-billed Hornbill *T. e. erythrorhynchus* has brown eyes and pale yellow to pink facial skin. **4b.** Damaraland Red-billed Hornbill *T. e. damarensis* has brown eyes and bright pink facial skin. **4c.** South African Red-billed Hornbill *T. e. rufirostris* has yellow eyes and pink facial skin. **4d.** Populations with black circumorbital skin of at least two undescribed taxa exist in W. Tanzania (yellow eyes, illustrated) and Senegambia (brown eyes). See species account for further details. Display with wings closed, but in some N and W African populations wings are held partly open.

5. **Von der Decken's Hornbill**
Tockus deckeni p. 142
Savannas of NE Africa, with 2 subspp. Ad ♀ and juv with black bill, subad and ad ♂ with orange bill. **5a.** Ad Von der Decken's Hornbill *T. d. deckeni* with plain black wing-coverts, and ad ♂ with bicoloured orange and yellow bill. **5b.** Ad Jackson's Hornbill *T. d. jacksoni* with white-spotted wing-coverts, like juv, and ad ♂ with mainly orange bill. Display with the wings wide open and fanned above the back.

Plate 6
Indian grey hornbills, genus *Ocyceros*, and small African forest hornbills, genus *Tockus* but subgenus uncertain

Indian grey hornbills, genus *Ocyceros*. With pale silvery-grey plumage and white tips to wing and tail.

1. Indian Grey Hornbill
Ocyceros birostris p. 157
Savannas of the Indian subcontinent. Long tail with dark subterminal band and projecting casque are distinctive. Ad ♀ with smaller, more yellow bill, similar to juv. Usually in family parties, flying with fast beats of the stubby wings or uttering loud, high piping calls.

2. Malabar Grey Hornbill
Ocyceros griseus p. 153
Forests of W India. Pale stripe above eye and rufous wash on undertail-coverts distinctive. Ad ♂ with deep yellow bill, ad ♀ with black marks at base of bill, juv with plain dull yellow bill. Usually in noisy family parties, with various cackling calls.

3. Sri Lankan Grey Hornbill
Ocyceros gingalensis p. 155
Forests of Sri Lanka. Pale stripe above eye, pale underparts, and white outer tail feathers distinctive. Bill of ad ♂ mainly yellow, of ad ♀ black with yellow mark, and of juv pale yellow. Occur in pairs or family parties, with loud clucking calls uttered at intervals.

Small African forest hornbills, genus *Tockus* but subgenus uncertain.

4. Long-tailed Hornbill
Tockus albocristatus p. 149
Moist lowland forests of W and central Africa, with 3 subspp. White crest and long, graduated white-tipped tail distinctive. **4a.** Western Guinea Long-tailed Hornbill *T. a. albocristatus* with only the crest white. **4b.** Eastern Guinea Long-tailed Hornbill *T. a. macrourus* with white head and neck. **4c.** Lower Guinea Long-tailed Hornbill *T. a. cassini* with white crown, plus white tips to wing-coverts and flight feathers. Sexes and ages similar, ad ♂ with largest bill, juv with smallest. Occupy dense tangles in and around forest, moving with dexterity through the foliage, and utter loud double crowing note.

5. Dwarf Black Hornbill
Tockus hartlaubi p. 145
Moist lowland forests of W and central Africa, with 2 subspp. Small size, black plumage, white underparts, and white stripe above eye distinctive. Bill black, with red tip and yellow base in ad ♂. **5a.** Upper Guinea Dwarf Black Hornbill *T. h. hartlaubi* with plain black wing-coverts, and little red on bill in ad ♂. **5b.** Lower Guinea Dwarf Black Hornbill *T. h. granti* with white spots on wing-coverts, and more red on bill in ad ♂. Solitary and secretive in middle layers of forest, with a high trisyllabic piping call.

6. Dwarf Red-billed Hornbill
Tockus camurus p. 147
Moist lowland forests of W and central Africa. Small size, red-brown colour, red bill, and white edges to wing feathers distinctive. Ad ♀ with black tip to bill. Often low down in forest, usually in small noisy social groups, uttering a distinctive series of descending hooting notes.

Plate 7
Indomalayan pied hornbills, genus *Anthracoceros*

Indomalayan pied hornbills, genus *Anthracoceros*. Solid areas of white and black plumage, the latter with a notable green gloss.

1. **Palawan Hornbill**
Anthracoceros marchei p. 170
Forests of Palawan archipelago in the Philippines. The only hornbill on these islands; black body, plain yellow bill, and all-white tail distinctive. Bill of ad ♂ larger than ♀. Favours forest edge, and calls with raucous crowing notes.

2. **Sulu Hornbill**
Anthracoceros montani p. 175
Forests of Sulu archipelago in the Philippines. The only hornbill on these islands; black body and all-white tail distinctive. Bill black in both sexes, larger in ad ♂, which has white eye (red in ad ♀). Utter loud squawking calls.

3. **Indian Pied Hornbill**
Anthracoceros coronatus p. 161
Forests of India and Sri Lanka. White outer tail feathers, white underparts, and pink throat patches distinctive. Ad ♂ with more black on bill than ♀ and blue (not pink) circumorbital skin, juv with small plain yellow bill. Occupy forest and forest edge; noisy with loud squealing and cackling calls, roost communally.

4. **Oriental Pied Hornbill**
Anthracoceros albirostris p. 164
Forests of SE Asian mainland and Sunda Shelf landmasses, with 2 subspp. Pale blue throat-skin patches, white underparts, and white tips to black tail feathers distinctive. Ad ♀ with more black on end of bill than ♂. **4a**. Asian Pied Hornbill *A. a. albirostris* slightly smaller, and ad ♀ with less black on bill, than **4b**. Sunda Pied Hornbill *A. a. convexus*, but much individual variation. Inhabit forest and forest edge, usually as small family parties, with noisy cackling calls.

5. **Malay Black Hornbill**
Anthracoceros malayanus p. 172
Forests of the Sunda Shelf landmasses. All black, with white tips to outer tail feathers. Ad ♂ with large yellow bill and high casque, ♀ with black bill and smaller casque, juv with pale green bill. Some individuals with grey or white stripe above eye. Favour dense stands of large trees, forming core of territory, from where utter harsh braying calls.

Plate 8
Indomalayan great hornbills, genus *Buceros*

Indomalayan great hornbills, genus *Buceros*. Large birds, with pied plumage and a broad black band across the white tail in most spp.

1. Great Pied Hornbill
Buceros bicornis p. 178
Forests of India, SE Asian mainland, and W Sunda landmasses. Massive yellow bill with double-pronged casque, white neck, white underparts, and white tips to wing-coverts and flight feathers distinctive. Ad ♂ with black markings on bill and casque and with red eyes, ad ♀ with bill unmarked and white eyes, juv with no casque. Occupy forest canopy, fly buoyantly, and utter loud honking calls.

3. Great Philippine Hornbill
Buceros hydrocorax p. 189
Forests of large islands in the Philippines, with 3 subspp. Rufous plumage with black face and underparts, together with all-white tail (in ads), distinctive. **3a**. Great Luzon Hornbill *B. h. hydrocorax* with large casque and all-red bill. **3b**. Great Mindanao Hornbill *B. h. mindanensis* with yellow tip to bill and smaller casque. **3c**. Great Samar Hornbill *B. h. semigaleatus* with yellow tip to bill and only slight casque. Both sexes with pale blue eyes (may be red in ad ♂ of Luzon subsp.), but exact details of soft-part colours uncertain. Juv quite unlike ads, mainly white with black bill and black band across outer tail feathers. Occur in social groups and utter loud honking calls.

2. Great Rhinoceros Hornbill
Buceros rhinoceros p. 184
Forests of the Sunda Shelf landmasses, with 3 subspp. Large orange bill with upturned orange casque, and black plumage with white abdomen, are distinctive. **2a.** Malayan Great Rhinoceros Hornbill *B. r. rhinoceros* has a long recurved casque. **2b.** Bornean Great Rhinoceros Hornbill *B. r. borneoensis* has a stouter and even more recurved casque. **2c.** Javan Great Rhinoceros Hornbill *B. r. sylvestris* has a long pointed casque and broader black tail band. Ad ♂ with black markings on bill and casque and with red eyes, ad ♀ with bill unmarked and white eyes, juv with no casque. Pairs occupy canopy of forest and often utter duets of loud honking calls.

4. Great Helmeted Hornbill
Buceros vigil p. 192
Forests of the Sunda Shelf landmasses. Dark brown plumage with naked head and neck, rufous face, white edges to wings, and elongated central pair of tail feathers all distinctive. Ad ♂ with red head and neck skin, ♀ with pale green. Favour canopy of tall foothill forest, from where utter unique hooting notes that break into peals of shrill laughter.

Plate 9
Indonesian and Philippine tarictic hornbills, genus *Penelopides*

Small Indonesian and Philippine tarictic hornbills, genus *Penelopides*. Shrill bleating calls; in most spp. ad ♀ with all-dark plumage, juvs of both sexes like ad ♂.

1. **Sulawesi Tarictic Hornbill**
Penelopides exarhatus p. 198
Forests of Sulawesi and nearby islands, with 2 subspp. Plumage all black, but for white throat and face in ad ♂ and juvs. Ad ♀ with black face, bill smaller than ad ♂, and facial skin black (not pink). **1a**. North Sulawesi Tarictic Hornbill *P. e. exarhatus* with brown casque and plain black patch at base of lower mandible. **1b**. South Sulawesi Tarictic Hornbill *P. e. sanfordi* with red-brown casque and yellow stripes across base of lower mandible. Found mainly in subcanopy in small social groups, which utter shrill piping calls.

2. **Visayan Tarictic Hornbill**
Penelopides panini p. 200
Forests of Panay and neighbouring islands in the Philippiness, with 2 subspp. Tail with broad black tip, tail-coverts rufous. **2a**. Visayan Tarictic Hornbill *P. p. panini* with tail ochre and underparts rufous. **2b**. Ticao Tarictic Hornbill *P. p. ticaensis* with tail white and underparts and rump pale rufous. Ad ♀ mainly black, ad ♂ and juvs with cream neck and breast. Favour lower levels of forest and occur as small social groups. Utter high bleating *tarictic* call.

3. **Luzon Tarictic Hornbill**
Penelopides manillae p. 202
Forests of Luzon and neighbouring islands in the Philippines, with 2 subspp. Tail with pale band of cream or rufous across centre. **3a**. Luzon Tarictic Hornbill *P. m. manillae* with upperparts dark brown and narrow tail band. **3b**. Polillo Tarictic Hornbill *P. m. subnigra* with upperparts black and glossy and a broad tail band. Ad ♀ with brown or black underparts, ad ♂ with cream neck and underparts, juv like ad of respective sex, not like ad ♂ as in other *Penelopides* spp. Occur as small social groups in lower levels of forest, calling with high squeaky calls.

4. **Mindanao Tarictic Hornbill**
Penelopides affinis p. 204
Mindanao and neighbouring islands in the Philippines, with 3 subspp. Tail with black base and tip surrounding ochre centre. **4a**. Mindanao Tarictic Hornbill *P. a. affinis* with black rump; other subspp. with rufous rump, and ochre patch in tail more extensive in **4b**. Samar Tarictic Hornbill *P. a. samarensis*, and even replacing black base in **4c**. Basilan Tarictic Hornbill *P. a. basilanica*. Ad ♀ with black plumage and dark blue facial skin, ad ♂ and juvs with pale head and underparts. Occupy lower levels of forest in small groups.

5. **Mindoro Tarictic Hornbill**
Penelopides mindorensis p. 207
Forest of Mindoro I. in the Philippines. No obvious sexual plumage differences, unlike all other *Penelopides* spp. Ad ♂ with pink facial skin, which dark blue in ad ♀. Little studied.

Plate 10
Indomalayan white-crowned and wrinkled hornbills, genus *Aceros* and subgenera *Berenicornis* and *Aceros*

Indomalayan white-crowned and wrinkled hornbills, genus *Aceros* and subgenera *Berenicornis* and *Aceros*. Adult ♀ with black head, neck, and underparts, ad ♂ and juvs of both sexes with white or rufous on these areas.

Subgenus *Berenicornis*. With long, white, graduated tail and high spiky crown feathers.

1. White-crowned Hornbill
Aceros comatus p. 210
Forests of Sunda Shelf landmasses. Ad ♀ with white crest, ♂ and juvs with white head and neck, flecked with black in juvs. Secretive in lower levels of forest and dense tangles, uttering mellow double or triple cooing notes.

Subgenus *Aceros*. With raised blade-like casque, variously wrinkled in different species.

2. Rufous-necked Hornbill
Aceros nipalensis p. 213
Forests of SE Asian mainland. Broad white tips of tail and wings, inflated naked red throat, and heavy yellow bill distinctive. Ad ♀ mainly black, ad ♂ and juvs with rufous head and neck. Utter sharp barking notes, less often loud roaring cries during display with head raised and tail jerked above the back.

4. Sunda Wrinkled Hornbill
Aceros corrugatus p. 218
Forests of the Sunda Shelf landmasses. White tail with black base, and bright colours of bill, casque, and facial skin distinctive. Ad ♀ with black face and neck, white in ad ♂ and juvs, and sexes differ in throat colour. White areas stained yellow or ochre with preen-gland oils. Usually in pairs or family groups, feeding in the forest canopy and flying long distances between trees. Utter rather soft coughing notes.

6. Mindanao Wrinkled Hornbill
Aceros leucocephalus p. 221
Forests of Mindanao and neighbouring islands in the Philippines. White tail with broad black tip and red bill distinctive. Ad ♀ with black head and neck, ad ♂ and juvs pale rufous. Usually in pairs or small groups, silent for much of the time, or utter low croaking notes.

3. Sulawesi Wrinkled Hornbill
Aceros cassidix p. 215
Forests of Sulawesi and nearby islands. All-white tail, and bright colours of casque, bill, and facial skin distinctive. Ad ♀ with black head and neck, ad ♂ and juvs with rufous. Spend much time in upper canopy and fly buoyantly across long distances. Usually as pairs, forming small flocks when not breeding. Utter gruff barking notes.

5. Visayan Wrinkled Hornbill
Aceros waldeni p. 223
Forests of Panay and neighbouring islands in the Philippines. Tail with broad white band across centre (often stained yellow) and red bill distinctive. Ad ♀ with black head and neck, ad ♂ and juvs deep rufous.

Plate 11
Indomalayan wreathed hornbills, genus *Aceros* and subgenus *Rhyticeros*

Indomalayan wreathed hornbills, genus *Aceros* and subgenus *Rhyticeros*. All with casque formed of low flat wreaths.

1. Sumba Wreathed Hornbill
Aceros everetti p. 239
Forests of Sumba I. in the Lesser Sunda archipelago. The only hornbill on the island and the only member of the subgenus with a black tail and body. Ad ♀ with head and neck also black, ad ♂ and juv with rufous. Usually in pairs or small groups, utter harsh clucking notes.

3. Plain-pouched Wreathed Hornbill
Aceros subruficollis p. 230
Forests of SE Asian mainland. The black body and all-white tail distinctive. Ad ♀ with black head and neck, ad ♂ and juvs with rufous. Ad ♀ with blue throat skin, ad ♂ and juvs with yellow. Occur with and very similar to Bar-throated Wreathed Hornbill, but lack throat bar and slightly smaller. Often in wandering flocks, uttering barking call-notes at perch or in passage.

5. Narcondam Wreathed Hornbill
Aceros narcondami p. 227
Forests of Narcondam I. near the Andaman Archipelago. The only hornbill on the island, the black body and all-white tail distinctive. Ad ♀ with head and neck black, ad ♂ and juvs with rufous. Strong, dextrous flyers, congregating at fruiting trees with noisy cackling calls.

2. Bar-pouched Wreathed Hornbill
Aceros undulatus p. 232
Forest of SE Asian mainland and Sunda Shelf landmasses. The black body and all-white tail distinctive. Ad ♀ with black head and neck, ad ♂ and juvs with rufous. Ad ♀ with blue throat skin, ad ♂ and juvs with yellow, all with an incomplete dark bar across the naked skin. Occurs with and very similar to Plain-pouched Wreathed Hornbill, but has throat bar and is slightly larger. Often in wandering flocks, uttering barking call-notes at perch or in passage.

4. Papuan Wreathed Hornbill
Aceros plicatus p. 224
Forests of Moluccas, New Guinea, and Solomon Is. The only hornbill within its range, the black body and all-white tail distinctive. Ad ♀ with head and neck black, ad ♂ and juvs with varying shades of rufous. Usually in small flocks, flying buoyantly above the forest, and sometimes uttering harsh barking notes.

Plate 12
African white-rumped and wattled hornbills, genus *Ceratogymna* and subgenera *Bycanistes*, *Baryrhynchodes*, and *Ceratogymna*

African white-rumped and wattled hornbills, genus *Ceratogymna* and subgenera *Bycanistes*, *Baryrhynchodes*, and *Ceratogymna*. Adult ♂ with much larger casque than ad ♀, juv with more or less brown on face and neck.

Subgenus *Ceratogymna*. With blue pendulous throat wattles, lack white rump, ad ♀ and juvs of both sexes with rufous head and neck.

1. **Black-casqued Wattled Hornbill**   *Ceratogymna atrata*   p. 260
   Moist lowland forests of W and central Africa. All-black plumage and white tips to outer tail feathers distinctive. Ad ♂ with black bill and huge casque, ad ♀ and juvs with pale bill and rufous head and neck. Often in family parties with ad ♀♀ and juvs predominating; powerful flyers with noisy nasal braying calls.

2. **Yellow-casqued Wattled Hornbill**   *Ceratogymna elata*   p. 263
   Moist lowland forests of W Africa. All-black plumage with white scaling on neck and all-white outer tail feathers distinctive. Ad ♂ with high, abruptly ending casque, ad ♀ and juvs with rufous head and neck. Often in family parties with ad ♀♀ and juvs predominating; powerful flyers with noisy nasal braying calls.

Subgenus *Baryrhynchodes*. With extensive white rump and back, but only abdomen white below, pale grey or brown facial feathers, and large more-or-less yellow casque in ad ♂.

3. **Brown-cheeked Hornbill**   *Ceratogymna cylindricus*   p. 249
   Moist lowland forests of W and central Africa, with 2 subspp. Large pale casque, black band across white tail, and broad white wing tips distinctive. **3a.** Upper Guinea Brown-cheeked Hornbill *C. c. cylindricus* with black thighs and pale yellow bill and casque. **3b.** White-thighed Hornbill *C. c. albotibialis* with extensive white underparts and black bill. Usually in pairs or family parties; generally quiet, but do utter harsh coughing calls.

4. **Grey-cheeked Hornbill**   *Ceratogymna subcylindricus*   p. 252
   Moist lowland forests of W and central Africa, with 2 subspp. Broad white wing tips and outer tail feathers and white rump and back all distinctive. Black and white casque of ad ♂, black bill and casque of ad ♀. **4a.** Upper Guinea Grey-cheeked Hornbill *C. s. subcylindricus* with less yellow on casque of ad ♂ than **4b.** Lower Guinea Grey-cheeked Hornbill *C. s. subquadratus*. Often at forest edge in pairs or small groups, with direct flight and loud hooting and quacking calls.

5. **Silvery-cheeked Hornbill**   *Ceratogymna brevis*   p. 256
   Montane and coastal forests of NE and E Africa. Large pale casque, blue circumorbital skin, all-black wings, white tips to outer tail feathers, and white rump and back distinctive. Ad ♀ and juv with small dark bill and casque. Usually fly as pairs or family trios, with direct flight and loud braying calls.

Subgenus *Bycanistes*. With white rump and underparts, the smallest members of the genus.

6. **Piping Hornbill**   *Ceratogymna fistulator*   p. 242
   Moist lowland forests of W and central Africa, with 3 subspp. Small bill and casque, together with white on wings and tail, distinctive. **6a.** Upper Guinea Piping Hornbill *C. f. fistulator* with ridges on bill and least white on wings and tail. **6b.** Sharpe's Piping Hornbill *C. f. sharpii* and **6c.** Dubois's Piping Hornbill *C. f. duboisi* with extensive white on wing tips and outer tail feathers, the latter with the casque well developed. Usually in small flocks in forest canopy and edges. Flight direct, and utter high piping and cackling calls.

7. **Trumpeter Hornbill**   *Ceratogymna bucinator*   p. 245
   Montane and riverine forests of E and S Africa. The large black bill and casque, white tips to wings and tail, and red circumorbital skin distinctive. Ad ♀ and juv with notably smaller casques. Often in small groups, moving with direct flight, and making loud wailing and crying calls.

Plate 13
Appearance in flight of African hornbills, genera *Bucorvus*, *Tockus*, and *Ceratogymna*

Adult males only (not to scale)

1. **Piping Hornbill**   *Ceratogymna fistulator*
   White edges to wing and tail, white rump, small bill.
   1a. Dubois's Piping Hornbill   *C. f. duboisi*
   Extensive white edges to wing and tail, pale bill.
   1b. Upper Guinea Piping Hornbill   *C. f. fistulator*
   White edges only to secondaries and tail tips, dark bill.
   1c. Sharpe's Piping Hornbill   *C. f. sharpii*
   White edges to wing and tail, dark bill.

2. **Northern Ground Hornbill**   *Bucorvus abyssinicus*
   White primaries, short tail, high casque, blue face.

3. **Southern Ground Hornbill**   *Bucorvus leadbeateri*
   White primaries, short tail, low casque, red face.

4. **Brown-cheeked Hornbill**   *Ceratogymna cylindricus*
   Black bar on white tail, white rump, broad white edges to wing.
   4a. White-thighed Hornbill   *C. c. albotibialis*
   Bill dark, casque pale.
   4b. Upper Guinea Brown-cheeked Hornbill   *C. c. cylindricus*
   Bill and casque pale.

5. **Grey-cheeked Hornbill**   *Ceratogymna subcylindricus*
   White edges to wings, tail, and wing coverts, white rump and back, casque black and white.

6. **Silvery-cheeked Hornbill**   *Ceratogymna brevis*
   White tips to tail, white rump and back, black wings, pale casque, dark bill.

7. **Trumpeter Hornbill**   *Ceratogymna bucinator*
   All-black bill with large casque, narrow white wing edges, white tail tips, white rump.

8. **Long-tailed Hornbill**   *Tockus albocristatus*
   Long graduated tail with white tips, white crest.
   8a. Eastern Guinea Long-tailed Hornbill   *T. a. macrourus*
   White head and neck, black wings.
   8b. Lower Guinea Long-tailed Hornbill   *T. a. cassini*
   White head and neck, white tips to coverts and primaries.
   8c. Western Guinea Long-tailed Hornbill   *T. a. albocristatus*
   White crown only, black wings.

9. **Yellow-casqued Wattled Hornbill**   *Ceratogymna elata*
   White outer tail feathers, pale cut-off casque, black wings.

10. **Black-casqued Wattled Hornbill**   *Ceratogymna atrata*
    All black but for white outer tail tips, protruding casque.

11. **Hemprich's Hornbill** *Tockus hemprichii*
    White outer tail stripes, dusky orange bill, sooty plumage.

12. **Bradfield's Hornbill** *Tockus bradfieldi*
    Pale brown plumage, white tail tips, orange bill.

13. **African Crowned Hornbill** *Tockus alboterminatus*
    Charcoal plumage, orange bill, prominent casque.

14. **Dwarf Red-billed Hornbill** *Tockus camurus*
    Red bill, pale edges to coverts, white tail tips.

15. **Dwarf Black Hornbill** *Tockus hartlaubi*
    Dark bill, white stripe above eye, white tail tips.

16. **African Pied Hornbill** *Tockus fasciatus*
    Yellow bill, all-black wings.
       16a. Upper Guinea Pied Hornbill   *T. f. semifasciatus*
    Black marks on bill, white outer tail tips.
       16b. Lower Guinea Pied Hornbill   *T. f. fasciatus*
    Red marks on bill, all-white outer tail feathers.

17. **Monteiro's Hornbill** *Tockus monteiri*
    White secondaries and outer tail feathers, red bill.

18. **Pale-billed Hornbill** *Tockus pallidirostris*
    Brown, with white outer tail tips, pale yellow bill.

19. **African Grey Hornbill** *Tockus nasutus*
    Dark brown with white eye and back stripes, black bill with white stripe.

20. **Southern Yellow-billed Hornbill** *Tockus leucomelas*
    White-spotted coverts, yellow bill, pink facial skin.

21. **Eastern Yellow-billed Hornbill** *Tockus flavirostris*
    White-spotted coverts, yellow bill, black facial skin.

22. **African Red-billed Hornbill** *Tockus erythrorhynchus*
    White-spotted coverts, slender red bill with black base.
       22a. Damaraland Red-billed Hornbill   *T. e. damarensis*
    White face, brown eye.
       22b. South African Red-billed Hornbill   *T. e. rufirostris*
    Grey face, white stripe over eye, yellow eye.
       22c. North African Red-billed Hornbill   *T. e. erythrorhynchus*
    Pale grey face, white stripe over eye, brown eye.

23. **Von der Decken's Hornbill** *Tockus deckeni*
    Pied plumage, white secondaries and outer tail feathers.
       23a. Jackson's Hornbill   *T. d. jacksoni*
    White-spotted wing-coverts, orange bill.
       23b. Von der Decken's Hornbill   *T. d. deckeni*
    Unspotted wing-coverts, red and yellow bill.

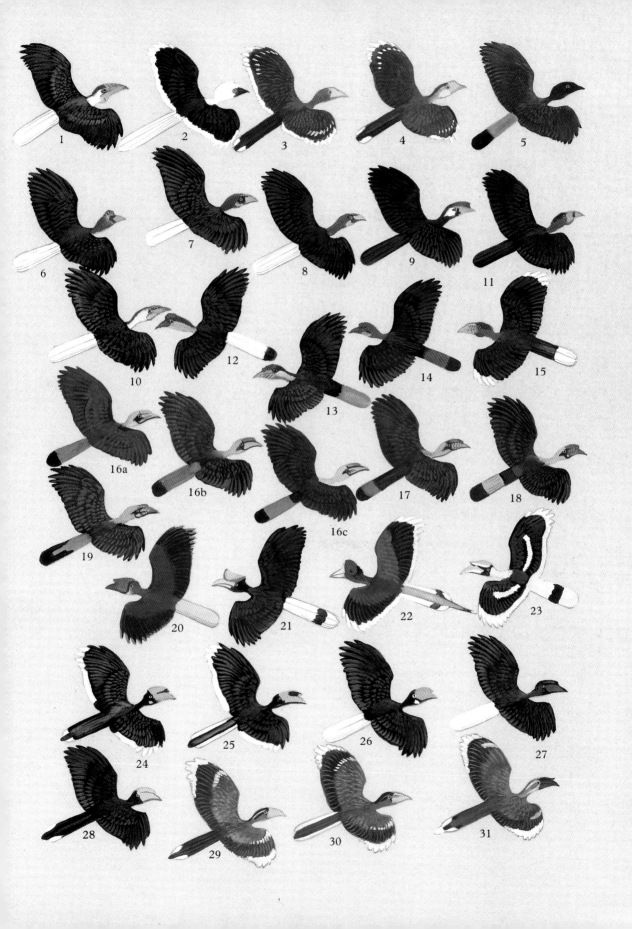

Plate 14
Appearance in flight of Indomalayan hornbills, genera *Anorrhinus*, *Ocyceros*, *Anthracoceros*, *Buceros*, *Penelopides*, and *Aceros*

Adult males only (not to scale)

1. **Bar-pouched Wreathed Hornbill**   *Aceros undulatus*
   White tail and neck, rufous crown, yellow throat with dark bar, low casque.

2. **White-crowned Hornbill**   *Aceros comatus*
   Long white tail, white wing edges, white head and neck, black bill.

3. **Tickell's Brown Hornbill**   *Anorrhinus tickelli*
   White wing edges, rufous head and neck, pale bill.

4. **Austen's Brown Hornbill**   *Anorrhinus austeni*
   White wing edges, white throat and neck, pale bill.

5. **Bushy-crested Hornbill**   *Anorrhinus galeritus*
   Dark head and bill, pale grey tail with black tip, dark wings.

6. **Sulawesi Wrinkled Hornbill**   *Aceros cassidix*
   White tail, high casque, colourful face, rufous head and neck.

7. **Papuan Wreathed Hornbill**   *Aceros plicatus*
   White tail, blue face, yellow bill, rufous head and neck, low casque.

8. **Narcondam Wreathed Hornbill**   *Aceros narcondami*
   White tail, red base to yellow bill, rufous head and neck, low casque.

9. **Sulawesi Tarictic Hornbill**   *Penelopides exarhatus*
   All-black plumage but for white face, dark casque.

10. **Plain-pouched Wreathed Hornbill**   *Aceros subruficollis*
    White tail and neck, rufous crown, plain yellow throat, low casque.

11. **Sumba Wreathed Hornbill**   *Aceros everetti*
    Black tail, rufous head and neck, low casque.

12. **Mindanao Wrinkled Hornbill**   *Aceros leucocephalus*
    White tail with broad black tip, red bill, rufous head and neck.

13. **Sunda Wrinkled Hornbill**   *Aceros corrugatus*
    Pale yellow tail with black base, colourful face and casque, black crown and nape, rufous head and neck.

14. **Visayan Wrinkled Hornbill**   *Aceros waldeni*
    Rufous tail with broad black base and tip, rufous head and neck, red bill.

15. **Rufous-necked Hornbill**   *Aceros nipalensis*
    Broad white wing and tail tips, rufous head and neck, low casque.

16. **Mindanao Tarictic Hornbill**   *Penelopides affinis*

16a. Basilan Tarictic Hornbill   *P. a. basilanica*
Rufous tail with black tip, cream head and neck.
16b. Samar Tarictic Hornbill   *P. a. samarensis*
Pale rufous rump, tail with white edges and black tip, cream head and neck.
16c. Mindanao Tarictic Hornbill   *P. a. affinis*
Black base and tip to rufous tail, cream head and neck.

17. **Luzon Tarictic Hornbill**   *Penelopides manillae*
    Dark tail with pale rufous band, cream head and neck.

18. **Visayan Tarictic Hornbill**   *Penelopides panini*
    Rufous rump, black base and tip to pale rufous tail, cream head and neck.

19. **Mindoro Tarictic Hornbill**   *Penelopides mindorensis*
    Rufous rump and tail with black tip and edges, cream head and neck.

20. **Great Philippine Hornbill**   *Buceros hydrocorax*
    Overall brown, with pale yellow tail and wing tips, crimson bill, large casque, rufous head and neck.

21. **Great Rhinoceros Hornbill**   *Buceros rhinoceros*
    White tail with black band, orange bill with high casque.

22. **Great Helmeted Hornbill**   *Buceros vigil*
    White wing edges, white tail with black band and long central feathers, high red casque, naked neck.

23. **Great Pied Hornbill**   *Buceros bicornis*
    White edges to wings and coverts, white tail with black band, white neck, yellow bill, large casque, black face.

24. **Oriental Pied Hornbill**   *Anthracoceros albirostris*
    White wing edges and outer tail tips, yellow bill with black marks.

25. **Indian Pied Hornbill**   *Anthracoceros coronatus*
    White wing edges and outer tail feathers, yellow bill with black marks.

26. **Palawan Hornbill**   *Anthracoceros marchei*
    All-white tail, yellow bill without dark markings.

27. **Sulu Hornbill**   *Anthracoceros montani*
    White tail, rest (including bill) all black.

28. **Malay Black Hornbill**   *Anthracoceros malayanus*
    Long pointed tail with white outer tips, wings all black, bill yellow.

29. **Malabar Grey Hornbill**   *Ocyceros griseus*
    White wing and outer tail tips, yellow bill, pale grey.

30. **Sri Lankan Grey Hornbill**   *Ocyceros gingalensis*
    White wing tips and outer tail feathers, yellow bill, pale grey.

31. **Indian Grey Hornbill**   *Ocyceros birostris*
    Long pointed tail with white tip, white wing tips, black bill with pointed casque.

to begin daily walk in search of food. On moonlit nights, rarely calls and even descends to the ground to walk around before returning to roost (R. Randall *in litt.* 1973). Spends on average 70% of day walking, but amount varies considerable from day to day; may cover up to 11 km in one day. Though prefers walking, flies to cross unsuitable habitat, to pursue intruders (when may rise to 300 m to cover several km), or to reach roost. Flight speed about 30 km/hr.

Eats any animals it can overpower, up to the size of hares, squirrels, large tortoises, and snakes. Individuals band together to hunt large, fast, or dangerous prey. Reported to distract large snakes either with wing quills outspread (Ayres in Gurney 1861) or by covering them with grass (Roberts 1924). Arthropods, especially termites, beetles, and grasshoppers, toads, *Achatina* snails, lizards, and snakes the principal diet in Transvaal (Kemp and Kemp 1978), chameleons and scorpions in Kenya (Hofer 1982), frogs in Natal (Knight 1990), and earthworms in E Cape Province (G. Ranger unpublished notes). A lame bird was observed eating carrion from an antelope carcass (Anonymous 1982*a*), another picking scraps from bones in a drought (Babich 1983), but regularly fed at sheep carcasses in farmland (G. Ranger unpublished notes) and once at the remains of a lion kill (P. Beaumont pers. comm. 1990). Often ingests small pieces of bark and sticks but vegetable food very rarely recorded, e.g. maize (Aurelian 1957). Undigested remains regurgitated as loose pellets.

Most food is simply picked up from the ground or low vegetation, sometimes after a short ground or aerial pursuit, favouring areas of short ground cover (Knight 1990). In Transvaal, 38 per cent of items dug or scratched out with heavy bill, from excavations that may be 40 cm deep (Kemp and Kemp 1978), especially in newly hoed ground (Benson and Benson 1977). Digging especially prevalent during dry season and often done in and around piles of elephant dung (Verheyen 1953; pers. obs.), but also includes excavation of wasp and bee nests and removal of comb and honey (Burton 1984). Occasionally ascends trees to pry off loose bark or pursue prey through branches.

Pursues large eagles for their prey, scavenges in or preys on their nests, and chases them off when breeding; a group of 5 even flew down and killed a 2.9-kg ad ♂ Black Eagle *Aquila verreauxii* which passed by their nest in still air at dusk (L. Brown *in litt.* 1982). However, its own food is often pirated by Tawny Eagles *A. rapax*, to which the hornbills react by group defence and, once robbed, by walking around with bills raised and wings partly open. Attends grass fires for prey flushed by flames. Sometimes feeds in association with other animals, especially game mammals, apparently for the arthropods they flush (Mlingwa 1990), or Carmine Bee-eaters *Merops nubicoides* or Dark Chanting Goshawks *Melierax metabates* attracted by their own activities.

When not foraging may sunbathe, lying flat on ground with wings outspread, sometimes being groomed by other group-members, or may foliage-bathe after rain. Removes thorns from feet with bill when necessary. Daily activities limited by high temperatures (Fig. 3.7, p. 29), strong wind, or rain. Birds fly into trees at end of day, just before dark, and usually after bout of play involving mainly imms in bill-wrestling, chasing, or pouncing on one another. Do not have a regular roost site, just roosting crouched on tree limb, but often make a second long, secretive flight to their final roost just as darkness falls. A group of three reported roosting in the nest-chamber with the ♀ (Behr 1970) is most unusual.

Displays and breeding behaviour

Monogamous but breeds co-operatively (Hutchinson in Sclater 1902; van Someren 1922; Ranger 1931), usually as a dominant pair with several, mainly ♂, helpers.

Loud call uttered with head bowed (Fig. 3.9, p. 29), even in flight (Kemp and Crowe 1985). Only one pair breeds per territory, with up to 9 ad and imm helpers. All members of a group, except 1–2-year-old imms, take part in territorial calling and defence. All ad ♂♂ courtship-feed the dominant breeding ♀, rarely an imm also (Hofer 1982), but only the dominant ♂ copulates with her. Copulation is almost always attempted on a branch, where the ♂ can bring his tail under the ♀ after preening her hard on the head and neck and forcing her to crouch (Fig. 3.16, p. 33).

The nest site is visited almost daily before breeding, the ♀ sitting inside for up to 5 hr and often pecking loose wood off the interior. A deep lining of dry leaves is brought in bundles by the ♂♂ before and during the nesting cycle, always with a food item included in the bundle (Kemp and Kemp 1980; Hofer 1982; Fig. 5.1, p. 52), and ♂♂ also bring food alone in courtship-feeding. ♂♂ will sometimes crouch by the nest with bill raised and uttering a squeaky call, possibly soliciting the ♀ to enter (Ranger 1931; Kemp and Kemp 1980).

Breeding and life cycle

NESTING CYCLE: about 126–129 days: 37–43 days incubation, 86 days nestling period. ESTIMATED LAYING DATES: E Africa (rainfall region A: see Brown and Britton 1980), Feb–Mar, (region C) Aug, (region D) Jan–Feb, Apr, Oct–Nov; Tanzania, June–Aug; Zaïre, Oct; Angola, Aug; Zambia, July–Dec; Malawi, Sept–Nov; Zimbabwe, Aug–Jan, peak Oct–Nov; South Africa (Transvaal), Sept–Jan, (Natal) Sept–Nov, (Cape Province) Oct. Breeds in early rainy season in most areas (Kemp 1976b), with onset of egg-laying usually after first heavy falls, spread over 7 weeks and laying over 11 weeks in different seasons in Transvaal (Kemp and Kemp 1991). c/1–2, rarely 3 (Benson and Benson 1977; G. Knight and K. S. Begg pers. comm. 1992/3); in Transvaal, 85% c/2 ($n = 34$) (Kemp 1988a). EGG SIZES: ($n = 19$, Transvaal) 61.6–79.0 × 45.6–54.0 (72.2 × 50.3); ($n = 15$, Zimbabwe) 68.8–77.9 × 46.9–53.9 (73.9 × 51.5). Eggs with white, rough and pitted shells, rather pointed at one end, and with considerable variation in size between and within clutches (Kemp 1988a). The second-laid egg is usually smaller, and the hatching interval suggests that it is laid 3–5 days after the first.

Most nests are in natural holes in dead or live trees of DBH ($n = 25$), 55–500 (119) cm including spp. of *Adansonia*, *Sclerocarya*, *Schotia*, *Lannea*, *Acacia*, *Kirkia*, *Podocarpus*, *Diospyros*, *Terminalia*, *Kigelia*, *Brachystegia*, *Ficus*, *Combretum*, *Sterculia*, and *Raphia* palm stumps. Nest heights ($n = 25$) 1.8–8.5 (4.6) m, depth below rim 9–150 (61) cm, entrance width 13–65 (30) cm and height 25–110 (51) cm, cavity diameter 26–70 (37) cm. Some are in rock faces or earth banks, in cavities of about 1 × 1.2 m with a 35-cm-wide entrance (van Someren 1922; Jones 1969; Behr 1970), and an old stick nest of another bird was used in two successive years (Woodward and Woodward 1875). Often the nests are near watercourses, owing to the predominance of large trees, and the same sites may be used over at least 17–36 seasons. Sites often also used by Barn Owls *Tyto alba*, and seen chasing Pel's Fishing Owl *Scotopelia peli* from an area where both have nested (Carr and Carr 1984). Also provide retreats for monitor lizards, pythons, genets, and leopards.

Incubation begins with the first egg. Only the dominant ♀ incubates and is fed on the nest by most group-members, including sometimes other ad ♀♀ and 2-year-old and older imms. The group usually visits the nest at dawn, and the incubating ♀ emerges to call with them. The group also visits the ♀ on average 3–4 more times per day with food (Kemp and Kemp 1980; Hofer 1982). She may leave the nest for up to 45 mins to defecate, preen, dustbathe, and, occasionally, to catch some food for herself, returning to nest when group moves off again to forage. ♀ usually utters the low acceptance-bray when taking food (Hofer 1982). One group was reported to roost with the ♀ in the rock cavity at the top of a cliff which formed the nest (Behr 1970), another to roost in trees adjacent to nest (G. Ranger unpublished notes).

The chicks hatch 3–5 days apart (Tramontana and Rider 1988), but the younger has always died of starvation within a few days (with one exception, where two chicks were raised to a month old before one succumbed to starvation, K. S. Begg *in litt.* 1993). One egg, rarely both, may also fail to hatch. The older chick may weigh 250 g by the time the younger hatches, at about 60 g, and easily dominates it in any competition for food. Each chick hatches blind, naked, and with pink skin (Fig 5.4, p. 55). Within 3 days the skin turns dark purple, while the throat and mouth lining remain pale pink. After 7 days the eyes begin to open, the first feather-quills show through the skin, and the dorsal air sac is fully developed. By 14 days the eyes are fully open, the body covered in short spiny feather-quills, and the bare throat partially inflated. By 21 days the quills are well out and on the point of breaking open, and the legs are strong and well developed. By 30 days the quills have opened and feathers cover most of body; thereafter, for the remaining 7 weeks until fledging, the principal development is in growth of wing and tail feathers.

Food is carried in the bill to the nest as multiple items (Fig. 5.1, p. 52) collected by placing those already caught on the ground while securing additional prey. The group delivers 4–9 billfuls of food per day to the nest (Plate 2d). Young birds are the least successful foragers, especially when digging or probing, but helpers can supply 25% of food delivered to a nest and reduce the time spent off by the ♀ (Knight 1990). The quantity delivered depends on group size and number of helpers, frequency of visits, and size of each billful. The largest billfuls may consist of one or two large items, such as snakes, mice, and lizards, together with as many as 12 smaller items. Food is dropped on the nest-floor and fed by the ♀ to the chick until the latter can see to feed itself. The time spent in the nest by the ♀ before she leaves the chick and emerges to forage with the group is unknown. She may leave the chick for intervals when it is only 10 days old, and by 30 days it may be left totally alone. The ♀ returns regularly with the group to feed the chick, and at one nest the ♂♂ reduced deliveries after her emergence (Knight 1990).

The chick defecates on the copious dry lining, the nest becoming odorous only late in the cycle, if at all. Chicks remain with the group at least until mature at about 4–6 years old. Initially they are fed by all group-members, following them around with incessant begging, even stealing items left on the ground by the latter while pursuing other prey. Although the chick can feed itself after six months or a year, it may still be given food when at least 2 years old if it solicits an ad. After leaving the nest site, the group does not normally return until the following breeding season.

Transvaal data suggest low productivity and long lifespan (Kemp 1988a; Kemp *et al.* 1989). Groups may not attempt to breed every year, with a higher proportion laying after a good start to the summer rains. Breeding success, as reflected by increase in mean group size recorded during aerial counts, also increases after good summer rains. Some groups breed much more regularly and successfully than others, and nests started early in a season are more successful (Kemp and Kemp 1991). Groups fledge, on average, one chick every 9.3 years (n = 215 group breeding seasons), or 49 per cent of breeding attempts; 31 per cent of fledged chicks survive to maturity, but ad turnover is estimated at only 1.5 per cent per annum. Leopards are suspected of preying on ads and chicks; one chick was killed by a genet (L. Brown *in litt.* 1971), and another was probably killed by a still-live puff-adder brought to the nest as food. Adults with deformed bills or broken and reset legs are occasionally recorded. Bred less regularly in captivity than congener *B. abyssinicus* (Tramontana and Rider 1988), and captive program with second-hatched chicks harvested from the wild under way (Marais 1993).

Bucerotidae

Medium-sized to large true hornbills. All true hornbills have 14 cervical vertebrae and short tarsi, even the few spp. which walk rather than hop when on the ground. Only left carotid artery functional. Vary from small insectivores to large frugivores, from social to solitary, with considerable variation in basic biology from genus to genus (Kemp 1979). Details of anatomy, behaviour, and biology differ between genera and subgenera and are described under these taxa.

Exhibit unique breeding biology for which all true hornbills are renowned. Basically, ♀ seals herself into the nest-hole, leaving a narrow vertical slit, moults her rectrices and remiges almost simultaneously, and is fed throughout by the ♂. Details vary between genera and spp. Laying actually begins only some days after final entry, and this, together with the closed nest, makes estimation of laying dates and other breeding details difficult. Long tail often raised over back to conserve space in nest. Early development of legs enables chicks to sit upright, reach for food from entrance slit, and, like ♀, reverse to squirt droppings out through entrance and reseal nest where necessary.

Family contains 52 spp. in 8 genera, distributed across Afrotropical (21 spp., 2 genera), Indomalayan (30 spp., 6 genera), and Australasian (1 sp.) biogeographical realms. Most spp. occupy forest habitats, but for 11 spp. in African and 1 sp. in Indian savannas.

Genus *Anorrhinus* Reichenbach, 1849

Medium-sized hornbills, with head feathers long and broad, forming an obvious crest. Casque only slightly developed, and unfeathered throat skin barely to obviously present. Sexes differentiated by bill colour and plumage of head and neck, imm either quite different from ad or resembles ad ♂. No emargination of primaries. Utter loud cackling calls. All spp. live in groups and breed co-operatively. The only birds infested with mallophagan feather-lice of the genus *Bucerocolpocephalum*.

Breed in tree holes, the entrance sealed, ♀ moulting rectrices and remiges simultaneously during breeding, and ♂ and group-helpers delivering food in gullet, from where regurgitated and passed into nest. Chick's skin remains pinky yellow throughout development. ♀ emerges with chick at end of nesting cycle. Juvs of both sexes resemble ad ♂ in plumage.

Three spp. in forest habitats from Myanmar to Sumatra and Borneo. Now placed in one genus on the basis of common morphology, behaviour, and louse parasites, with the spp. separated by differences in ad soft-part colours and head plumage (Kemp 1988*b*). Previously 2 spp. in monotypic genera *Anorrhinus* and *Ptilolaemus* Ogilvie-Grant, 1892. The White-crowned Hornbill *Aceros* (*Berenicornis*) *comatus* shows some affinities with this genus in crest development and social organization, but is little known and currently appears best placed as an offshoot of *Aceros* hornbills.

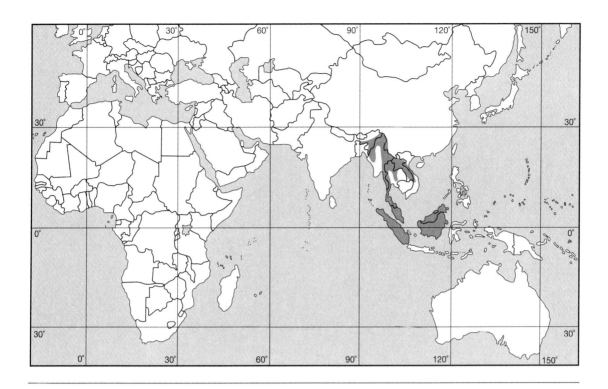

Austen's Brown Hornbill   *Anorrhinus austeni*

Anorhinus [sic] *austeni* Jerdon, 1872. *Ibis*, **(3)2**, 6. Asalu, Cachar Hills, Assam, India.

PLATES 2c, 3, 14

Other English names: Brown Hornbill, White-throated Brown Hornbill, White-throated Brown-backed Hornbill, Brown-backed Hornbill

Description

ADULT ♂: upperparts dark brown and underparts rust-brown, but for white cheeks and throat. Flight feathers dark brown, primaries with white tips and white spots on the outer web. Tail dark brown, with white tips to all but the central pair of feathers. Bill and low casque ridge ivory-white with orange wash at base. Bare circumorbital and throat skin pale to bright blue. Eyes brown, legs and feet dull greenish brown.

ADULT ♀: like ♂ in overall coloration, but smaller; cheeks, throat and underside grey-brown. Facial skin pale pink with yellow wash below eye.

IMMATURE: both sexes like ad ♂ but undersides paler, and neck, wing-coverts and flight feathers with broad pale brown tips. Bill pale yellow to horn-coloured, bare facial skin yellowish. Eyes grey-brown, legs and feet brown.

SUBADULT: imm ♀ develops ad plumage at first moult when about 1 year old, bill then ivory-yellow with greenish wash at base, especially on lower mandible, and facial skin becomes darker (Frith and Douglas 1978).

SUBSPECIES

This sp. has been combined previously with *A. tickelli* as a subsp. of a single sp. in the monotypic genus *Ptilolaemus*.

102 Austen's Brown Hornbill *Anorrhinus austeni*

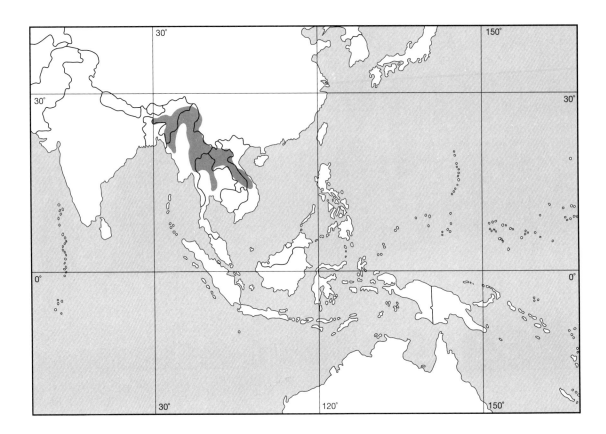

MEASUREMENTS AND WEIGHTS
SIZE: wing, ♂ ($n = 9$) 310–345 (329), ♀ ($n = 5$) 300–312 (306); tail, ♂ ($n = 6$) 290–302 (295), ♀ ($n = 4$) 260–270 (265); bill, ♂ ($n = 7$) 112–135 (125), ♀ ($n = 4$) 105–117 (110); tarsus, ♂ ($n = 6$) 48–51 (50), ♀ ($n = 4$) 44–46 (45). WEIGHT: unrecorded.

Field characters
Length 60–65 cm. Dark brown above and rufous below with ivory-white bill and sharp-edged casque in both sexes; ♀ noticeably smaller. Long, graduated, white-tipped tail notable, with dark pair of central feathers. Face and neck white to pale rufous in ad ♂ and juv of both sexes, underparts more rufous than on ♀, and imm with pale yellow bill. Flight smooth and deliberate, without such obvious flap-and-glide of other spp. Generally noisy, with loud calls.

Voice
Loud, plaintive scream repeated, together with loud croaks, chuckles, and screams. Similar to *Anthracoceros coronatus* but less harsh (Ali and Ripley 1970). Nasal *ank-ank-ank* (Deignan 1945). Tape recordings available.

Range, habitat, and status
India, at Barail Range and Patkai Hills in Assam; Myanmar, in N and E; China, S part of Xishuangbanna in Yunnan Province (Tso-Hsin 1987); Thailand, at Phu Lom Lo, Doi Chiang Dao, Khao Yai, and Tonkin on Laos border (Bangs and van Tyne 1931); Vietnam, at Keng Tung, Muong Moun, Bana, Phou Kong Ntoul, and on slopes of the central chain (Wildash 1968); Laos, at Tranninh (Beaulieu 1944).

Favours dense evergreen forest from plains up to 1500 m. Uncommon in deciduous forest and only locally common in evergreen forest (Ali and Ripley 1970), but was very common on plains of E Assam (Baker 1927). Was rare on higher hills of NW Thailand, at edge of evergreen forest or in open pine and oak forest (Deignan 1945). Apparently not sympatric with *A. tickelli* in Thailand, where population in Khao Yai now isolated (Sheppard and Worth 1992).

Feeding and general habits
Territorial in pairs or groups of 5–15, rarely up to 20. Noisy, often contact-calling with loud plaintive screams or, when feeding, with low cackles. Often joins with *Anthracoceros albirostris* and other frugivores in the tops of forest trees. Very restless, always moving about in the branches or from tree to tree, but without really noisy wingbeats (Ali and Ripley 1970). Sometimes flies high above forest in follow-my-leader fashion, as when crossing valleys, but usually makes only short flights, travelling within the canopy. Forages from the treetops to the lower canopy, rarely on the ground. Sunbathes on branch or ground with wings and tail spread (Frith and Douglas 1978).

Fruit and animal food taken in similar proportions throughout the year. Most carnivorous of 4 spp. at Khao Yai, Thailand, with average delivered to nests by weight of 22 per cent figs, 38 per cent non-fig fruit, and 40 per cent animal food items (Poonswad *et al.* 1983, 1987, 1988). Fruit included 32 spp. of 15 families in the 21 genera *Polyalthia, Uvaria, Canarium, Elaegnus, Elaeocarpus, Cinnamomum, Litsea, Amoora, Aphanamixis, Chisocheton, Aglaia, Ficus, Horsfieldia, Knema, Eugenia, Connarus, Piper, Podocarpus, Symplocus, Strombosia,* and *Artocarpus*. Animals included bats (Chiroptera), reptiles (Viperidae, Agamidae, Scincidae), molluscs (Cyclophoridae), arthropods (Psychidae, Cossidae, Orthoptera, Tettigoniidae, Phasmatidae, Gryllidae, Blattidae, Lepidoptera larvae), and annelid earthworms, together with eggs or nestlings of various bird spp. such as doves and bulbuls.

Displays and breeding behaviour
Monogamous, co-operative breeder. Of 22 nesting pairs, 18 had 1–5 helpers, in all cases ♂♂ of various ages. ♀♀ in a group were not permitted by the helpers or the breeding ♂ to feed the breeding ♀. Very noisy during visits to nests, attracting conspecifics and often resulting in vigorous fighting (Poonswad *et al.* 1983, 1987, 1988).

Breeding and life cycle
NESTING CYCLE: estimated at 92 days: 30 days incubation, 62 days nestling period (Poonswad *et al.* 1987). Incubation period reported by local Nagas as 24 days (Ali and Ripley 1970). ESTIMATED LAYING DATES: India (N Cachar) May, (Assam) Mar–June, at beginning of rainy season; Thailand, Feb (5) and Mar (3), so chicks emerge at beginning of the monsoon rains. c/2–3, rarely 4–5. EGG SIZES: ($n = 24$) 46.0–57.0 × 33.0–35.4 (48.8 × 34.2) (Baker 1927). Eggs white with pitted shells.

Nests in cavities in trees, in India at height of 5–8 m, a few much higher, and the ♀ sealed in with droppings and food remains. One nest was 18 m up in a 35-m tree of DBH 80 cm (B. Beehler *in litt.* 1989). Twelve nest sites in Thailand were all closely aggregated in higher areas of dense forest, with a mean nearest-neighbour distance of 565 m (Poonswad *et al.* 1983, 1987, 1988). All were in trunks of living trees, *Eugenia* (2), *Dipterocarpus* (6), *Altingia excelsa* (2), *Sterculia* (1), *Ficus* (1), *Tetrameles nudiflora* (2), and unknown spp. (8). Two nest-cavities measured 40–46 cm deep, 32–40 cm wide, with a ceiling 110–200 cm high and the floor up to 17 cm below the entrance lip. Entrances were round to oval in shape, and most sealing material was food debris and pieces of wood.

Most nests were in natural cavities, but also used old nest-holes of the Great Slaty Woodpecker *Mulleripicus pulverulentus*. Nests were reused regularly and were defended frequently against other sympatric hornbill spp. and conspecifics, one intruder ♀ breaking into an active nest despite fights with defending ♂ (Poonswad *et al.* 1987). Interchanged cavities with *Aceros undulatus* and *Anthracoceros albirostris*.

Food-delivery rate to the nest was 16.4–25.4 g/hr (mean 14.6 g/hr). Co-operative feeders included ♂♂ of all ages but no ♀♀. There were two permanent feeders in one group of 5 and an infrequent third: one, probably the *alpha* ♂, brought more food, removed droppings at the nest, and attended the ♀ on emergence, while the rest attended only to the chicks. Groups forage and return to the nest together, being very noisy when at the nest and each taking its own position while feeding; often regurgitate digested seeds, which are not fed to the ♀. There is an increase in the feeding rate and in delivery of non-fig and animal foods at hatching, this decreasing as the chicks develop.

A radio-tracked ♂ foraged close to nest over a home-range of about 3 km², of which 0.8 km² was grassland. It changed roosts every night, but another two breeding flocks roosted at the same site every night, 0.5 and 1.0 km from their respective nests, splitting up only in the morning on their way to feed. The ♂ remained around the nest area for 7–9 days after the chick fledged, and then both moved away (Tsuji *et al.* 1987). Mean home range of ♂♂ was 4.3 km² when breeding (Poonswad and Tsuji 1994).

Of 24 nesting pairs that sealed the nests, 18 produced some fledged young. Nests with few or no helpers tended to be less successful, but more helpers did not mean more food delivered to the nest although they did help keep off intruders and predators (Poonswad *et al.* 1987). The average was 2.3 chicks fledged per season and 3.4 helpers. A marten preyed on two chicks in one nest from which the ♂ had emerged unusually early.

Tickell's Brown Hornbill *Anorrhinus tickelli*

Buceros Tickelli Blyth, 1855. *Journal of the Asiatic Society of Bengal,* **24,** 266. Tenasserim, Myanmar.

PLATES 3, 14

Other English names: Brown Hornbill, Assam Brown-backed Hornbill

Description
ADULT ♂: upperparts dark brown. Underparts, including cheeks and throat, rich rust-brown. Flight feathers dark brown with white edging and tips, primaries with white spots on the outer web. Tail feathers dark brown, with white tips to all but the central pair. Bill and low casque ridge pale yellow. Bare circumorbital and throat skin pale blue. Eyes brown, legs and feet black.

ADULT ♀: like ♂ in overall coloration, but darker brown, smaller, and bill horn-coloured to black.

IMMATURE: both sexes like adult ♂ in overall coloration, but paler.

SUBSPECIES
This sp. has been combined previously with *A. austeni* and placed as a subsp. of a single sp. in the monotypic genus *Ptilolaemus*.

MEASUREMENTS AND WEIGHTS
SIZE: wing, ♂ ($n = 11$) 314–343 (326), ♀ ($n = 12$) 300–320 (308); tail, ♂ ($n = 5$) 280–295 (288), ♀ ($n = 5$) 258–279 (270); bill, ♂ ($n = 7$) 122–132 (127), ♀ ($n = 9$) 110–122 (116); tarsus, ♂ ($n = 7$) 46–52 (48), ♀ ($n = 7$) 45–47 (46). WEIGHT: ♂ ($n = 4$) 854–912 (869); ♀ ($n = 7$) 683–797 (723).

Field characters
Length 60–65 cm. A medium-sized hornbill with rich red-brown plumage and white tips to the flight feathers and tail. Bill pale yellow in ♂ and dark brown in ♀, not yellow in both sexes as in *A. austeni* and without white throat of ♂ and imm of that sp. Usually in small groups, moving below the canopy and not very noisy, but keeps up steady soft calling.

Voice
Loud, plaintive scream, *whey-wheyo*, repeated (Tickell 1864). No tape recordings located.

Tickell's Brown Hornbill *Anorrhinus tickelli*

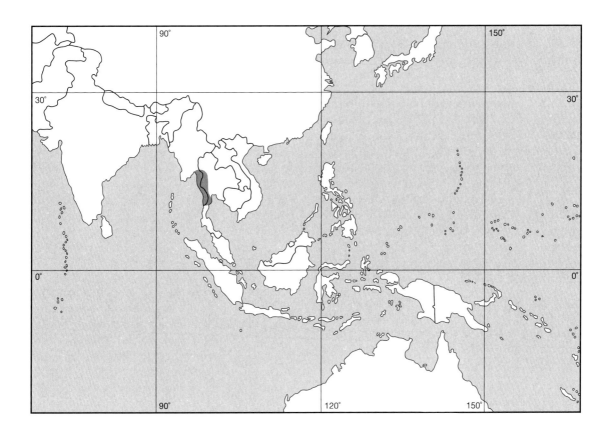

Range, habitat, and status
S Myanmar, in mountainous areas of Tenasserim; SE Thailand, at Huai Kha Khaeng and historically S from Hue Nya Pla to Petchaburi R., with recent sighting in Cumporn Province (P. Poonswad *in litt.* 1992).

Dense evergreen and deciduous forest from foothills to 1500 m, generally uncommon. Favours the tallest primary forest, including stands of *Hopea odorata*, and was commonest on Thai side of peninsula ridge. Does not occur on S Thailand plains (Worth *et al.* 1994), endangered elsewhere in country.

Feeding and general habits
Unrecorded, but probably similar to *A. austeni*.

Displays and breeding behaviour
Unrecorded. Probably monogamous and breeding co-operatively.

Breeding and life cycle
NESTING CYCLE: unrecorded. ESTIMATED LAYING DATES: Myanmar (Tenasserim), Feb–Apr. EGG SIZES: ($n = 22$) 42.3–51.2 × 32.2–35.5 (46.4 × 33.8) (Baker 1927). Eggs taken from one nest were white, slightly glossy, with the shell finely pitted, but without the tubercles found on eggs of larger spp.

Three nests in Tenasserim were at height of 3.5–6 m, the entrance sealed with droppings; one was in a *Xylia dolabriformis* and all were in relatively small trees (Bingham 1879). Other nests were above 4 m, but most below 8 m (Baker 1927). At least 5 out of a group of 8–10 ad ♂♂ fed at a nest near Three Pagodas Pass, Myanmar (H. Bartels *in litt.* 1982).

Bushy-crested Hornbill   *Anorrhinus galeritus*

Buceros galeritus Temminck, 1831. *Planches colorées*, **8,** Plate 520   PLATES 3, 14
Sumatra and W Borneo, restricted to Pontianak, Kalimantan, Indonesia.

Description
ADULT ♂: upperparts dark brown with metallic green sheen, cheeks grey, and head feathers large and elongate forming loose crest. Underparts dark brown at neck and on breast, becoming brown to grey-brown on abdomen and undertail-coverts. Flight feathers dark brown, tail grey-brown with distal third black. Bill and low casque ridge black. Bare circumorbital and throat skin blue, with white area behind eyes and on angle of jaw, and black eyelids forming ring around eye. Eyes red, legs and feet dark brown.

ADULT ♀: like ♂ in overall coloration, but smaller, bill with broad yellow stripe along upper mandible and tip to lower mandible, legs and feet blue-grey.

IMMATURE: plumage like ad ♂ but undersides paler, neck, wing-coverts and flight feathers with broad pale brown tips, bill and gape pale olive, bare facial skin pale yellow with pink ring around eye, eyes blue-grey.

SUBADULT: eye colour hazel by 1 year old, same red as ad by 2 years. Facial skin begins to change from yellow to blue at 1 year old (Frith and Douglas 1978).

MOULT
Not obviously confined to one season, as shown by study skins (Sanft 1960).

MEASUREMENTS AND WEIGHTS
SIZE: wing, ♂ (n = 82) 320–375 (351), ♀ (n = 62) 308–355 (329); tail, ♂ (n = 39) 256–295 (276), ♀ (n = 31) 241–275 (260); bill, ♂ (n = 26) 132–155 (144), ♀ (n = 17) 118–142 (134); tarsus, ♂ (n = 27) 47–53 (50), ♀ (n = 17) 44–50 (48). WEIGHT: ♂ 1134, 1247 (Riley 1938); unsexed 1134, mean 1172.

Field characters
Length 60–65 cm. All-dark plumage together with dense droopy crest are diagnostic; bill dark in ♂ but cream-coloured in ♀. Confusable with *Anthracoceros malayanus* when tail of latter appears all dark, and both have sexes with black and cream bills, but colours are reversed for each sp. Usually in noisy groups, flying with whooshing wingbeats, giving fast loud calls, and usually encountered well below forest canopy.

Voice
Quiet *wah wah wohawaha* rising and falling in volume, or alternating with strident squawking of rapidly repeated notes lasting for several min. High-pitched, gull-like, communal chorus audible at least 1.6 km away (Madge 1969). Uttered when perched in loose groups and always given during visits to nest, with ♀ joining in from within. Alarm a short, high-pitched *aak aak aak*. Chicks beg with loud high *kwee-wee-wee-wee-weah* for minutes on end (Leighton 1982). Tape recordings available.

Range, habitat, and status
Myanmar, S from Nwalabo; Thailand, S from Lam Pi waterfall; Peninsula Malaysia, including the island of Pinang; Indonesia, on Sumatra and Kalimantan; Sarawak; Sabah, and the islands of Bunguran Utara; Brunei.

Favours dense forest, at up to 1800 m on Sumatra (van Marle and Voous 1988), mostly around 750 m. Most common, with largest groups, in medium- to low-elevation primary forest with good fig populations, including peat swamp forests; scarce in more open coastal forests. Also remains and breeds in selectively logged dipterocarp forests (Johns 1987), and can generally subsist among smaller trees than other hornbills (van Marle and Voous 1988). Resident, with densities of 1.2–5.6 km^2 per group depending on habitat quality (Kemp and Kemp 1974; Leighton 1982; Johns 1987).

Feeding and general habits

Territorial, in groups of 2–20 or rarely even more; median group size 7 in primary forest. Average group composed of dominant *alpha* breeding pair with 4–6 helpers, the sexes approximately equal (Leighton 1986). Favours patches of dense forest and tangled growth, as along rivers and streams, edges of clearings, bases of hills, in gorges, or secondary growth. Moves systematically through foliage, taking fruit in passing and then moving on, never into emergent canopy but often down to ground level. If threatened utters alarm-call, leading to silence, alertness, and gliding down into cover. Has fixed roost sites, usually in isolated trees near a stream, where group resorts to the tips of long slender twigs (Kemp and Kemp 1974; Leighton 1982).

Eats fruits at various stages of ripeness during relatively short visits to trees compared with larger spp. (mean 3.1 mins, max 11, $n = 24$: Johns 1987), visits to fruit trees interspersed with bouts of hunting animals. Prises off bark, digs among epiphytes, and splits husks and capsules, the last especially important when fruiting poor. Capsular, large-seeded, dehiscent, and drupaceous lipid-rich fruits most important in diet (especially Meliaceae and Myristicaceae: Leighton and Leighton 1983), followed by sugar-rich figs (only 10% of diet), and then animals. Groups arrive early at fruiting trees within their own territory, but only about half of group feeds at any one patch per visit. Regular feeder on at least nine spp. of fig in Malaysia (Lambert 1989). Animal-hunting absorbed 24% of time but yielded only 38 items per 700 hrs of observation (Leighton 1986). Group-members may unite to chase agile prey, which goes to the captor, and were successful in 10 out of 19 attempts (Leighton 1982).

Members gang up to drive larger spp. of hornbill or similar-sized *Anthracoceros malayanus* from fruit, and larger groups drive off smaller ones. Groups are larger during fruit-rich

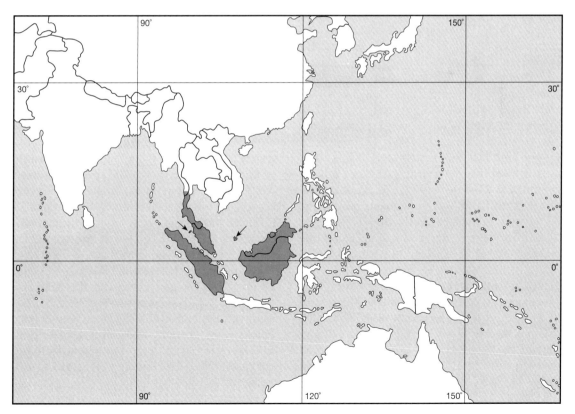

times, breaking up with helpers splitting off for all or part of time when fruits become scarce. Helpers move off in various associations (all ♀, all ♂, or mixed ages and sexes), during which some inter-territorial display takes place between subgroups. Normally ♂♂ and ♀♀ intermingle freely (Leighton 1986). A *Spizaetus* hawk-eagle spent 3.5 hrs following a group and chasing small animals which they flushed (Leighton 1982).

Food delivered to nests was mainly fruit, but at one included at least 32 cicadas, a 10-cm-long item of animal food, a stick insect, and a *Draco* lizard, and at another 15 lizards, 6 arthropods (cicadas, mantids, myriapods), and 4 frogs. Seeds thrown out of one nest, many of which germinated before nesting was over, included Rutaceae (*Evodia*, 2), Leguminosae (*Entada* sp., 1, the largest at 4.5 × 2.5 cm; other sp., 1), Fagaceae (*Lithocarpus cyclophorus*, 1), Sterculiaceae (*Sterculia foetida*, 4), Sapindaceae (*Sapindus rarak*, 1), Connaraceae (*Connarus* sp., 1), Palmae (1 fruit), and unidentified (4). At another nest, 21 spp. of seed included *Canarium denticulatum*, *C. ordontophyllum*, *Alangium evebaceum*, and *Dacryodes*, *Lithocarpus*, *Horsfieldia*, *Eugenia*, and *Dehasia* spp. Also recorded to eat *Ficus glabella* and *F. sumatrana* figs (McClure 1966) and *Lanraecae litsa* (de Silva 1981).

Fruit pulp and a cockroach recorded from one stomach (Smythies 1960). A captive ♀ caught an adult robin *Copsychus saularis* and took live fish from near the surface of a pond (Frith and Douglas 1978). It also bathed under a sprinkler and in standing water, the only one of 10 captive spp. to do so. Sunbathed with wings and tail spread while perched.

Displays and breeding behaviour

Monogamous, co-operative breeder (Whitehead in Sharpe 1890). Pairs call at intervals during the day, and both sexes and helpers call and defend the territory. Group-members perch shoulder to shoulder in display, calling in unison with rising and falling gobbling notes, followed by sharp cries that rise to loud high screams. They may advance on their neighbours, only to retreat with a special harsh call, and then advance again (Leighton 1982).

Only the *alpha* ♀ breeds, chasing off all ♀♀ not members of the group, sometimes very aggressively, especially when fruiting is poor. The *alpha* pair travels separately from the group before laying, this possibly due to mate-guarding (Leighton 1986).

Breeding and life cycle

NESTING CYCLE: at least 75 days: estimated 30 days incubation, 60 days nestling period. ESTIMATED LAYING DATES: Peninsula Malaysia, Feb, May; Sarawak, May, Aug (2), Sept (2), Dec; Sabah, June; Indonesia (Kalimantan), Jan, Mar. Bred and fledged chicks during one Jan–May fruiting peak and not again in two years (Leighton and Leighton 1983). At another nest, apparently bred successfully from layings in Feb and then May of same year (Madge 1969). No obvious breeding season. c/2–3 from number of chicks fledged. Eggs undescribed.

One nest was 10 m up in the mid-trunk of a live *Shorea* of 75 cm DBH, the entrance 30 cm long × 6 cm wide. Another was 13–15 m up in the trunk of a live tree (DBH 175 cm, DNH about 150 cm, cavity 50 cm diameter, entrance estimated at 30 cm long × 10 cm wide). A third was 25 m up in the trunk of a dead tree in secondary forest logged five years earlier, the entrance also about 30 × 10 cm.

There was a group of at least 10 birds (incubating ♀, 4 ♂♂, 2 ♀♀, 2 juvs, and one other) at one nest (Madge 1969) and 7 (4 ♂♂, 3 ♀♀) at another (Walton and Watt 1987). Not all members fed at each visit to the nest, and the whole group may not visit each time, but numbers attending and feeds increased once chicks were present, and both sexes and all ages were involved. The number of fruits regurgitated by each bird may depend on fruit size; rarely, two small fruits may be regurgitated together.

At one nest (Madge 1969), most feeds were recorded in mornings and especially evenings, with fruit loads per bird of 4, 2, a few, 2, 17, 16, 14, 5, and a few singles (total 56 at one visit by 6 birds). Items were rarely delivered carried in the bill and most were fed to the ♀, but once probably to the chicks. The two chicks were first seen at the nest-entrance when about 2 weeks old. At another nest (Walton and Watt 1987), with three chicks close to fledging, all members usually came on each visit but on average fed on only 46% of visits ($n = 402$). Some came more than others (range 8–21%) and fed more (range 7–20%), a total of 288 food items being delivered over 402 visits. Food-delivery dropped temporarily after each chick fledged. Of 306 items, 80% were fruit, 8% animals, and 11% unidentified.

♀ and chicks squirted their droppings out of the sealed nest. At one nest the ♀ and one chick emerged at least four days before the other, which remained in the opened hole with its 7.5 cm-wide exit. The two eggs were estimated to have been laid about five days apart, judging from the size of the chicks. At another nest, the ♀ apparently emerged at least a week before the three chicks, which each fledged a day apart in the mornings (Walton and Watt 1987). The chicks remained in undergrowth near the nest, at separate localities, for several days after emerging, although they could already fly well. The chick's skin remains pinkish yellow throughout (E. McClure pers. comm. 1974).

A maximum of 3 chicks recorded per nesting attempt, usually 2, which are fed for at least 6 months after fledging by all older members of the group and supplied with all their food requirements for the first few months. One juv ♂ left the group 5 months after fledging, but a young ♀ remained for over 12 months. May breed again immediately after successful brood if food adequate (Leighton 1986).

Genus *Tockus* Lesson, 1830

Small to medium-sized hornbills. Only the skin around the eye and a small patch at base of throat unfeathered. Outer primaries not emarginated, and casque usually only a small ridge, developed into small cylinder in a few spp. Territorial calls accompanied by conspicuous display movements in most spp.

Monogamous, and only 1 sp. possibly breeding co-operatively. Only the ♀ seals the nest, although in some spp. the ♂ helps to bring lumps of mud carried in the bill. Chicks retain the pink skin throughout development, and ♀ emerges half-way through nestling period, after undergoing a simultaneous rectrix and remex moult. ♂, and later ♀, bring food to nest as single items held in bill. Mates do not appear to allopreen regularly.

Fourteen spp. endemic to Africa, but for one which extends just into the Arabian Peninsula. The genus forms at least four separate clades, separated as subgenera, which share distinctive calls, displays, and ecological and behavioural characteristics (Kemp 1976a; Kemp and Crowe 1985). At least 2 spp. within each clade are sympatric, so cannot be considered as superspp. Most spp. are hosts to feather-lice in the genus *Buceroemersonia* (Elbel 1977b).

African Crowned Hornbill *Tockus (Rhynchaceros) alboterminatus*

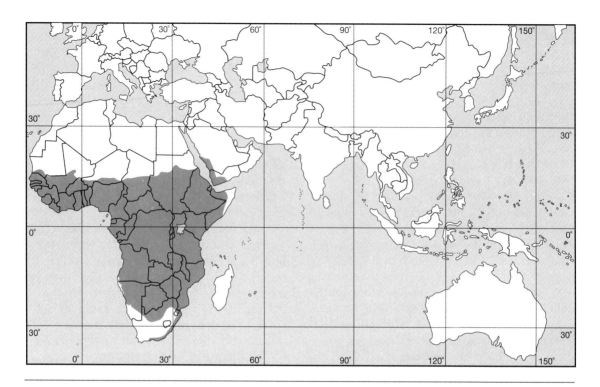

Subgenus *Rhynchaceros* Gloger, 1841

One of two mainly arboreal-living clades within the genus which, together with the subgenus *Lophoceros*, is distinguished by head-up posture and uttering of whistling calls during territorial display. In this subgenus, *Rhynchaceros*, the wings are not opened during display. Both clades prefer to line their nest with flakes of bark, and both share certain species of *Chapinia*, *Bucerocophorus*, and *Buceronirmus* feather-lice (Elbel 1967, 1969a, b, 1976; Table 6.1).

Three spp. are probably allopatric, at least when breeding, and form a clear supersp. group (*T. fasciatus*, *T. alboterminatus*, *T. bradfieldi*). The fourth, *T. hemprichii*, is sympatric in part of the range of *T. alboterminatus*, and has a distinctive ending to its display in which the tail is fanned over the back (Fig. 3.10, p. 30).

African Crowned Hornbill *Tockus (Rhynchaceros) alboterminatus*

Lophoceros alboterminatus Büttikofer, 1889. *Notes from the Leyden Museum*, 11, 67. Gambos, upper Cunene district, Angola.

PLATES 4, 13

Other English name: Crowned Hornbill

Description
ADULT ♂: head, neck, back, wings, and tail dark sooty brown, but for the broad white stripe over the eye, pale grey streaking on the cheeks, and white tips to all but central and outermost pairs of tail feathers. Upperwing-coverts and flight feathers edged with pale brown. Throat and upper breast dusky brown, streaked with grey; rest of underparts white. Bill, with prominent casque ridge ending in short protrusion,

red to deep orange with yellow band at base. Bare circumorbital and throat skin black. Eyes yellow, legs and feet black.

ADULT ♀: like ♂ in overall coloration, but smaller, with casque shorter and less developed; throat skin blue-green and becoming more prominent when breeding.

IMMATURE: like ad but for white specks on wing-coverts, yellow bill lacking casque and notches on cutting edges; throat skin dull yellow, eyes grey.

SUBADULT: bill, bare throat skin, and eye colour like ad by 8–10 weeks after fledging, when post-juvenile moult of fluffy head and neck feathers begins. Point of casque separate from bill by 16 weeks (Ranger 1941).

SUBSPECIES

Birds from the main range in eastern Africa, including Ethiopia (described above, with measurements and weights below) have been separated as the subsp. *T. a. geloensis* (Neumann), 1905; larger and greyer birds from W have been described as *T. a. alboterminatus* (wing, ♂ ($n = 6$) 235–254 (243), ♀ ($n = 8$) 211–234 (221)); smaller and paler brown birds with a prominent supercilium from eastern coastal forests as *T. a. suahelicus* (Neumann), 1905 (wing, ♂ ($n = 67$) 207–247 (236), ♀ ($n = 48$) 209–237 (222)); leading to larger and darker birds to the S as *T. a. australis* (Roberts), 1911 (wing, ♂ ($n = 30$) 237–273 (252), ♀ ($n = 14$) 226–247 (237)). See Sanft (1960). However, the forms each grade into one another and are not considered to warrant subspecific status (Kemp and Crowe 1985).

MEASUREMENTS AND WEIGHT

SIZE: wing, ♂ ($n = 115$) 237–273 (255), ♀ ($n = 56$) 220–252 (234); tail, ♂ ($n-24$) 216–248 (229), ♀ ($n = 11$) 199–234 (220); bill, ♂ ($n = 95$) 86–109 (95), ♀ ($n = 26$) 76–91 (82); tarsus, ♂ ($n = 29$) 31–37 (34), ♀ ($n = 14$)

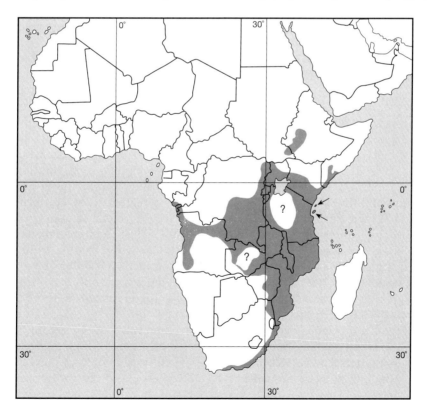

30–36 (33). WEIGHT: ♂ (*n* = 31) 191–332 (244); ♀ (*n* = 13) 180–249 (205).

Field characters
Length 50 cm. Dusky-brown plumage contrasting with white undersides and tail tips, red bill with yellow stripe at base, and yellow eyes are all characteristic. Slender build, buoyant flight, and high piping calls also aid identification, but similar to those of other arboreal spp., such as African Pied Hornbill *T. fasciatus* (which differs in pied plumage, cream and black bill, brown eyes), Bradfield's Hornbill *T. bradfieldi* (lighter brown, lacking pale supercilium, more slender paler orange bill), African Grey Hornbill *T. nasutus* (paler grey-brown plumage, black and white (♂) or maroon and cream (♀) bill), and Pale-billed Hornbill *T. pallidirostris* (cream bill, pale grey-brown plumage).

Voice
Various loud, high, whistling calls. Contact-call a long loud *kew*, repeated rapidly in alarm. Series of notes, rising and falling in pitch, serve also for contact or, together with display, for territorial maintenance. Low growl of threat or alarm, or loud piping alarm-cry leading up to screech of fear. Tape recordings available.

Range, habitat, and status
SW Ethiopia, as an isolated population; S Somalia; SE Sudan, in Boma Hills; Uganda, N to Kabalega Falls; Kenya, on W highlands and also along Tana R.; Rwanda; Burundi; E and SE Zaïre; Tanzania, S along Rift Valley highlands, W lowland forests, and also islands of Zanzibar and Pemba; Malawi; Zambia, excluding dense central woodland; Angola, N to Cabinda and S to Huila along the escarpment; Namibia, in E Caprivi Strip; Zimbabwe, excluding dense woodland in NE and drier SW; Mozambique; E South Africa, S to Knysna forests and once W to Robertson.

Found in evergreen montane, coastal, riparian, and secondary forest patches, extending to some areas of dense deciduous woodland and up to 3000 m. Absent from moist lowland evergreen forest, where replaced by African Pied Hornbill. Local, but usually common where it does occur. Spacing and density vary with the amount of wooded habitat, which is often fragmented. Pairs breed within a defended territory of about 3–5 km^2 of wooded habitat (S Africa: Ranger 1941, 1949–52; pers. obs.). A forest patch as small as 7 ha will support a pair, but more usually in patches of 10–50 ha, or 2 pairs in a 90-ha patch (Malawi: F. Dowsett-Lemaire *in litt.* 1989). Total of 28 pairs bred in 440 ha of riverine forest in E Cape Province, S Africa (M. du Plessis *in litt.* 1992).

Feeding and general habits
An arboreal sp. which lives in pairs or small family groups of up to 7. In marginal habitat, conditions deteriorate in dry or cold seasons and produce seasonal movements, with groups of up to 80 wandering far from their normal forested haunts (e.g. E Cape Province: Anonymous 1982*b*, 1983; Vernon 1988; Claasen 1992). In parts of E Africa, movements resemble a regular migration, but in S Africa they are linked to drought conditions and have no regular timing or direction.

Forages mainly in foliage, while perched or by plucking off food items while hovering. Hawks insects in mid-air and swoops to the ground to snatch up other food items. Less commonly, descends to the ground, mainly in the dry season, to hop about in search of food or to dustbathe. Also foliage-bathes regularly (G. Ranger unpublished notes).

Roosts on long, thin branches or vines with open space around, above, and below them. Up to 6 regular roost sites per territory, used in irregular sequence and for varying lengths of time. They enter this safe refuge cautiously, perching silently nearby, flying out calling or dashing for cover if a hawk passes, slipping in to the final perch only at the very last light and remaining crouched there until after sunrise (Ranger 1941, 1949–52). The roost is usually sited in a valley over a stream, where it is sheltered from weather and predators.

Feeds mainly on invertebrates and fruit, the latter especially important during the dry season, but both always a regular part of the diet (Ranger 1950; Irwin 1981). Various fruits and nuts are eaten, including figs (*F. sansibarica*), *Trichilia emetica, Rhoicissus capensis, Burchelia capensis, Allophylus abyssinicus, Commiphora caryaefolia, Royena pallens, R. licioides, Cassine, Aberia, Canthium, Scutia indica, Encephalartos* cycad seeds (Ranger 1950), exotic fruits such as brambles, maize, bananas (Haagner and Ivy 1908), peanuts, and oil-palm fruits (Dean 1973).

Consumes a wide range of animal food, especially arthropods such as beetles (Buprestidae, Elateridae, Tenebrionidae, Cerambycidae), termite and ant alates, locusts, caterpillars and wasps, mantids, bugs, cicadas, butterflies, moths and their eggs, fly maggots and pupae from carcasses, and centipedes and spiders, augmented by snails, eggs and chicks of small birds such as doves, *Ploceus bicolor*, and *Terpsiphone viridis* (Moreau 1935), lizards, chameleons, and a golden mole. Regularly eats the aposematic grasshopper *Phymateus leprosus* in S Africa, especially during the dry season, but crushes and wipes this and hairy caterpillars repeatedly before swallowing to remove irritating fluids and hairs (Ranger 1950). Undigested remains are regurgitated whole, such as seeds, or as fragments formed into a pellet, usually at night. Finer fragments are expelled in faeces.

Displays and breeding behaviour
The nest site within a territory is often used season after season, and visited throughout the year. Visits increase prior to breeding, when the birds perch more conspicuously near the nest, the ♀ begins to seal external cracks to the nest, and the ♂ begins courtship-feeding of the ♀. Imms from previous season are driven out by ads at the end of the winter dry season (Ranger 1941, 1949–52). The pair-members are vocal in defence of their territory, calling with the bill skywards and rocking back and forth with each phrase of the whistling call. They also call loudly in flight on visits to the nest.

Breeding and life cycle
NESTING CYCLE: about 83 days: pre-laying period 7–14 days, incubation 25–27 days (mean 26, n = 18), nestling period 46–55 days (mean 50, n = 13) (G. Ranger unpublished notes). ESTIMATED LAYING DATES: E Africa (rainfall region A: Brown and Britton 1980), Feb, Apr, July, (region B) Feb, July, (region D) Feb, Sept, Nov, (region E) Sept, Oct; Kenya, Mar; Tanzania, Sept, Nov; Zaïre, Nov–Dec; Angola Sept–Nov; Zambia, Oct, Feb; Malawi, Aug–Nov; Zimbabwe, Oct–Nov, Jan; South Africa (Transvaal), Nov, (Natal) Nov–Dec, (Cape Province) Oct–Jan with peak Nov (17) to early Dec (3). Breeds at beginning of rains in S, less seasonally in E Africa (Kemp 1976b). c/2–5, usually 3 or 4. EGG SIZES: (n = 24) 36.5–41.5 × 25.5–30.4 (39.3 × 27.8). Eggs white with pitted shells.

The nest is placed in a natural hole in a tree, usually within the trunk or a large branch, 1.2–12 m above the ground. Tree spp. include *Adansonia digitata, Acacia nigrescens, A. albida, Piptadenia, Sideroxylon inerme, Euphorbia grandidens, Erythrina caffra, Commiphora caryaefolia, Schotia latifolia, S. speciosa, Harpephyllum, Calodendron, Vepris lanceolata, Cassine croeceum, Olinia capensis, Grevillea robusta* (Schönland 1895; Ranger 1949–52, unpublished notes; pers. obs.). A cavity of about 20 cm diameter, with an entrance as small as 6.5 × 4.8 cm or as large as 40 × 9 cm, is sufficient for a ♀ and 4 chicks. Artificial nest-chambers with entrances 18 cm high and 6–9 cm wide are also used (Büttiker 1960). One nest was in the same tree as that of an African Red-billed Hornbill *Tockus erythrorhynchus* (Greenberg 1975).

All sealing is by the ♀, the entrance closed to a narrow vertical slit with her droppings, nest debris, and food items such as millipedes and sticky fruits. The ♂ brings nest lining throughout the nesting period but especially at the beginning, mainly bark flakes but including flowers and snail shells, the latter possibly also containing calcium for egg-formation. Up to 1500 bark flakes accumulate in a nest, raising the level of the floor considerably (M. du Plessis *in litt.* 1992).

The eggs are laid at intervals of 2 days, but up to 4 days for last egg of clutch. Most laying between midnight and dawn. Incubation, by the ♀, begins with the first egg. The ♀ moults first her rectrices and then her remiges more or less simultaneously, starting 3–5 days after completion of the clutch and taking 3–7 days to complete; in one case, the moult of a sick ♀ was incomplete and suspended (G. Ranger unpublished notes). ♂ feeds the ♀ throughout the nesting cycle with single food items brought in the bill, making 40–80 visits per day. A small chick which died was consumed by a sibling. A second ♀ may have assisted at one nest (Siemens 1981).

Eggshells are often thrown out on the ground below at hatching. The chicks hatch asynchronously, and the ♀ leaves the nest 25–30 days after hatching of the eldest chick, after enclosure of 61–69 days. Her feathers are then regrown, and she helps the ♂ to feed the chicks. ♂'s moult usually starts only after ♀ emerges, or towards end of nestling period, and proceeds for at least 4 months. The chicks reseal the entrance alone and emerge from 15 days after the ♀, flying well and not returning to nest.

Mean clutch size in E Cape, S Africa, 3.6 (range 2–5, $n = 20$); 85% of nestings fledged some young ($n = 34$), mean number of chicks fledged 2.5 (range 1–4, $n = 29$) (G. Ranger unpublished notes). Addled eggs laid late in clutch and starvation of later chicks are main causes of losses, but the nest is sometime taken over by bees, or flooded, ♀ suffers from a mouth infection or becomes egg-bound, or the parents are killed outside the nest (Haagner and Ivy 1908; Moreau 1935; Siemens 1981; G. Ranger unpublished notes). Breeding ♀ ♀ regularly beheaded by genets in riverine forests of E Cape, S Africa (M. du Plessis *in litt.* 1992). One ♀ was replaced within a month. Juvs may remain in the parental territory for 6–8 months until the following breeding season, but feed themselves within a month of fledging.

Bradfield's Hornbill *Tockus (Rhynchaceros) bradfieldi*

Rhynchaceros bradfieldi Roberts, 1930. *Ostrich*, **1**, 65. Designated as Waterberg, Namibia.

PLATES 4, 13

Description

ADULT ♂: head and neck grey, streaked with white in broad band from above eye to nape. Rest of upperparts grey-brown, darker on wing feathers, which are edged with pale brown. White tips to tail-feather pairs 3 and 4, only a narrow white rim to rest of tail. Underparts off-white. Bill, with casque a low ridge merging into middle, deep orange with yellow line across base and brown cutting edges. Circumorbital and throat skin dark grey. Eyes pale orange, legs and feet black.

ADULT ♀: like ♂ in overall coloration, but smaller, casque even shorter, bare throat skin dull turquoise.

IMMATURE: like ad, but bill smaller and paler, eye grey; facial skin colours undescribed.

MOULT
Nov–Mar, even while breeding (Hoesch 1937).

MEASUREMENTS AND WEIGHTS
SIZE: wing, ♂ ($n = 17$) 247–271 (259), ♀ ($n = 12$) 231–250 (237); tail, ♂ ($n = 17$) 210–235 (219), ♀ ($n = 12$) 194–213 (200); bill, ♂ ($n = 17$) 85–111 (99), ♀ ($n = 12$) 80–91 (86); tarsus, ♂ ($n = 11$) 37–43 (39), ♀ ($n = 8$) 35–39 (37). WEIGHT: ♂ ($n = 1$) 200; ♀ ($n = 4$) 170–204 (193); unsexed ($n = 15$) 180–395 (da Rosa Pinto 1983).

Field characters

Length 50 cm. Recognizable by deep orange bill, grey-brown plumage with narrow white tip to tail, pale orange eye, buoyant flight, slender build, and whistling calls. Most like African Crowned Hornbill *T. alboterminatus*,

Bradfield's Hornbill *Tockus (Rhynchaceros) brafieldi*

with which parapatric in extreme NE of range. Differs in being pale brown, not sooty-brown, with the bill orange, not red, the casque of ♂ not well developed, and the pale supercilium not prominent. Sympatric with African Grey Hornbill *T. nasutus*, the plumage of which is more patterned, darker brown, with clear white tail tips, the bill black and white (♂) or maroon and cream (♀), the calls more drawn out, and the display includes wing-flicking.

Voice
Loud, high, whistling calls uttered as single notes or in series. Loud, long, single alarm-note, also fast series of notes when anxious. Series of notes rising and falling in pitch and volume, but increasing in tempo, *pi-pi-pi pi-pi-pi pi-pi-pi-pieeu*, accompany display. Tape recordings available.

Range, habitat, and status
NW Zimbabwe, E to Silobela and Nkai, even wandering S to Bulawayo (Chadwick 1984); SW Zambia, NE to Namushakende and Mulobezi; N Botswana, S to Maun; NE Namibia, SW to the isolated Waterberg population; S Angola, NW to Humbe.

Occurs mainly in *Baikaea–Guibourtia–Pterocarpus* woodlands on sandy soils, extending to adjacent drier woodlands, especially *Colophospermum* mopane. Confined to wooded and watered gorges in the Waterberg, from which it ranges to adjoining dry *Acacia* to forage. Locally common, especially when congregating into flocks during the May–Sept dry season. Overlaps marginally with the African Crowned Hornbill, mainly during the dry winter months when feeding more terrestrially (Irwin 1982).

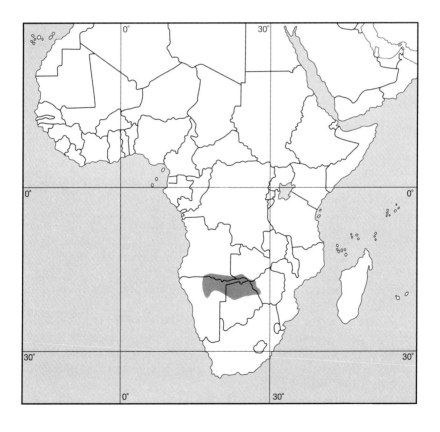

Feeding and general habits

Usually found in pairs or small groups in the wet and early dry seasons; later in the dry season forms into nomadic flocks of up to 62 birds, which often concentrate on open areas rich in harvester *Hodotermes* and other termites, seeds, or litter and droppings, as around waterholes (J. Rushworth *in litt.* 1971), even *Croton megalobothrys* thickets on grassland (Penny 1987).

Much food is taken from foliage or hawked in the air. Often descends to the ground to feed, much more so than its close relative the African Crowned Hornbill, hopping about to dig and scratch for food with the bill. Terrestrial foraging occurs mainly during the dry season.

Feeds mainly on invertebrates, including locusts, mantids, beetles, ants, bugs, maggots, and termites. Also takes small reptiles, some fruit, and grass seeds. During the dry season, termites and the red seeds of *Guibourtia coleosperma* are especially important elements in the diet.

Displays and breeding behaviour

Monogamous, and in most areas may be territorial only during the summer breeding season. Calls with bill skywards, rocking back and forth with each phrase, probably for territorial proclamation and mate attraction.

Breeding and life cycle

Little known. ESTIMATED LAYING DATES: Namibia, Dec; Zimbabwe, Sept–Dec, peak Nov. Breeds at beginning of wet season. EGG SIZES: ($n = 1$) 39.0×26.7 (Hoesch 1937). Egg white with pitted shell.

Nest placed in a natural cavity in a rock or tree hole, lined with bark flakes, and the entrance sealed to a narrow vertical slit. One hole in a cliff was about 40 cm deep (Hoesch 1937).

♀ seals herself into the nest at least 10 days before starting to lay. She may moult the rectrices and remiges simultaneously by time that the first egg is laid, and a few of these feathers may be moulted before enclosure (Hoesch 1937). ♂ feeds the ♀ until she emerges when the chicks are half grown, whereafter both parents feed the young until they fledge. Food is delivered to the incubating ♀ about every 1–2.5 hrs (Hoesch 1937).

African Pied Hornbill   *Tockus (Rhynchaceros) fasciatus*

Buceros fasciatus Shaw, 1811. *General Zoology*, **8**, 34. Angola, restricted to Malimbe in Cabinda Province.

PLATES 4, 13

Other English names: Zande Hornbill, Allied Hornbill

Description

T. f. fasciatus (Shaw), 1811. Lower Guinea Pied Hornbill.

ADULT ♂: head, upper breast, and entire upperparts black, mantle and wing-coverts with metallic sheen. Tail black, but for almost all-white outer tail-feather pairs 3 and 4 forming two white stripes down the sides. Underparts white. Primaries and secondaries black, latter with buff edges. Bill: upper mandible, with raised casque ridge attenuating sharply near tip, pale yellow but for black of tip extending back as broad black lines along top of casque and cutting edges; lower mandible yellow, with dark red of tip extending back along underside. Bare circumorbital and separate throat-skin patches dark blue. Eyes brown, legs and feet greenish brown to black.

ADULT ♀: like ♂ in overall coloration, but smaller, casque smaller and shorter, lower mandible with black instead of red at tip and underneath, throat skin orange.

IMMATURE: like ad, but bill smaller, uniform pale yellow, and without any casque development. Only tips of outer rectrix pairs 3 and 4

African Pied Hornbill *Tockus (Rhynchaceros) fasciatus*

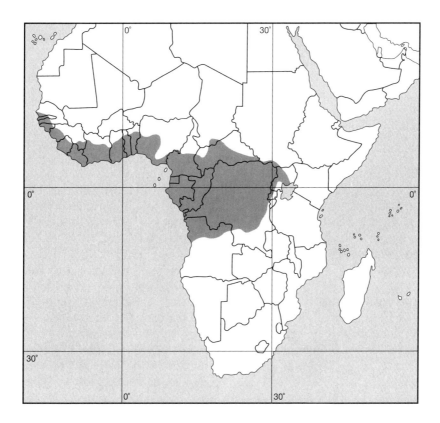

white, bare facial skin flesh-coloured, eyes dark grey.

SUBSPECIES

T. f. semifasciatus (Hartlaub), 1855. Upper Guinea Pied Hornbill. Slightly smaller, only tips of outer rectrix pairs 3 and 4 white in both ad and imm (only in imm of *T. f. fasciatus*). Bill, lacking ridges along sides, with more extensive black areas over tip and casque, broad black lines across lower mandible, and little or no red.

MOULT

Two in moult in Apr (Uganda: Friedmann 1966), Aug (Guinea: Hald-Mortensen 1971), and Feb, June, and Sept (Liberia: Rand 1951; Colston and Curry-Lindahl 1986).

MEASUREMENTS AND WEIGHTS

T. f. fasciatus SIZE: wing, ♂ (n = 39) 253–285 (265), ♀ (n = 34) 232–256 (243); tail, ♂ (n = 39) 217–254 (236), ♀ (n = 30) 200–230 (211); bill, ♂ (n = 43) 90–117 (98), ♀ (n = 32) 78–95 (85); tarsus, ♂ (n = 41) 32–38 (35), ♀ (n = 32) 30–35 (32). WEIGHT: ♂ (n = 15) 250–316 (278); ♀ (n = 7) 227–260 (241).

T. f. semifasciatus SIZE: wing, ♂ (n = 25) 240–267 (254), ♀ (n = 16) 220–250 (234); tail, ♂ (n = 20) 212–250 (226), ♀ (n = 13) 190–216 (203); bill, ♂ (n = 19) 95–110 (101), ♀ (n = 14) 76–98 (88); tarsus, ♂ (n = 19) 31–37 (34), ♀ (n = 13) 30–35 (32). WEIGHT: ♂ (n = 1) 271; ♀ (n = 4) 191–228 (215); unsexed (n = 50) mean 262 (Thiollay 1937).

Field characters

Length 45 cm. Boldly marked with all-black upperparts and contrasting white underparts, and pale yellow bill marked with black lines and dark tip. Ads (but not imms) of each subsp. separable by extent of white in outer tail feathers. Distinguished most easily from black-and-white *Ceratogymna* forest hornbills by small size, small casque, slender build, and buoyant flight. Overlaps with very similar African Crowned Hornbill *T. alboterminatus* in S and E of range, both having the same buoyant flight, whistling calls, and displays. African Crowned Hornbill differs in having an orange bill, yellow eye, broad white stripe behind the eye, and browner plumage.

Voice

Shrill, high, whistling notes uttered singly, as when used to maintain contact, or in series of varying length and differing tempo. A series of phrases of 3–4 notes, rising and falling in pitch and tempo, *pi-pi-pi pi-pi-pi pi-pi-pieeu*, accompanies territorial display. Tape recordings available.

Range, habitat, and status

T. f. fasciatus: Nigeria, from E of Niger R.; Cameroon; Rio Muni; Gabon; Congo; Zaïre, throughout Congo R. basin; S Central African Republic; S Chad, N to 100 km SE of Bozoum (Blancou 1939); S Sudan, at Fada N'gourma; Uganda, N to Budongo, S to Bwamba and E to Sese I. of Lake Victoria (Short *et al.* 1990); N Angola, S to Luanda and E to Lunda. *T. f. semifasciatus*: The Gambia; SW Senegal; Guinea-Bissau; Guinea; Sierra Leone; Liberia; Côte d'Ivoire; Ghana; Togo; Benin; Nigeria, W of Niger R.

Found in most forms of evergreen forest, along the edge of primary forest, in secondary and even in tertiary forest. Also enters savanna along bands of riparian forest growth, occupies patches of dense deciduous woodland, and moves into oil-palm plantations. Common to abundant, not often above 900 m in Liberia (Colston and Curry-Lindahl 1986).

Feeding and general habits

Resident in small family groups of 3–5 over most of its range, and probably territorial for much of the time. At the edges of its range and in changeable habitat, however, may form flocks of up to 70 outside the breeding season (Rand 1951), more usually 10–20, and these flocks range widely in search of food, with several gathering at food sources such as fruiting oil palms.

Primarily an arboreal sp., foraging among foliage or hawking on the wing with buoyant flight, snatching prey either in mid-air (termite alates), from tree trunks, or from the ground (lizards: Young 1946). Often perches low within vegetation or settles on fallen logs under cover, even hopping about on the ground to take some food, as during regular dustbathing sessions. Roosts in trees, sometimes communally with up to 22 birds together (Gore 1990), often on palm leaves from which it clears the surrounding fronds.

Feeds mainly on fruit and insects. Of 48 stomachs, 43 had fruit (27 from oil palms, 20 unidentified), and those with insects included grasshoppers (13), cetoniid and buprestid beetles (7), mantids (6), sphingid moths (5), Hymenoptera (2), and termites, cicada, and a bug (each 1), and also the only vertebrate, a single tree frog (Germain *et al.* 1973). Of 16 stomachs from Zaïre, 13 had fruit, including oil palm, and 8 held insects (Chapin 1939). A regular predator at emergences of ant and termite alates, consuming 41–86 at a time, besides many other insects and half the stomachs also containing small fruits ($n = 6$, Thiollay 1970).

Ten genera of fruits are reported in the diet from Gabon: *Musanga*, *Dacryodes*, *Morinda*, *Xylopia*, *Ficus*, *Heisteria*, *Coelocaryon*, *Guibourtia*, *Pycanthus*, and *Tricalysia* (Brosset and Erard 1986). Also takes *Tetrorchidium didymostemon* (Happel 1986), plus oil-palm fruit (Colston and Curry-Lindahl 1986) and the large hairy caterpillars that feed on them (Young 1946).

Food delivered to large chicks in a nest ($n = 147$ items) included caterpillars (47), grasshoppers (11), mantids (4), beetles (7), wasps (4), flies (1), bugs (8), spiders (1), butterflies (10), larvae and pupae (6) unidentified insects (17) and a small lizard. Full stomach contains an average 16 g of food ($n = 50$) (Thiollay 1976).

Beetles are important animal food, but also takes ants, grasshoppers, caterpillars, termite alates, bugs, and lizards (Chapin 1939), plus saturniid moths and their larvae (Marchant 1953), small mice, and *Agama* lizards (van Someren and van Someren 1949). Recorded robbing nests of a small sunbird and a woodpecker, and often mobbed by small birds (Bates 1930).

Displays and breeding behaviour
Pairs or small family groups call to one another with the bill skywards and head rocked back with each 3-note phrase.

Breeding and life cycle
Little known. ESTIMATED LAYING DATES: Senegambia, Aug; Liberia, Dec–Feb; Côte d'Ivoire, Nov–Jan; Ghana, Jan–Apr; Nigeria, Jan–Apr; Cameroon Feb–Mar, ♀ with active ovaries Feb, Mar, May, Dec; Uganda, Mar–Apr, Sept; Gabon, Sept–Feb, in drier season; Zaïre, July–Sept, ♂ in breeding condition Dec. c/4 maximum recorded. EGG SIZES: *T. f. fasciatus*: ($n = 3$) 40.1×28.6. *T. f. semifasciatus*: ($n = 2$) $33.5–35.0 \times 25.0–25.5$. Eggs white with pitted shells.

Nests are in natural cavities 9–24 m up in large trees, the entrance sealed to narrow vertical slit by the ♀ using her own droppings. One was in a trunk below an active Crowned Eagle *Stephanoaetus coronatus* nest.

The ♀, while immured in the nest and incubating, is fed by the ♂, mainly on grasshoppers and fruit supplied at rates of 6–12 times/hr (Brosset and Erard 1986). The ♀ moults her rectrices and remiges during enclosure. The ♀ emerges before the chicks to help the ♂ in feeding them (Voisin 1953). A captive survived for at least 22.3 years (K. Brouwer *in litt.* 1992).

Hemprich's Hornbill   *Tockus (Rhynchaceros) hemprichii*

Buceros (Lophoceros) Hemprichii Ehrenberg, 1833.
Symbolae Physiologae Aves, sign. aa. Coastal Abyssinia, restricted to Archiko in mountains about 40 km S of Massaua, Ethiopia

PLATES 4, 13
FIGURE 3.10

Description
ADULT ♂: head, neck, upper breast, and back dark brown. Tail sooty brown, except outer tail-feather pairs 3 and 4 which are all white to form two white stripes down the length of the tail. Underparts white. Remiges and wing-coverts sooty brown with cream edges and tips. Bill dark red, small casque ridge merging into bill and ridges at base of mandibles brownish. Bare circumorbital and throat skin black. Eyes dark brown, legs and feet black.

ADULT ♀: like ♂ in overall coloration, but smaller, with smaller casque and blackish base to lower mandible; circumorbital skin undescribed, probably grey; throat skin pale greenish-yellow.

IMMATURE: similar to ad ♀, but bases of white rectrices flecked black. Bill sooty brown, with yellow tip to lower mandible (Bannerman 1910).

MEASUREMENTS AND WEIGHTS
SIZE: wing, ♂ ($n = 20$) 279–305 (289), ♀ ($n = 7$) 255–270 (262); tail, ♂ ($n = 18$) 237–267 (255), ♀ ($n = 5$) 214–238 (227); bill, ♂ ($n = 19$) 106–133 (121), ♀ ($n = 6$) 96–108 (102); tarsus, ♂ ($n = 14$) 38–42 (40), ♀ ($n = 5$) 38–40 (39). WEIGHT: ♀ ($n = 1$), in breeding condition, 297 g.

Field characters
Length 50 cm. Dark brown colour, deep red bill, and white feathers on each side of tail distinctive, as are buoyant flight, slender build, and high whistling calls. Sympatric with African Crowned Hornbill *T. alboterminatus* in some areas: separable from latter by brown (not sooty) plumage, lack of pale supercilium, brown (not yellow) eye, deep red (not orange-red) bill, and lack of prominent casque. Larger than African Grey Hornbill *T. nasutus*, which has dark black and white (♂) or maroon and cream (♀) bill. Display climax, with tail

Hemprich's Hornbill *Tockus (Rhynchaceros) hemprichii*

fanned over back showing white stripes down sides, is unique.

Voice
High, whistling notes uttered singly or in series. Single-note contact-call, also 2-syllable piping call. During display repeats long series of piping notes, *pi-pi-pi-pioh-pioh-pioh*, ending with faster, shriller calls. Young give loud yelping calls while begging in nest. Tape recordings available.

Range, habitat, and status
Ethiopia; Djibouti; N Somalia; SE Sudan, in Didinga Mts and Boma Hills; NW Kenya, rarely as far S as Nakuru (Hegner 1980); NE Uganda, at Mt Moroto.

Occupies rocky habitat in hills and gorges at up to 4300 m, favouring wooded gorges and watercourses, especially where *Euphorbia* or sycamore trees abound. Commonest in Ethiopian highlands, but elsewhere local and uncommon. Resident in main part of range, but peripherally descends into lowlands after rains (Somalia: Archer and Godman 1961) or moves far from normal range after breeding (Kenya: Hegner 1980), sometimes entering range of African Crowned Hornbill *T. alboterminatus* (Short et al. 1990).

Feeding and general habits
Usually in pairs or small family parties, sometimes up to 14 together when moving locally during dispersal or drought (Hegner 1980; Duckworth et al. 1992). Makes use of buoyant flight to move between forest patches on the mountains, or through scattered trees and bushes in more arid habitats in the S parts of its range.

Takes much food arboreally, but also at times descends to the ground for termite alates

or searches in rock crevices for prey. Catches carpenter bees in the air, which it then carefully devenoms.

Omnivorous, but feeds principally on insects and other small animal food, including chameleons, skinks, bees, beetles, caterpillars, and grasshoppers. Fruits eaten include figs, juniper berries, *Trichilia emetica*, and possibly even *Euphorbia* fruits.

Displays and breeding behaviour
Appears to be monogamous and territorial. During territorial display, calls with bill skywards, wings partly spread, and body bobbing up and down, ending with faster calls during which the tail is raised vertically and partly spread over the back (Brown 1976; Fig. 3.10, p. 30). Some wing-flapping also reported (Phillips 1913).

Breeding and life cycle
Little known. ESTIMATED LAYING DATES: Ethiopia, Jan, Mar–Apr, Aug–Oct; Kenya, Mar–May. c/3. EGG SIZES: ($n = 3$) 37.6–40.0 × 26.3–27.6 (39.1 × 27.1). Eggs white with pitted shells.

Most nests have been found in holes in rock faces, only one in a natural hole in an *Erythrina* tree, despite the presence of trees at most sites (Urban *et al*. 1970; Brown 1976; Williams 1978). Most nests were placed on the sides of gorges and ravines. One entrance to a rock hole was 60 × 30 cm, sealed to form a slit 7.2 × 2.2 cm; at another, the cavity was 20 × 20 cm and 40 cm deep, with the entrance sealed to 14.5 × 2.5 cm. The nest is lined with wood chips and bark. The ♀ seals the entrance, and the ♂ feeds her with food taken to the nest as single items held in the bill tip.

Subgenus *Lophoceros* Hemprich and Ehrenberg, 1833

Two arboreal spp. that display with head-up postures and whistling calls but flip their wings open with each call-note. Wing-coverts and flight feathers edged with paler brown to produce scalloped effect. Most closely allied to the subgenus *Rhynchaceros*, with which they share several morphological and behavioural characters such as buoyant flight, hopping when on the ground, and roosting in open sites.

Pale-billed Hornbill *Tockus (Lophoceros) pallidirostris*

Buceros pallidirostris Hartlaub and Finsch in Finsch and Hartlaub, 1870. *Vögel Ost-Afrika*, p. 871, from von der Decken *Reisen in Ost-Afrika*, **4**, Caconda, Benguella, Angola.

PLATES 4, 13

Description
T. p. pallidirostris (Finsch and Hartlaub), 1870. Western Pale-billed Hornbill.

ADULT ♂: head and neck dark grey, with broad white stripe from above eye to nape. Back brown, tail sooty brown with white tips to all but central pair of rectrices. Underparts white. Primaries, secondaries, and wing-coverts sooty brown with buff borders. Bill creamy yellow with some grey blotching and tip, casque ridge terminating sharply at about one-third of length. Facial skin colours not recorded. Eyes red or red-brown, legs and feet sooty brown.

ADULT ♀: like ♂ in overall coloration, but smaller, casque ridge shorter, eyes brown; facial skin colours not recorded.

IMMATURE: like ad, but bill smaller, casque undeveloped. Soft-part colours not recorded.

SUBSPECIES
T. p. neumanni (Reichenow), 1894. Eastern Pale-billed Hornbill. Smaller, with head and

Pale-billed Hornbill *Tockus (Lophoceros) pallidirostris*

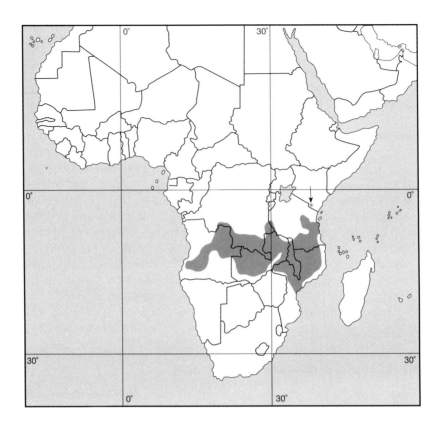

neck paler than *T. p. pallidirostris*, bill with reddish tip extending back along casque and cutting edges; facial skin colours unrecorded; eyes brown in both sexes. Hybrid specimens with *T. p. pallidirostris* reported W of Luangwa valley, Zambia (Sanft 1960).

MEASUREMENTS AND WEIGHTS
T. p. pallidirostris SIZE: wing, ♂ (*n* = 18) 235–262 (247), ♀ (*n* = 15) 220–240 (233); tail, ♂ (*n* = 9) 218–238 (224), ♀ (*n* = 4) 210–223 (215); bill, ♂ (*n* = 14) 73–100 (83), ♀ (*n* = 6) 69–80 (74); tarsus, ♂ (*n* = 10) 38–40 (39), ♀ (*n* = 5) 35–38 (37). WEIGHT: ♂ (*n* = 3) 248–266 (256); ♀ (*n* = 2) 206–217.

T. p. neumanni SIZE: wing, ♂ (*n* = 18) 215–237 (227), ♀ (*n* = 6) 197–217 (205); tail, ♂ (*n* = 14) 188–213 (201), ♀ (*n* = 5) 178–194 (184); bill, ♂ (*n* = 14) 73–85 (82), ♀ (*n* = 5) 70–77 (74); tarsus, ♂ (*n* = 15) 33–37 (35), ♀ (*n* = 4) 32–34 (33). WEIGHT: ♀ (*n* = 1) 170, unsexed, 200–325 (8♂♂, 2♀♀: da Rosa Pinto 1983).

Field characters
Length 50 cm. Grey and brown plumage, contrasting with cream bill and white stripe behind eye distinctive, and buoyant flight and whistling calls also useful. In some areas overlaps with African Grey Hornbill *T. nasutus*, which has very similar plumage, flight, and calls, but black and white (♂) or maroon and cream (♀) bill. All other spp. black or dark brown.

Voice
A series of drawn-out piping whistles very similar to, but slightly deeper and more mellow than, those of African Grey Hornbill, including accelerated phrases that accompany display, *pipipieu pipipipieu pipipipieu* (C. Vernon *in litt.* 1973). Tape recordings available.

Range, habitat, and status
T. p. pallidirostris: Angola, S to Kasinga; S Zaïre, with one record from W shore of Lake Tanzania; Zambia, W of and excluding Luangwa valley. *T. p. neumanni*: Zambia, E of Luangwa valley; Malawi; Mozambique, north of Zambezi R.; S Tanzania, ranging N to Kidugallo, Mgera and Kilosa (Short *et al.* 1990) and also S Kenya, as rare visitor at Taveta.

Favours dense, extensive stands of tall *Brachystegia* miombo woodland below 1372 m. Interdigitates but is rarely strictly sympatric with African Grey Hornbill, which avoids denser woodlands. Frequent, but uncommon in Tanzania.

Feeding and general habits
Feeds mainly in trees, but known to descend to the ground to take food, chiefly in the dry season, as well as to dustbathe. Usually encountered in pairs or parties of 5–8.

Diet of plant and animal matter reported, but few details. Known to eat small seeds and gum oozing from damaged pods (Roberts 1912).

Displays and breeding behaviour
Displays with bill pointing skywards, wings flicking, rocking back and forth, and calling with fast whistles, very like African Grey Hornbill (C. Vernon *in litt.* 1973).

Breeding and life cycle
Little known. ESTIMATED LAYING DATES: Angola, Aug–Sept, Nov; Zambia, Sept; Zaïre, Aug–Nov; Malawi, Oct; Mozambique, Oct–Nov. Breeds at the beginning of the rainy season, starting even in the dry period just before the rains when many miombo trees come into leaf. c/4–5. EGG SIZES: ($n = 3$) 39.3–39.9 × 29.0–29.7 (Mozambique: Roberts 1912), about 36 × 26 (Angola: da Rosa Pinto 1983). Eggs white with pitted shells.

Nests in a natural hole in a tree with the entrance sealed to a narrow vertical slit. A funkhole existed above one nest (Roberts 1912). Difference in sizes of chicks in one brood suggests incubation from the first egg and laying at intervals of at least 1 day. Incubation is by the ♀, fed by the ♂ with food items brought singly in the bill. ♀ moults her rectrices and remiges simultaneously during enclosure.

African Grey Hornbill *Tockus (Lophoceros) nasutus*

Buceros nasutus Linnaeus, 1766. *Systemae Naturae*, edition 12, 1, 154. Senegal, restricted to environs of Dakar.

PLATES 4, 13
FIGURES 3.3, 3.11, 5.6

Other English name: Grey Hornbill

Description
T. n. nasutus (Linnaeus), 1766. North African Grey Hornbill.

ADULT ♂: head and neck dark grey, except for broad white stripe from above eye to nape. Back pale brown with white stripe down centre. Tail blackish-brown, all feathers except central pair tipped white. Underparts white with faint brown wash, especially on upper breast. Flight feathers sooty brown with buff tips and outer edges; wing-coverts dark brown, edged buff. Bill black, with white patch at base of upper mandible and white ridges across base of lower mandible, casque a small ridge merging into middle of bill. Bare circumorbital and throat skin dark grey. Eyes red-brown, legs and feet sooty brown.

ADULT ♀: like ♂ in overall coloration, but smaller, casque ridge shorter; bill dark red, with casque and basal half of upper mandible pale yellow and pale yellow ridges across base of lower mandible. Bare throat skin pale green.

African Grey Hornbill *Tockus (Lophoceros) nasutus*

IMMATURE: most like ad ♂, but bill smaller and uniform dark grey.

SUBSPECIES

T. n. epirhinus (Sundevall), 1850. South African Grey Hornbill. Smaller than *T. n. nasutus*, with narrower pale edges to the wings and tail. Ad ♂ has a discrete tubular casque on the bill which ends in a short projection, ad ♀ with less pronounced ridges at the base of the lower mandible. Bill of juv resembles that of ad ♂, being black with a white patch at the base of the lower mandible.

T. n. forskalii (Hemprich and Ehrenberg), 1833, from the Arabian Peninsula, has been separated from *T. n. nasutus* on its darker upperparts and larger size: wing, ♂ (n = 12) 231–256 (241), ♀ (n = 5) 212–226 (219).

T. n. dorsalis Sanft, 1954, of arid SW Africa, has been separated from *T. n. epirhinus* on account of its having paler upperparts, although of similar size: wing, ♂ (n = 10) 214–228 (224), ♀ (n = 9) 200–212 (206).

Neither of the last 2 subspp. is recognized as distinct (Kemp and Crowe 1985).

MOULT

Main period Nov–May (Transvaal: Kemp 1976a).

MEASUREMENTS AND WEIGHTS

T. n. nasutus SIZE: wing, ♂ (n = 137) 210–250 (225), ♀ (n = 47) 187–225 (206); tail, ♂ (n = 61) 184–218 (201), ♀ (n = 36) 167–200 (183); bill, ♂ (n = 68) 89–120 (99), ♀ (n = 40) 69–97 (78); tarsus, ♂ (n = 67) 34–42 (38), ♀ (n = 35) 32–39 (35). WEIGHT: ♂ (n = 5) 220–258 (233); ♀ (n = 4) 163–215 (183).

T. n. epirhinus SIZE: wing, ♂ (n = 55) 196–237 (217), ♀ (n = 14) 186–218 (201); tail, ♂ (n = 26) 172–209 (195), ♀ (n = 11) 163–185 (178);

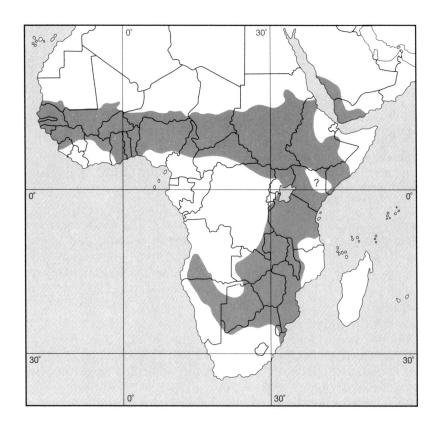

bill, ♂ (n = 32) 73–102 (82), ♀ (n = 13) 66–80 (72); tarsus, ♂ (n = 17) 31–38 (35), ♀ (n = 11) 32–39 (33). WEIGHT: ♂ (n = 34) 159–234 (167); ♀ (n = 29) 122–175 (146).

Field characters

Length 45 cm. Grey and brown plumage and bill colour, black and white in ♂ or maroon and cream in ♀, most distinctive characters. Slender build, buoyant flight, piping calls, and wing-flicking display also notable. Most like Pale-billed Hornbill *T. pallidirostris* in colour, calls, flight, and display, but where they occur together, in central Africa, Pale-billed Hornbill identified by pale cream bill and lack of white stripe down centre of back. African Crowned Hornbill *T. alboterminatus* separable by yellow eye, red bill, and dark sooty plumage with pale supercilium, and Bradfield's Hornbill *T. bradfieldi* by orange bill and pale eye stripe.

Voice

High, rather drawn-out, whistling and piping notes, uttered either singly or in a series, depending on the intensity of alarm or contact. Shorter notes, at faster tempo, precede display, when a series of notes is run together into phrases to accompany the visual signals, *pi pi pi pi pipipieu pipipieu pipipieu*. Tape recordings available.

Range, habitat, and status

T. n. nasutus: S Mauritania; Senegal; The Gambia; Guinea-Bissau; Guinea; Mali, along Niger R.; Ghana; Burkina Faso; Togo; Benin; Nigeria; S Niger, N to 17°N; N Cameroon; S Chad; Central African Republic; Sudan, N to 20°N along Nile R.; Ethiopia; SW Saudi Arabia; Yemen; S Somalia; Uganda, S to Kigezi; Kenya, S to Rabai but excluding arid NE. *T. n. epirhinus*: S Uganda; SE Kenya; Rwanda; Burundi; Tanzania; S Zaïre; Malawi; Mozambique, in W and S; Zambia, in E and S; S Angola, N to Malanje in winter; NE Namibia; Botswana; Zimbabwe; South Africa, in N Transvaal and N Natal.

Occupies any wooded habitat, from scattered thorn bushes with few trees in semi-desert to tall open woodland with stands of large trees. Replaced in denser and taller central African woodlands by very similar Pale-billed Hornbill, and does not enter moist evergreen forest. Widespread and common, including in S Saudi Arabia (Silsby 1980). Pairs maintain a territory of 22–63 ha (Transvaal: Kemp 1976a; Tarboton et al. 1987), much larger than that of sympatric ground-foraging congeners.

Feeding and general habits

Forages mainly arboreally in woodland, taking only 20–30% of food items from ground cover (Transvaal: Kemp 1976a), but this may vary regionally, with 8 out of 10 items taken terrestrially in Kenya (Lack 1985) or many large insects in Ethiopia (Duckworth et al. 1992). Relies on aerial agility to obtain about half its food, catching it in flight in the air, as with termite alates, or off vegetation; may snatch rodents and solifugids from the ground in flight. Follows animals such as baboons, monkeys, zebras, and large birds for insects they flush (Gore 1981a; Lewis 1984; Ryall 1984; Duckworth et al. 1992). Rarely, digs for insect larvae (Roberts 1940) or eats scavenging insects around carcasses (Komen 1984). Also robs weaver nests (Tree 1979; Beasley 1985) and hence sometimes attracts mobbing from small birds (Alston 1937).

Ranges widely in search of food, moving with floating, dipping flight from tree to tree at speed of about 30 km/hr (Brooke 1956). At the end of the day it roosts on a thin branch tip, often at a regular site. Since the vegetation over much of its range is deciduous, it is often forced to make seasonal movements in search of leafy foraging habitat, then often congregating in loose flocks of up to 100. Makes regular N–S movements S of the Sahara, e.g. moving N in Ghana and Nigeria in May–June after breeding and S in parties of up to 50 in Oct–Nov (Elgood 1982; Grimes 1987), and migrants may even pass over more luxuriant southern habitats containing residents. In southern Africa, irregular irruptive movements on to grasslands increase in drought years (Nuttall 1992), and regular altitudinal movements are suspected in parts of Zimbabwe (Zambezi valley) and South Africa (E Transvaal escarpment).

Feeds mainly on small animals, especially insects such as grasshoppers, mantids, scale insects, caterpillars, and longhorn and buprestid beetles. Small vertebrates, including tree frogs, chameleons, and lizards, are important by weight in the diet (Kemp 1976a). Catches large bees in flight and devenoms them (Cheesman and Sclater 1935). Regularly takes fruit, especially early in the dry season, including fig and *Commiphora* fruits, acacia seeds, and peanuts (Scholtz 1972). Undigested food remains are regurgitated, sometimes as pellets and usually at the roost (Cunningham-van Someren 1977). Full stomach contains on average 11 g of food ($n = 5$) (Thiollay 1976).

Displays and breeding behaviour

Monogamous, and territorial when breeding. Pairs display by calling with the bill pointing skywards, rocking back and forth, and flicking the wings in and out with each phrase (Fig. 3.11, p. 30). Also displays alone, ♂ more often than ♀. Pairs inspect nest-holes early in the season; later the ♂ courtship-feeds the ♀, carrying single items in bill. ♀ does all preliminary sealing of nest, often briefly entering cavity.

Breeding and life cycle

NESTING CYCLE: 72–86 days: pre-laying period 5–11 days, incubation period 24–26 days, nestling period 43–49 days (Kemp 1976a). ESTIMATED LAYING DATES: Senegambia, July–Aug, Oct–Apr; Mali, Apr–June; Niger, Mar, Dec; Ghana, Dec–Jan; Nigeria, Jan–Apr in S, Aug–Sept in N, mainly Feb–July; Sudan, Jan–Feb, Apr, June–Sept; Ethiopia, Apr–Sept; Somalia, May; E Africa (rainfall region A: Brown and Britton 1980), Jan–Feb, (region D) Nov–Mar; Uganda, Jan–Mar; Kenya, Jan, Apr; Zaïre, Jan–Feb, Aug; Angola, Sept–Oct; Zambia, Aug–Dec; Malawi, Aug–Nov; Namibia, Oct–Mar, peak Dec; Zimbabwe, Sept–Dec, peak Oct–Nov; South Africa (Transvaal), Sept–Dec, peak Oct–Nov. Usually lays during or even before first rains when vegetation comes into new leaf (Kemp 1973), 55% of nests started before as little as 25 mm of rain in Namibia (Riekert 1988). Captives, England, California (Mace 1992) laid Apr–May. c/2–5 (Transvaal, mean 4.3, $n = 18$; Namibia, mean 3.9, $n = 7$). EGG SIZES: *T. n. nasutus*: 37.0 × 25.5. *T. n. epirhinus*: ($n = 22$) 34.2–40.0 × 25.0–28.3 (37.3 × 26.2). Eggs white with pitted shells.

Nests are placed in a natural tree hole or large barbet nest-hole, also in rock holes in more arid areas (Hoesch 1937; Kemp and Kemp 1972) or in ground holes of Red-and-Yellow Barbets *Trachyphonus erythrocephalus* (Cunningham-van Someren 1977). Territory size probably varies with density of woody vegetation, and nests are often sited among larger stands of trees. Uses trees such as *Sclerocarya, Acacia, Combretum, Lonchocarpus, Lannea, Diospyros*, and *Colophospermum*, and will use artificial nestboxes (Riekert and Clinning 1985, Mace 1992). Mean nest height 4 m (range 70 cm to 9 m, $n = 31$). Nest-cavity with mean diameter 23 cm, floor 8 cm below entrance, entrance 9 cm high × 4 cm wide, 85% of nests with funk-hole above cavity. Same hole often used in consecutive years (Kemp 1976a).

Entrance sealed to narrow vertical slit by ♀, using her droppings and the remains of food, especially millipedes, brought by ♂. Lining of bark flakes and some dry leaves brought by ♂ throughout nesting period, especially early in cycle, when 26% of visits ($n = 119$) were with lining rather than food (Brown 1940).

After sealing herself into the nest, ♀ waits for 5–11 days before laying the first egg, the rest of the clutch being laid at intervals of 1–7 days, usually on alternate days (Mace 1992), but intervals becoming longer between later eggs. The ♂ supplies the incubating ♀ with food, and is often very wary when approaching the nest. Chicks hatch at about the same intervals as the eggs are laid, and a brood of 4 may take up to 8 days to hatch.

The chicks hatch naked, blind, and pink-skinned (Kemp 1976a, Mace 1992). The eyes begin to open and feather-quills emerge by 5 days old. By 15 days the quills are well out but not unsheathed, giving a prickly appearance. Feathers first emerge on the head, neck, and underparts, later on the back and wing–coverts. Appears well feathered at 30 days old, and fully feathered at fledging except for a few partly encased head, neck, and flight feathers. Leg growth is complete at 16 days, body growth at 30 days, but wing and tail feathers are still growing at fledging.

The chick weighs 12–14 g at hatching, increasingly by 7–10 g per day for 40 days, by which time the young are about ad weight. The weight then oscillates, dropping slightly before the young fledge at 43–49 days. The chicks hatch asynchronously and staggered in size (Fig. 5.6, p. 58), and younger chicks generally take longest to fledge (Kemp 1976a).

The ♀ breaks out of the nest when the eldest chick is 19–34 days old and the youngest sometimes only 13. The chicks reseal the entrance alone. The ♀ moults rectrices and remiges just before or at the start of laying, with regrowth virtually complete at emergence (Mace 1992). A captive moulted twice during successive breeding attempts in the same season (Wilkinson and McLeod 1990). The ♀ feeds herself after emergence and assists the ♂ in feeding chicks. Chicks perch within cover near the nest after emerging, and later join the ads to forage. Younger chicks reseal the nest until ready to emerge. Juvs in captivity first fed themselves within 10 days of fledging (Wilkinson and McLeod 1990).

In the Transvaal, 56% of eggs laid produced fledged young ($n = 16$ nests), 91% of all nesting attempts producing some young ($n = 22$), and 2.4 young fledged per successful attempt ($n = 20$) (Kemp 1976a). Less subject to predation by raptors than are its sympatric ground-feeding congeners, although recorded as prey of Lanner Falcon *Falco biarmicus*. The main factor limiting breeding success was growth of tree bark, which closed the nest-entrance before emergence of ♀ or chicks. May sometimes be double-brooded in nature (Grimes 1987). Recorded as prey of Lanner Falcon (Tree 1963), Ayres' Eagle *Hieraaetus ayresii* (Steyn 1982), Wahlberg's Eagle *Aquila wahlbergi* (Tarboton and Allan 1984). A pair recorded chasing off an African Gymnogene *Polyboroides typus*, a potential nest-predator (R. Brooke *in litt.* 1973) even mobbing a dead hawk (Osborne and Colebrook-Robjent 1988), but the hawk not always able to break into a sealed nest (Steyn 1982). A captive survived at least 20 years (K. Brouwer *in litt.* 1992).

Subgenus *Tockus* Lesson, 1830

This subgenus represents a radiation of ground-dwelling spp. within the genus. All but *T. monteiri* run rather than hop along the ground on relatively long tarsi, and all fly with direct flap-and-glide flight on relatively short wings. They also have large, irregular areas of white in the tail and secondaries, and wing-coverts spotted with white. They bow the head and utter clucking calls during territorial display, several spp. sharing similar postures, with the wings either closed or fanned during display. Roosts are in concealed positions, and contact-calls during the day are not much employed. Grass, leaves, and twigs form main material for lining the nest. All share distinctive spp. of mallophagan feather-lice, *Chapinia lophocerus* and *Buceroemersonia brelihi*.

At least 5 and possibly 8 spp. The less derived spp., *T. monteiri* and the *T. erythrorhynchus* complex, keep the wings more or less closed during display; the more derived *T. leucomelas*, *T. flavirostris*, and *T. deckeni* fan the wings. Spp. are sympatric in both groups, so superspp. are not recognized.

Monteiro's Hornbill *Tockus (Tockus) monteiri*

Toccus monteiri Hartlaub, 1865. *Proceedings of the Zoological Society, London,* **1865**, 87. Benguella, restricted to environs of Benguela, Angola.

PLATES 1c, 1d, 5, 13

Description

ADULT ♂: head, neck, and upper breast dark grey, throat and sides of head streaked white. Back brown, uppertail-coverts sooty brown. Central two pairs of rectrices sooty brown, outer pairs white with brown bases. Underparts white. Primaries black with cream tips, widest on inner feathers, primaries 4–8 with a white spot on the outer web. Outer secondaries white with black bases, inner secondaries brown with paler edges. Extent of white on remiges variable. Wing-coverts brown with large cream spots. Bill red with yellowish base, large, scored with grooves along its length, and surmounted by a low, grooved casque ridge. Circumorbital skin black, throat skin dark grey. Eyes brown, legs and feet black.

ADULT ♀: like ♂ in overall coloration, but smaller, bill notably shorter, casque ridge terminating well before tip, bare throat patches blue-green.

IMMATURE: like ad, but bill much smaller and paler orange with dusky patch at base of lower mandible, bare facial skin pale flesh-coloured, eyes brown, legs dark grey. Feathers, especially wing-coverts, tipped light brown.

MOULT

♂ and ♀ often begin moult before breeding and complete it while breeding (Kemp and Kemp 1972).

MEASUREMENTS AND WEIGHTS

SIZE: wing, ♂ ($n = 28$) 204–224 (215), ♀ ($n = 10$) 181–203 (194); tail, ♂ ($n = 19$) 219–245 (231), ♀ ($n = 8$) 192–215 (207); bill, ♂ ($n = 23$) 103–127 (115), ♀ ($n = 11$) 86–105 (95); tarsus, ♂ ($n = 17$) 44–50 (47), ♀ ($n = 6$) 43–46 (44). WEIGHT: ♂ unrecorded; ♀ ($n = 3$) 323–350 (340), ♀ in nest ($n = 8$; J. Mendelsohn *in litt.* 1992) 269–423 (321); unsexed ($n = 15$) 210–400 (da Rosa Pinto 1983).

Field characters

Length 50 cm. Large dark red bill, grey and brown plumage, and white outer tail feathers readily distinguish it from much smaller sympatric congeners. African Red-billed Hornbill *T. erythrorhynchus* and Southern Yellow-billed Hornbill *T. leucomelas* have pied plumage, different bill colours, and higher-pitched clucking calls. African Grey Hornbill *T. nasutus* has muted grey and brown plumage, black and white (♂) or maroon and cream (♀) bill, and whistling calls.

Voice

Deep, hoarse, clucking notes, uttered singly as alarm- or contact-calls. Also given as a series with varying tempo, faster when taking flight in alarm, slower when making nervous contact from afar. During territorial calling, an accelerating series of notes breaks into double notes as display commences, *kok-kok-kok-kokok-kokok-kokok*, accompanied by rise in volume. Sequence similar to that of related African Red-billed Hornbill, but tempo slower. Tape recordings available.

Range, habitat, and status

SW Angola, N to Catumbela; N Namibia, mainly along the escarpment inland of the Namib Desert and E of Etosha Pan, extending E to the Waterberg Mts and S to Rehoboth.

Occupies the driest habitat of any hornbill (Plate 1c), including areas with <100 mm average annual rainfall. Prefers hilly country with stony substrate, but also found on level ground with sandy soils, usually with only scattered thorn trees. Shy and wary in many areas, but generally common. Mean territory size of 15 ha ($n = 12$, range 10–21, Namibia; Kemp and Kemp 1972).

Feeding and general habits

Resident as pairs within a defended territory in those parts of its range with higher rainfall.

Monteiro's Hornbill *Tockus (Tockus) monteiri*

The rains, however, are unpredictable throughout its distribution, and during the dry season it regularly forms wandering flocks of up to 47 birds, which often range out on to flat country adjacent to the hills.

It has the direct flap-and-glide flight characteristic of ground-foraging spp. in this subgenus, and like them feeds mainly on the ground. Unlike them, however, rather than running, it progresses on the ground by hopping, almost bounding, which may best suit its stony habitat. Does much digging with the strong bill, excavating trenches up to 30 cm long and 5 cm deep in search of bulbs. Takes some food from the scrubby vegetation. Roosts in trees, often on rock faces, or may use a rock ledge if no trees are available.

Feeds mainly on insects, also taking some fruits, seeds, shoots, flowers, and bulbs. Bulbs may be important during the dry season, but takes any small animals it can capture, including centipedes, millipedes, moths, Lepidoptera caterpillars and pupae, buprestid, tenebrionid, and meloid beetles, and long-horned and short-horned grasshoppers, and also raids wasp nests. *Acanthoplus* long-horned crickets are especially favoured during summer, forming 70% of food items delivered to nests ($n = 95$) (Kemp and Kemp 1972).

Displays and breeding behaviour

Monogamous, and territorial at least during the breeding season. Territories are maintained by calling and display, with fighting resorted to when necessary. Pair-members often call together, giving clucking notes which rise in tempo and volume until they break into double notes, at which point the head is lowered and the body bobbed up and down with each phrase. A similar display is given by single birds, and even by imms before leaving the parental territory (Kemp and Kemp 1972).

Pairs inspect nest sites, usually in the early morning, spending more and more time sealing external fissures and starting to close up the entrance. ♀ does most sealing, often from within the nest-cavity, but the ♂ does some sealing and sometimes brings lumps of mud to the ♀ in his bill. Areas as extensive as 600 cm² may be closed over (294 cm² required 14 days of preliminary sealing before ♀ entered), but same site often used year after year and may require only closure of small exit hole of the previous year. ♂ often brings food to the ♀ in courtship-feeding, both at and away from the nest.

Breeding and life cycle
NESTING CYCLE: 72–84 days: pre-laying period 5–11 days, incubation ($n = 6$) 24–27 days, nestling period ($n = 5$) 43–46 days. ESTIMATED LAYING DATES: Angola Nov–Dec; Namibia, Oct–Mar, peak Jan–Mar. Lays about one month after first heavy fall (>25 mm) of summer rains (Kemp and Kemp 1972; Riekert 1988). c/2–8. Mean clutch sizes in different seasons 4.0 ($n = 8$; J. Mendelsohn *in litt.* 1992), 4.4 ($n = 8$; Kemp and Kemp 1972), 5.0 ($n = 4$; Riekert 1988, over 6 (J. Mendelsohn *in litt.* 1993). Clutch size apparently related to amount of rain falling before laying. EGG SIZES: ($n = 31$) 37.2–43.9 × 25.3–29.2 (40.4 × 27.0). Eggs white with pitted shells.

Usually nests in rock face (Plate 1d) along watercourses in areas of few trees, foraging on flats and slopes away from the riverbed. May, however, nest far from watercourses where there are trees large enough to provide nest-cavities (Kemp and Kemp 1972). Nests are from ground level, among tree roots or at the base of a cliff, to 39 m up on a cliff. Will readily use artificial nestboxes (Riekert 1988).

Large cavities of 32 cm mean diameter are used in rock faces, usually without the funk-hole typical of most hornbill nests. Entrance finally sealed to narrow vertical slit by ♀ using her droppings, usually within the day of final entry. ♂ brings some mud and millipedes as material, but only rarely helps to seal. The cavity is lined with dry leaves, pods, bark, and grass stems that are brought by the ♂, both before and after laying. Snail shells are also brought, either for lining or as a calcium source for the laying ♀.

The eggs are usually laid at intervals of 2–3 days (range 1–5), the interval increasing on average as the clutch proceeds, with the last egg laid up to 23 days after the ♀ enters. Incubation begins with the first egg. ♀ moults her old remiges and rectrices simultaneously at or just before commencement of laying, and some body feathers during enclosure.

The ♂ usually feeds the ♀ with single food items carried in the bill, but sometimes puts down one to catch another while foraging, then picks up both and may carry up to 3 items per visit, a habit otherwise shown only by *Bucorvus* hornbills. It may be an adaptation to the harshness of the environment, conserving energy by reducing the number of visits to the nest necessary. Feeding rates of 1–3.5 feeds/hr are recorded (Kemp and Kemp 1972).

Hatching is spread over 4 (c/3) to 10 (c/5) days. The chicks hatch naked and blind, and retain the pink skin throughout development. Feather-quills are just visible through the abdominal skin at hatching. At 9 days old, they are emerging all over, especially on the underparts and head, colour is just coming to the bill and legs, and the eyes just opened. At 16 days all quills are out, giving a prickly appearance, and the feathers are just breaking from the quills on the underparts. The legs are fully grown at 22 days old. At 26 days the chick appears well feathered, but its rectrices and remiges are still growing and will not be complete even by the time the chick fledges. ♀ breaks out of the nest 19–25 days after hatching of the first chick, and the chicks reseal the nest alone. ♀ helps the ♂ feed the chicks, the feeding rate rising to 5.9 visits/hr.

Breeding success during different seasons in Namibia showed 30% of eggs laid producing fledged young ($n = 8$ nests, mean clutch size 4.4, mean brood at hatching 3.8, mean brood fledged 1.5; Kemp and Kemp 1972), or 69% of eggs laid producing fledged young ($n = 8$ nests, mean clutch size 4.0, mean brood at hatching 3.3, mean brood fledged 2.8;

J. Mendelsohn *in litt.* 1992), when a sample of 15 pairs reared a mean of 0.63 chicks per adult.

Broods of 5 are known to have been fledged successfully, the overall mean brood size for successful nests in Namibia being 2.3 ($n = 23$). Nearly all losses were chicks that starved soon after hatching, usually the youngest in staggered broods. Annual losses probably vary considerably in a habitat with such erratic rainfall. Pairs laying earliest fledged most young, and fledging success was correlated with rainfall prior to laying, as well as during the nesting cycle.

African Red-billed Hornbill   *Tockus (Tockus) erythrorhynchus*

Buceros erythrorhynchus Temminck, 1823.
Planches colorées, **36**, sp. 19. Senegal and Guinea, restricted to Podor about 175 km N St Louis, Senegal.

PLATES 5, 13
FIGURES 1.5, 3.5, 3.12, 3.13, 5.3

Other English name: Red-billed Hornbill

Description
T. e. erythrorhynchus (Temminck), 1823. North African Red-billed Hornbill.

ADULT ♂: crown and nape dark grey, neck and face white but for ear-coverts, which are faintly streaked grey and separated from crown by broad white post-orbital stripe. Back sooty brown, darkening towards tail, and with white stripe down centre. Two central pairs of rectrices black, rest black basally with increasing amounts of white distally and irregular black bar across centre. Underparts white. Primaries black with white spots in centre. Outer secondaries black, 2–4 from outer with white spot in centre, 5–7 almost all white, and inner secondaries dark brown with cream edging. Upperwing-coverts sooty brown with large white spot in centre, greater secondary coverts almost all white. Bill red with narrow yellow base, basal half of lower mandible black, casque only a slight ridge. Circumorbital and throat skin creamy yellow to pale pink. Eyes brown, legs and feet sooty brown.

ADULT ♀: like ♂ in overall coloration, but smaller, bill red with only black fleck on lower mandible.

IMMATURE: like ad, but bill smaller and plain brownish yellow; eyes grey, soon turning brown.

SUBSPECIES
T. e. rufirostris (Sundevall), 1850. South African Red-billed Hornbill. Darker and slightly larger than *T. e. erythrorhynchus*, neck and upper breast flecked grey, basal half of secondaries 5–7 dark brown, tail-bar broader. Bill of ♀ without black fleck, of imm with black patch at base resembling adult ♂. Eyes yellow when ad; bare facial skin flesh-coloured, flushing in breeding season to deep pink.

T. e. damarensis (Shelley), 1888. Damaraland Red-billed Hornbill. Like *T. e. erythrorhynchus*, but larger, forehead and ear-coverts white, bare facial skin more extensive and flesh-coloured, and eyes brown. Plumage characters reported to intergrade with those of *T. e. rufirostris* along zone of overlap (Sanft 1960), but may involve imm birds, which have grey on face like *T. e. rufirostris*: the eye colour would be distinctive.

T. e. rufirostris will probably prove to be a distinct sp., and *T. e. damarensis* may be either an isolate of the nominate form or a distinct sp. The nominate N subsp. and SW *T. e. damarensis* seem to inhabit drier thorn scrub than *T. e. rufirostris* (Kemp 1990). There are also at least two other discrete and undescribed populations that have been observed in captivity, among museum specimens, and on photographs. Both have black rather than pink or yellow circumorbital skin when ad, and

African Red-billed Hornbill   *Tockus (Tockus) erythrorhynchus*

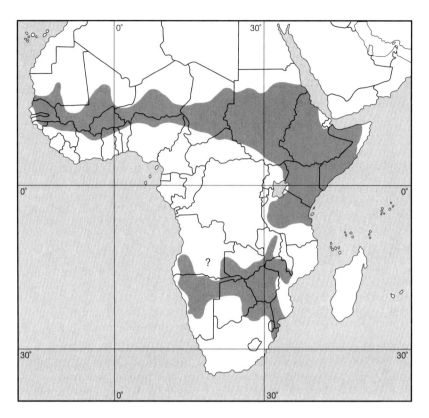

both occur within the range of *T. e. erythrorhynchus*. One in Senegambia has similar brown eyes but is smaller (wing, ♂ (*n* = 6) 170–183 (176), ♀ (*n* = 5) 157–164 (162)), while the other in W Tanzania has yellow eyes but is of similar size (wing, ♂ (*n* = 5) 177–185 (180), ♀ (*n* = 3) 165–168 (167)). Photos of the Senegambian form show the facial skin to be pink in imms (B. Tréca *in litt.* 1993) and suggest a different display from *T. e. rufirostris*, with the wings half-opened (T. Boon *in litt.* 1992) and most like the display of the nominate form in E Africa (Jones 1992). Smaller specimens in E of range of *T. e. rufirostris* have also been separated as *T. e. degens* Clancey, 1964; wing, ♂ (*n* = 10) 166–179 (172), ♀ (*n* = 5) 160–163 (161) (Clancey 1964). A detailed review of this species complex is clearly indicated, especially since the Senegambian population has been implicated in transmission of Crimean–Congo hemorrhagic fever virus (Zeller *et al.* 1992)

MOULT
Main period Dec–May (Transvaal: Kemp 1976a). A partial albino has been photographed in the Transvaal.

MEASUREMENTS AND WEIGHTS
T. e. erythrorhynchus SIZE: wing, ♂ (*n* = 78) 170–197 (182), ♀ (*n* = 62) 154–175 (167); tail, ♂ (*n* = 53) 183–214 (199), ♀ (*n* = 55) 166–198 (180); bill, ♂ (*n* = 60) 73–97 (83), ♀ (*n* = 58) 59–74 (67); tarsus, ♂ (*n* = 50) 36–42 (39), ♀ (*n* = 57) 34–39 (36). WEIGHT: ♂ (*n* = 4) 172–185 (177); ♀ (*n* = 7) 120–140 (131).

T. e. rufirostris SIZE: wing, ♂ (*n* = 32) 177–202 (188), ♀ (*n* = 27) 170–188 (177); tail, ♂ (*n* = 27) 187–214 (200), ♀ (*n* = 24) 172–208 (186); bill, ♂ (*n* = 28) 68–93 (79), ♀ (*n* = 24) 54–76 (65); tarsus, ♂ (*n* = 26) 38–42 (41), ♀ (*n* = 17) 34–40 (38). WEIGHT: ♂ (*n* = 75) 124–172 (150); ♀ (*n* = 75) 90–151 (128).

T. e. damarensis SIZE: wing, ♂ (*n* = 13) 186–203 (195), ♀ (*n* = 4) 181–186 (183); tail, ♂ (*n* = 8) 190–212 (199), ♀ (*n* = 2) 175–195 (183); bill, ♂ (*n* = 10) 83–99 (90), ♀ (*n* = 4) 73–77 (75); tarsus, ♂ (*n* = 5) 41–43 (42), ♀ (*n* = 2) 36–40 (38). WEIGHT: ♂ 200, 220; ♀ 150, 200.

Field characters

Length 35 cm. Slender red bill, small size, spotted wing-coverts, and patterned wing and tail distinctive, together with clucking calls and bobbing display. Most similar to Monteiro's Hornbill *T. monteiri* with which sympatric in SW Africa, but smaller, without dark head and neck, less white in tail, and higher-pitched calls. Similar in plumage to yellow-billed hornbills *T. leucomelas* and *T. flavirostris* but, apart from different bill colours, is smaller, with primary spots forming white line across open wing in flight. Von der Decken's Hornbill *T. deckeni* has little or no spotting on wing-coverts and has orange (♂) or black (♀, imm) bill.

Voice

High-pitched clucking notes uttered singly for low-intensity contact or alarm, or in series of varying tempo when more intense. Same call leads into double notes, *kok-kok-kok-kokok-kokok-kokok*, that accompany display. Screeches when frightened by aerial predators or when handled. Chicks beg with high piping notes and take food with harsh squawk of acceptance, as does nesting ♀. Tape recordings available.

Range, habitat, and status

T. e. erythrorhynchus: S Mauritania, Senegal; The Gambia; Guinea-Bissau; Guinea; S Mali, along Niger R.; Burkina Faso; Niger; N Côte d'Ivoire; N Ghana; Benin; Togo; Nigeria; Chad; N Central African Republic; Sudan, to 18°N along Nile R.; Ethiopia; Somalia; Kenya; E Uganda; Tanzania. *T. e. rufirostris*: Malawi; Zambia, in Luangwa and Zambezi valleys; Zimbabwe; N Botswana; S Angola, as far north as Catumbela; NE Namibia; SE Mozambique; South Africa, in Transvaal and N Natal Provinces. *T. e. damarensis*: Namibia, in Damaraland and N and central highlands, possibly intergrading with *rufirostris* in NE around Otjiwarongo.

Favours open savanna, woodland, or thorn scrub, especially *Acacia* and *Combretum* scrub in S and central Africa where there is heavy grazing and limited ground cover. Hence often rather local, but usually common where occurs. *T. e. erythrorhynchus* and *T. e. damarensis*, which inhabit drier thorn scrub and extend into more arid and mountainous areas, appear to undertake regular movements or visit more arid areas only sporadically. Ascends to 2100 m in N Ethiopia (Smith 1957).

Territory size about 10 ha (range 5–14, *n* = 8), with overall density of 50 ha/pair (Transvaal: Kemp 1976a; Tarboton *et al.* 1987). Four *T. e. damarensis* nests with mean inter-nest distance of 208 m (plus 2 *T. monteiri* and 1 *T. leucomelas* in same area along river, Otjongoro, Namibia, pers. obs.).

Feeding and general habits

Usually found as pairs or small family parties, often forming core of bird parties (Schifter 1986), but during the dry season may congregate in flocks of several hundred at good feeding areas, such as around waterholes. Birds move to feeding areas in the early morning, travelling out of their territory when necessary in the dry season, but usually return to a regular roost site within the territory at night. In drier areas, however, flocks appear to undertake local migrations in search of food. Roost in a tree close to a trunk or main limb.

Obtains almost all food on the ground (94% of items, Kenya: Lack 1985), searching actively while running about. Obtains 35% (wet season) to 65% (dry season) of food items by digging in loose ground, litter, and game droppings (Transvaal: Kemp 1976a). Rarely actively pursues or takes flight after prey. Eats mainly invertebrates and fruit in summer and grain in winter in Senegal (Diop 1993).

In Transvaal feeds largely on insects, especially beetles (mainly dung beetles) and their larvae, with grasshoppers, termites (6 spp.), ants, and fly larvae also important. Takes a wide range of small arthropod prey, such as antlions, butterflies, and crickets, with solifugids, scorpions, centipedes, and grasshoppers among the larger items (Kemp 1976a; Wambuguh 1988). Also takes geckos (Wambuguh 1988), raids Red-billed Quelea *Quelea quelea* colonies for eggs and chicks, robs nests of Crimson-breasted Shrikes *Laniarius atrococcineus* (Hoesch 1933), and scavenges dead rodents (Gérard Morel *in litt.* 1984). A few fruits and seeds are eaten, such as *Boscia senegalensis* and *Commiphora holtziana*, and in some areas and seasons, especially in E and W Africa, these may be frequent and important in the diet (Elliott 1972; Gérard Morel *in litt.* 1984; Lack 1985; Wambuguh 1988). Once reported to drink water in captivity (Bigalke 1951).

Displays and breeding behaviour
Territorial, at least during the breeding season. Pairs maintain territories by calling and displaying together, especially in early morning. When clucking notes accelerate and break into double notes, head is bowed down, the body bobbed up and down, and in the subsp. *T. e. rufirostris* the wings are held slightly away from the body but not opened (Fig. 3.12, p. 31): exactly the same movements as in Monteiro's Hornbill. In *T. e. erythrorhynchus* and *T. e. damarensis*, however, the wings are partly opened during display, even more so in the black-faced Senegambian population (Fig. 3.13, p. 31). The display is usually performed on a prominent perch but may occur anywhere, including on the ground. The ♂ courtship-feeds the ♀ single items carried in the bill tip, and ♀ accepts food with a loud acceptance-squawk.

Breeding and life cycle
NESTING CYCLE: 65–99 days: pre-laying period 3–24 (6) days, incubation period 23–25 (24) days, nestling period 39–50 (43) days (Kemp 1976a; Wambuguh 1987). ESTIMATED LAYING DATES: Senegambia, Aug–Nov, peak Sept (Guichard 1947; Morel and Morel 1990); Niger, Mar, Aug, Dec; Nigeria, Apr, June, Aug–Nov; Sudan, Aug–Sept; Ethiopia, Feb–Aug, peak Feb–Apr; Somalia, Apr, May; E Africa (rainfall region A: Brown and Britton 1980), May, July–Aug, (region D) Oct–June; Kenya, Dec–Feb, July–Aug; Zambia, Jan–Mar; Malawi, Jan–Mar; Namibia, Feb–Mar; Zimbabwe, Sept–Feb, peaks Oct–Nov and Jan–Feb; South Africa (Transvaal), Oct–Mar, peak Nov–Dec, one second brood started May, (Natal) Nov–Dec. Nesting usually begins 4–7 weeks after commencement of rains, after short rains in Kenya (Kemp 1973, 1976b; Wambuguh 1987). Captives: Germany, Apr, July (von Wieschke 1928; Encke 1970); Britain, pair re–laid within 6 weeks of fledging chicks. c/2–7, usually 3–5, mean clutch size 3.0 (n = 80, Transvaal, Kemp 1976a; n = 9, Kenya, Wambuguh 1987) to 5.0 (n = 4, Namibia, J. Mendelsohn *in litt.* 1992). Seasonal variation in clutch size is related to rainfall prior to breeding. EGG SIZES: *T. e. erythrorhynchus* (n = 26) 31.5–37.0 × 22.0–25.5 (33.9 × 24.0). *T. e. rufirostris* (n = 12) 31.8–35.2 × 23.0–24.7 (33.6 × 24.1). *T. e. damarensis* (n = 5) 34.5–37.3 × 24.1–26.0 (36.1 × 25.0) (Schönwetter 1967; J Mendelsohn *in litt.* 1992). Egg size within clutches appears to decrease as laying proceeds (Wambuguh 1987). Eggs white with pitted shells. The breeding biology of the black-faced Senegambian population has recently been studied in detail (Diop 1993), with no obvious differences from the other populations.

Nests usually in a natural hole in a tree, but also uses old barbet or woodpecker holes and even hollow beehive logs (Williams 1978). Trees used include *Adansonia, Sclerocarya, Acacia, Combretum, Lonchocarpus, Diospyros, Albizzia, Colophospermum, Lannea, Schotia, Delonix, Commiphora*, and *Ehretia* spp. In S Africa, mean nest height 3.2 m above ground (range 30 cm to 9.1 m, n = 104), with internal diameter of cavity about 23 cm, floor 9 cm down, and narrow entrance about 8 cm high × 4 cm wide, 70% of nests with long funk-hole above the cavity (Kemp 1976a). In Kenya, nests averaged 1.9 m above ground (range 0.5–3 m, n = 12) and all had a long funk-hole (mean 60 cm) (Wambuguh 1987). In Namibia and South Africa, readily uses nestboxes without funk-hole (Büttiker 1960; J. Mendelsohn *in litt.* 1992). Nests usually close to or within area of sparse ground cover, and cavities used repeatedly in successive seasons, often also by different hornbill or other bird spp.

All cracks and holes leading into the cavity, other than the main entrance, are sealed prior to enclosure. ♀ does nearly all sealing of the narrow vertical entrance slit, although ♂ attends closely, feeds her (Fig. 5.3, p. 54), sometimes brings lumps of mud in his bill, and assists with sealing (Diop 1993). Latterly, ♀ uses her droppings and juicy insects (Wambuguh 1987). Lining, brought by ♂ throughout the nesting cycle but especially at beginning, is mainly green leaves with some bark, dry grass, flowers, and snail shells. It may be used to raise the level of the nest-floor (Roots 1968).

Eggs are laid at intervals of 1–6 days, the whole clutch sometimes being laid at daily intervals, but the interval between eggs often increasing as the clutch increases: a fertile c/5 was laid at intervals of 5–7 days (Moreau 1938). ♀ incubates from laying of the first egg and is fed by the ♂. In captivity, one ♂ continued to feed a brood after the original ♂ had died (Azua and Azua 1989). ♀ moults rectrices and remiges simultaneously once laying starts, moult usually completed within 1–3 days of laying first egg.

Individual eggs usually hatch in the sequence laid, but not always (Wambuguh 1987). The nestling hatches naked, blind, and with pink skin retained throughout. Air sacs spread over the dorsal surfaces within a week. Leg growth is complete in 15 days (whereas ulna takes 25–30 days), giving the chick early mobility in the nest. Feather-quills appear in 5 days, initially on crown, breast, thighs, and wings, later on back, face, and neck. Eyes partly open at 7 days. Feathers begin to break from quills at about 20 days. Rectrices and remiges almost fully grown when chick emerges from nest. Main weight increase in first 30 days, to peak at 28–38 days, weight thereafter fluctuating and falling off some 12% before emergence (Kemp 1976a; Wambuguh 1987).

Single food items are delivered to the nest at mean rates of <1–17/hr, highest when chicks are half grown and very vocal (Transvaal: Kemp 1976a), and rising to 50/hr in morning (Wambuguh 1988). In Kenya, delivers mainly insects of mean length 1.3–4.8 cm at different nests, especially Lepidoptera (adults and larvae), Orthoptera, and Coleoptera, with fruits such as *Premna resinosa*, *Boscia coriaceae*, *Commiphora holtziana*, *C. riparia*, and *Dictyoptera* forming a lesser proportion of the diet (Wambuguh 1987). Feeding rates and food size delivered to nest were highest soon after ♀ had entered and was laying (5.1 feeds/hr) and after hatching (5.5–9.4 feeds/hr). Later, ♀ usually fed chicks less often and with items of smaller mean length than ♂. Mean feeding rate per inmate was 6.0 feeds/hr, but at some nests as low as 2.2–2.5 feeds/hr (Wambuguh 1987).

♀ breaks out of the nest when eldest chick 21–22 days old (range 16–24; youngest chick may be only 13 days), but up to 38 days in captivity (Roots 1968). Chicks reseal entrance alone, using their droppings and food remains, but some help given by ad in captivity (Roots 1968). Moulted feathers regrown by time of emergence; ♂♂ (and non-breeding ♀♀) undergo normal gradual moult during same summer breeding period. ♀ and chicks readily defend nest against visits of other animals, including other hornbills (Wambuguh 1988).

Chicks fly well on first emergence from nest, never returning. Remain hidden in foliage nearby for first few days, then accompany ad. Later help in territorial defence, the basic display being shown even when still in the nest-hole. Still receive food from parents when 6 weeks out of nest, although easily able to feed themselves; may remain with parents for 6 months. Can breed in first season after fledging (Moreau 1938).

In Transvaal, 45% of eggs laid produced fledged chicks ($n = 26$ nests); 90% of nesting attempts produce some chicks, with an average of 1.5 young reared per breeding attempt ($n = 73$) (Kemp 1976a). In Kenya, 59% of eggs hatched and 94% of hatched chicks fledged, most losses being through eggs failing to hatch, with overall productivity of 1.4 chicks per nest ($n = 11$ nests) (Wambuguh 1987). Main causes of failure are predation of ♂, infertile or cracked eggs (20–41%), starvation of younger chicks in large staggered broods, flooding of nest, or elephant damage to nest tree.

Full-grown birds regularly fall prey to large raptors, some become hooked in thorns, and several cases reported of predation by puff-adders (Rasa in Wambuguh 1987): can lead to death of nest-inmates by starvation or make them vulnerable to predators. Recorded as prey of Bateleur *Terathopius ecaudatus*, African Gymnogene *Polyboroides typus*, Tawny Eagle *Aquila rapax*, Wahlberg's Eagle *A. wahlbergi*, Martial Eagle *Polemaetus bellicosus*, especially African Hawk-eagle *H. spilogaster*, Dark Chanting Goshawk *Melierax metabates*, and Gabar Goshawk *Micronisus gabar* (Snelling 1969, 1971; Steyn 1982; Tarboton and Allan 1984; Osborne and Colebrooke-Robjent 1988; Searle 1993). Large raptor also snatched large chick from nest-hole (Parker 1984). Probably the first hornbill bred in captivity (von Wieschke 1928; Neunzig 1930), and repeatedly so thereafter. Captive ♀ *T. e. rufirostris* hybridized with ♂ *T. leucomelas* to produce viable young with calls intermediate in pitch and display intermediate in wing-fanning (cf. Figs 3.12 and 3.14, p. 31) at National Zoological Gardens, Pretoria (pers. obs.). A captive ♂ survived for at least 18.1 years (K. Brouwer *in litt.* 1992).

Southern Yellow-billed Hornbill *Tockus (Tockus) leucomelas*

Buceros leucomelas H. K. Lichtenstein, 1842.
Verz. Sammlung der Vögel Kaffernland, **17**. Kaffirland, restricted to Vaal R. valley between Bloemhof and Commando Drift, Transvaal Province, South Africa.

PLATES 5, 13
FIGURES 3.14, 5.2

Other English name: Yellow-billed Hornbill

TAXONOMY: previously considered conspecific with NE African *T. flavirostris* (Kemp and Crowe 1985).

Description
ADULT ♂: crown and nape dark grey, broad white stripe from forehead over eye to nape, ear-coverts streaked with grey. Back black with broad white stripe down centre. Central two pairs of tail feathers all black, next pair with terminal third white, outer two pairs white with black bases and narrow black band across centre. Throat and upper-breast feathers white, edged grey to give streaked effect, abdomen pure white. Outer primaries black with white spot in centre of outer web, rest of remiges black but for secondaries 4–6 from outer being white with black bases; inner secondaries and scapulars dark grey, edged white. Upperwing-coverts black with large white spots. Bill yellow, cutting edge of mandibles and tip dark horn; casque, a low ridge along entire length of bill, darker yellow than bill. Circumorbital and throat skin deep flesh-coloured, especially when breeding. Eyes yellow, legs and feet black.

ADULT ♀: like ♂ in overall coloration, but smaller, casque ridge extending only half-way down bill.

IMMATURE: like ad, but bill smaller and uniform pale yellow mottled with dusky patches, eyes grey.

SUBSPECIES
Specimens at extreme W and E edges of range are smaller and have been described as separate subspp. (Sanft 1960; Kemp and Crowe 1985). Angolan birds (wing, ♂ ($n = 13$) 173–195 (182), ♀ ($n = 3$) 166–173 (169)), de-

scribed as *T. l. elegans* (Hartlaub), 1865, have the crown paler grey, more extensive areas of white in wings and tail, and ♀ has an orange base to the lower mandible; eye colour described as brown, not yellow (Büttikofer 1889). Eastern birds have been separated as *T. l. parvior* Clancey, 1959 mainly on size: wing, ♂ ($n = 15$) 186–204 (194), ♀ ($n = 10$) 173–185 (179) (Clancey 1959). Further details of soft-part colours and displays are necessary to decide exact status.

MOULT
Main period Dec–May (Transvaal: Kemp 1976a).

MEASUREMENTS AND WEIGHTS
SIZE: wing, ♂ ($n = 25$) 193–215 (205), ♀ ($n = 16$) 182–203 (191); tail, ♂ ($n = 18$) 203–235 (219), ♀ ($n = 14$) 192–215 (201); bill, ♂ ($n = 21$) 81–99 (90), ♀ ($n = 15$) 67–80 (74); tarsus, ♂ ($n = 13$) 41–44 (43), ♀ ($n = 12$) 39–41 (40). WEIGHT: ♂ ($n = 75$) 153–242 (211); ♀ ($n = 75$) 138–211 (168).

Field characters
Length 40 cm. Yellow bill, pied plumage with spotted upperparts, streaked breast, white underparts, and fanned-wing display distinctive. Widely sympatric with African Red-billed Hornbill *T. erythrorhynchus*, differing in bill colour, larger size, and less obvious primary spots. Sympatric with Monteiro's Hornbill *T. monteiri* in SW Africa, but smaller, less white in tail, lacks dark head and neck, and bill yellow, not red.

Voice
Clucking notes, uttered singly in low-intensity alarm or contact, increasing in number and tempo when more intense. Series of notes breaks into continuous bubbling call, *kok-kok-kok-kok-korkorkorkorkor*, that accompanies display. Slightly lower-pitched, more liquid,

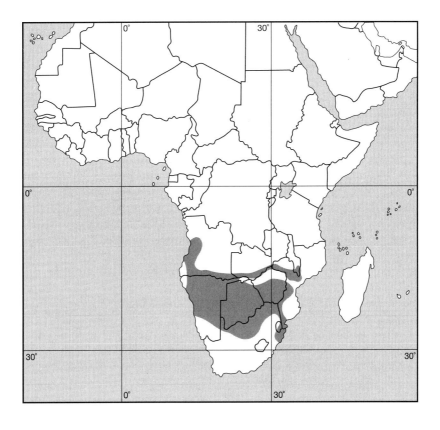

and notes more run together than in smaller African Red-billed Hornbill. Chicks utter high squeak and later a louder squawk when handled, also a shrill *kirr* when insecure (Prozesky 1965). Tape recordings available.

Range, habitat, and status
SW Angola, N to Luanda; Namibia; S Zambia; S Malawi, in Shire valley; Botswana; Zimbabwe, excluding N woodlands; N and E South Africa; SW Mozambique.

Occupies different types of savanna, from semi-arid, where it usually occurs among denser vegetation along watercourses, to open woodland. Widespread and common, especially in open thorn savanna. Local altitudinal seasonal movements during dry season known (Transvaal: Kemp 1976*a*), from deciduous lowveld to base of more evergreen and productive escarpment, and some movement in arid Kalahari (Herholdt 1989). Territory size about 17 ha (range 14–27, $n = 8$), and overall density 20 ha/pair (Transvaal: Kemp 1976*a*; Tarboton *et al.* 1987).

Feeding and general habits
Resident over most of range and usually found in pairs or small family parties, its generalized feeding habits enabling it to remain on territory throughout the year. Moves to feeding areas at dawn, returning at dusk to regular roost site, often against the main limb of a tree. Makes local movements in some areas, congregating at food sources such as waste-tips or tourist camps, often returning some distance to roost back within its territory.

Most food is simply picked from the ground or low vegetation, and does not dig in loose soil or litter to the same extent as the African Red-billed Hornbill (Kemp 1976*a*). Other techniques, such as hawking, digging, or turning over objects using the bill as a lever, are used when applicable, especially in the dry season. Favours areas where ground cover not too dense or tall, since forages while running about on the ground, but prefers some herb and grass cover as this supports invertebrate food.

In Transvaal, feeds mainly on wide variety of small animals. Termites (4 spp.), beetles and their larvae, grasshoppers, and caterpillars form the bulk of the diet, but rodents, solifugids, centipedes, scorpions, and ants may be seasonally important (Kemp 1976*a*), as during rodent plagues (Sharp 1985). Includes bird's eggs in diet. Takes some fruit and seeds, such as *Trichilia emetica* fruit and legume seeds broken from their pods (Stark and Sclater 1903), also oil-palm fruit (Dean 1973), and young leaves of *Boscia albitrunca* (Hines 1990).

Displays and breeding behaviour
Monogamous and territorial. Displays by calling with head bowed, open wings fanned above the back, and rocking back, and forth with each phrase (Fig. 3.14, p. 31). Pairs or single birds usually display from a prominent perch in the early morning, but in disputes with neighbours may display at any time or place. Rarely leads to bill-to-bill fighting or pursuit. Only intraspecifically exclusive, since allows at least African Red-billed Hornbill and African Grey Hornbill *T. nasutus* to nest close by without aggression.

Pairs inspect trees for nest-holes, then use mud for sealing of external cracks and crevices. ♂ carries some lumps of mud in his bill to the ♀, who enters the nest-hole for periods, does most of sealing, and also uses her own droppings. ♂ courtship-feeds ♀ for up to a month before laying and latterly supplies all her food. Food is carried as single items in the bill, very rarely regurgitated from within the gullet. Copulation is infrequently recorded, often being performed on a perch away from nest with no ritual or call.

Breeding and life cycle
NESTING CYCLE: 70–76 days: pre-laying period 4–5 days, incubation period 24 days, nestling period 42–47 (44) days (Kemp 1976*a*). ESTIMATED LAYING DATES: Angola, Oct; Namibia, Oct–Mar, peak Jan–Feb; Botswana, active nest Nov; Zimbabwe, Sept–Mar, peak Oct–Dec; South Africa (Transvaal), Oct–Dec, single records Jan–Mar probably second

broods, (Natal) Oct–Nov, c/2–6 (mean 3.7, n = 68) (Transvaal: Kemp 1976a). EGG SIZES: (n = 52) 31.5–41.0 × 21.3–28.5 (36.9 × 26.0). Eggs white with pitted shells.

Nests in a natural cavity 76 cm to 12.2 m up (mean 3.9 m, n = 88) in a tree, the entrance often facing NE (Hoesch 1937). Trees used include spp. of *Sclerocarya, Acacia, Combretum, Lannea, Peltophorum, Diospyros, Kigelia, Terminalia, Albizzia,* and *Colophospermum.* The nest-chamber is about 20 cm in diameter, and almost 80% of nests have a long funk-hole for escape in the roof of the cavity (Kemp 1976a). The nest-floor is about 11 cm below the average 7 × 4-cm entrance, which is sealed to a narrow vertical slit about 5 mm wide, the original hole being as small as possible and as narrow as 25 mm. ♀ does the final sealing in alone, using her own droppings, and the ♂ brings dry lining of grass, leaves, and bark. Also bred in captivity using artificial nest-chamber (Büttiker 1960).

Incubation begins with the first egg and the eggs are laid at intervals of 1–4 days, later eggs in the clutch being laid at longer intervals. Rectrices and remiges are moulted simultaneously by the ♀ (Fig. 5.2, p. 53), just before laying (Hoesch 1937) or more usually when she starts to lay, and all are dropped within 1–3 days of laying the first egg. In one case, a ♀ had lost most of her claws as well (Prozesky 1965).

The eggs hatch at about the same intervals as laid, and the whole clutch may take as long as 9 days. The chicks hatch naked, blind, and with the skin pink. The eyes begin to open and the feather-quills emerge at about 5 days, the feather remaining in quill until about 20 days, giving the chicks a prickly appearance. Thereafter the quills break open and feathers emerge first on the head, wing-coverts, and underparts (Prozesky 1965). The bill becomes pale yellow by 15 days, and is mottled with dusky brown by time of fledging. Weight increases from 12 g at hatching, by 5–10 g daily to about ad weight at 35 days; rate of gain depends on weather conditions and food availability (Prozesky 1965; Kemp 1976a).

The ♂ feeds the ♀ and chicks, the ♀ passing food to the chicks until they are 10–15 days old, when their rapidly growing legs enable them to reach the slit. A lone ♂ was once seen to take food to chicks of *T. nasutus* in their nest (Mendelsohn 1990). The chicks regurgitate a pellet daily from soon after hatching, usually at night; this includes dry leaves from the nest lining, the largest pellet recorded being 3 × 5 mm and weighing 4 g (Prozesky 1965). The ♀ emerges from the nest when the eldest chick is about 20 days old (range 19– 27), and the chicks reseal the nest alone. The stimulus of an open nest at this stage seems important in the ontogeny of the chick's behaviour (Prozesky 1965). The ♀ then helps the ♂ to feed the chicks and feeds herself. Food is brought to the nest at rates of 3–11 items/hr, highest when the chicks are about 10–20 days old, the rate rising abruptly after hatching (Kemp 1976a). The ♀ removes soiled lining from the nest, throwing it out of the slit, but by 10–15 days the chicks are able, like the ♀, to squirt their droppings out through the entrance.

On fledging, the chicks fly sufficiently well to reach perches in nearby trees, where they remain, being fed by parents, for a few days. Younger chicks left behind reseal the entrance and may emerge up to 10 days after the eldest. Fledglings soon accompany ads to feed, foraging more and more for themselves, but still taking food from parents 6 weeks after leaving the nest. Chicks give rudimentary territorial displays while still in the nest, and later assist parents in border encounters.

Ads are preyed upon by mongooses and raptors, especially large eagles and falcons. Recorded as prey of Secretarybird *Sagittarius serpentarius*, Lanner Falcon *Falco biarmicus*, Bateleur *Terathopius ecaudatus*, African Gymnogene *Polyboroides typus*, Tawny Eagle *Aquila rapax*, Wahlberg's Eagle *A. wahlbergi*, Martial Eagle *Polemaetus bellicosus*, especially African Hawk-eagle *H. spilogaster*, Dark Chanting Goshawk *Melierax metabates*, Black Sparrowhawk *Accipiter melanoleucus*, Spotted Eagle Owl *Bubo africanus*, and Milky Eagle

Owl *B. lacteus* (Dixon 1968; Snelling 1969, 1971; Steyn 1980, 1982; Tarboton and Allan 1984; Thornhill 1986). About 38% of eggs laid led to fledged young ($n = 29$), 92% of nesting attempts led to some production, with 1.4 young fledged per nesting attempt ($n = 54$) (Transvaal: Kemp 1976a). Mean of 2.4 chicks (range 1–5) fledged per nest over several seasons in Namibia (Riekert 1988). Loss of parents through predation, infertile eggs, and starvation of younger members of staggered broods are main causes of failure. Bred in captivity (van Ee 1963; Marshall 1984; Azua and Azua 1989), and one pair raised 43 chicks over a decade. Captive ♂ hybridized with *T. erythrorhynchus* to produce viable young with intermediate calls and displays (National Zoological Gardens, Pretoria, pers. obs.).

Eastern Yellow-billed Hornbill *Tockus (Tockus) flavirostris*

Buceros flavirostris Rüppell, 1835. *Neue Wirbeltiere zu der Fauna der Abyssinia gehörig, Vögel,* **6**, table 2, figure 2. Valleys at base of Taranta Mts, near Masawa, Ethiopia.

PLATES 5, 13

Other English name: Yellow-billed Hornbill

TAXONOMY: previously considered conspecific with S African *T. leucomelas* (Kemp and Crowe 1985).

Description
ADULT ♂: crown and nape black, broad white stripe from forehead over eye to nape. Back black with broad white stripe down centre. Central two pairs of rectrices all black; rest black at base, then white, with broad black band across centre. Throat and underparts white. Primaries black with white spot in centre, spots increasing in size on inner primaries and outer secondaries, central secondaries white with black base and tip. Upperwing-coverts sooty brown with large white spots. Bill, with casque a low ridge along entire length of bill, deep yellow with an orange base, and the tips and cutting edges of the mandibles black. Circumorbital skin black, bare throat patches rose. Eyes yellow, legs and feet sooty brown.

ADULT ♀: like ♂ in overall coloration, but smaller, throat skin black, casque ridge extending only half-way down bill.

IMMATURE: like ad, but bill smaller and uniform pale yellow mottled with dusky patches, upper breast streaked dark grey, and eyes dull grey.

SUBSPECIES
Specimens from extreme NE Somalia are slightly smaller (wing, ♂ ($n = 2$) 180, 205, ♀ ($n = 2$) 178, 186, with deep orange extending across the base of bill (♂) or right along the lower mandible (♀). They have been separated as *T. f. somaliensis* (Reichenow), 1894, but the single character seems somewhat variable, and more information on soft-part colours and behaviour is required to resolve their status (Sanft 1960; Kemp and Crowe 1985).

MEASUREMENTS AND WEIGHTS
SIZE: wing, ♂ ($n = 28$) 187–211 (200), ♀ ($n = 27$) 178–193 (183); tail, ♂ ($n = 22$) 203–235 (221), ♀ ($n = 19$) 180–227 (194); bill, ♂ ($n = 22$) 79–94 (89), ♀ ($n = 19$) 62–80 (74); tarsus, ♂ ($n = 16$) 38–42 (40), ♀ ($n = 13$) 35–41 (38). WEIGHT: ♂ ($n = 3$) 225–275 (258); ♀ ($n = 3$) 170–191 (182).

Field characters
Length 40 cm. Deep yellow bill, spotted upperparts, and white underparts distinctive. Clucking calls and fanned-wing display, showing off patterned wing and tail, aid identification, but shared with sympatric Von der Decken's Hornbill *T. deckeni*. Differs from latter in heavily spotted wing-coverts, and uniform orange rather than orange and cream

Eastern Yellow-billed Hornbill *Tockus (Tockus) flavirostris*

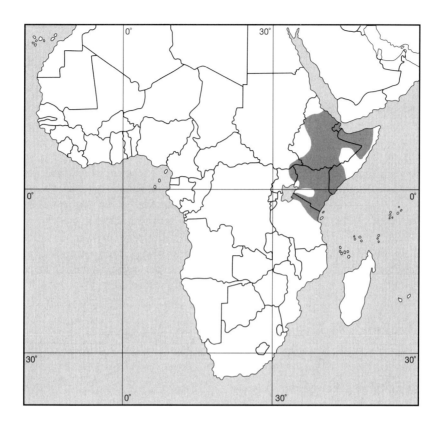

(♂) or black (♀, imm) bill. Widely sympatric with African Red-billed Hornbill *T. erythrorhynchus*, differing in bill colour, larger size, and less obvious primary spots.

Voice
Clucking notes, usually given in a series, and often breaking into the continuous bubbling *kok-kok-kok-kok-korkorkorkorkor* that accompanies display. Calls notably lower-pitched than those of Southern Yellow-billed Hornbill *T. leucomelas*. Tape recordings available.

Range, habitat, and status
Somalia, in N and E; Ethiopia, mainly in N, E, and S highlands; SE Sudan; NE Uganda, in Kidepo valley; Kenya, W to Loiya R.; NE Tanzania, S to Singida.

Generally uncommon (Short *et al.* 1990), but most abundant in open thorn or *Commiphora* savanna and woodland. Resident over most of range, but some altitudinal seasonal movements suggested (Somalia: Archer and Godman 1961), from arid lowlands to more productive escarpments during the dry season.

Feeding and general habits
Forages while running on the ground, and most food picked from the ground (97% of items: Lack 1985) or off low vegetation. Enjoys a special mutual relationship with Dwarf Mongooses *Helogale undulata*, feeding on insects, especially locusts, flushed by the mammals in return for giving warnings of specific avian predators, including those relevant only to mongooses (Rasa 1981, 1983). Mongooses delay departure and hornbills have special 'chivvying' and 'waking' behaviours to facilitate co-operation. This behaviour is shared with Von der Decken's Hornbill.

Feeds mainly on insects, such as locusts and termites, together with some fruits, including figs and especially *Commiphora*. Seen to dig regularly into a rotten stump for beetle larvae and also to eat a scorpion. Recorded to drink regularly (Cunningham-van Someren 1986), an unusual behaviour for a hornbill if normal.

Displays and breeding behaviour
Monogamous and territorial. Displays by calling with head bowed, fanning wings over back, and rocking back and forth with each phrase. Courtship food carried by ♂ to ♀ as single items in the bill.

Breeding and life cycle
Little known. Likely to be similar to Southern Yellow-billed Hornbill. ESTIMATED LAYING DATES: Somalia, Apr–May; Ethiopia, Mar–May, possibly Oct–Nov; E Africa (rainfall region A: Brown and Britton 1980), Feb–Mar, (region D) Nov. c/2–3. EGG SIZES: (n = 8, Somalia) 36.0–38.0 × 26.0–27.5 (36.8 × 26.6); 34.0–40.1 × 25.0–27.7 (37.9 × 26.8) (Schönwetter 1967). Eggs white with pitted shells.

Recorded to nest in natural cavities in *Acacia* and 'hagar' trees, 1.5–4.5 m up usually with a funk-hole above (North 1942). An entrance 6 cm across was sealed to a narrow vertical slit 5.7 cm × 17 mm wide, another to a slit measuring 10.5 cm × 18 mm. One cavity was only 12 cm wide, and the nest-floor was lined with wood chips and bark.

Incubation is by the ♀ alone, her rectrices and remiges being moulted simultaneously when she starts to lay. The ♂ feeds the ♀ during incubation.

Von der Decken's Hornbill   *Tockus (Tockus) deckeni*

Buceros (Rhynchaceros) Deckeni Cabanis, in von der Decken *Reisen in Ost-Afrikas*, **3(1)**, 37, plate 6. E Africa, restricted to Seyidie Province, Kenya

PLATES 5, 13
FIGURE 1.1

Description
T. d. deckeni (Cabanis), 1869. Von der Decken's Hornbill.

ADULT ♂: crown black; sides of head, neck, centre of back, and underparts white, but for some grey streaking on ear-coverts. Sides of back, rump, and central two pairs of rectrices black, the outer rectrices white with only a black base. Wings all black, except for white central secondaries and their coverts. Bill, with low casque ridge along entire length, red on basal half, with outer half and patch below nostril yellow, cutting edges black. Circumorbital skin black, throat patches flesh-coloured. Eyes brown, legs and feet black.

ADULT ♀: like ♂ in overall coloration, but smaller, bill black, casque ridge terminating half-way down bill.

IMMATURE: most similar to ad ♀, the bill smaller and blackish horn with yellowish patches (possibly more orange in juv ♂: Moreau 1935), wing-coverts flecked white, eyes grey-brown (A. Rasa pers. comm. 1987).

A semi-albino recorded with black areas only a pale grey (Duckworth *et al.* 1992).

SUBSPECIES
T. d. jacksoni (Ogilvie-Grant), 1891. Jackson's Hornbill. Smaller, ad ♂ with uniform deep orange bill. Shows conspicuous white spots on wing-coverts of both ad and imm. Allopatric to the nominate subsp. and previously considered a separate sp. or an imm form (Sanft 1954; Ellis 1993). Known to breed in this plumage (Brown and Britton 1980) but not to differ in soft-part colours or displays, so probably warrants only subspecific (Kemp and Crowe 1985) or megasubspecific (Short *et al.* 1990) status. Deserves more study where the two forms meet in E Turkana basin.

MEASUREMENTS AND WEIGHTS
T. d. deckeni SIZE: wing, ♂ (n = 20) 175–203 (188), ♀ (n = 14) 158–180 (169); tail, ♂ (n = 20) 206–232 (221), ♀ (n = 14) 187–210 (196);

bill, ♂ (*n* = 20) 76–94 (87), ♀ (*n* = 14) 60–72 (68); tarsus, ♂ (*n* = 5) 37–41 (39), ♀ (*n* = 5) 32–36 (34). WEIGHT: not separated by subsp.

T. d. jacksoni SIZE: wing, ♂ (*n* = 16) 168–187 (181), ♀ (*n* = 6) 152–167 (161); tail, ♂ (*n* = 16) 195–214 (206), ♀ (*n* = 6) 174–185 (178); bill, ♂ (*n* = 15) 73–87 (82), ♀ (*n* = 6) 65–73 (68); tarsus, ♂ (*n* = 5) 38–40 (39), ♀ (*n* = 3) (37). WEIGHT: not separated by subsp.

Both subspp. combined (Sanft 1960) SIZE: wing, ♂ (*n* = 50) 171–200 (183), ♀ (*n* = 46) 152–185 (167); tail, ♂ (*n* = 44) 196–240 (218), ♀ (*n* = 39) 176–215 (196); bill, ♂ (*n* = 42) 75–95 (84), ♀ (*n* = 38) 62–74 (69); tarsus, ♂ (*n* = 20) 38–43 (40), ♀ (*n* = 19) 34–39 (36). WEIGHT: ♂ (*n* = 7) 165–212 (194); ♀ (*n* = 5) 120–155 (145).

Field characters
Length 35 cm. Solid black and white areas of plumage, especially white face and underparts, black wings with white secondary patch, and white outer tail feathers, prevent confusion with other small sympatric *Tockus* spp., all of which have more spotted plumage. Red-and-yellow or all-red bill of ♂ and black bill of ♀ and imm are distinct from bills of African Red-billed Hornbill *T. erythrorhynchus* and Eastern Yellow-billed Hornbill *T. flavirostris*. Shares fanned-wing display with latter, but has higher-pitched clucking calls, more like those of African Red-billed Hornbill. All 3 spp. have direct flap-and-glide flight.

Voice
Clucking notes uttered either singly or in series. Calls of *T. d. jacksoni* may be harsher than those of the nominate subsp. (North and McChesney 1964). Display-call an accelerating series of notes breaking into a continuous bubbling. Often calls for long periods during the breeding season. Tape recordings available.

Range, habitat, and status
T. d. deckeni: S Ethiopia, including highland areas; S Somalia, with single record from N; Kenya, W to Lokitaung and Loita Hills; Tanzania, S to Tabora and Iringa lowlands. *T. d. jacksoni*: NE Uganda, at Karamoja and Teso; NE Kenya, W to Kerio valley, E to Lake Turkana and Samburu district, S to Lake Bogoria.

Occurs in semi-arid savanna with scattered trees and bushes, especially *Commiphora* woodland. Generally common; frequent along Omo R. in Ethiopia (Toschi 1959), where up to 1 pair/100 m recorded on transects (Duckworth *et al*. 1992).

Feeding and general habits
Forages mainly on the ground (93% of items: Lack 1985), where it runs about in search of food or drops down repeatedly to collect items (Jackson 1938). Enjoys special mutual relationship with Dwarf Mongooses *Helogale undulata*, feeding on insects they flush, especially locusts, in return for providing warning of specific avian predators, including those relevant only to mongooses (Rasa 1981, 1983). Mongooses delay departure and hornbills have special 'chivvying' and 'waking' behaviours to facilitate co-operation. This behaviour shared with Eastern Yellow-billed Hornbill.

Feeds mainly on small invertebrates, as well as berries, especially *Commiphora* and *Cissus* fruit in season, seeds such as *Dicrostachys*, and some buds. Insects are especially important in the diet, including grasshoppers, mantids, crickets, beetles and their larvae, termites, ants, and cicadas. Other small animals eaten include snails, mice, nestling birds, lizards, and tree frogs.

Displays and breeding behaviour
Pairs maintain a territory by displaying with the head down and wings fanned (Moreau 1935), like the Eastern Yellow-billed Hornbill. Displays appear similar in both subspp. (Kemp 1976*a*).

Breeding and life cycle
NESTING CYCLE: 80–82 days: pre-laying plus incubation period 33 days, nestling period 47–50 days. ESTIMATED LAYING DATES: *T. d. deckeni*: Ethiopia, Feb–July; E Africa (rainfall region D: Brown and Britton 1980),

Von der Decken's Hornbill *Tockus (Tockus) deckeni*

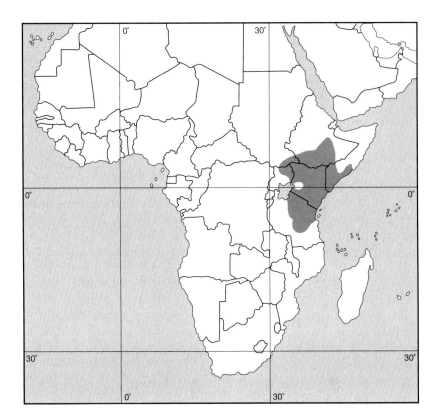

Nov–Mar, peak in short rains: Tanzania, Nov. Captives: England, May, Sept (Paterson 1992). *T. d. jacksoni*: E Africa (region A), Feb–Mar; Kenya, Apr–May. c/2, rarely 3; clutch small compared with congeners. EGG SIZES: *T. d. jacksoni* (n = 2) 34.3– 35.0 × 22.4–23.9. Eggs white with pitted shells.

Nests are in natural holes in trees, such as *Acacia tortilis* or palm stems, or in large woodpecker holes, from as low as 50 cm up to 5 m (Williams 1978). The entrance, as small as 5 cm in diameter, is sealed to a narrow vertical slit with droppings and food remains by the ♀ alone, and the final closure may be done within a day (Moreau 1935; Paterson 1992). One nest was lined with bark flakes.

The eggs appear to be laid at intervals of 2–4 days, judging from different sizes of chicks within broods. Laying was earlier in a season of early rainfall than in one of drought (Moreau 1935). Incubation begins with the first egg, judging by the asynchronous hatching of chicks, and is performed solely by the ♀, who spends about 8 weeks in nest and is fed throughout by the ♂. The incubation period is unrecorded, but hatching occurred 33–34 days after the ♀ sealed into the nest (Moreau 1935; Paterson 1992).

The ♂ feeds the chicks in the nest with single items brought in the bill. The ♀ breaks out of nest when the chicks are 21–28 days old. The chicks reseal the entrance and emerge 22–28 days later (Moreau 1935; Paterson 1992). A second brood was started immediately after fledging and removal of the chicks in captivity (Paterson 1992). The ♀ moults the rectrices and remiges simultaneously during enclosure. The ♂ (and non-breeders) moult gradually during the breeding season. Recorded as prey of Eastern Chanting Goshawk *Melierax poliopterus* (A. Rasa pers. comm. 1987). A captive-bred ♂ (*T. d. jacksoni*) survived 11.9 years (K. Brouwer *in litt.* 1992).

Subgenus *incertae sedis* (uncertain)

Relationships of the two dwarf hornbill spp. *T. hartlaubi* and *T. camurus* and of the long-tailed, bushy-crested *T. albocristatus* (previously in the monotypic genus *Tropicranus* P. Sclater, 1992) are indeterminate through lack of study in their African moist evergreen forest habitats. All are typical members of the genus *Tockus*, with their feeding at the nest with single items carried in the bill, the ♀ emerging well before the chicks, and the chicks being pink-skinned. They apparently lack the elaborate territorial displays of other species in the genus, and are host to some of the same but also some different feather-lice (Elbel 1967).

T. albocristatus is distinct in structure and lice, but in proportions is most like arboreal-foraging spp. in the subgenus *Rhynchaceros*. *T. hartlaubi* also resembles this subgenus in lice and proportions. *T. camurus* is more similar to the terrestrial-foraging subgenus *Tockus* in proportions and the red bill, but resembles arboreal spp. in at least one of its lice; it is the only member of the genus thought to breed co-operatively.

Dwarf Black Hornbill *Tockus hartlaubi*

Tockus hartlaubi Gould, 1860. *Proceedings of the Zoological Society, London*, **1860**, 380. W Africa, restricted to Fanti, Ghana.

PLATES 6, 13

Other English names: Black Dwarf Hornbill, Congo Hornbill

Description
T. h. hartlaubi Gould, 1860. Upper Guinea Dwarf Black Hornbill.

ADULT ♂: head and neck black, with broad white stripe from above eye to nape. Foreneck dark grey with white tips to feathers. Back and tail black, all but central two pairs of rectrices tipped white. Underparts pale grey, lightest towards abdomen. Wings black with metallic sheen, outer primaries with small white spot in centre. Bill black with dark red tip, low casque ridge attenuating sharply half-way down bill. Circumorbital skin grey to black, throat patches flesh-coloured. Eyes red-brown with grey rims, legs and feet black.

ADULT ♀: like ♂ in overall coloration, but smaller, bill all black, eyes deep red with grey rims, casque only a slight ridge at base.

IMMATURE: similar to ad ♀ in bill colour and lack of casque ridge.

SUBSPECIES
T. h. granti (Hartert), 1895. Lower Guinea Dwarf Black Hornbill. Similar to *T. h. hartlaubi*, but wing-coverts and inner secondaries tipped white, red tip to bill of ♂ extending back to include most of casque, bill of ♀ with brown tip, and facial skin with bluish tinge.

MOULT
♂ with wing moult in July (Liberia: Colston and Curry-Lindahl 1986).

MEASUREMENTS AND WEIGHTS
T. h. hartlaubi SIZE: wing, ♂ ($n = 24$) 146–167 (154), ♀ ($n = 13$) 137–148 (141); tail, ♂ ($n = 18$) 152–167 (160), ♀ ($n = 10$) 136–151 (142); bill, ♂ ($n = 19$) 59–68 (63), ♀ ($n = 11$) 48–53 (50); tarsus, ♂ ($n = 18$) 22–25 (24), ♀ ($n = 8$) 22–23 (22). WEIGHT: ♂ ($n = 7$) 96–135 (109); ♀ ($n = 2$) 83, 83.

T. h. granti SIZE: wing, ♂ ($n = 18$) 150–165 (156), ♀ ($n = 6$) 140–150 (144); tail, ♂ ($n =$

Dwarf Black Hornbill *Tockus hartlaubi*

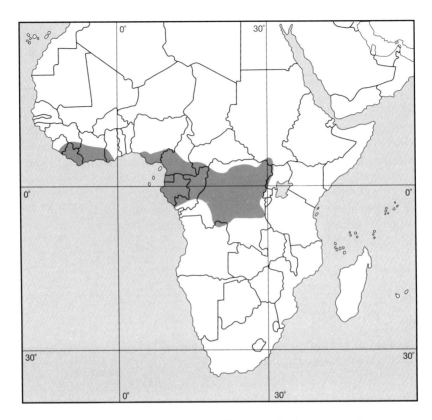

14) 151–177 (161), ♀ (*n* = 6) 142–156 (148); bill, ♂ (*n* = 12) 58–69 (64), ♀ (*n* = 6) 48–54 (52); tarsus, ♂ (*n* = 13) 23–27 (25), ♀ (*n* = 6) 23–27 (24). WEIGHT: ♂ (*n* = 2) 102, 135; ♀ (*n* = 2) 88, 99.

Field characters
Length 32 cm. Small size, black plumage with clear white supercilium, dark bill, and calls distinctive. Rather silent compared with vocal Dwarf Red-billed Hornbill *T. camurus*, which has red bill and red-brown plumage. Much smaller than African Pied Hornbill *T. fasciatus*, which has white underparts, areas of cream colour on bill, and more extensive white in tail.

Voice
High piping notes uttered singly, in a series of 2–3 rhythmic whistling notes, a plaintive *eep*, *eep*, *eep*, quite unlike other hornbills. May be followed by a low howl, with both calls regularly repeated. Tape recordings available.

Range, habitat, and status
T. h. hartlaubi: S Sierra Leone; S Guinea; Liberia; Côte d'Ivoire; W Ghana; Togo; Benin; S Nigeria; Cameroon; Rio Muni; Gabon; Congo; Zaïre, W of Congo R. *T. h. granti*: Zaïre, Congo R. basin E of river; S Sudan, at Bengengai; W Uganda, in Bwamba forest; Central African Republic; N Angola, in Cabinda province.

Inhabits mature evergreen lowland forest, including gallery forest, where prefers more tangled areas and only rarely enters secondary forest. Home-range of one pair 20–30 ha (Gabon: Brosset and Erard 1986). Inconspicuous, and generally considered uncommon.

Feeding and general habits
Usually found in pairs or small family parties of up to 8. Occurs mainly in upper boughs and middle-storey canopy, especially where lianas abound (Brosset and Erard 1986). Spends

much time perched on a horizontal and open site, searching for prey.

Feeds arboreally at 10–35 m above ground and hawks many insects in mid-air (Friedmann 1966, 1978; Grimes 1987). Less often descends to ground level to feed on grasshoppers disturbed by army ants (Brosset and Erard 1986). Mobs birds of prey, but often follows monkeys for the insects they disturb (Chapin 1939; Germain *et al.* 1973).

Diet mainly relatively large insects, especially beetles and Orthoptera, but only rarely takes fruit. Ten stomachs all contained insects, but two also had some fruit (van Someren and van Someren 1949; Berlioz and Roche 1960; Germain *et al.* 1973). Buprestid and other beetles, beetle larvae, ants, grasshoppers, mantids, cicadas, bugs, butterflies, crickets, ants, termite alates, spiders, caterpillars, and small lizards recorded in diet.

Displays and breeding behaviour
Unrecorded. Apparently resident and territorial.

Breeding and life cycle
Little known. ESTIMATED LAYING DATES: Ghana, ♀ close to laying Feb, ♀ in full breeding condition Dec; Nigeria, June, Oct, feeding chicks at nest Dec; Cameroon, July–Aug, Nov; Gabon, July, Sept–Nov; Congo, ♂ calling Nov; Central African Republic, June; Zaïre, June, Aug, Dec; Uganda, ♂ in breeding condition July. A nest which fledged a chick in Jan was active the following Sept (Gabon: Brosset and Erard 1986). Probably lays up to c/4, one ♀ containing 4 well-developed oocytes. Eggs unrecorded.

Breeds in natural holes in trees, often in knotholes, 9–25 m above ground, the entrance sealed to form a narrow vertical slit. ♂ feeds the ♀ during her enclosure, delivering arthropods as single items carried in the bill and collecting food often within only 100 m of nest (Chapin 1939). Acceptance-shriek on receipt of food often reveals an occupied nest.

♀ emerges before the chicks fledge to help the ♂ feed them. The chicks remain with their parents for at least 1 month after fledging. One brood consisted of 2 chicks, but in three others only a single chick was fledged (Brosset and Erard 1986).

Dwarf Red-billed Hornbill   *Tockus camurus*

Tockus camurus Cassin, 1857. Proceedings of the Academy of Natural Sciences, Philadelphia, **8**, 319. Cape Lopez, Gabon.

PLATES 6, 13
FIGURE 1.1

Other English name: Dwarf Hornbill

Description
ADULT ♂: head, neck, upper breast, and back red-brown, with white tips to feathers of breast. Tail dark brown, all but central pair of feathers tipped white. Underparts white. Primaries and secondaries dark brown with white tips and edges, primaries with white median spot. Wing-coverts dark brown, tipped white, broadest tipping on median and greater coverts. Bill red, with low casque ridge merging midway into bill. Circumorbital skin matt brown, throat skin undescribed. Eyes white or pale yellow, legs and feet dark brown.

ADULT ♀: like ♂ in overall coloration, but smaller, with black tip to bill and with casque ridge only at base.

IMMATURE: like ad, but iris grey, bill orange-red, facial skin pale grey.

SUBSPECIES
Birds in W African forest patches, from Sierra Leone and Guinea to lower Niger R., smaller (wing, ♂ (*n* = 5) 143–157 (148), ♀ (*n* = 5) 140–148 (143)) and generally darker above, except for crown. Previously considered as

Dwarf Red-billed Hornbill *Tockus camurus*

subsp. *T. c. pulchrirostris* (Schlegel), 1862, but overlap in size and colour with nominate form (Kemp and Crowe 1985).

MOULT
Wing moult May–July (Liberia: Colston and Curry-Lindahl 1986).

MEASUREMENTS AND WEIGHTS
SIZE: wing, ♂ ($n = 40$) 147–167 (157), ♀ ($n = 21$) 140–160 (150); tail, ♂ ($n = 20$) 138–159 (148), ♀ ($n = 10$) 134–150 (142); bill, ♂ ($n = 24$) 59–76 (66), ♀ ($n = 14$) 54–66 (59); tarsus, ♂ ($n = 24$) 29–32 (31), ♀ ($n = 11$) 28–31 (30). WEIGHT: ♂ ($n = 8$) 101–122 (111); ♀ ($n = 7$) 84–115 (97).

Field characters
Length 30 cm. Small size, red bill, red-brown plumage, and loud distinctive calls readily distinguish this sp. from other forest hornbills, including the (equally small) Dwarf Black Hornbill *T. hartlaubi*, which is mainly black with a broad white supercilium.

Voice
Main call some 15 loud, mellow notes, starting with a short note and following with longer double notes, falling in pitch and volume, *hoo hoo-oo hoo-oo hoo-oo* ...... Also utters a fluting double-noted *kolong*, given with a pause between each call. Calls are audible 500 m away, uttered regularly throughout the year, and often with the whole group calling together. Tape recordings available.

Range, habitat, and status
S Sierra Leone; S Guinea; Liberia; Côte d'Ivoire; Ghana; Togo; Benin; Nigeria; Cameroon; Rio Muni; Gabon; Congo; Zaïre; W Uganda, in Bwamba forest; Rwanda; S Central African Republic; S Sudan; N Angola, in Cabinda province.

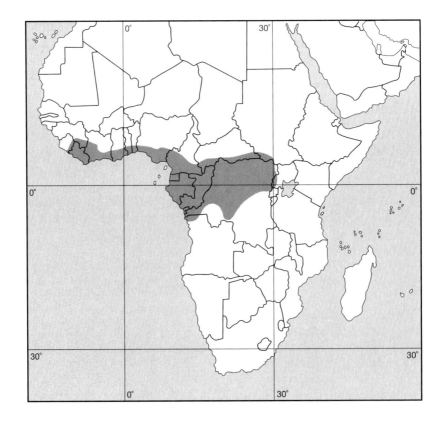

Inhabits dense moist lowland evergreen forest, especially swampy areas, forest edge, and around glades, extending into secondary forest, riverine gallery forest, and isolated forest patches. Ascends to 1350 m in Zaïre. Home-range of one group 20–25 ha (Gabon: Brosset and Erard 1986). Widespread and frequent in suitable habitat.

Feeding and general habits

Occurs in pairs but more often in groups of 4–6, sometimes as many as 12: probably exists in territorial social groups. Inconspicuous, except for the loud calls in which the whole group participates, but very active. May retreat into the canopy at midday, and most active morning and evening (van Someren and van Someren 1949).

Usually found in lower and middle levels of the forest, foraging at times on the forest floor. Often associates with bird parties, squirrels, and army ants *Anomma nigricans* (Bouet 1961), for the insects they disturb (Bates 1930). Sometimes mobbed by small birds (Maclatchy 1937).

Feeds mainly on insects, including beetles, mantids, crickets, grasshoppers, and caterpillars (Chapin 1939; Colston and Curry-Lindahl 1986). Three stomachs contained only insects, but one also had fruit (Germain *et al.* 1973), and recorded taking *Musanga* fruits (Brosset and Erard 1986). Full stomachs contain an average 6 g of food ($n = 4$) (Thiollay 1976).

Display and breeding behaviour

Probably breeds co-operatively, with a small group helping a dominant pair.

Breeding and life cycle

Little known. ESTIMATED LAYING DATES: Liberia, ♀ in breeding condition June; Burkina Faso, feeding at nest Aug; Ghana, ♂ in breeding condition Feb; Nigeria, ♂ in breeding condition Sept; Cameroon, pair in breeding condition Dec; Zaïre, nest contained 2 chicks in Nov (probably from eggs laid in Sept), ♂ in breeding condition Oct; Gabon, copulation observed Dec; Uganda, none in breeding condition Apr–Dec. Clutch sizes and eggs unrecorded.

Nests in natural tree holes, the entrance sealed to a narrow vertical slit. Chicks have pink skin throughout nestling period.

Long-tailed Hornbill   *Tockus albocristatus*

Buceros albocristatus Cassin, 1848. *Proceedings of the Academy of Natural Sciences, Philadelphia*, **3**, 330. St Paul's River, Liberia.

PLATES 6, 13
FIGURE 1.1

Other English names: White-crested Hornbill, African White-crested Hornbill

Description

T. a. albocristatus (Cassin), 1848. Western Guinea Long-tailed Hornbill.

ADULT ♂: head, including crown and nape feathers, white with black shafts and tips to feathers, giving grizzled effect. Rest of plumage black with metallic sheen, except for white tips to elongated feathers of the graduated tail. Bill dark grey, with black spot leading to cream patch at base of upper mandible, some with red-brown stripe forward of nostril; casque a discrete blade-like ridge attenuating sharply near tip of bill. Circumorbital skin blue, throat patches flesh-coloured. Eyes cream; legs and feet dark blue-grey, turning black after death.

ADULT ♀: like ♂ in overall coloration, but smaller, with shorter casque ridge.

IMMATURE: like ad, but bill greenish, darker at base, and smaller; without casque initially, later like ad ♀; eyes pale blue, feet greyish horn-coloured.

150 Long-tailed Hornbill *Tockus albocristatus*

SUBSPECIES

T. a. macrourus (Bonaparte), 1850. Eastern Guinea Long-tailed Hornbill. Like *T. a. albocristatus* and of similar size, but neck, as well as head, white flecked with grey.

T. a. cassini (Finsch), 1903. Lower Guinea Long-tailed Hornbill. Like *T. a. albocristatus* but larger, and remiges, greater wing-coverts, and scapulars tipped white. Sides of head black and only the crest white. Pale patch at base of bill indistinct or absent, but bare throat patches more extensive.

MOULT

♂ with arrested wing moult in Aug, ♂ in moult in July (Liberia: Rand 1951; Colston and Curry-Lindahl 1986); ♀ with incomplete tail moult June (Guinea: Holgersen 1956).

MEASUREMENTS AND WEIGHTS

T. a. albocristatus SIZE: wing, ♂ (n = 9) 230–247 (239), ♀ (n = 8) 201–210 (205); tail, ♂ (n = 8) 410–445 (427), ♀ (n = 6) 360–375 (369); bill, ♂ (n = 7) 91–103 (96), ♀ (n = 8) 67–74 (70); tarsus, ♂ (n = 6) 37–40 (38), ♀ (n = 5) 34–36 (35). WEIGHT: ♂ (n = 4) 279–315 (297); ♀ 276, 288.

T. a. macrourus SIZE: wing, ♂ (n = 7) 225–251 (240), ♀ (n = 4) 197–230 (213); tail, ♂ (n = 5) 430–455 (439), ♀ (n = 3) 345–405 (377); bill, ♂ (n = 7) 90–100 (94), ♀ (n =4) 70–86 (75); tarsus, ♂ (n = 5) 36–41 (39), ♀ (n = 4) (34). WEIGHT: unrecorded.

T. a. cassini SIZE: wing, ♂ (n = 39) 240–277 (253), ♀ (n = 20) 194–224 (209); tail, ♂ (n = 23) 435–505 (465), ♀ (n = 10) 390–420 (406); bill, ♂ (n = 33) 93–119 (103), ♀ (n = 17) 67–80 (74); tarsus, ♂ (n = 36) 36–41 (38), ♀ (n = 19) 33–38 (35). WEIGHT: ♂ 310.

Field characters

Length 70 cm. Only African hornbill with white head and crest. Slender shape and long, graduated, white-tipped tail more suggestive of large cuckoo than hornbill, especially when bird is seen from behind as it slips through thick vegetation. Voice also distinctive.

Voice

Not very noisy, but calls have distinctive hollow quality. Main call a series of crowing notes followed by a rasping howl, *uoo-uoo-uoo-aah-aah*, rather like human imitation of a hyena whooping. Contact-note a short *squark*, given as birds move through forest. Short cough given in threat, accompanied by swinging tail back and forth. Loud *raw-crow* or *krouic* alarm uttered at avian and mammalian predators and heeded by monkeys (Willis 1983; Brosset and Erard 1986). Chicks beg with harsh, grating squeal. Tape recordings available.

Range, habitat, and status

T. a. albocristatus: S Guinea; Sierra Leone; Liberia; W Côte d'Ivoire. *T. a. macrourus*: E Côte d'Ivoire; Ghana; Togo; Benin. *T. a. cassini*: Nigeria; Cameroon; Rio Muni; Gabon; Congo; Zaïre; NW Uganda, in Bwamba forest; Central African Republic; N Angola, in Cabinda and Lunda provinces.

Prefers dense moist primary evergreen forest, but also readily enters small patches of adjacent secondary forest, strips of riparian forest, and even deciduous parkland. Ascends to 900 m on Mt Nimba and to 1500 m on Mt Kivu. Generally common. Most common where troops of monkeys occur, with which it is closely associated, but remains common even after monkeys eliminated (Dowsett-Lemaire and Dowsett 1991).

Feeding and general habits

Occurs in pairs accompanied by 1–3 young, under natural conditions each territory apparently coinciding with that of a monkey troop (Brosset and Erard 1986). Monkeys are alerted by hornbill alarm-calls and the association is mutually beneficial, since the hornbills habitually follow monkeys, army ants, or bird parties for the small animals they disturb (Bates 1930; Maclatchy 1937; Brosset and Dragesco 1967;

Long-tailed Hornbill *Tockus albocristatus*

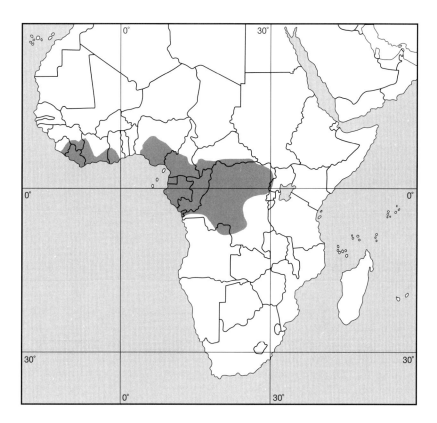

Willis 1983). This trait is shared by the squirrel *Protoxerus stangeri* (Chapin 1939).

Plucks most food in flight off subcanopy foliage and stems (89%, $n = 65$), regularly hunting among lower branches 5–30 m above ground. For the rest (11%), descends to the ground to snatch food (Willis 1983). Moves through the branches and trunks with graceful swerving flight. Often hunts in and moves through poorly lit areas, to which notably large eyes may be suited. Often hawks insects on the wing, especially those disturbed by monkeys, after a dextrous pursuit steered by the long tail. May remain perched very still and well concealed between flights, for as long as 30 mins and usually on medium-sized horizontal perch, slowly turning the head in search of food (Willis 1983). May also bound along branches (Willis 1983) or chase one another in flight (Bates 1930).

Diet largely insects, often rather small ones, but also small vertebrates such as ad and nestling birds, shrews, lizards, snakes, and frogs. Large cicadas, grasshoppers, and mantids are taken most regularly, also beetles, caterpillars, bugs, butterflies, moths, spiders, and slugs. Attacks small birds trapped in nets (Dowsett-Lemaire and Dowsett 1991). Of 27 stomachs, 17 contained only animal food, 2 only fruit, and 11 both (Chapin 1939; Germain *et al.* 1973). Also known to eat oil-palm fruits (Colston and Curry-Lindahl 1986), arils of *Trichilia heudelotii* (Dowsett-Lemaire and Dowsett 1991), and to descend to ground for fallen *Dacryodes* fruit (Brosset and Erard 1986).

Displays and breeding behaviour

Usually found in pairs, suggesting territoriality, or as small family groups. Noted to grapple with the bill in combat (Willis 1983).

Breeding and life cycle

Little known. ESTIMATED LAYING DATES: Liberia, Nov, Jan–Feb, ♂ in breeding condition Sept, Oct; Côte d'Ivoire, Apr; Ghana, Jan–Feb, Oct; Gabon, Oct–Dec; Congo, Oct; Zaïre, Jan–Feb, June–Aug, Dec. c/2. EGG SIZE: unrecorded. Eggs oval, white, and with pitted shells.

Five nest-holes were in natural cavities 10–16 m above the ground (Brosset and Erard 1986; Dowsett-Lemaire and Dowsett 1991), and one in a palm stem about 15 m up. The ♀ seals the entrance with her own droppings.

A ♂ was seen catching insects on the wing and taking them singly to a nest in the bill, another returning repeatedly to the nest with food from the direction of a nearby monkey troop (Brosset and Erard 1986). ♀ moults rectrices and remiges simultaneously while breeding (Chapin 1939). Usually only 1 chick reared, but brood of 2 recorded (Büttikofer 1885). An adult ♂ recorded as prey of Red-chested Goshawk *Accipiter toussenelii* (Chapin 1932; Brosset 1973). A captive survived at least 11.8 years (K. Brouwer *in litt.* 1992).

Genus *Ocyceros* Hume, 1873

Small hornbills, previously placed with 14 Afrotropical species in the genus *Tockus*. Differ from latter in lack of distinctive territorial display, in delivering food to the nest in the gullet and regurgitating at the nest, and in having a different sp. of *Chapinia* mallophagan louse as a parasite. Do share the same sp. of *Buceroemersonia* louse with *Tockus*. One species, *O. birostris*, has a distinctive pointed casque, short wings giving an unusual parrot-like flight, elongated pair of central tail feathers, an unusual wing:tarsus ratio, and shows

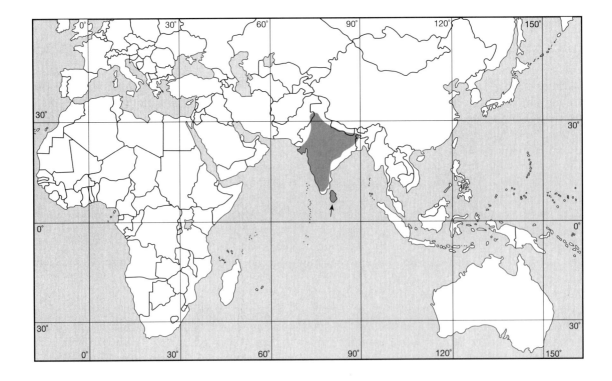

some similarities to species of *Anthracoceros*. Previously assigned erroneously to the genus *Meniceros* Bonaparte, 1854 (Kemp and Crowe 1985; Browning 1992).

Reported to breed as pairs alone, but at least *O. gingalensis* and *O. birostris* may be co-operative breeders at times. Otherwise resemble *Tockus* in most aspects of biology, but are much more frugivorous.

Total of 3 spp., all on Indian subcontinent: 1 widespread, 1 confined to W India, and 1 on Sri Lanka.

Malabar Grey Hornbill   *Ocyceros griseus*

Buceros griseus Latham, 1790. *Indian Ornithology*, **1**, 147. New Holland, restricted to environs of Bombay, India.

PLATES 6, 14

TAXONOMY: considered previously to be conspecific with *O. gingalensis*, although latter's differences in nostril shape and aspects of plumage, such as more grey-brown than slate, and soft-part coloration, such as purplish orbital skin, suggest specific status (Ali 1936; Legge 1983; Kemp 1988*b*).

Description
ADULT ♂: head, neck, and upperparts dark grey with white superciliary stripe and shafts of head and neck feathers. Underparts lighter grey, with white abdomen and fulvous undertail-coverts. Primaries and secondaries blackish grey, the former with white tips and bases. Tail blackish grey, all but the central two pairs of feathers with white tips. Bill, with oval nostrils, and low casque ridge yellow with orange at the base. Bare circumorbital and throat skin black. Eyes red-brown, legs and feet dark grey.

ADULT ♀: like ♂ in overall coloration, but smaller, especially casque; bill pale yellow with blackish patches on casque and sides of the lower mandibles.

IMMATURE: bill like ad ♀ but with less extensive blackish areas, plumage paler with only fulvous tint to undertail-coverts, fulvous edging to upperwing-coverts, and white edging to flight feathers. Bare facial skin undescribed. Eye grey, turning dirty yellow then brown; legs and feet grey-green. Remiges and rectrices more pointed than on ad.

MEASUREMENTS AND WEIGHTS
SIZE: wing, ♂ ($n = 14$) 205–220 (212), ♀ ($n = 7$) 185–201 (195); tail, ♂ ($n = 9$) 208–235 (222), ♀ ($n = 6$) 188–205 (196); bill, ♂ ($n = 11$) 97–110 (104), ♀ ($n = 6$) 72–87 (81); tarsus, ♂ ($n = 7$) 39–42 (40), ♀ ($n = 6$) 36–40 (38). WEIGHT: ♂ ($n = 4$) 238–340 (292).

Field characters
Length 45 cm. Slaty grey-brown hornbill without prominent casque; head and underparts streaked with white, broad white band from bill to behind eyes, white tips to primaries and outer tail feathers. Yellow bill, shorter tail, and more buoyant flight distinguish it from parapatric Indian Grey Hornbill *O. birostris*.

Voice
Incessant variety of loud, harsh croaks, chuckles, and mock laughter, varied by raucous cackling, reminiscent of pigs or of squawks of domestic chicken when held by legs (Ali 1936; Ali and Ripley 1970). Small chicks utter nasal *tain-tain-tain* all the time (Abdulali 1951). Tape recordings available.

Range, habitat, and status
India, on W Ghats from Kandala S to Palni and Cardamom Hills and on Salsette I. at Bombay. Bred as far N as Kanara (Baker 1927).

Open moist evergreen and deciduous forests from plains to 1600 m. Most common above 600 m, especially where figs abound, and also extends into gardens, tea plantations, and tall

154 Malabar Grey Hornbill *Ocyceros griseus*

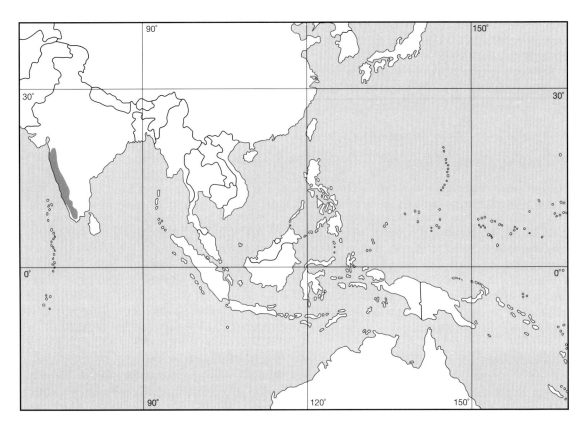

shade trees in cardamom plantations. Historically probably favoured coastal, valley, and ravine forests. In parapatry with *O. birostris* where forests interdigitate with drier deciduous woodland in a few areas (e.g. E of Sirsi: Davidson 1891, 1898). Declining in northern half of range (S. A. Hussain *in litt.* 1991).

Feeding and general habits
Found in flocks of 5–20, moving in follow-my-leader fashion, forming larger congregations with other frugivores at fruiting fig trees, where keeps up incessant calling. Diet includes figs, drupes, berries, insects, and lizards (Ali and Ripley 1970).

Displays and breeding behaviour
Monogamous, apparently without helpers. Tail somewhat fanned to display white tips while calling (Kemp 1988*b*).

Breeding and life cycle
Little known. ESTIMATED LAYING DATES: India, Jan–May, at end of dry season. c/2–4. EGG SIZES: (n = 50) 35.5–46.0 × 27.0–31.8 (41.8 × 30.3) (Baker 1934). Eggs white with pitted shells.

A nest at Kandala was 12 m up in a natural hole in a tree (Abdulali 1942, 1951). ♂ carried dragonfly to nest in bill and regurgitated two red berries, tapping with bill on arrival (when shot, had about 15 *Ixora* berries in gullet, remains of 40 more in stomach). ♀ in the nest was brooding 3 chicks and an addled egg. Its flight feathers had been moulted simultaneously, but for 4 rectrices which remained (probably new feathers), but there was no body moult. Faeces were ejected through the sealed entrance slit.

Sri Lankan Grey Hornbill *Ocyceros gingalensis*

Buceros Gingalensis Shaw, 1811. *General Zoology*, **8**, 37. Ceylon, restricted to Colombo, Sri Lanka.

PLATES 2a, 6, 14

Other English name: Ceylon Grey Hornbill

TAXONOMY: previously considered conspecific with *O. griseus*; differs in nostril shape and in plumage and soft-part coloration, but further details necessary. Possible specific status long recognized (Ali 1936).

Description

ADULT ♂: head, neck, and upperparts dark grey with white shafts to head and neck feathers. Underparts white. Primaries and secondaries blackish grey, the former with white tips and bases. Tail blackish grey, all but the central pair of feathers with broad white tips, outer 3 pairs becoming almost pure white in older birds (Ali and Ripley 1970). Bill, with round nostrils, and low casque ridge cream with black patches at the base of the upper mandible and underside of lower mandible. Bare circumorbital and throat skin dark purple. Eyes red, legs and feet greenish grey.

ADULT ♀: like ♂ in overall coloration, but smaller, especially casque; black bill with cream stripe along base of upper mandible. Bare facial skin undescribed; eyes brown.

IMMATURE: bill pale greenish, narrower white tips to rectrices than ads. Bare facial skin undescribed. Eye blue-grey just before fledging.

MOULT

Commencing primary moult in Jan (Whistler 1944).

MEASUREMENTS AND WEIGHTS

SIZE: wing, ♂ (n = 15) 200–220 (210), ♀ (n = 13) 191–203 (197); tail, ♂ (n = 13) 208–236 (220), ♀ (n = 11) 186–213 (200); bill, ♂ (n = 13) 92–110 (103), ♀ (n = 12) 79–90 (84); tarsus, ♂ (n = 15) 38–42 (40), ♀ (n = 8) 37–40 (39). WEIGHT: ♂ 238 (Ali and Ripley 1970).

Field characters

Length 45 cm. The only small, grey-brown hornbill on Sri Lanka, the white tips to the flight feathers and loud voice distinctive. Occurs alongside the much larger Indian Pied Hornbill *Anthracoceros coronatus*.

Voice

Loud *kaa ... kaa ... kakakaka* by ♂, or *kuk ... kuk ... kuk-kuk-kuk ko ko kokoko* answered by mate in similar tones (Henry 1978). Main call a hard *ka* repeated (de Zylva 1984), or harsh laughing *kaa* syllable repeated at accelerating rate. No tape recordings located.

Range, habitat, and status

Sri Lanka. Lowland evergreen forest and deciduous woodland, up to 1200 m in S. Common breeding resident in moist and dry lowland forest, moving seasonally (Sept–Oct) to hills after breeding (Phillips 1978).

Feeding and general habits

Occurs as pairs or family parties, but small flocks may form at fruiting trees. Shy, with slow dipping flight. Often sits quietly and upright among foliage, only moving the head in search of food. Prefers middle levels of forest where creepers abound.

Diet mainly fruit, such as small banyan and bo figs, wild cinnamon, wild nutmeg, and dawata *Carallia integerrima*. Also takes small animals such as green lizards, tree frogs, scorpions, and insects such as mantids (Henry 1978; Legge 1983).

156 Sri Lankan Grey Hornbill *Ocyceros gingalensis*

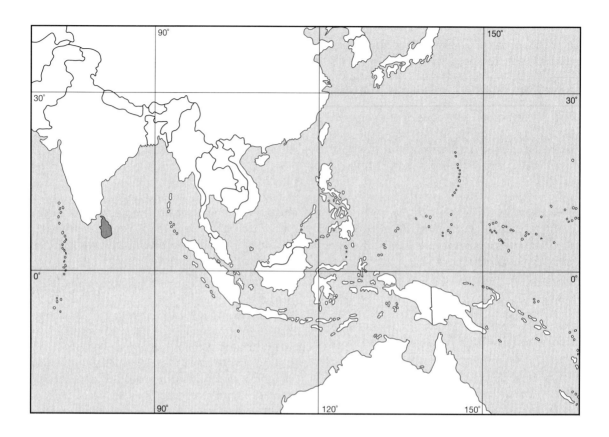

Displays and breeding behaviour
Monogamous, but possibly with helpers at times (based on distinctive coloration of juv).

Breeding and life cycle
Little known. Incubation period 28–30 days (Phillips 1979). ESTIMATED LAYING DATES: Sri Lanka, mid-Mar to early July, some pairs as late as Oct (Phillips 1979), with peak in May (45 %, n = 16). c/1–3. EGG SIZES: (n = 11) 37.4–44.5 × 21.0–34.8 (39.9 × 28.7). Eggs white with pitted shells.

Nests 2–21 m up in natural holes in trees, most at 4.5–10.5 m. Floor up to 50 cm below entrance but usually much less (Phillips 1979). One nest was 9 m up in top of trunk of lone kumbuk tree and used in two successive years; the cavity was about 46 cm in diameter, the chicks disappearing, possibly up a funk-hole, when the nest was examined (Punchihewa 1968). Also uses old cavities of Crimson-backed Woodpecker *Chrysocolaptes lucidus* as nest-holes (Phillips 1979), the entrance often just large enough for the ♀ to enter.

One ♂ fed incubating ♀ at 06.30, 09.30, 11.00, 12.30, 15.00, 17.00, and 18.00 hr with gulletfuls of fruit and small insects, regurgitated over about 2 min per visit (Punchihewa 1968). Also seen to present a live garden lizard. ♂ carries large animal food items to nest in bill, but regurgitates fruits. ♀ emerged about a month after chicks hatched, but did not help ♂ to feed chicks until about a month after emergence. ♀ undergoes simultaneous moult of flight feathers when breeding (Henry 1978).

Indian Grey Hornbill *Ocyceros birostris*

Buceros birostris Scopoli, 1786. *Del. Flor. Faun. Insubr.*, **2**, 87. Coromandel, restricted to Pondichery, India.

PLATES 6, 14
FIGURE 1.1

Other English names: Grey Hornbill, Common Grey Hornbill

Description

ADULT ♂: upperparts medium grey. Head, neck, and underparts pale grey, with dark grey patch on cheeks. Primaries, secondaries, and tail dark grey, primaries and all but 5 inner secondaries with white tips, outer 5 primaries with additional narrow white stripe in centre of leading edge. Tail with darker grey band preceding white tips, and central pair of feathers elongated. Bill, with serrations along cutting edge of upper mandible and casque a slender cylinder with tip protruding above bill, dark brown with pale yellow tip and underside. Bare circumorbital skin lead-coloured. Eyes red-brown to orange, legs and feet grey-brown.

ADULT ♀: like ♂ in overall coloration, but smaller, with less protruding casque; primaries without white tips but with broader white stripes. Circumorbital skin undescribed; eyes brown.

IMMATURE: like ad ♀ in plumage, including lacking white tips to primaries. Bill pale yellow; circumorbital skin and eye colours undescribed.

SUBSPECIES

Birds from N India are reported to be consistently greyer than those on the S peninsula and have been recognized as *O. b. pergriseus* Koelz, 1939, a distinction supported by Biswas (1961).

MEASUREMENTS AND WEIGHTS

SIZE: wing, ♂ ($n = 17$) 209–228 (220), ♀ ($n = 10$) 195–211 (203); tail, ♂ ($n = 14$) 250–295 (270), ♀ ($n = 7$) 246–280 (259); bill, ♂ ($n = 17$) 96–115 (102), ♀ ($n = 8$) 80–97 (88); tarsus, ♂ ($n = 15$) 42–46 (44), ♀ ($n = 8$) 38–44 (41). WEIGHT: ♂ 375.

Field characters

Length 50 cm. Grey-brown bird with dark brown bill and protruding casque; long graduated tail, with pattern of black band and white tip exposed on landing. Flies clumsily with fast, short, parrot-like wingbeats, with tips upturned, and interspersed with bouts of gliding. ♀ with smaller casque and imm with yellow bill, and both lacking white tips to primaries. Flight, calls, long tail, and dark bill distinguish it from parapatric Malabar Grey Hornbill *O. griseus* on W of peninsula.

Voice

Territorial call described as loud cackling and squealing notes *k-k-k-ka-e* (Ali and Ripley 1970), or short, rapid, piping notes *pi-pi-pi-pi-pipipieu-pipipieu-pipipieu* (Roberts 1991). Also shrill monotonous squeal, various loud cackling and squealing notes (Ali and Ripley 1970), or weird squeals in alarm if flushed. Normal contact-call a shrill, monotonous, kite-like squeal, *wheee*, repeated. Also hoarse growling call like a griffon vulture. Noisy in early morning, calling when flying or perched. ♀ and chicks utter screech of acceptance when taking food (Hall 1918; Lowther 1942). Wings produce low buzzing noise in flight (Roberts 1991). Tape recordings available (Roberts 1991).

Range, habitat, and status

NE Pakistan (Roberts 1991); India, W to Kathiawar, Mt Abu, N along base of Himalayas up to 1400 m in Kumaon (Bhim Tal), E to Rajmahal Hills and Singhbhum, S throughout except arid parts of Rajasthan and Gujarat (Kutch) and moist areas of Kerala; S Nepal, in lowlands below 500 m (Inskipp and Inskipp 1985); NW Bangladesh (Sanft 1960). Parapatric with and then replaced in extreme SW by *O. griseus*.

Indian Grey Hornbill *Ocyceros birostris*

Resident and often common, but with local movements during summer to follow fruiting trees, which may also lead to breeding areas. Occupies deciduous parkland and open thorn forest, favouring open woodland with figs, as in many rural areas of cultivation (Ali and Ripley 1970). Regular breeder in gardens, generally on open plains, around topes and avenues, and in villages (Lowther 1942). Extinct in Kathiawar, including Gir Forest. Common breeding resident in Delhi (MacDonald 1960).

One ♀ in heavy moult was taken for medicinal purposes (Horne 1869). Ad and chicks killed for food and medicine, broth of entire bird, including feathers, administered internally or as lotion to cure post-childbirth pains in women.

Feeding and general habits

Occurs as pairs or family groups of up to 8. Forms assemblages of up to 30 at fruiting banyan or peepul figs, together with many other frugivorous birds, but departs in original small groups. Moves short distances from tree to tree, flying with fast beats of the relatively short wings; often utters harsh note in flight. Uses the bill to assist with climbing to reach fruit (Horne 1869). Mainly arboreal, but descends to ground for fallen fruit or termite alates, or to dust-bathe (Santharam 1990), and then hops clumsily about with tail cocked (Ali and Ripley 1970). Sometimes leaps up from perch to snatch flying insects.

Small figs the staple diet, but also eats other berries, poisonous fruit of Yellow Oleander *Thevetia neriifolia*, *Bauhinia* petals, together with beetles, mantids, dragonflies, grasshoppers, wasps, lizards, mice, and parakeet nestlings extracted from tree hole (Newnham 1911). In captivity, takes live fish.

Displays and breeding behaviour

Monogamous, possibly with helpers at times (judging from distinctive colours of juv). At

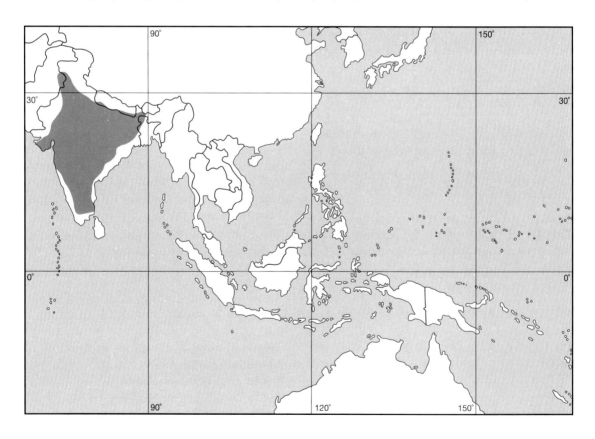

one nest, 2 ♂♂ were present but only one took food to the nest (Horne 1869). Territorial display conducted with loud calling, the tail depressed under the perch and swung slowly back and forth, and the bill held skywards. ♂ courtship-feeds ♀, sometimes after a hopping approach along a branch (MacDonald 1960).

Breeding and life cycle
ESTIMATED NESTING CYCLE: at least 66 days (incubation period at least 21 days, ♀ emerges after at least 45 days) (Lowther 1942). ESTIMATED LAYING DATES: India (S, at Sirsi), Mar, (N, at Lahore) May–June, with different local peaks; Nepal, Mar–June; Pakistan, late Apr. Most laying at end of dry season. c/2–5, usually 3, rarely more. EGG SIZES: ($n = 30$) 39.1–46.0 × 27.5–32.0 (41.9 × 30.0) (Baker 1927). Eggs matt white with pitted shells, becoming dirty in nest. Like eggs of *O. griseus*, but latter usually only c/2–3 and laid a month earlier (Davidson 1891, 1898).

Nests in natural tree holes 3–8 m up, including in semul *Bombax heptaphyllum* (2, both used previously by parrots), aroo *Ailanthes excelse*, peepul *Ficus religiosa*, and sisso *Dahlbergia sissoo* (Horne 1869), and exotic trees such as mangoes, jaman, and nim. Sometimes enlarges entrance hole slightly and often selects cavities with a funk-hole present. Sites often used in successive seasons, one over at least seven. Some conflict with other hole-nesters, such as rollers, common mynahs, and parrots *Psittacula krameri* (Hall 1918). Two nests were found 29 km apart in Punjab, where the sp. was described as rare (Finlay 1928), another two within close proximity of each other (Hall 1918).

One nest with cavity 25 cm deep and entrance 12 × 8 cm (Horne 1869), another with entrance 15 × 9 cm (Hume and Oates 1890), another sealed to 10 × 1.2-cm slit and lined with bark flakes (Finlay 1928). ♀ seals hole and ♂ delivers wet earth, but ♀ also uses droppings and food remains. At two nests the sealing was completed in 2–3 days (Horne 1869; Hall 1918).

Pre-laying interval of 7–10 days; clutch incubated from laying of first egg, and difference in chick size of up to 14 days suggests variable laying interval (Lowther 1942; Roberts 1991). ♂ regurgitates to ♀ from gullet, including figs, berries, green leaves, tamarind and bean pods, locusts without appendages, ants, lizards, dove's egg, and nestlings. Up to 24 peepul figs and a leaf, 22 nim berries, or 7 locusts in one load. During a 14-hr watch when small chicks in the nest, a ♂ delivered 12 times (Lowther 1942).

Breeding ♀ moults flight feathers almost simultaneously, but not the body feathers (Osborne 1904). ♀ leaves nest when chicks estimated to be 7–14 days old. One ♀ took 2 hours to break out of the sealing, making deliberate wing- and leg-stretching movements once out and then flying off with ♂ for several hours (Lowther 1942). Nest resealed, apparently by chicks as sealing less neatly applied. On emergence of ♀, both parents feed the chicks, and later families leave the nest area immediately after fledging (Hall 1918). ♀ recorded as thin after laying and fat at emergence, ♂ declining in condition during nesting cycle (Lowther 1942).

Genus *Anthracoceros* Reichenbach, 1849

Medium-sized hornbills, the genus recently reviewed in detail with detailed plumage descriptions, measurements, and distributions (Frith and Frith 1983a). Both sexes have a well-developed casque, a helmet-like curved cylinder becoming laterally flattened towards tip, with longitudinal grooving in one sp. Imms resemble ads but for less developed bill and casque. Outer two primaries (9 and 10) emarginated at tip. Extensive bare patches around eyes and on

Genus *Anthraceros*

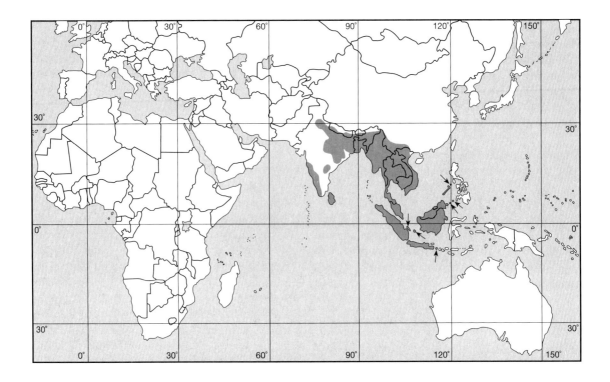

throat in all but one sp., *A. montani*. The latter is also the only sp. in the genus with sexually dimorphic eye colours. It most resembles in form the other most aberrant sp., *A. malayanus*, which has the central pair of tail feathers slightly elongated, a character shared with *A. coronatus* and *A. albirostris*.

Breed monogamously without helpers, but biology of two isolated island endemics (*A. marchei, A. montani*) poorly known. Most spp. occur in relatively open forest and along forest edge. Breed in tree holes as pairs, despite often flocking when feeding. ♀ seals nest, with some sealing material delivered by ♂. ♂ delivers food to nest in gullet, regurgitating before passing items in to inmates, although *A. malayanus* often delivers single items carried in bill tip and other spp. bring sealing material in the bill. ♀ undergoes simultaneous moult of rectrices and remiges during incubation. ♀ emerges during nestling period to assist ♂ in feeding young (*A. coronatus*) or, more often, at end of nesting cycle (*A. albirostris, A. malayanus*). Chick retains pink or pinky yellow skin after hatching.

Genus contains 5 spp., on the Indian subcontinent and on islands of the Sunda basin and of the Palawan and Sulu archipelagos. Of these, 3 spp. form an obvious supersp., the Indomalayan Pied Hornbills (*A. coronatus*, *A. albirostris*, and *A. marchei*), which also share similar calls and *Chapinia acutovulvata* feather-lice. The other 2 spp. are quite different and possibly not even congeneric. Not a very cohesive genus, overlapping with *Ocyceros* in elongated central tail feathers and related lice, or with *Buceros* in limited and similar sexual dimorphism in casque form and colour and with sexually dimorphic eye colour (*A. montani*, which also has its own *Chapinia* louse sp.).

Indian Pied Hornbill   *Anthracoceros coronatus*

Buceros coronatus Boddaert, 1783. *Table des Planches Enluminées d'Histoire Naturelle, de M. D'Aubenton*, p. 53 (after Daubenton, 1788, *Planches Enluminées*, plate 873). Philippines, designated as Mahé, Calicut Province, India.

PLATES 7, 14

Other English names: Malabar Pied Hornbill, Pied Hornbill

TAXONOMY: Was combined with *A. albirostris* in complex synonymy of common and scientific names, but is much larger, less sexually dimorphic, with different throat-skin colour, and about half as many individuals show black marks in white outer tail feathers (Kemp 1979; Frith and Frith 1983*a*).

Description

ADULT ♂: head, neck, upper breast, back, and wings black, glossed with greenish blue. Breast, thighs, abdomen, and underwing-coverts white, underparts with black feather bases. Primaries, except outer two, and secondaries, except basal three, white with black bases. Tail white but for central pair of rectrices, which are black and 25–40 mm longer than the rest; individual variation in extent of slight black markings in white outer tail feathers. Bill, with casque an axe-like structure ridged along the sides that originates as a cylinder above the forehead and ends as projecting blade near end of bill, yellow with black base to casque and bill and extensive black patch over end of casque: extent of black patch individually variable and increases with age (Frith and Frith 1983*a*). Bare circumorbital skin blue-black and throat patches flesh-coloured, eyelids black. Eyes red to orange-red, legs and feet dark greenish grey.

ADULT ♀: like ♂ in overall coloration, but smaller, although only slightly so in bill:casque proportions. No black base to casque, and black patch on casque smaller. Circumorbital skin white with pink tinge, gular skin flesh-coloured, eyes brown.

IMMATURE: like ad but for black bases to white outer tail feathers, and more often with black markings in the tail than ad (46% against 14%), but some with all-white outer tail feathers as in ad. Bill smaller, with casque undeveloped, and with less black; bare facial skin like ad of respective sex, eyes grey-brown. Black on casque increases and that on outer tail feathers decreases with age.

MEASUREMENTS AND WEIGHTS
SIZE: (Frith and Frith 1983*a*) wing, ♂ ($n = 28$) 320–360 (343), ♀ ($n = 16$) 309–333 (320); tail, ♂ ($n = 26$) 270–336 (314), ♀ ($n = 15$) 272–324 (304); bill, ♂ ($n = 29$) 171–221 (198), ♀ ($n = 17$) 154–189 (175); tarsus, ♂ ($n = 28$) 54–67 (61), ♀ ($n = 15$) 48–63 (57). WEIGHT: unrecorded. No variation in size between 3 discrete populations (E India, W India, and Sri Lanka), but ♀♀ of latter with longer tails relative to ♂♂ than in other populations.

Field characters

Length 65 cm. Black, with white underparts and entire outer tail feathers. Extensive white tips and trailing edges to wings obvious in flight. Large yellow and black bill, and casque with flattened projecting tip, distinctive. Notably larger, less sexually dimorphic, and with different throat-skin colour than parapatric *A. albirostris* in NE India, and only about half as many individuals with black marks in white outer tail feathers. Much larger than drab sympatric *Ocyceros* spp., the Indian Grey and Malabar Grey Hornbills.

Voice

A variety of loud, shrill squeals and raucous cackles, or loud cackling *kleng-keng, kek-kek-kek-kek-kek-kek* rising in pitch and volume, or rapid piercing *kak-kak, kak-kak*, similar to *A. albirostris* but deeper in tone and less modulated

(Ali and Ripley 1970). Utters loud whistle on arrival at roost, or in response to others, *kak-kak-kak-kaa-kaa-kaa*, uttered with exaggerated raising and lowering of bill. At dawn calls fast series of up 163 clucks ending with few deep whistles, clucks heard rarely for rest of day (Modse 1988). Tape recordings available.

Range, habitat, and status
India; Sri Lanka. Occurs in 3 separate populations. One in E Indian states of Bihar, Orissa, Madhya Pradesh, and Andhra Pradesh; a second on W side of peninsular India, along foothills of Western Ghats from Aroli and Ratnagiri, S of Bombay, through Goa, W Mysore (Malnaad), W Tamil Nadu (Nilgiri District), and Kerala; the third on Sri Lanka and Jaffa Is (Sanft 1960; Frith and Frith 1983*a*). Status of further populations now isolated in intervening habitat requires investigation.

Occupies periphery of moist evergreen and deciduous forests, also less dense forest on foothills and ridges and including bamboo forest, plantations, and groves of figs and mangoes. Not in hills of extreme S where *Buceros bicornis* found (Legge 1983). Mainly on lowlands in Sri Lanka, but at up to 750 m at NE Monsum. Overlaps with *A. albirostris* in E Chota Nagpur (S Bihar) and adjacent hill forest of Orissa and W Bengal; no hybrids known.

Common in a few areas but frequently reduced through habitat destruction, especially in W India and Sri Lanka (Modse 1988; R. Kannan *in litt*. 1992). On Sri Lanka is now confined to more secluded areas, not nearly so common as 20–50 years ago, with few near hill country, and uncertainty if it still ascends into the hills after breeding (Phillips 1979; Legge 1983). In Orissa, large numbers of chicks are removed annually from nests for medicinal purposes, to the extent that a captive-breeding programme has been initiated for the species' conservation (Dev 1992).

Feeding and general habits
Studied in detail only in N Kanara, W Ghats, India (Modse 1988). Occurs in flocks of 4–58, in mixed deciduous forests at 300–600 m. Recorded in all monthly transects, with peak number of 74 recorded in Mar and most common in Dandeli forest, where fruiting trees also most common, with peak fruiting in Nov–Apr, especially *Strychnos nux-vomica* (but for *Ficus glomerata* that fruited throughout the year).

Roosted communally throughout the year in dense riverine foliage, usually in inner branches over water in an evergreen *Holigarna arnotiana*, a bamboo patch, or a deciduous *Terminalia arjuna*, the roost being over huts and by a busy temple and bridge. Gathered both pre- and post-roosting in trees on the opposite bank, some 120 m away. The roost had been in use for at least 20 years and the same perch sites were used over two years, with only minor local movement if monkeys in trees at roost time. Peak numbers of 44 in Apr, dropping to 3–12 in Jun–Sept when breeding and with only non-breeders and subads coming to roost. Breeding ♂♂ apparently roost near their nests.

Leave roost in groups of 1–3 immediately on waking, all following leader, usually in pairs. Depart 36 mins before to 6 mins after sunrise: general but not specific correlation with light intensity, but earlier on hot mornings with low humidity. Post-roost gathering for 2–13 mins, pre-roost only 2–4 mins, when spend time preening, allopreening, scratching, stretching (especially mornings), bill-cleaning, bill-banging, billing one another (especially afternoons and between pairs in Feb mating period), and allofeeding (especially afternoons, with growling during food-pass). Fly into top of roost tree and hop down to roosts at dark, entering after sunset with more variation than at dawn emergence. Mean period of diurnal activity at about 15°N: 11.32–13.32 hrs.

Forages in dispersed small flocks, some gathering together at fruiting trees. Foraging linked to sunrise–sunset times, unless delayed by rains, with termination less fixed than onset. Starts later and ends earlier with increasing daylength. Forages mainly arboreally, but will descend to ground for small animal food. Seasonal variation in foraging sites, area, and distances moved from roost, moving out 0.4–4 km over 3.97 km² in summer, when feeding mainly on *Strychnos* and *Ficus* along river banks, but 1–6 km over 8.31 km² in monsoon and winter, when few fruits along river. On Sri Lanka also favours foraging along riparian trees, where eats fruits of ironwood, even taking them from the ground (Legge 1983).

Feeds in various ways (Modse 1988): plucking and twisting off fruits when perched; pecking small insects such as termites out of runways on branches or on the ground; breaking orange-sized drupes (e.g. *Strychnos*) and capsules (e.g. *Swietenia*) by cracking in or hammering with the bill; swooping down on *Calotes* lizards, grasshoppers, beetles, insect larvae, and snails on ground; hawking flying insects such as cicadas; hopping in pursuit along the ground (but taking no fallen fruits); and hovering for 3–5 sec to take inaccessible fruits. Forages most actively in morning and during the monsoon (63% of time, versus winter 47% and summer 40%). May also pluck fruits in flight, sometimes hanging by bill if fruit too well attached, or using underside of bill as grapple to pull itself onto branches.

During 12–13 Feb, a marked ♀ in a pair fed almost solely on cherry-sized fruits of *Machilus macrantha*, with little animal food: fed 42% and rested 58% of 11-hr day, eating 166 fruits (97% of items) in 47% of foraging time and 4–5 small animals (3% of items) in 53% of time (Modse 1988). Preens, bill-cleans and stretches when resting, moving into denser foliage during the heat of day, gaping, and sometimes having to suspend foraging owing to heat.

Overall frugivorous, with animal food taken mainly when fruits limited. Total of 17 fruit spp. (for which moisture, lipid, protein, and carbohydrate measured), leaves of *Melia composita*, and 9 types of animal food recorded in

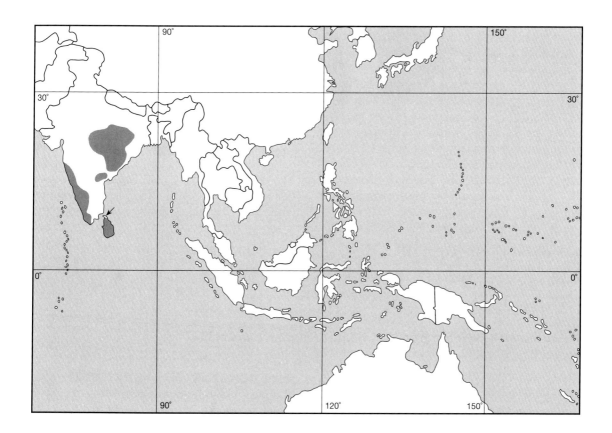

diet (Modse 1988). Plants (68% by items) included: *Grewia tiliifolia, Mallotus philippensis, Ficus* (20% of items: *F. infectoria, F. glomerata, F. religiosa, F. bengalensis, F. rumphi*), *Strychnos nux-vomica, Polyalthia fragrans, P. longifolia, Schleichera oleosa, Machilus macrantha, Caryota urens, Melia composita* (fruit and leaves), *Swietenia mahagoni* (placenta of fruit only), *Syzygium cuminii, Buchanania* spp., and also reported to eat *Ardina cordifolia* and to raid chilli fields. Animals (32% by items) included: terrestrial snails, beetles, Lepidoptera larvae, grasshoppers, cicadas, *Mabuya* skinks, *Calotes* lizards, and termites. Bird's egg in one stomach (Mason and Lefroy 1912). Never seen drinking.

Ficus and *Strychnos* were the major fruit items in SW India, the pulp of the latter (which contains alkaloids poisonous to humans) being eaten. Most fruits of diameter 9–29 mm: *Strychnos* (51 mm) and *Melia* (24 mm) are broken open only to obtain pulp and placenta respectively. Max number of fruits taken per visit to a tree varies with size (*Strychnos* 5, *Grewia* 79), ranging in weight from about 6 to 118 g per bout (highest recorded 118 g for *Ficus glomerata*, 102 g for *Strychnos*, 98 g for *Machilus*). High *Ficus* consumption may also relate to water balance since have high water content. Least favoured, with low water and large seeds, are *Melia, P. longifolia, Caryota*, and *Mallotus*. Dry placenta, with low lipid and protein, of *Swietenia* is countered by high carbohydrate content.

Displays and breeding behaviour
Monogamous in separate pairs, otherwise undescribed.

Breeding and life cycle
ESTIMATED NESTING CYCLE: 80 days: 29–30 days incubation, 49 days (5–6 weeks) nestling period. ESTIMATED LAYING DATES: E India (Travancore), Mar; W India, Mar–Apr (Ali and Ripley 1970), Jun–Sept (Modse 1988); Sri Lanka, Apr–July (Ali and Ripley 1970), peak mid-Apr to early May (Phillips 1979); laying during rainy season. Captives: S Carolina, May. c/2–4, c/2 normal for Sri Lanka (Phillips 1979). EGG SIZES: (n = 7) 49.7–56.2 × 36.2–41.3 (52.5 × 38.2) (Sanft 1960). Egg white with pitted shell.

Nests in natural cavities in trees, entrance 3.5–15 m above ground. Eggs may be laid on floor of cavity or on a ledge near the entrance, the ♀ climbing up to take food if the nest-floor too deep. ♀ seals nest with little help from ♂ (Phillips 1979). ♀ undergoes simultaneous flight-feather moult (Osmaston 1913).

Apparently some variation in timing of ♀'s emergence. A wild ♀ broke out of the sealed nest when the two chicks were estimated at only 10–14 days old. The ♀ made yapping calls from the entrance when the ♂ was nearby, emerging stiffly to flap her wings and stretch, before flying off with the ♂ for several hours. The chicks probably resealed the nest alone (Lowther 1942). At another nest, the chicks emerged 1–2 weeks after the ♀ (Phillips 1979).

Oriental Pied Hornbill *Anthracoceros albirostris*

Buceros albirostris Shaw and Nodder, 1790. *Nat. Misc.,* **19**, plate 809 (after Levaillant, 1801, *Histoire Naturelle d'Oiseaux Americaine* **1**, 29, plate 14). Chandernagore, Bengal, India.

PLATES 7, 14
FIGURES 1.1, 3.4

Other English names: Northern Pied Hornbill, Sunda Pied Hornbill, Indian Pied Hornbill, Malaysian Pied Hornbill

TAXONOMY: Species was combined with *A. coronatus* in complex synonymy of common and scientific names, but is much smaller, more sexually dimorphic, with different throat-skin colour, and with far fewer than half as many individuals showing black marks in white outer tail feathers (Kemp 1979; Frith and Frith 1983*a*).

Oriental Pied Hornbill *Anthracoceros albirostris*

Description

A. a. albirostris (Shaw and Nodder), 1790. Asian Pied Hornbill.

ADULT ♂: head, neck, upper breast, back, and wings black with greenish gloss. Breast, thighs, abdomen, underwing-coverts, and tips to all but outer two primaries and basal three secondaries white. Tail black with broad white tips to all but central pair of feathers, which protrude up to 30 mm and which in only a few individuals have narrow white tips (Frith and Frith 1983*a*). Bill, with casque a cylinder from above base and tapering to blade midway along bill, yellow, with black base to both casque and bill and extensive black patch over end of casque and extending on to bill (extent of black variable, and increases with age). Bare circumorbital skin white with black spot in front of eye, and throat patches white with blue tinge. Eyes dark red, legs and feet dull greenish grey.

ADULT ♀: like ♂ in overall coloration, but smaller, casque less convex without projecting tip; bill and casque yellow, with distal half black, and brown patches at sides and base of lower mandible (dark areas more individually variable than on ♂). Eyes brown to grey-brown.

IMMATURE: like ad but without gloss on upperparts; tail feathers often with decreasing white tips towards centre, and more often with black marks on white (20% versus 3%). Bill smaller, with casque undeveloped, and all pale yellow; circumorbital skin white, tinged pink; throat patches fleshy blue-white; eyes dark brown.

SUBADULT: casque development begins 1–2 months after fledging and is basically complete by 12–14 months old, but still some growth at 3 years old (Gee 1933*b*; Frith and Frith 1978). White areas on the tail increase with age.

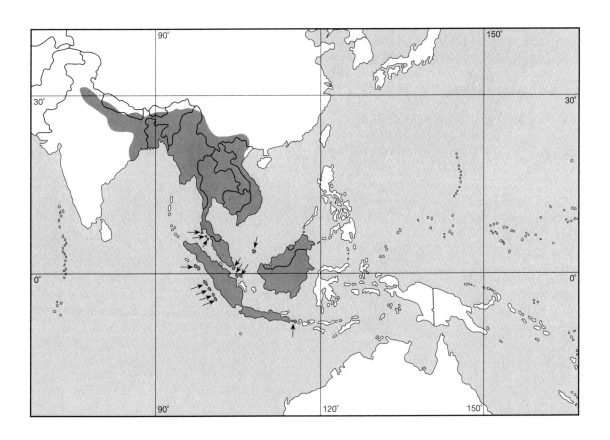

Oriental Pied Hornbill *Anthracoceros albirostris*

SUBSPECIES

A. a. convexus (Temminck), 1831. Sunda Pied Hornbill. Plumage same as *T. a. albirostris*, but more frequent flecks of black in white areas of tail of many specimens (23% ads, 88% imms) and asymmetrical junction of black and white. Also slightly larger, especially in bill and casque (Frith and Frith 1983a). Bill colour and casque form like nominate subsp., but generally less black on bill of ad ♀ despite considerable local and geographical variation, some of the latter showing discrete island forms. Circumorbital skin of ad apparently lacks black spots before eye. Eye colour of both sexes light cinnamon-brown (Bartels and Bartels 1937), of juv grey-brown, and leg and foot colour of both sexes and ages dull greyish green or black (Bartels and Bartels 1937; Blasius 1882). Probably age variation in decreasing black in tail and deepening eye colour, as in nominate subsp. (Frith and Frith 1983a).

MOULT

Twenty specimens from Burma and Indo-China with no moult Nov–Jan in dry season, when not breeding. Juv still in postnatal moult in June and ♀ in moult in July, in Thailand (Deignan 1945). Juv tail feathers narrower than ad.

MEASUREMENTS AND WEIGHTS

A. a. albirostris SIZE: (Frith and Frith 1983a) wing, ♂ (n = 101) 234–358 (299), ♀ (n = 85) 249–323 (279); tail, ♂ (n = 99) 209–316 (265), ♀ (n = 84) 210–305 (249); bill, ♂ (n = 101) 113–187 (154), ♀ (n = 86) 110–156 (129); tarsus, ♂ (n = 99) 42–61 (52), ♀ (n = 78) 38–57 (49). WEIGHT: ♂ 680–795; F 567–680. Decrease in size in cline from N to S.

A. a. convexus SIZE: wing, ♂ (n = 82) 288–340, ♀ (n = 39) 267–298 (281); tail, ♂ (n = 79) 240–310 (273), ♀ (n = 38) 219–285 (252); bill, ♂ (n = 82) 155–197 (175), ♀ (n = 39) 124–153 (138); tarsus, ♂ (n = 81) 45–61 (55), ♀ (n = 38) 42–58 (51). WEIGHT: (Sumatra: Ripley 1944) ♂ 907, ♀ 879. Some inter-island variation, but also considerable local individual variation.

Field characters

Length 55–60 cm. Black, with white underparts and tip of tail. Bill, with large cylindrical casque, yellow with black patch at end of casque in both sexes. ♀ with more black on bill and brown patches on sides. Imm with casque undeveloped. White trailing edge of wing and tail conspicuous in flight. Flaps with short, shallow wingbeats, then dips as glides, but not over long distances like larger spp., moves more in short flights over habitat. Audible in flight, but not noisy like larger spp.

Voice

Variety of loud, shrill, nasal squeals and raucous chuckles, including loud cackling *kleng-keng*, *kek-kek-kek-kek-kek* rising and falling in pitch and volume (Medway and Wells 1976) and uttered by both sexes. Also low, soft clucking *kuck-kuck-kuck* or still softer *kiew-kiew-kiew* (Frith and Douglas 1978), and low continuous growling during food-presentation. Loud alarm-cries when nest examined, and ♂ makes soft clucking notes, *tuuk tuuk tuuk*, while ♀ working on nest-hole (Bartels and Bartels 1937). Tape recordings available.

Range, habitat, and status

A. a. albirostris: India, in Himalayan foothills from Dehra Dun in Bihar, E Bengal to Assam; S Nepal: S Bhutan; Nepal; N Bangladesh; Myanmar, plus islands in Mergui Archipelago of Salanga, Kadan Kyun, and Thayawthadangyi Kyun; China, in W Yunnan and S Xishuangbanna provinces plus Guangxi and S Zhuang Aca regions (Tso-Hsin 1987); Vietnam, including islands of Hon Thom and Quan Phu Quoc; Laos; Cambodia; Thailand, to Phatthalung plus islands of Ko Chang, Phuket, and Tarutao; NE Peninsula Malaysia, along W coastal plain to Kedah at Alor Star and Mabek and including islands of Langkawi and Pulau Butang (Sanft 1960; Frith and Frith 1983a).

Forest edge and open moist deciduous and evergreen forests, at up to 670 m in N Thailand (Deignan 1945). Includes coastal, riverine, secondary, and selectively logged forests, offshore islands, bamboo brakes, forested islands, wooded gardens, rice-paddies, and irrigated fields. In S Nepal, moves locally with animal and fruit availability, common but declining owing to deforestation (Inskipp and Inskipp 1985). Overlaps with *A. coronatus* in E Chota Nagpur (S Bihar) and adjacent hill forest of Orissa and W Bengal; no hybrids known. Only hornbill in Yunnan (Smythies 1986), and, historically, one or two pets in every village in Arakan, Tenasserim, Myanmar (Tickell 1864). Commonest hornbill throughout S Vietnam (Wildash 1968). Locally common in Peninsula Malaysia, with 1.0–1.6 km^2/pair (Johns 1988*a*).

A. a. convexus: S Thailand, at Pattani, Yala, and Narathiwat; Peninsula Malaysia, historically including Pinang; Singapore, now as a vagrant; Indonesia, on Sumatra and islands of Riau Archipelago (Batam, Bintan, Merah, Durian, Kurimon, Bulan, Galang, Mapor), Linnga Archipelago, Mansalar Is, Nias, Batu Is (Telo, Tanahmasa, Tanahbala), Mentawi Is (Siberut, Sipura, Pagai Utari, and Utara Selantan) and Legundi, Java and island of Panaitan, W Bali, islands between Malaya and Borneo of Natuna groups (Bunguran, Lingung, Laut, Pandak) and Tambelan group (Tambelan Besar, Benua, Gilla, Uwi), Kalimantan; Sarawak; Sabah and islands to NE of Gaya and associated islets; Brunei (Sanft 1960; Frith and Frith 1983*a*).

Commonest along coast and on offshore islands; possible exclusion by Malay Black Hornbill *A. malayanus* in parts of range, such as from islands of Banka and Belitung. Flies between mainland and islands, e.g. Noesa to Java mainland (Bartels and Bartels 1937). On Sumatra, found on forest edge or in lightly wooded country in lowlands, near coast and larger rivers, to 700 m at Aech (van Marle and Voous 1988). Still common in remote areas of Java (Ujung Kulon, Pangandaran, Bakuran: Holmes and Nash 1989). Radio-tagged ♀ in flock of 50 moved up to 4 km from capture site, another covering about 5.6 km^2 daily of which 2.0 km^2 was unused grassland (Tsuji *et al.* 1987).

Feeding and general habits

Occurs in pairs or family groups of 4–6, but up to 50 may congregate at fruit trees with other frugivorous birds, or 30–50 in loose flocks after breeding (Wells 1975; Poonswad *et al.* 1983, 1987). Mainly arboreal, but regularly descends to ground to feed, hopping about alone or in groups, especially along streams. Normally occurs in noisy flocks, but quiet and secretive when breeding. May roost communally, with up to 130 recorded at one roost in Thailand (Tsuji *et al.* 1987), but numbers vary and roosts change repeatedly during year, most on forest edge but with no regular routes to and fro. Radio-tracked birds preferred to forage in remnant forests, as families or flocks of up to 30, avoiding open areas and ranging over 15 km^2 (of which 5 km^2 was grassland) (Tsuji *et al.* 1987).

Diet mainly fruit, but also insects and small vertebrates. Fruits eaten, besides figs and drupes, include *Corypha utan* palm fruit, *Melanoxylon* berries, small fruits on quick-growing lianas, and papaw/papaya. One of few hornbills to eat very large *Ficus aurantiacea* fruits, but not a regular fig-eater (Lambert 1989). Also eats exotic oil-palm fruit (Pan 1987*b*). Hawks insects in flight, and a tame bird caught passing swallows and munias in flight (Primrose 1921). Fish remains found in one stomach, and seen taking live fish in shallow pools (Bingham 1897). Other animal prey includes *Oriolus chinensis* nestlings, nestling birds from tree holes, bats, frogs, lizards, snakes, scorpions, snails (Baker 1927), beetles, cicadas, cockroach, moths, butterflies, grasshoppers, termites from mounds, and termite alates hawked in air. No drinking recorded.

At nests in Thailand, diet 35% figs, 45% non-figs, and 20% animal items by weight, with much variation from nest to nest probably depending on availability of animals (15–44%) (Poonswad *et al.* 1983, 1987). Fruits included 35 spp. from 16 families and in the 24 genera *Polyalthia, Uvaria, Canarium, Tricosanthes, Elaegnus, Sloanea, Cinnamomum, Litsea, Amoora, Aphanamixis, Chisocheton, Aglaia, Ficus, Horsfieldia, Knema, Eugenia, Connarus, Piper,*

Jasminum, *Lepisanthes*, *Symplocos*, *Strombosia*, *Artocarpus*, and *Rourea* (Poonswad *et al.* 1987).

Animals included beetles (Buprestidae, Scarabidae, Cerambycidae, Lucanidae, Passalidae, Cetoniidae, Elateridae), bugs (Reduviidae, Cicadidae), ants, wasps (Vespidae) and their nests, butterflies, moths and their larvae (Psychidae, Cossidae), lacewings, grasshoppers, leaf insects, longhorns, mantid eggs (Orthoptera), stick insects, crickets, dragonflies, cockroaches, centipedes, diplopod millipedes (Julidae, Sphaerotheriidae, Polydesmidae), scorpions, spiders, crabs, river- and land-snail shells, earthworms, frogs, lizards, geckos, blind snakes, eggs and chicks of several Pycnonotidae and Columbidae spp., a Banded Kingfisher *Lacedo pulchella*, and rats.

Frequently sunbathes with head to one side, one or both wings lowered, and head, neck, and rump feathers raised. At highest intensity or with sudden exposure of sun, will adopt full spread posture of wings and tail (Fig. 3.4, p. 27). Foliage-bathes, flapping about while grasping twig in bill, and even under running tap or on wet grass in captivity. Usually followed by sunbathing, and most waterbathing done during showers on sunny days (Frith and Douglas 1978). Dustbathing reported as regular in wild (Primrose 1921), but not seen in captives.

Displays and breeding behaviour

Monogamous as separate pairs without helpers. Allofeeding is common throughout the year by both sexes, usually involving unwanted food exchanged with loud growling during presentation. Also allopreens frequently, soliciting attention to head and neck not elsewhere, even as nestlings; allopreening also precedes copulation (Hutchins 1976; Frith and Douglas 1978). May tilt back head when preened (Tickell 1864). Individuals show an assertive display with head and neck lowered towards one another and then repeated sharp upward-flicking motion of bill, often when one trying to displace another from perch (Frith and Douglas 1978).

Only ♀ enlarged nest-hole and threw out humus and leaves, although ♂ present uttering soft clucking notes (Bartels and Bartels 1937). Courtship-feeding frequent during nest visits, and copulation noted between ♀'s entrances into nest (Poonswad *et al.* 1987), once 12 days before sealing in. ♂ will also bring earth for sealing to the nest, up to 20 times per day, taken from nearby streams or among roots, and still delivered well into the incubation period. Copulation recorded 4 days before sealing in in captivity, lasting only 3–4 sec, the ♂ making a loud cackling before and after mounting. Both sexes took part in sealing, although mainly the ♀, and the ♂ would offer food either to the nest-entrance or to the ♀ (Hutchins 1976).

Breeding and life cycle

NESTING CYCLE: 79–89 days: incubation period 25–27 days (or 29 days in captivity), nestling period 6–7 weeks (49–55 days in captivity) (Hutchins 1976; Poonswad *et al.* 1987; Seibels 1989). ♀ emergence variable, from 20 days before chicks to after last chick. ESTIMATED LAYING DATES: *A. a. albirostris*: India, Mar–Apr, with most laid Mar; Myanmar (Tenasserim), Feb–Mar ($n = 17$); Thailand, 20 Feb to 23 Mar ($n = 14$; Poonswad *et al.* 1987); Peninsula Malaysia, Jan–Mar. *A. a. convexus*: Peninsula Malaysia, Jan–Mar; Borneo, Jan (2), Feb, Mar; Sumatra, Dec–Mar, May; central Java, Sept, Nov. Most at start of or in dry season. Captives: Washington, May. c/1–3, usually 3 (4 in captivity: Hutchins 1976). EGG SIZES: *A. a. albirostris* ($n = 65$) 43.6–54.0 × 30.0–38.0 (47.8 × 34.4); *A. a. convexus* ($n = 6$) 39.5–53.3 × 31.8–35.5 (48.1 × 33.9). Java ($n = 2$) 52.1 × 33.7, 51.8 × 34.1; Malacca ($n = 1$) 42.5 × 35.5 (Hoogerwerf 1949). Eggs with smaller end obvious, strong, glossy, but with rough grey-noduled shell.

A. a. albirostris: Nests in natural cavities in trees, such as dead *Dipterocarpus alatus* in cultivation patches (2, 15, and 30 m up); in dead *Koompassia malacensis* of 50 m with 8-cm-diameter entrance (Pan 1987*b*); in *Lagerstroemia flos-reginae* at 15 m in stump of dead branch, with entrance 25 × 19 cm and cavity 60 cm wide × 25 cm high; others only 5.5 m up in swamp jungle and 15 m up on edge of paddy-field (Madoc 1947). One possible record of nesting in hole in limestone cliff (Yang Lan pers. comm. 1992).

In Khao Yai National Park, Thailand, 24 nests were spaced 400–3200 m apart (mean 667 m): 4 in open forest, 5 on the forest edge, and 15 in dense forest (Poonswad et al. 1983, 1986, 1987), at a mean density of 1.67 km²/nest in a 34-km² study area. On average, half of the nest-holes used in a previous season were reused subsequently, but few were used repeatedly by the same sp. and some were alternated with Austen's Brown Hornbill *Anorrhinus austeni*. Nests were in live tree trunks, but for one in a large branch and another in a dead tree.

Tree spp. used were *Eugenia* (2), *Dipterocarpus* (6, one dead), *Altingia excelsa* (2), *Sterculia* (1), *Ficus* (1), *Tetrameles nudiflora* (2), *Pterospermum* (2), and unidentified (6). Tree heights were 15–35 m ($n = 9$, mean 25.9, SD 6.6); nest heights 5–20 m (12.6, 5.6); DBH 46–164 cm (104.4, 42.7); DNH 40–110 cm (56.7, 21.9). Mean nest-entrance dimensions were 25 × 10 cm, and old nest-holes of the Great Slaty Woodpecker *Mulleripicus pulverulentus* were also used. The range of interior dimensions for three nests was length 40–45 cm, width 23–30 cm, height 30–200 cm, and the floor 0–7 cm below the entrance. The nest-holes faced in no particular direction. Mud and food debris were the principal sealing materials. Part of the trunk from a fallen nest was cut out, nailed to another tree, and reused.

Start of delivery of animal food to the nest coincides with expected hatching (Poonswad et al. 1983), especially in captivity (Seibels 1989). One ♂ made 3–5 feeding visits per day, of 0.5–10 min duration, during which up to 25 food items were delivered; 114 fruits and 6 animal food items were delivered (Pan 1987b). ♂ feeds ♀ from gullet by regurgitation, delivering food at a mean rate of 14.2 g/hr, increasing as chicks grow and with no obvious diurnal rhythm (Poonswad et al. 1983, 1987). Radio-tracked breeding ♂ foraged up to 1.5 km from nest over an area of 0.9 km² (Tsuji et al. 1987).

♀ undergoes complete flight-feather moult (Sanft 1960). Nestlings emerged 13 May to 11 June at end of dry season in Thailand (Poonswad et al. 1987), ♀ emerged up to 20 days before the chicks but more usually with them, but chicks made no attempt to reseal nest, and departure of brood may extend over 1–14 days. ♂ seen to chase away hawk near nest with calling and attack, as with conspecific neighbours. Competition for nest-holes with Austen's Brown Hornbill and Bar-pouched Wreathed Hornbill *Aceros undulatus* was recorded.

Broods of 1–2 chicks fledged in Thailand, 90% of 42 nests that were sealed producing fledglings (Poonswad et al. 1987). Twice chicks were killed in the nest by ants, twice chicks were found dead in the nest, and once a chick was killed by a human. Sealing behaviour was first shown 8–9 months after fledging (Frith and Douglas 1978).

Captives bred for several successive years in a nestbox 50 × 50 × 100 cm, the entrance to which was 20 × 30 cm before being sealed to a slit of 17 × 2 cm (Hutchins 1976). Single brood per season, laid and hatched up to 4 eggs, but reared only 2–3 chicks. Amount of protein in diet important for success, with up to 10 mice being eaten per day when chicks about a month old. Nestbox removed between seasons and replaced as stimulus: courtship-feeding and billing started 58 days after replacement and 18 days before sealing in. ♂ regurgitated food at the nest, but carried sealing material in the bill tip. A captive ♀ used small pebbles to reline its nestbox (Seibels 1989).

Laying started 4 days after sealing in on one occasion. The first two eggs of a clutch were laid 2 days apart, the ♀ sitting with wings drooped at the side before laying but tight at side thereafter. The eggs were kicked about as the ♀ moved in nest, besides being turned intentionally. ♀ and chicks squirt out their droppings and throw out some debris. Chick's skin pink (E. McClure, C. B. Frith *in litt.* 1975). ♀ underwent a rapid and almost simultaneous moult of its flight feathers by 12 days after the first egg was laid: the feathers were kept in a pile at various places in nest, but there was little or no moult of contour feathers.

Chicks were first heard 29 days after enclosure, and the chicks and ♀ emerged 49 days thereafter (Seibels 1989). Chicks emerged a day apart after nestling period respectively of 55 and 54 days, and ♀ emerged with the second chick after 89 days in the nest, as in wild (Gee 1933b), to be greeted by calling of the ♂ and chicks (Hutchins 1976). Sexual difference in size of the chicks apparent at emergence, and ♀♀ predominated (8:3) among chicks raised by the captive pair (Seibels 1989).

♂ fed the chicks with food carried in the bill until they began to feed themselves 21 days after emergence. Another captive fledgling was fed for 3 weeks by its free-flying mother, mainly on insects (Sanft 1960). First bred in captivity in W hemisphere at Honolulu Zoo in 1953 (Lint 1972). Captive had lifespan of 12.5 years (Flower 1925); another in a Naga village said to be over 30 (Baker 1927).

A. a. convexus: Four nests in Sumatra all 20–40 m up in emergent live large forest trees, one in a *Koompassia*, another 38 m up with a 30 × 10-cm entrance, and a third in a natural rot-hole where a branch had broken off leaving a 13 × 4.7-cm entrance (Bartels and Bartels 1937). The latter hole was sealed to a 2.2-cm-wide slit, mostly with mud that must have been brought by the ♂, also with food or droppings including the remains of seeds, beetles, and a large cockroach. The cavity was more than 1 m deep, 30 cm wide and 20–25 cm high. The nest-floor was lined with dry earth and food remains, including several types of seeds, and a remex and rectrix of the ♀. ♀ and chick retreated out of reach to the back of the nest, and the ♀ emerged and flew off soon after the first visit to nest.

The ♂ fed by regurgitation of single items from gullet, once a load of some 20 cherry-sized fruits. ♀ uttered loud alarm-cries in nest when attacked by pair of Great Helmeted Hornbills *Buceros vigil*. Once a small flock visited the active hole of Malay Black Hornbill, and they probably exchange holes with other spp. owing to limited availability.

A captive ♀ survived at least 20.8 years (K. Brouwer *in litt.* 1992).

Palawan Hornbill   *Anthracoceros marchei*

Anthracoceros marchei Oustalet, 1885. *Le Naturaliste, Paris,* **7**, **15 July**, 108. Puerto Princesa, Palawan I., Philippines.

PLATES 7, 14

Description

ADULT ♂: black, with upperparts glossed green, but for all-white tail (a few individuals, especially imms (79% versus 26%), with black marks or shaft sections in tail, and others with rudimentary white tips to some wing flight feathers: Frith and Frith 1983a). Bases of primaries with narrow area of pale grey. Bill and casque, the latter forming a low cylinder above the forehead and ending in a pointed projection above the bill tip, pale yellow, with black base to lower mandible. Bare circumorbital and throat-skin patches white with blue tinge, and black ring around eye. Eyes red-brown, legs and feet dark leaden black.

ADULT ♀: like ♂ in overall coloration, but smaller; eyes brown.

IMMATURE: like ad, but bill smaller and casque undeveloped; often with black flecks on white tail feathers; bill cream with grey base, bare facial skin white with blue tinge, but lacking black ring around eye; eyes grey.

Palawan Hornbill *Anthracoceros marchei*

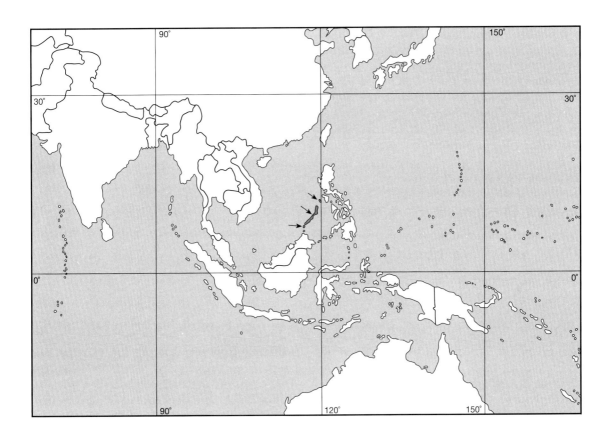

MOULT
Flight feathers moulting on ♂ and ♀ in July, the rainy season, while no moult on six other specimens taken in Jan, June, Aug, Sept, Dec (Sanft 1960).

MEASUREMENTS AND WEIGHTS
SIZE: (Frith and Frith 1983a) wing, ♂ (n = 11) 291–316 (300), ♀ (n = 14) 270–311 (283); tail, ♂ (n = 9) 221–242 (231), ♀ (n = 14) 197–237 (218); bill, ♂ (n = 11) 142–172 (158), ♀ (n = 14) 126–157 (142); tarsus, ♂ (n = 11) 49–55 (51), ♀ (n = 14) 44–54 (50).
WEIGHT: unrecorded.

Field characters
The only hornbill on the Palawan Archipelago. The black body and wings, white tail, and pale yellow bill distinctive.

Voice
Call a raucous *caaaww* or more usually *kreek-kreek*, given most often in early morning or late evening (Gonzales and Rees 1988). Also a soft *kiew* note repeated, similar to quiet contact-call of *A. albirostris* (Frith and Douglas 1978). No tape recordings located.

Range, habitat, and status
Philippines, restricted to islands of Busuanga in the Calamian Group and Palawan and Balabac in the Palawan Archipelago. Recently reported from Calawit and Culion in the Calamian Group, but records unsubstantiated (Dickinson *et al.* 1991).

Occupies primary and secondary evergreen forest, as well as mangrove swamp and areas of cultivation.

Used to be common and tame on Culion and Palawan, feeding at times in low fruit trees within a metre or two of the ground, but wary and remaining in high canopy of forest on Palawan (Whitehead 1890; McGregor 1909). Still common in St Paul National Park, Palawan (D. Yong *in litt.* 1987), but islands of Busuanga, Culion, and Balabac now largely deforested (Worth *et al.* 1994).

Feeding and general habits
Found solitarily or in small groups. Ranges from the ground up to the canopy when feeding, foraging groups audible by their calls and wing-flapping. Largely frugivorous, but includes other plant material, and animals such as insects and lizards, in its diet (Gonzales and Rees 1988). The trematode *Brachylaima fuscata* has been found in the small intestine of this hornbill (Fischthal and Kuntz 1973).

Displays and breeding behaviour
Unrecorded.

Breeding and life cycle
Unrecorded.

Malay Black Hornbill *Anthracoceros malayanus*

Buceros malayanus Raffles, 1822. *Transaction of the Linnean Society, London* **13(2)**, 292. Malacca, Malaysia.

PLATES 7, 14

Other English names: Malaysian Black Hornbill, Asian Black Hornbill, Black Hornbill

Description
ADULT ♂: all black, with upperparts glossed dark greenish blue. Central pair of tail feathers black (some specimens with narrow white tips) and projecting well past other tail feathers, outer pairs graduated and with broad white tips. A broad superciliary stripe, in both sexes and from their first plumage (Dunselman 1937), varies individually from being absent (black, 2%), to dark grey (13%), pale grey (43%), or white (42%), with no regional or sexual trends evident other than that pale grey stripes appear more abundant in Borneo (Frith and Frith 1983a). Bill, with casque a large cylinder from above forehead ending as projecting blade over tip of bill and ridged along sides, light yellow, with narrow black base to bill and casque. Bare circumorbital skin and small throat-skin patches dark blue-black, often with yellow patch below eye. Eyes dark red, legs and feet black.

ADULT ♀: like ♂ in overall coloration, but smaller, bill black, and casque a small cylinder tapering to point midway along bill; circumorbital skin flesh-coloured, throat-skin patches yellowish buff, eyes red-brown.

IMMATURE: plumage like ad, but white tail tips more often flecked with black (67% versus 16%). Casque undeveloped, bill pale greenish yellow with some dark underpigment in very young birds. Circumorbital and throat skin dull yellow with orange surround to eye, eyes dark brown, legs and feet grey.

SUBADULT: bill and casque development of ♂ still far from complete at 18 months old. Circumorbital skin of ♂ same black as ad by 20 months (Frith and Douglas 1978).

SUBSPECIES
Wing of Bornean population is on average 5.6% shorter than in populations of Peninsula Malaysia and Sumatra, and the tail is also shorter. Bornean population was named *A. m. deminutus* Sanft, 1960: wing, ♂ ($n = 28$) 289–330 (307), ♀ ($n = 16$) 262–286 (275). Bills and casques of Sumatran birds are larger than the rest (Frith and Frith 1983a), giving typical mosaic of characters for island populations. ♂ has a longer tail in relation to the wing than the ♀, but the ratio is unity in the Sumatran population.

Malay Black Hornbill *Anthracoceros malayanus*

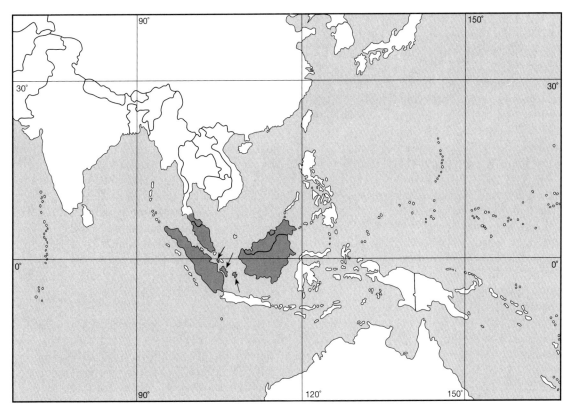

MOULT
Specimens in moult recorded throughout the year (Sanft 1960).

MEASUREMENTS AND WEIGHTS
SIZE: wing, ♂ (n = 34) 310–339 (324), ♀ (n = 21) 277–300 (288); tail, ♂ (n = 12) 294–345 (323), ♀ (n = 8) 263–294 (279); bill, ♂ (n = 7) 161–185 (170), ♀ (n = 6) 114–133 (123); tarsus, ♂ (n = 8) 49–54 (52), ♀ (n = 6) 44–49 (46). WEIGHT: ♂ 1050. See Frith and Frith (1983a) for detailed regional measurements.

Field characters
Length 60–65 cm. Black hornbill with cream (♂) or black (♀) bill and long white-tipped tail. Rather shy, with quiet flight, and spends much time in the canopy. The very loud calls, however, attract attention, and pale yellow bill of ♂ often conspicuous. Possibly confusable with Bushy-crested Hornbill *Anorrhinus galeritus* when white tips of tail not easily visible: both spp. have the bill black or whitish, but with sexual differences in this respect reversed in Bushy-crested (Duckett 1986).

Voice
Harsh, rasping, prolonged cawing and braying notes, rising and falling in pitch and volume before ending abruptly with a brief second note (Kemp 1988b), generally crow-like but harsher, louder, and more drawn out (Bartels and Bartels 1937). Tape recordings available.

Range, habitat, and status
Thailand, S from Trang; Peninsula Malaysia, S to Johore; Indonesia, on Sumatra plus islands of Bangka, Belitung, and Singkep in Linnga Archipelago, Kalimantan; Sarawak; Sabah; Brunei.

Occupies primary evergreen lowland forest, at up to 600 m but uncommon above 200 m. Found more in centre of forests than sympatric Oriental Pied Hornbill *A. albirostris*, but does occupy riverine gallery, tidal swamp, mixed deciduous, and even tall secondary or selectively logged forest.

Generally uncommon but often most frequently heard hornbill species in an area, at density of just over 1 km²/pair in Kalimantan (Leighton 1986), 4 km² pair in W Malaysia (Johns 1987), 0.5–5 km²/pair in Sabah (Johns 1988a), or 0.6 km²/pair in Sarawak (Kemp and Kemp 1974). Population most under pressure in W Sumatra (Worth et al. 1994).

Feeding and general habits
Resident in year-round territories, favouring a central stand of large trees with adjacent tangles, where occupies canopy and middle storey of forest. Enters ribbons of secondary forest along rivers for small fruits, as on quick-growing lianas (Johns 1987). Usually encountered in pairs, sometimes accompanied by a juv, but may join with neighbours in groups of 5–6 and subads may also form small groups (Leighton 1986; Johns 1987). Rarely, up to 33 together (Malaysia: Medway and Wells 1976; Sumatra: Silvius and Verheught 1985).

Early morning visitor to fruiting trees to extract fruit from newly dehisced capsules, staying for 0.5–1.0 hr, and trees may be adapted to supply this important disperser of their seeds (Becker and Wong 1985). Spends much time resting, preening, and calling in the forest canopy, only occasionally flying in the open, e.g. to isolated fruiting fig trees. Forages mostly in dense tangles, where it hops and creeps about, down to ground level, only some feeding in canopy as when taking figs and other fruits. Uses bill to prise off bark and split husks of capsules when foraging. Shows typical spreading of wings and tail when sunbathing and after foliage-bathing (Frith and Douglas 1978).

Diet of diverse, often large fruits, especially large-seeded dehiscent Meliaceae and Myristicaceae (Leighton and Leighton 1983). Includes *Polyalthia*, red arils of *Myristica* spp., and a major disperser of seeds of *Aglaia* spp. (Meliaceae) (Becker and Wong 1985; Wong 1986). Also eats odd triangular fruits of *Amorphophallus titanum*, which must be bird-dispersed but which taste bittersweet to humans, with burning aftertaste (Bartels and Bartels 1937). Feeds regularly on figs, including the very large fruits of *Ficus auranticacea* among 17 other fig spp. (Lambert 1989). Sugar-rich figs form 40% of diet, capsular and drupaceous lipid-rich fruits becoming especially important when other fruits are limited. Feeds on fruit of exotic oil palm (Worth et al. 1994). A captive reported to drink water after swallowing a bat (Harrison in Smythies 1960).

Visits to fruit trees interspersed with bouts of hunting animals, but animals taken only occasionally (hunting as against fruit-eating on 25% of first encounters, $n = 72$) with low hunting returns per unit time (Leighton 1986). Animal food delivered to a nest included beetles (Dynastidae, Cerambycidae), tettigoniids, bombycid butterflies, and eggs of *Pericrocotus flammeus* minivet (Bartels and Bartels 1937). Subordinate to all other hornbill spp. at fruiting trees, even similar-sized *Anorrhinus galeritus*, whose larger groups gang up to drive off other spp.; tends to arrive late at some shared fruiting trees (Leighton 1986). Also regularly displaced by some squirrels (Becker and Wong 1985).

Displays and breeding behaviour
Monogamous in separate pairs without helpers. Six ads inspected a nest-cavity together, and once a subad ♂ assisted a pair, probably its parents (Leighton 1982, 1986), so possibly co-operative breeding does occur under some circumstances. Pairs call at intervals during the day, with long territorial calling disputes along borders by both sexes.

Breeding and life cycle
Little known. ESTIMATED NESTING CYCLE: at least 80 days: about 30 days incubation, 50 days nestling period. ESTIMATED LAYING DATES: Sumatra, Feb, Apr, Nov; Borneo, Jan (2), Aug (3), Dec (2). No obvious breeding season: in Kalimantan pairs bred and fledged chicks during Jan–May, over a peak of fruit

availability, but did not breed again over a 2-year period (Leighton and Leighton 1983). c/2–3, but often only one chick raised and usually only single chick recorded with parents (but may be biased because of chicks each staying with only one parent). EGG SIZES: Borneo (n = 3) 49.5–46 × 32–33 to nearest 0.5 mm (mean 46.8 × 32.6) (Gibson-Hill 1950).

One nest, found 4.5 m up in live medium-sized forest tree, was apparently sealed; the ♀ defecated forcibly out of entrance, throwing out seeds, and the ♂ delivered dry leaves and twigs, probably as lining. Seen visiting active Great Slaty Woodpecker *Mulleripicus pulverulentus* nest (Short 1973), possibly as potential breeding site. The ♂ made brief feeding visits to the nest, often with only a single fruit or insect carried in the bill tip. Eggshells were thrown out of the nest after hatching. Many insects were fed to the chick, and also a bird's egg (Bartels and Bartels 1937). The ♂ was very cautious during visits to the nest, flying silently in and out and watching carefully before feeding. The ♂ rarely called during breeding, but began again as soon as the chick fledged. The same nest was reoccupied within 2 months of the chick fledging, but the ♂ parent was not the same, suggesting either a change of mate or a new pair in residence.

Chick with pinkish-yellow skin throughout nestling life (E. McClure *in litt.* 1975). Juv fed for at least 6 months after fledging, being supplied with all food for the first few months. Each parent travels separately with and gathers food for one of the two chicks, all four only rarely coming together. Subad may remain in parental territory but stay apart from ad for up to 18 months after fledging (Leighton 1986). A captive juv showed practice nest-sealing behaviour when still under 1 year old (Frith and Douglas 1978). Three captives survived for at least 12 years and two females 19.2 and 21 years respectively (K. Brouwer *in litt.* 1992, 1993).

Sulu Hornbill *Anthracoceros montani*

Buceros montani Oustalet, 1880. *Bulletin hebdomadaireda, Association Scientifique de France, Paris,* (2)2, 205. Sulu Archipelago, Philippines.

PLATES 7, 14

Other English names: Montano's Hornbill

Description
ADULT ♂: all black, with glossy upperparts, except for white tail. Bill and casque black, the latter a high blade originating above the forehead and ending abruptly two-thirds of the way along the bill. Bare circumorbital skin and small patches at base of bill (meeting under throat) black. Eyes cream, legs and feet dull lead-coloured.

ADULT ♀: like ♂ in overall coloration, but smaller, eyes dark brown.

IMMATURE: bill tip white or pale horn and casque undeveloped. A few individuals with dull white tips to some primaries.

MEASUREMENTS AND WEIGHTS
SIZE: (Frith and Frith 1983a) wing, ♂ (n = 3) 293–310 (304), ♀ (n = 2) 276, 290; tail, ♂ (n = 3) 240–243 (242), ♀ (n = 2) 226, 260; bill, ♂ (n = 2) 147, 150, ♀ unrecorded; tarsus, ♂ (n = 3) 50–54 (52), ♀ (n = 2) 49, 52. WEIGHT: unrecorded.

Field characters
The only hornbill on the Sulu Archipelago, the overall black appearance and white tail distinctive. Sexes separable on eye colour.

Voice
Call 'begins with series of notes precisely like the "song" of a common hen magnified about

fifty-fold, and ends with an indescribable combination of cackles and shrieks' (McGregor 1909). Nasal cackle, *ga-ga-ga-ga-GA-GA* (J. Hornskov *in litt.* 1992). Starts calling about 09.00 hrs with distinctive cackling and shrieking typical of the genus (R. Krupa *in litt.* 1988). No tape recordings located.

Range, habitat, and status
Philippines, restricted to islands of Jolo and Tawitawi in the Sulu Archipelago (Sanft 1960), and recently found on Sanga Sanga (Dickinson *et al.* 1991).

Historically common on hills behinds town of Sulu (McGregor 1909) but had not been recorded subsequently since the 1930s and was thought extinct! Fairly common in evergreen dipterocarp forest on Tawitawi (du Pont and Rabor 1973*b*), including Balabak Forest (R. Krupa *in litt.* 1988).

Forests were still quite extensive, with relatively little logging under way, and human emigration rather than immigration taking place (Worth *et al.* 1994). On Tawitawi, however, logging is accelerating so fast that there is grave concern for the species (F. Lambert to K. Brouwer *in litt.* 1993). Local name 'teugsi'.

Feeding and general habits
A pair or two recorded feeding together in tall fruiting trees, in primary forest and often on mountain slopes, moving from one part of the tree to another, but shy and if flushed moved a good distance before again congregating (du Pont and Rabor 1973*b*). Four birds visited a fruiting tree over 1 km from the nearest forest (J. Hornskov *in litt.* 1992).

Displays and breeding behaviour
Undescribed.

Breeding and life cycle
Undescribed.

Genus *Buceros* Linnaeus, 1758

Very large hornbills. Sexes similar in size and form of casque, which is large and of distinctive design in each sp. Casque and eye colour differ between sexes in two, possibly three spp. Imm most similar to ad ♀ but for undeveloped casque and eye colour, except in one sp., *B. hydrocorax*, where quite differently coloured. Tail white with black subterminal band, except in *B. hydrocorax*, where imm has black band but ad an all-white tail. Bare facial skin only around eye, except for small throat patches in some *B. hydrocorax* subspp. and in *B. vigil*, where the whole neck and most of head is bare.

Casque, bill, and some white areas of plumage coloured red, orange, or yellow by preen-gland oils applied in cosmetic coloration from the well-tufted uropygial gland. All spp. share this tufted gland and mallophagan feather-lice in the genus *Bucerophagus* with ground hornbills of the family Bucorvidae. Utter loud honking calls with bill held skywards, one sp., *B. vigil*, with a complex series of hooting and then laughing notes. The latter also differs in having greatly elongated central tail feathers and solid block of hornbill 'ivory' at front of casque; it was placed previously in the monotypic genus *Rhinoplax* Gloger, 1841. Noise of wingbeats in flight very loud, aided by emarginated outer (9 and 10) primary feathers.

Nest in tree holes, the ♀ sealing the entrance, partly with mud delivered by the ♂. The ♂ delivers food, mainly fruits, in its gullet and regurgitates them one at a time to the ♀ and chicks at the nest. c/2 normal, but often only one chick reared. Nesting cycle about 19 weeks. The chick retains pink skin, changing from pink to dark purple-black only in areas where black plumage will emerge. The ♀ emerges from the nest about 90 days into the nestling period, after which the chick reseals the nest itself. The ♀ undergoes a simultaneous moult of its rectrices and remiges while incubating. Take about 4 years to mature. Mates

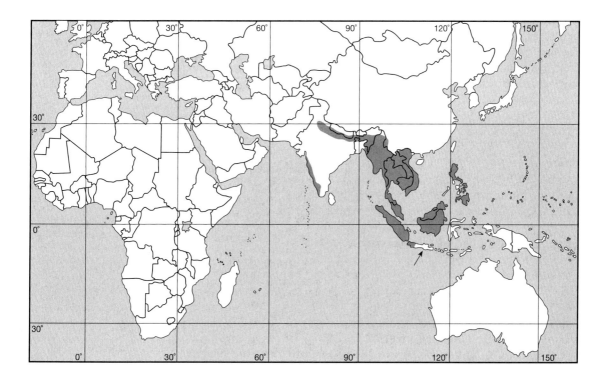

do not appear to employ allopreening extensively, but regularly nibble with bill tips.

Four, maybe 5, spp., ranging from W India to Java and the Philippines. Occur in moist primary forests on the Indian subcontinent, Asian mainland, and larger islands of the Sunda basin and Philippines. No supersp. groups indicated. *B. bicornis* and *B. rhinoceros* very similar, but they and *B. vigil* are sympatric in parts of range, and *B. hydrocorax* separated on the Philippines is unusual in breeding in co-operative groups.

Great Pied Hornbill   *Buceros bicornis*

Buceros bicornis Linnaeus, 1758. *Systemae Naturae*, edition 10, **1**, 104. China, designated as Sumatra, restricted to Benkulen, Sumatra, Indonesia.

PLATES 2b, 8, 14
FIGURES 3.1, 3.4, 3.6, 5.5

Other English names: Great Indian Hornbill, Concave-casqued Hornbill, Great Hornbill, Indian Hornbill, Homrai Hornbill, Indian Concave-casqued Hornbill

Description

ADULT ♂: crown, neck, and upper breast white; cheeks, chin, back, and lower breast black; thighs, abdomen, and uppertail- and undertail-coverts white. Minor upperwing-coverts black, with same metallic green gloss as back. Primaries, secondaries, and their greater and median coverts with broad white tips, forming white bands across wing which, together with the neck, are often stained yellow with preen-gland oils. Tail white with broad black band across centre. Bill, with large flat block-like casque bifurcated at the front, deep yellow from preen-gland oil applied over base, becoming orange to red towards tip, and with black line along bill–casque junction and at base of casque and mandibles. Bare circumorbital skin black. Eyes red, legs and feet olive-green.

ADULT ♀: same overall coloration as ♂, but smaller, casque yellow to orange without black marks; circumorbital skin red with black eyelids forming ring around eyes, eyes white.

IMMATURE: same overall coloration as ad, but bill much smaller and casque barely developed; circumorbital skin fleshy pink, bill cream, eyes pale blue-grey, legs and feet greenish yellow.

SUBADULT: casque begins to develop at about 6 months (148 days old in captive: Choy 1980) (Fig. 5.5, p. 57); by 2 years far from complete (Frith and Douglas 1978) but bill becomes serrated on inside cutting edge (Tickell 1864), by third year casque formed into horns but not concave on top (Phipson 1897), and only by fifth year appears fully ad (Tickell 1864). Eye colour of ♂ changes through grey, brown, and chestnut to red of ad by third year (Tickell 1864; Phipson 1897). Orbital skin turns pale grey (with lower eyelid still pink), to grey, then black. Casque formation and shape is suspected to depend to some extent on patterns of wear (Hodgson 1832; Prater 1921).

SUBSPECIES

The populations of continental Asia, described above, are larger than those of the Malay Peninsula and Sumatra (Deignan 1945) and have been separated as the form *B. b. homrai* Hodgson, 1832. The pattern suggests a cline towards the Equator with a step at the Isthmus of Kra, but the smaller 'nominate' peninsular form (wing, ♂ ($n = 18$) 462–494 (474), ♀ ($n = 16$) 428–472 (447)) is not recognized here as a subsp. At least three specimens of both sexes that are possibly hybrids with *B. rhinoceros* have been received into the Singapore bird trade from an unknown locality (L. Kuah *in litt.* 1992).

MOULT

♂ moulting flight feathers and tail in Sept (Thailand: Deignan 1945).

Great Pied Hornbill *Buceros bicornis*

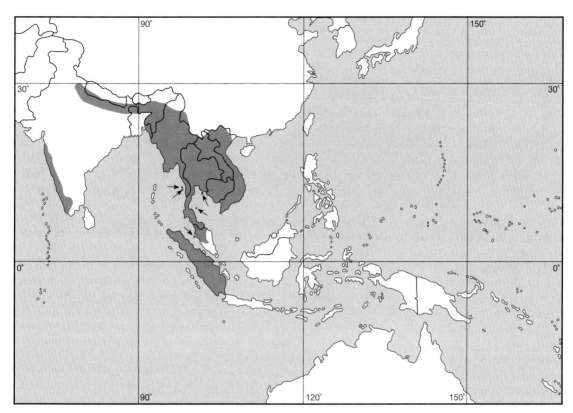

MEASUREMENTS AND WEIGHTS
SIZE: wing, ♂ (*n* = 28) 470–580 (527), ♀ (*n* = 11) 453–508 (480); tail, ♂ (*n* = 11) 400–463 (435), ♀ (*n* = 8) 350–437 (402); bill, ♂ (*n* = 5) 290–340 (319), ♀ (*n* = 6) 245–295 (268); tarsus, ♂ (*n* = 11) 60–79 (72), ♀ (*n* = 7) 59–73 (67). Largest casque dimensions: 192 long, 106 wide, 56 high. Wingspan 162 cm (Bangs and van Tyne 1931). WEIGHT: ♂ (*n* = 7) 2600–3400 (3006), imm ♂ 2390; ♀ (*n* = 3) 2155–3350 (2591) (Riley 1938; Preuss and Preuss 1973).

Field characters
Length 95–105 cm. Very large black and white hornbill with long, heavy, deep yellow bill and large concave-topped casque. Face, upperparts, and wings black, the latter with conspicuous white bars across them and white trailing edges. Neck, abdomen, and tail white, the latter with a black band across the centre. ♀ smaller, with pale eye and no black markings on bill, imm without casque and with smaller bill. Great Rhinoceros Hornbill *B. rhinoceros* of similar size, but has orange bill with upturned casque and lacks white on wings, neck, and crown. Loud calls distinctive, and particularly noisy at start of breeding season and when flying to roost.

Voice
Very loud reverberating *kok* repeated at slow regular intervals (Ali and Ripley 1970); Malayali name is Malamorakki or 'mountain-shaker'. Also coarse guttural deep *who* ♂ and *whaa* (♀) (Frith and Douglas 1978), similar to alarm-call of langur (Fleming 1968); or double *who-whaa* at take-off and in flight, which is a duet of ♂ then ♀ call. Calls on both inhalation and exhalation (Tickell 1864), and audible to 800 m through forest. Small nestlings make constant feeble croaking, alternating with soft, piping, whistling noise (Tickell 1864); ♀ and older chicks utter harsh churr as begging and food-acceptance calls; 70-day-old chick utters guttural squeal when handled (Choy 1980). ♀

makes loud growling when defending nest (Davidson 1891). Utter nasal wheezing noise when allofeeding (Frith and Douglas 1978). Young captives uttered only low murmuring grunt, never full ad roar, but may beg incessantly with loud shrieking calls (Ellison 1923). Tape recordings available.

Range, habitat, and status
India, with disjunct distribution, an isolated population along W Ghats of the peninsula, from Kolaba to Tenmalai, with the main area along the base of the Himalayas from Uttah Pradesh (Garhwal Terai, Kumaon) to Assam (Tezu); S Nepal; N Bangladesh; Myanmar, including islands of Kadan and Thayawthadangyi in Mergui Archipelago; China, in Yunnan province and W Yingjiang and S Xishuangbanna Districts (Tso-Hsin 1987); Vietnam; Laos; Cambodia; Thailand, including SE islands of Ko Chang and Ko Kram; Peninsula Malaysia, S to Bukit Fraser in Pahang, Gombak valley and Genting Simpah, plus W offshore islands of Salangar, Lontar, Telibun, Terutau, Langkawi, Pinang, Dingings, and Pangkor; Indonesia, an isolated population on Sumatra (Sanft 1960).

Inhabits primary evergreen dipterocarp and moist deciduous forest, mainly on lowland plains, especially between 600 and 1000 m, but along Himalayan foothills and in N Thailand extends up to 2000 m. Occupies selectively logged forests, but not so widely as smaller spp. (Johns 1987). Where sympatric with *B. rhinoceros*, favours more highland forest in Malaysia but is more common in coastal swamp forest on Sumatra.

Common in larger forest tracts, but uncommon to scarce in fragmented habitats, such as W Ghats, even historically (Davidson 1891, 1898). Local resident in SE Nepal in thick forest, but declining as a result of deforestation (Inskipp and Inskipp 1985) and probably long extinct further west (Fleming 1968). Very rare in Yunnan province of China (Tso-Hsin 1987), but present throughout S Vietnam (Wildash 1968). Rare and local on mainland Sumatra, up to 1000 m in N (van Marle and Voous 1988), also rare in Jambi province where *B. rhinoceros* common (Silvius and Verheught 1985). Passes over urban habitats *en route* to forests (Deignan 1945).

Has long been hunted for food and medicine in India, especially ♀ and chicks in nests, whose blood is thought to comfort the soul of the deceased by Kadar tribesmen of Kerala (R. Kannan *in litt.* 1992). Considered 'far superior to any fowl or pheasant' (Hume and Oates 1890), besides contributing their medicinal properties to 'the numerous tribe of children inhabiting the village' (Davidson 1898). Hunting was noted as making the birds wary and contributing to their decline along the W Ghats of India almost a century ago. Nesting unfortunately also coincides with the honey-gathering season on W Ghats, which attracts attention to cavities in large trees. Cutting of fig branches to feed domestic elephants and increase trees (which bear fruit eaten by hornbills) for their sap is also being addressed (James and Kannan 1993)

Feeding and general habits
Occurs in pairs, small family parties, or occasionally flocks of up to 40 in monsoon season or gathering at fruiting trees; once at least 200 at roost in huge peepul *Ficus religiosa* tree (Baker 1927). Usually as resident pairs, but some stable groups of up to 20, possibly non-breeders since mostly subads, often make local movements after fruit (Ali and Ripley 1970), especially tracking fig crops (R. Kannan *in litt.* 1992). Covers large area during daily feeding, with predictable and regular routine, but may spend whole day at particular fruiting trees (Bingham 1897). A family group in Thailand ranged over 608 km^2 (Tsuji *et al.* 1987) and mean home range of ♂♂ was 3.7 km^2 when breeding and 14.7 km^2 in the non-breeding season (Poonswad and Tsuji 1994). May leave forest to feed at isolated fruiting trees (Oates 1883), such as figs within bamboo forest (R. Kannan *in litt.* 1992). Break up into pairs to return to breeding territories, except for non-breeding flocks, and lone ♂♂ evident while feeding nesting ♀♀.

Flies with 3–4 quick, deep, noisy flaps, followed by a long whistling glide with wing tips upturned and tail spread, often high above the forest and far across valleys. Usually arboreal, hopping through branches with quick but ungainly buoyant sideways bounds, and also hops when on the ground, or rests on the 'knees', as when descending to feed on fallen fruit (Fig. 3.1, p. 25), to search for animals along stream banks, or to collect soil for sealing. Bathes in wet foliage and thereafter opens the wings to dry, sometimes sitting out in pouring rain for hours. Sunbathes with wings outstretched and tail spread (Fig. 3.4, p. 27), often preceded by trembling (Ellison 1923; Frith and Douglas 1978); may adopt heat-losing posture, gaping and with 'wrists' held away from the body (Fig. 3.6, p. 27). Often moves head and neck in short jerky movements, especially if inquisitive, but freezes with neck outstretched or moves very slowly if nervous.

Primarily frugivorous, but does hunt actively for small animals, including tearing and chipping off pieces of bark, and is a capable predator, especially during the breeding season. Usually dominant at fruiting trees, even chasing off smaller monkeys (Kannan 1992). Diet includes various fruits, especially figs *F. religiosa*, *F. indica* (Bingham 1897); *F. tsakela*, *F. mysorensis*, *F. callosa*, *F. talbotii*, *F. retusa* (R. Kannan *in litt.* 1992), nutmeg *Myristica*, and drupes of various spp., some very large. Also small animals such as locusts, beetles, lizards, snakes, and mynah nestlings taken from tree-hole nest (Wood 1927); larger prey battered against branches to kill it, and softened up in the bill if too large to swallow.

Diet when breeding in Thailand consisted of 57% figs, 29% non-figs, and 14% animal food by weight (Poonswad *et al.* 1983, 1987), and included fruits of 26 species in 14 families in the 19 genera *Polyalthia*, *Canarium*, *Elaegnus*, *Elaeocarpus*, *Cinnamomum*, *Litsea*, *Amoora*, *Aphanamixis*, *Chisocheton*, *Aglaia*, *Ficus*, *Horsfieldia*, *Knema*, *Eugenia*, *Connarus*, *Piper*, *Lepisanthes*, *Symplocus*, and *Strombosia*. Also takes some flowers and buds (Hume and Oates 1890), plantains (Mason and Lefroy 1912), and, in SW India, *Elaegnus conferta* and *Canarium strictum* (R. Kannan *in litt.* 1992). Diet at one nest in India comprised about 30% figs (*F. mysorensis*, *F. retusa*), 60% non-fig fruits (Lauraceae, *Persea*, *Alseodaphne*), 5% animal food, and 5% unidentified (Kannan 1992).

Animals in the breeding diet in Thailand were beetles (Scarabidae), Lucanidae, Passalidae, Cerambycidae, Cetoniidae, Elateridae), cicada (Cicadidae), digger wasp (Scoliidae), wasp nest (Vespidae), grasshoppers and leaf insects (Orthoptera), stick insect (Phasmatidae), crickets (Gryllidae), cockroaches (Blattidae), Lepidoptera larvae, centipedes, millipedes (Julidae, Sphaerotheriidae, Polydesmidae), giant scorpion, crabs, land snail, earthworm, frogs, lizards, flying lizard, skink, geckos, blind snakes, vipers, birds (*Megalaima* barbet, *Caprimulgus* nightjar, Greater Racquet-tailed Drongo *Dicrurus parodiscus*, Collared Scops Owl *Otus bakkamoena*), bird eggs and chicks (Pycnonotidae and Columbidae), insectivorous bat, common tree shrew, rat, squirrel, and flying squirrel. Also takes Barred Jungle Owlet *Glaucidium radiatum* (Kannan 1992). No drinking of water ever recorded.

Regularly uses communal roosts, arriving at sunset along same route each day, roosting among the topmost branches of large thinly foliaged trees, not clumped together but dispersed over neighbouring trees in a grove, with 3–4 per tree and with constant changes of position late into the dusk. Sleeps in typical posture, squatting on branch with head sunk between shoulders (Frith and Douglas 1978). Noisy just before and at dawn, in contrast to other communal-roosting spp. (*Anthracoceros coronatus*, *Aceros undulatus*), and possibly a form of communal display. Moves out in smaller parties than when returning at dusk (mean 5, $n = 13$ days, at dawn; mean 38, $n = 10$ days, at dusk), with total of up to 111 moving in same direction in non-breeding season (Thailand: McClure 1970). Applies yellow preen-gland oil during several bouts of preening in morning, whereafter rubs off during rest of day (Phipson 1897). Pair and chick, during seasons when present in the area, used same communal roost as about 30 other birds for 5 years, only sometimes using two subsidiary roosts on an erratic basis (Tsuji *et al.* 1987).

Nematode *Ascaradia galli* found in captive on post-mortem (Reddy and Rao 1983).

Displays and breeding behaviour

Monogamous in territorial pairs without helpers. All-♂♂ parties reported braying loudly near fruiting trees early in the breeding season (Myanmar, Feb: Bingham 1897). ♂♂ reported fighting by casque-to-casque aerial collisions (Lord Medway *in litt.* 1974). Bizarre mass lek-like courtship reported involving 21 birds, mainly pairs, calling, branch banging and hanging upside-down to break off orchid bulbs to present to mates (Hutton 1986). Calls with neck stretched vertically and bill pointing upwards at each loud *kok* note (Ali and Ripley 1970). Shows aggression by bouncing up and down on the perch, and indicates assertion by flicking movements of the bill (Frith and Douglas 1978).

Wild ♂ seen courtship-feeding successive fruits to ♀ outside nest, also regurgitating fruit into empty nest. Courtship-feeding recorded up to 5 months before breeding, sometimes with bill-grappling (Kannan 1992). ♂ rarely calls around nest once active, except when arrives and leaves, but ♀ may reply from within. Defends territory of about 100-m radius around nest, by calling and chasing off intruders, especially at start of the season, and may continue defence throughout season even if breeding not attempted. Copulation seen at nest after ♀ emerged from sealing (Poonswad *et al.* 1983, 1986, 1987).

A captive pair showed increase in calling, courtship-feeding, and restlessness before entering (Choy 1978, 1980), and sealing started few weeks before ♀ finally entered (Preuss and Preuss 1973; Golding and Williams 1986). Began bill-touching, especially ♂, 8 weeks before sealing in, leading to courtship-feeding 6 weeks before and first copulation 18 days before sealing, thereafter repeated frequently throughout that day. Much entering and leaving of the nest throughout, with copious sealing started by both sexes only 3 days before sealing in (Stott 1951). ♀ orbital skin may flush during courtship activities (Thormahlen and Healy 1989).

Breeding and life cycle

NESTING CYCLE: about 113–140 days: 1–4 days pre-laying, 38–40 days incubation, and 72–96 days nestling period. ♀ emerges 14–59 days before fledging of chick, after 114–134 (mean 121, $n = 15$) days in nest. Captive ♀ emerges when chick about 37–69 days old, after 77–112 days in nest, once emerging with chick after 124 days (Bohmke 1987); wild W Indian ♀ emerged when chick only 14–16 days old (Ali and Ripley 1987; Kannan 1992). ESTIMATED LAYING DATES: SW India, Jan–Apr; NE India, Jan–Apr, peak Feb–Mar; Nepal, Mar–Apr; Myanmar, Jan–Mar; Thailand, Jan 13 to Feb 11 ($n = 15$), Mar; S Sumatra, ♀ being fed in nest on 15 Dec (Nash and Nash 1985*b*); Singapore (captive), Jan. Lays at start of dry season so that chick emerges during SW monsoon. Captives: England, Mar; Germany, Mar; California, Apr (3); Missouri, Apr (2). c/1–3, usually 2 (a captive with third egg in ovary). Only 1 chick ever recorded fledged in wild, but 2 in captivity (Thormahlen and Healy 1989), and 3 small chicks reported from one nest (Hume and Oates 1890). EGG SIZES: ($n = 45$) 59.8–72.2 × 42.0–50.0 (65.1 × 46.3) (Baker 1927); eggs from c/1 larger than those from c/2 (Bingham 1879; Hume and Oates 1890). Eggs finely pitted, glossy, few with rough texture, and most become stained during incubation; oval or globular, but some variation in shape.

Nests in natural holes in trees and prefers entrance with elongate vertical slit. Nests 8–25 m above ground (Ali 1936; Sanft 1960; Ali and Ripley 1970; Nash and Nash 1985*b*; Kannan 1992, *in litt.* 1992), in trunk of a dead tree or live *Cullenia excelsa, Lagerstroemia flos-reginae, Tetrameles nudiflora, Eugenia* sp., *Palaguium ellipticum, Alseodaphne semecarpifolia, Homalium tomentosum, Hopea odorata*, primary limb of *Calophyllum tomentosum*, and two teak trees with strangling figs, usually close to fruiting figs (Bingham 1879). One nest tree DNH 1–1.2 m with entrance sealed to a 20 × 5-cm entrance slit (Ali 1936), another 72 cm DBH (Kannan 1992). A number of pairs have bred in captivity, in boxes, barrels, and logs, with cavities 45–100 cm in width, entrances 25–32 cm high × 18–25 cm wide, and floors up to 30 cm below entrance (Preuss and Preuss 1973; Choy 1978, 1980; Golding and Williams 1986; Bohmke 1987), despite such dimensions being quite different from cavities chosen in nature.

Of 25 nests found in Thailand, 20 were in dense forest (5 on forest edge) and 5 in open forest (Poonswad *et al.* 1983, 1986, 1987, 1988), at a density of about 2.3 km^2 per nest and a mean nearest-neighbour distance of 1002 m (range 200–2300), although some were only 30 m from nests of other hornbill spp. Tree spp. used were *Eugenia* (4), *Dipterocarpus* (7), *Cinnamomum glaucescens* (1), *Altingia excelsa* (1), *Lithocarpus* (1), Ulmaceae (1), and unidentified (8). Mean nest-tree DBH 111 cm (107–132) and height 25–53 m, cavity 12–22 m up in live trunks (but for two in branches), with no particular direction for entrance, but preferred elongate slits with the width just sufficient for ♀ casque to enter (8–9 cm); average entrance 40 × 14 cm, as for 8 of 10 nests in India (Kannan 1992). The range of interior dimensions of three nests was: diameter 42–70 cm, width 32–50 cm, height 39–200 cm, entrance lip to floor 0–7 cm. Only two nests did not have a high ceiling or funk-hole, and one of these was not used for four consecutive seasons. Nest-holes used regularly in successive seasons, even if robbed previously (Davidson 1898; R. Kannan *in litt.* 1992), including nests of other spp. (two taken over from *Aceros undulatus*: Poonswad *et al.* 1983).

Entrance sealed with droppings, chewed pieces of wood and bark, food and nest debris, but little if any soil: ♀ seals from within, ♂ from without and also delivers some material, often after swallowing material and then regurgitating pellets. Both sexes sealed in captivity (Poulsen 1970; Frith and Douglas 1978; Choy 1978, 1980; Golding and Williams 1986; Bohmke 1987; Thormahlen and Healy 1989), one pair taking 3 days to close the entrance and the ♀ a further 2 days after entering (see also Poonswad *et al.* 1987). One pair unable to seal circular entrance of 40-cm diameter but next year successful with 20 × 25-cm oval entrance (Bohmke 1987). ♀ does some banging on nest-walls while within, and spends some time in nest each day before sealing in.

♀'s call changes from honking to cackling sound at time of laying (Stott 1951). ♀ incubates the eggs and broods the chick between the legs, turning the eggs about hourly and raising her belly feathers before settling on them (Thormahlen and Healy 1989). At one nesting in captivity, the eggs were laid 5 days apart: one hatched after an unusually short incubation period of 28 days, and the ♀ emerged atypically with the chick at the end of the nestling period in this one season only (Thormahlen and Healy 1989). ♀ usually emerges well before the chick fledges, whereafter chick reseals nest alone. ♀ usually undergoes simultaneous flight-feather moult while nesting (Bourdillon in Hume and Oates 1890; Bohmke 1987; P. Poonswad *in litt.* 1992), but sometimes only a gradual moult (Preuss and Preuss 1973; Poonswad *et al.* 1983; Kannan 1992). One ♀ with an addled egg emerged about 2 months after sealing in (Poonswad *et al.* 1983).

♂ makes feeding visits of 15–20 mins every 2–3 hr, totalling about 3–5 per day, apart from some short visits with animal food. Number of items delivered, but not number of visits, rises about fivefold after chick hatches (Ali and Ripley 1970; Kannan 1992). Feeds by regurgitation from the gullet, which becomes distended with up to 50 grape-sized fruits (Stott 1951) or 185 pea-sized Lauraceae (Kannan 1992) per load. A last fruit, and usually any animal food item, is carried in the bill tip. Delivers about 26.5 g/hr of food to the incubating ♀, rising when chick present to peak of 43 g/hr and including similar rise in animal foods and non-fig (lipid-rich) fruits (Poonswad *et al.* 1987). Desertion occurred at a nest where only fruit was being fed, and at all nests snail shells were regularly supplied (Poonswad *et al.* 1987), possibly for calcium content. Radio-tracked ad foraged over 2.5–4.5 km² around nest in forest, another up to 3 km away over 7.5 km² of which 3.6 km² was grassland, with average daily over about 1.3 km² (Tsuji *et al.* 1987).

Feeding rate also increased in captivity after a chick hatched (Choy 1978, 1980), and ♂ then greatly favoured live-insect diet (grasshoppers, crickets, mealworms), swallowing 20–25 to form visible bolus in throat, regurgitating at nest, and taking other food only once all insects gone (Golding and Williams 1986). ♂ caught wild birds to feed ♀, and when chicks 18 days old began to feed day-old fowl chicks at nest, supplying more fruit only when chicks about 6 weeks old.

Chicks fledged from 80% of nests where ♀ sealed in (*n* = 25, Thailand: Poonswad *et al.* 1987). Captive juv began to take some food by 24 days after fledging, but still fed mainly by ad after 118–148 days, when virtually independent and casque started to develop (Choy 1978, 1980; Golding and Williams 1986). Wild juvs may also feed themselves well before 6 months after fledging (Benson 1968). A radio-tracked chick remained within 2 km of the nest for several months after fledging, almost to the next breeding season (Tsuji *et al.* 1987).

Long-lived in captivity, three for at least 31.6–33.5 years, and one ♀ now over 41 which laid its first egg at age of 39 (Flower 1938; K. Brouwer *in litt.* 1992, 1993). A Great Pied Hornbill, named 'William', taken as a nestling from the Karwar jungles in 1894, lived for over 25 years at the Bombay Natural History Society headquarters, which have since been named 'Hornbill House' (Ali 1976). First bred in captivity in 1953 (Lint 1972). A studbook of captives in Europe, including a detailed reference list, has recently been published (Brouwer 1993).

Great Rhinoceros Hornbill *Buceros rhinoceros*

Buceros Rhinoceros Linnaeus, 1758. *Systemae Naturae*, edition 10, **1**, 104. India, restricted to Malacca, Malaysia.

PLATES 8, 14
FIGURE 3.4

Other English name: Rhinoceros Hornbill

Description

B. r. rhinoceros Linnaeus, 1758. Malayan Great Rhinoceros Hornbill.

ADULT ♂: entire body and wings black, the upperparts with green metallic gloss, but for white thighs, abdomen, and undertail-coverts. Tail white with broad black band across centre. Bill, with large cylindrical casque laterally compressed and more or less turned back at tip, pale yellow or ivory-white with black bill–casque junction and base of casque and mandibles; orange preen-oil, with characteristic odour (Bartels and Bartels 1937), applied over all but for the tip of bill and casque. Bare circumorbital skin black. Eyes red, legs and feet olive-green, naked skin under wings and on abdomen dusky greenish (Riley 1938).

ADULT ♀: same overall coloration as ♂, but smaller, and casque without black marks; circumorbital skin red-orange with black eyelids forming ring around eyes, eyes white.

IMMATURE: same overall coloration as ads, but casque barely developed; circumorbital skin blue-grey, bill yellow with orange base, eyes light blue-grey at fledging. Some sexual variation and changes in colours of mouth lining (Reilly 1989).

SUBADULT: eyes brown within 2 months. Bill, casque, and soft-part colours far from fully developed when 3–4 years old (Frith and Douglas 1978). Casque development begins at 3–4 months (Reilly 1988), but still growing at 8.5 years old (Frith and Frith 1983*b*).

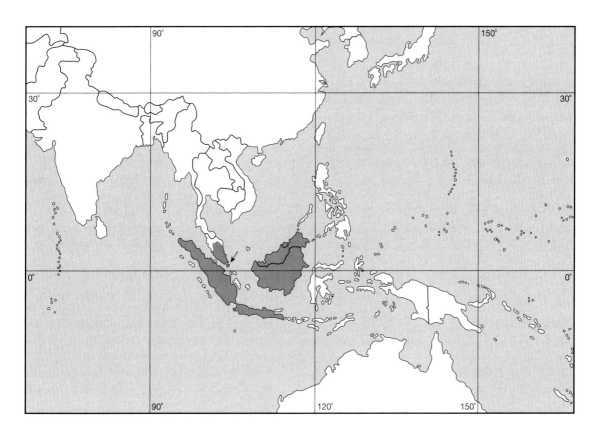

SUBSPECIES

B. r. borneoensis Schlegel and Müller, 1840. Bornean Great Rhinoceros Hornbill. Same overall coloration as nominate subsp., but smaller; casque normally shorter, broader, and more sharply rolled back at tip.

B. r. sylvestris Vieillot, 1816. Javan Great Rhinoceros Hornbill. Same overall coloration as nominate subsp., but black band across tail broader, and casque normally not rolled back at tip.

Local variation in casque form (the primary difference between subspp.), and the considerable differences developing among captive birds of known origin, suggest that recognition of formal subspp. may be superficial, especially as size differences appear as mozaic (as in other spp. of similar range). Casque form also shows much individual variation in development, partly the result of rubbing on perches (Bartels and Bartels 1937; Frith and Douglas 1978; P. Shannon *in litt*. 1990).

One ♂, in successive moults, retained rufous-tipped feathers on back of neck (just where black–white transition in *B. bicornis*) and in one moult grew a white-tipped P2 in each wing that was lost in the next moult (Frith and Douglas 1978). At least three specimens of both sexes that are possibly hybrids with *B. bicornis* have been received into the Singapore bird trade from an unknown locality (L. Kuah *in litt*. 1992).

MEASUREMENTS AND WEIGHTS

B. r. rhinoceros SIZE: wing, ♂ (n = 51) 460–530 (493), ♀ (n = 26) 415–490 (453); tail, ♂ (n = 19) 355–405 (377), ♀ (n = 17) 320–380 (352); bill, ♂ (n = 22) 270–340 (291), ♀ (n = 16) 225–270 (252); tarsus, ♂ (n = 19) 65–76

(69), ♀ (n = 15) 58–67 (63). WEIGHT: ♂ (n = 5) 2465–2960 (2580), ♀ (n = 3) 2040–2330 (2180).

B. r. borneoensis SIZE: wing, ♂ (n = 28) 440–500 (475), ♀ (n = 25) 420–450 (435); tail, ♂ (n = 17) 364–395 (368), ♀ (n = 17) 322–365 (343); bill, ♂ (n = 17) 221–270 (238), ♀ (n = 11) 180–215 (207); tarsus, ♂ (n = 18) 60–70 (64), ♀ (n = 18) 55–65 (61). WEIGHT: unrecorded.

B. r. sylvestris SIZE: wing, ♂ (n = 17) 465–550 (508), ♀ (n = 6) 445–470 (458); tail, ♂ (n = 9) 379–425 (406), ♀ (n = 2) 370, 383; bill, ♂ (n = 10) 212–245 (223), ♀ (n = 2) 195, 200; tarsus, ♂ (n = 8) 70–80 (75), ♀ (n = 2) 63, 70. WEIGHT: unrecorded.

Field characters

Length 80–90 cm. A very large black hornbill with white abdomen, and white tail with a broad black band across the centre. The large orange bill and recurved casque distinctive. Conspicuous and noisy, flying with rushing wingbeats and calling with loud braying notes. Parapatric in Peninsula Malaysia and Sumatra with *B. bicornis*, which has a flat, double-pronged yellow casque and white bar across centre and trailing edge of wings. Sympatric with *B. vigil*, which differs in having a short stout bill and casque, bare head and neck, elongated central tail feathers, white tips to the remiges, and a quite different hooting call.

Voice

♂ utters deep, short, forceful *hok* notes, usually followed by ♀ with higher *hak* notes, often in a duet *hok-hak hok-hak hok-hak ...*, sometimes even when ♀ still in nest (Hetharia 1941). ♂ almost always initiates calling, and often starts before and continues after ♀; largely silent when breeding. Calling usually starts before take-off, often after having sat quietly for long periods, and flight-call a disyllabic *ger-ronk*, higher, faster, more nasal, and immediately noticeable (Bartels and Bartels 1937). Flight-call from both sexes, often given simultaneously or antiphonally (Medway and Wells 1976; Leighton 1982), and audible over several km. Flight also audible, but less so than that of large *Aceros* spp. Food-acceptance call by ♀ a high continuous *eeee...*, uttered even before ♂ reaches nest; young chicks keep up a continuous high peeping while being fed. Threat-call by ♀ in nest a continuous harsh, high squealing, rising to a crescendo when something appears at the entrance. ♂ gives loud, harsh, nasal *hang* in alarm at nest (Kemp 1988b). Utter peculiar nasal coughing sound when handling food in allofeeding (Frith and Douglas 1978). ♂ occasionally makes quiet grunt when leaving nest (Johns 1982). Tape recordings available.

Range, habitat, and status

B. r. rhinoceros: Peninsula Malaysia, S of Songkhla, Bukit Maxwell in Perak, and Gunung Tahan in Kelantan; Singapore, up to about 1950; Indonesia, on Sumatra. *B. r. borneoensis*: Indonesia, in Kalimantan; Sarawak; Sabah; Brunei. *B. r. sylvestris*: Indonesia, on Java, E to Meru Betiri (H. Bartels *in litt.* 1982).

Inhabits dense lowland evergreen forest, excluding swamp forest (Holmes 1969), also hill dipterocarp forest, and occasionally up to 1200 m. Occupies and breeds in selectively logged forests (Johns 1987), and enters open areas between forests and suburbia (Frith and Frith 1983b), rarely even flying long distance over open country (Madoc 1947).

Generally common, widespread, and frequently heard within suitable habitat. Common in primary and disturbed forest on Sumatra at up to 800 m (van Marle and Voous 1988), but absent in S where *B. bicornis* common (Nash and Nash 1985a). Rare on Java except in Meru-Betiri area (Holmes and Nash 1989), extinct in Cibodas-Gunung Gede National Reserve (Andrew 1985). Locally reduced by hunting in areas of Borneo (Gore 1968) and numbers may fluctuate considerably depending on availability of fruit, with some individuals moving over long distances (Johns 1982).

Occurs at density of 0.5–2.5 km^2/pair in primary dipterocarp and secondary logged forest in Peninsula Malaysia (Johns 1987); 0.3–1.6 km^2/pair in Sabah, with up to 0.1 km^2/pair in an isolated forest patch (Johns 1988*a*); up to 1.2 km^2/pair in Sarawak (Kemp and Kemp 1974); 2–3 km^2/pair in Kalimantan (Leighton 1986). Population probably below 2500 on Borneo, 500 on Java, and declining fast through habitat destruction (Worth *et al.* 1994).

Emblem of the Malaysian state of Sarawak and important in the principal Iban ritual, the 'gawai kenyalang', involving elaborate carving of effigies (Freeman in Smythies 1960). Has long been collected by Bornean Dyaks for its tail feathers (Banks 1935) to use in headdresses and decorative capes, and this practice, together with keeping of preserved heads and casques, still continues (Kemp and Kemp 1974; Duckett 1985). Habit of rapid jumping up and down on branches in threat is imitated in Dyak dances (Hose in Shelford 1899). Also kept as pets, and one, reared on rice, was still being fed around a longhouse by its wild parents (Dunselman 1937).

Feeding and general habits

Most ads exist as resident, territorial pairs. Small flocks of up to 25, apparently subads and non-breeding ads, range widely, probably over hundreds of km^2, and move rapidly through an area, often as dispersed subgroups each several hundred metres apart, some associating in pairs and courtship-feeding prevalent (Leighton 1986; Wells 1974). Resident pairs call at intervals during the day, and both sexes call and defend their territory, but not against non-breeding flocks. Interact several times per day with neighbours, and very responsive to their calls or imitations thereof (Smythies 1960; Kemp and Kemp 1974). Travels with characteristic flap-and-glide flight, ♂ leading and third bird, if present, always a juv (Bartels and Bartels 1937). Sunbathes with fully spread wings and tail, also bathes in wet foliage (Frith and Douglas 1978; Fig. 3.4, p. 27).

Visits fruit trees between bouts of hunting small vertebrate and large arthropod animals. Sugar-rich figs form 40–60% of fruit diet, but capsular and drupaceous lipid-rich fruits are also important; especially active in prising off bark and splitting husks of capsules, the latter most important when fruiting poor. Actively hunts animals for about 50% of time ($n = 68$ first contacts), significantly more than smaller spp. but with same low rates of return. Members of pair remain up to 300 m apart when hunting animals, keeping in contact with calling (Leighton 1982). Arrives late at fruiting trees but dominant over all other hornbills, and most often seen with others at figs (Leighton 1986). Descends to ground in oil-palm plantations to feed on small animals and probably also fallen fruit (Johns 1987).

Fruit diet includes figs (*Urostigma, Ficus stupenda, F. sumatrana, F. glabella, F. benjamina*), *Santiria laevigata, Sterculia parvifolia*, Meliaceae, wild nutmeg, and oil-palm fruit, some of which have restricted fruiting seasons, others with multiple fruit crops (McClure 1966). A regular fig-eater taking at least 13 spp. in Malaysia (Lambert 1989). Able to break into 2-cm-diameter *Shorea xanthophylla* fruit but not larger fruits of five other spp. of dipterocarp (Gould and Andau 1989). Animal food includes lizards, tree frogs, bird eggs (*Hemiprocne longipennis*), spiders, and large insects (beetles, crickets) (Bartels and Bartels 1937). No drinking recorded. Attacked by drongos (Hetharia 1941), an indication of its having a predatory role.

Displays and breeding behaviour

Monogamous, possibly with helpers at times. Hose (in Shelford 1899) recorded several young ♂♂ feeding nesting ♀ after an ad ♂ had been shot at the nest, and imms have helped with provisioning in captivity (Reilly 1989; P. Shannon *in litt.* 1990), suggesting that co-operative breeding may occur at times.

Displays by jerking the head upwards with each note when calling, raising the bill skywards (Frith and Douglas 1978; Kemp 1988*b*). Performs perch-bouncing in threat, a very vigorous intimidating display, and bill-flicking also part of assertion (Hose in Shelford 1899; Frith and Douglas 1978).

Copulation sequence of captive pair started 16 days before sealing in and performed as frequently as every 30 min, involving courtship-feeding, then chasing and bill-fencing, and finally mounting for 5–15 s, accompanied by high-pitched nasal chatter (Reilly 1988, 1989). When chick scratched around head and neck, it raises its feathers and closes its eyes, but allopreening not recorded (Frith and Douglas 1978).

Once a ♀ was courtship-fed by the ♂ 1–1.5 months before it would normally enter, the ♀ giving the begging-call and fluttering its wings like chick; at the same visit the ♂ also tried to offer fruit to the empty hole while the ♀ was perched nearby, sometimes uttering a low guttural call (Reilly 1989). Visits to nests, including ♀ entering and being courtship-fed, recorded at least 6–12 weeks before sealing in (Hetharia 1941; Johns 1982). A captive ♀ slept in the nest for some days before sealing in (Reilly 1988).

Breeding and life cycle

NESTING CYCLE: in captivity 105–126 days: 37–46 days incubation, 78–80 days nestling period, ♀ emerging 86–97 days after layings (Reilly 1989). ESTIMATED LAYING DATES: Peninsula Malaysia, Mar, June, Nov; Thailand, June; Borneo, Jan, May–June, Sept; Sumatra, Jan; Java, Mar–Apr; no obvious seasonality. c/1–2, normally only one chick fledged, but two raised in captivity (Reilly 1988). EGG SIZES: *B. r. borneoensis* 63.0–64.8 × 63.0–44.6 (64.3 × 44.4) (Schöwetter 1967). Egg with pitted shell, leading to 'salt-and-pepper' staining.

Nests in natural hole in trunk of large forest trees, but seen inspecting holes in limestone cliff (Gunung Mulu, Sarawak: Kemp and Kemp 1974). Two of 7 nests in dead trees, one 15 m up in trunk of emergent *Shorea paucifolia* where branch had broken off (Johns 1982), several on top of ridge, one only 9 m up (Madoc 1947). Selects elongate entrance slit just small enough to admit ♀, and later digs around and within hole, possibly for sealing material or to improve shape. Two holes and entrances 33–35 cm high × 6.5–8 cm wide, and cavities 41–55 cm deep, 27–40 cm wide, and with the floor 7–15 cm below the sill. One cavity was 120 cm high, and at another the ♀ disappeared up a high funk-hole (Bartels and Bartels 1937). One artificial nest-barrel was 56 cm wide × 89 cm high, the entrance 25 × 25 cm with rope around rim and 52 cm from floor, but with 12-cm-deep lining (Reilly 1988).

Entrance sealed by ♀ alone, mainly with droppings from within nest (Bartels and Bartels 1937; Johns 1982), but some help from ♂ in captive pair (Reilly 1988). Pauses between sealing to collect and chew material from nest-floor, and sealing action audible from forest floor. Nests kept dry and clean, with faeces forcibly ejected by ♀ and larger chicks, faeces of small chicks and debris being thrown out, as with large seeds that are regurgitated not defecated. A captive chick ate most of its excreta from the nest-floor, which seemed to aid in redigesting certain food types (Frith and Douglas 1978). ♀ leaves the nest 4–7 weeks before the chick fledges, after about 3 months in the nest, and the half-grown chick reseals the nest alone. ♀ moult apparently not simultaneous in most instances.

♂, and later ♀, feed at nest. In one day, ♂ fed incubating ♀ 22 figs at 08.15 hrs, 4 fruits at 10.00, 8 at 15.30, and 22 in 2 min at 16.00 (Hetharia 1941). Another incubating ♀ was fed, from 06.45 to 18.41 hrs over eight visits lasting 1–15 min (mean 8), 43 figs, 3 *Litsea* fruits, and 1 insect grub (in loads of 12, 10, 1, 4, 0, 3, 12, 5); ♀ ate all offerings on four visits, but terminated four other visits by refusing food. The figs came from a nearby strangling fig, and nesting coincides with the fruiting season in the forest (Johns 1982). Another ♂ delivered food to ♀ and large chick about hourly, but breaks of up to 2.5 hr recorded and most visits during morning and evening with a midday break (Bartels and Bartels 1937). A captive ♂ switched from feeding mainly fruit to feeding crickets when the chicks hatched, but after two weeks began to add fruit and dog food to the diet (Reilly 1988).

Food, mainly fruits, are carried in the gullet, the swollen appearance of which is obvious. Few to many items are carried, depending on size (up to 58 small figs), and are fed one at a time after obvious neck-craning and regurgitation. ♂ recorded attacking and calling loudly at *B. vigil* near the nest (Hetharia 1941). ♂ roosts at night either near to or away from the nest.

Lay replacement clutch in captivity after death of nestling (Shannon 1993). Probably does not breed every year, although does reuse nest-holes annually at times (Bartels and Bartels 1937). Bred and fledged chicks during Jan–May in Kalimantan, during fruiting peak, and then not again in 2 years (Leighton and Leighton 1983). The smaller size of the ♀ chicks is apparent at fledging. Wild chicks were fed until at least 6 months after fledging, being supplied with all their food for the first few months. One juv remained in the parental territory for some months after it became independent, leaving only when a flock of subads passed through the area (Leighton 1986). A juv began trial plastering when 14 months old, using its own excreta (Frith and Douglas 1978). Captives survived at least 21 and 22.8 years (Schenker 1990; K. Brouwer *in litt.* 1993).

Great Philippine Hornbill *Buceros hydrocorax*

Buceros Hydrocorax Linnaeus, 1766. *Systemae Naturae*, edition 12, **1**, 153. Moluccas, designated as Manila, Luzon I., Philippines.

PLATES 8, 14
FIGURE 1.3

Other English names : Rufous Hornbill, Philippine Rufous Hornbill, Philippine Brown Hornbill

TAXONOMY: *B. h. hydrocorax* should probably be considered a distinct monotypic sp., given the differences in soft-part colours, lack of bare gular skin, and distinctive juv plumage, with *B. h. semigaleatus* and *B. h. mindanensis* as two subspp. of a further sp. Best resolved by field and molecular studies, and careful documentation of soft-part colours and imm plumage.

Description

B. h. hydrocorax Linnaeus, 1766. Great Luzon Hornbill.

ADULT ♂: forehead, cheeks, chin, and breast black; white band across throat. Head, neck, upper breast, thighs, abdomen, and undertail-coverts rufous. Wing-coverts, back, and uppertail-coverts dark brown. Remiges black with buff tips on the outer webs. Tail white, becoming stained yellowish. Bill, with large block-like casque extending forward into a point, crimson from oxidation of gamboge-yellow preen-gland oils (McGregor 1909), with black base to mandibles. Bare circumorbital skin yellow. Eyes red, feet red-brown.

ADULT ♀: same overall coloration as ♂, but smaller, bill and casque without black marks; circumorbital skin black, eyes white (specimen records; requires confirmation for both sexes). Back and shoulders of two specimens smeared with olive-yellow powder, probably from preen gland (Tweeddale 1877).

IMMATURE: unique among hornbills in being completely different from ad. Head, neck, and body white. Tail white, with broad black band across and grey-brown shading at base of all but the central two pairs of feathers (much individual variation in tail markings). Wings black with white tips to all coverts and edges to all remiges, giving mottled effect. Bill black, with red tip and base of lower mandible, and casque undeveloped. Circumorbital skin and eye colours undescribed; legs and feet black.

SUBADULT: change to ad coloration begins at 3–4 months, and moult of body plumage is complete within a year.

SUBSPECIES

B. h. mindanensis Tweeddale, 1877. Great Mindanao Hornbill. Ad coloured like nominate subsp., but slightly smaller, casque relatively larger, distal half of bill pale yellow (not all crimson), circumorbital skin black with yellow patch below eye (not all-yellow), small area of bare throat skin yellow, eyes pale blue-grey or green in both sexes (Witmer 1989; not red in ♂), and legs and feet light coral-red (not red-brown). Ad ♀ like ♂, but smaller, and circumorbital and throat skin yellow. Juv with bill all black, basal half of the outer three pairs of rectrices more or less black and without clear black band across centre, facial skin yellow, eyes brown, legs grey-green.

B. h. semigaleatus Tweeddale, 1878. Great Samar Hornbill. Same overall coloration at all stages as *B. h. mindanensis*, but casque only a shallow wedge. ♀ with circumorbital and throat skin greenish yellow. Juv with basal half of tail feathers more or less black like *B. h. mindanensis*.

MOULT

No obvious season (Sanft 1960).

MEASUREMENTS AND WEIGHTS

B. h. hydrocorax SIZE: wing, ♂ ($n = 8$) 395–421 (409), ♀ ($n = 12$) 362–400 (380); tail, ♂ ($n = 8$) 310–340 (334), ♀ ($n = 12$) 285–320 (302); bill, ♂ ($n = 6$) 195–215 (208), ♀ ($n = 8$) 169–205 (189); tarsus, ♂ ($n = 8$) 58–61 (59), ♀ ($n = 9$) 55–60 (57). WEIGHT: ♂ 1824.

B. h. mindanensis SIZE: wing, ♂ ($n = 21$) 375–405 (389), ♀ ($n = 21$) 355–375 (366); tail, ♂ ($n = 20$) 291–324 (308), ♀ ($n = 15$) 275–296 (291); bill, ♂ ($n = 16$) 180–210 (197), ♀ ($n = 10$) 155–205 (179); tarsus, ♂ ($n = 14$) 57–61

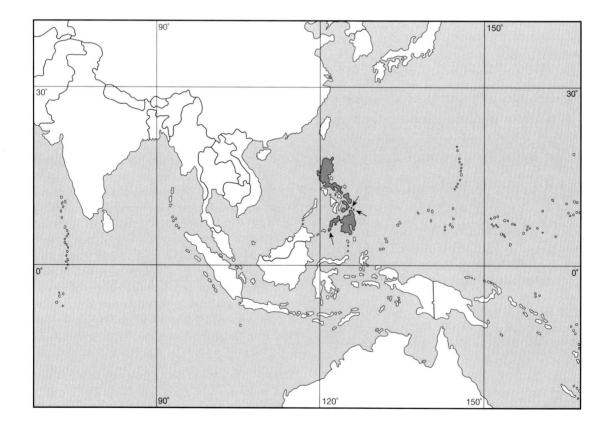

(58), ♀ (*n* = 11) 53–58 (56). WEIGHT: ♂ (*n* = 6) 1345–1612 (1574); ♀ (*n* = 6) 1413–1662 (1557) (Rand and Rabor 1960).

B. h. semigaleatus SIZE: wing, ♂ (*n* = 10) 378–397 (386), ♀ (*n* = 15) 350–365 (360); tail, ♂ (*n* = 10) 290–320 (302), ♀ (*n* = 9) 270–285 (280); bill, ♂ (*n* = 8) 177–190 (182), ♀ (*n* = 7) 160–180 (168); tarsus, ♂ (*n* = 10) 54–59 (57), ♀ (*n* = 10) 52–56 (54). WEIGHT: ♂ (*n* = 7) 1185–1550 (1357), ♀ (*n* = 3) 1017–1200 (1136) (Samar), and ♀ (*n* = 5) 1369–1652 (1517), ♀ (*n* = 6) 1171–1307 (1246) (Leyte) (Rand and Rabor 1960).

Field characters
Length 60–65 cm. The largest hornbill on the Philippines; the rich rufous body, brown wings, and black face, neck, and breast contrast with the white tail and crimson bill. Juv quite different, white with a black bill and mottled back and tail. In small noisy flocks, the loud honking calls and whooshing wingbeats also attract attention. Other spp. either much smaller (*Penelopides*) or with predominantly black and white plumage (*Aceros*).

Voice
Clear resonant *honk*, repeated (Stott 1947), ♀ at higher pitch. Captive utters harsh acceptance-squawk on taking food, and a short coarse bark when angry or frightened. No differences between subspp. documented. Tape recordings available.

Range, habitat, and status
Philippines: *B. h. hydrocorax* on islands of Luzon and Marinduque; *B. h. mindanensis* on Mindanao and Basilan (Sanft 1960), since recorded from Dinagat and Siargo, with sightings from Balut, Bucas, and Talicud (Dickinson *et al.* 1991); and *B. h. semigaleatus* on Samar, Leyte, Bohol, Panaon, and Buad (Parkes 1965), and recently traced to Calicoan and sighted on Biliran (Dickinson *et al.* 1991).

Confined to stands of primary evergreen dipterocarp forest at up to 400 m (Rand and Rabor 1960), also using some secondary forest (duPont and Rabor 1973*a*), and ascending to 760 m on Mt Isarog on Luzon I. (Goodman and Gonzales 1990) and to 2100 m on Mt Apo on Mindanao (Hachisuka 1934).

Used to be abundant on Samar in hills behind Catbalogan, also in mangrove swamps near the town (McGregor 1909). Up to 40 used to gather at fruiting trees on Mindanao (Kutter 1883). Habitat now much reduced, and absent from many areas of previous range. Primary (30%) and secondary forests still relatively extensive on Luzon and Mindanao, but reduced to less than 30% on Samar, Leyte, and Bohol, and on the remaining relatively small islands even further reduced (Worth *et al.* 1994). Skull may be used as part of decorative head-dress (Fig. 1.3, p. 4)

Feeding and general habits
Recorded in small groups of 3–7 that range through the canopy or fly small distances just above it (Stott 1947; Witmer 1989). Regular morning and evening calling from communal roosts, then, during passage from tree to tree and hillside to hillside, calls during alternating flapping and gliding flight. Known as 'clock of the mountains' for regular calling times, and audible at over 1.5 km (duPont and Rabor 1973*a*).

Rest of the day silent and relatively inactive in canopy of large forest trees. Feeds in tall fruiting trees, up to 12 gathering at one time, and noisy with much jumping about (Rand and Rabor 1960). Commonly joins larger flocks of Mindanao Wrinkled Hornbill *Aceros leucocephalus*, usually 20 or more, including feeding, on passage, and roosting together. Calling stimulates reply from neighbouring groups, suggesting territoriality, and also responds to imitation of calls (McGregor 1909).

Feeds mainly on seeds and fruit, including, on Mindanao, insects, wild figs, and feral guavas (Aug–Oct: Stott 1947). Rarely comes down to low bushes (McGregor 1909), but in early morning descends to the ground to scrape around tree roots and add insect larvae, centipedes, and grasshoppers to the diet (Tweeddale 1877). Captive ad ♂ preferred fruits high in sugar and red in colour; became very excited, wobbling head like begging juv, when offered animal food, first softening prey in bill. Crest raised as food is swallowed, and some insects even caught in mid-air (Stott 1947). Some specimens carry extensive subcutaneuous and abdominal fat long after the breeding season, while others collected at same time are lean (Curl 1911).

Leucocytozoon sp. (*andrewsi* type) reported in blood cells of three specimens (Manuel 1969).

Displays and breeding behaviour
Monogamous, breeding in co-operative groups with helpers (Witmer 1994). Courtship-feeding of *alpha* ♀ by other ad group-members recorded (Witmer 1994), including with both fruits and arthropods. Copulation twice recorded near the nest site. ♂ repeatedly rubbing his head under chin of ♀ and then jumping over her back, up to six times before mounting. ♀ also recorded allopreening with another ad.

Breeding and life cycle
Little known. ESTIMATED LAYING DATES: Leyte, May; Mindanao, Mar–Apr. Birds in juv plumage collected on Mindanao in Apr and on Luzon and Basilan in Nov (Dickinson *et al.* 1991). The timing of breeding at two nests suggested differences in timing of the breeding cycle for different groups (Witmer 1994). c/2 – 4. EGG SIZES: *B. h. mindanensis* ($n = 2$) 56.7 × 41.0 and 57.5 × 38.6, weights when fresh 480 and 465 g (Kutter 1883).

One nest on Mindanao was 30 m up in the top of the trunk of a huge 40-m emergent dipterocarp (DBH 220 cm), and another 14 m up in a virtually dead 20-m tree (DBH 102 cm) (Witmer 1994). One group comprised 4 ads and an imm, one presumably ad ♀ repeatedly entering the cavity. ♀ in nest excavated cavity and threw out debris, while others chased off *Aceros leucocephalus* or hammered with their bills around the nest-entrance. A neighbouring group consisted of 2 ads and an imm.

The second nest, in which the chick was close to fledging and the ♀ had already emerged, was apparently unsealed, with a long vertical slit-like entrance hole. It was visited by 2 ads, presumably the pair and both with blue-grey eyes, and an imm. The ♀, ♂ and imm each delivered food on 23%, 44% and 15% respectively of nest visits ($n = 72$). Feeds averaged every 1.35 hr (SE 0.13, range 0.25–3.00 hr), often by only one bird, although they usually arrived as a group. Once chased off intruding conspecific from nest area. Only a single chick fledged, and the nest tree was blown down before the next season.

Hybrid pair of nominate subsp. with another subsp. laid 4 fertile eggs that failed to hatch (K. Brouwer *in litt.* 1992). A captive ♂ survived at least 15.7 and a ♀ 13.9 years (K. Brouwer *in litt.* 1992).

Great Helmeted Hornbill   *Buceros vigil*

Buceros vigil J. R. Forster, 1781. *Indische Zoologie,* p. 40 (after Edwards, 1760, *Gleanings of Natural History* 2, plate 281, figure C). Designated as Sumatra, restricted to Benkulen, Sumatra, Indonesia.

PLATES 1a, 1c, 2f, 8, 14

FIGURE 1.1

Other English names: Solid-billed Hornbill, Helmeted Hornbill

Description
ADULT ♂: head and neck bare, but for dark brown feathers on crown and rufous feathers on cheeks. Body and wings dark brown but for white thighs, abdomen, and underwing- and undertail-coverts. Remiges black with white tips. Tail white with black subterminal band and white tip; the central pair of feathers are elongated by about 20 cm and silvery grey, but also with the black subterminal bar and a white tip. Bill, with broad high casque ridged along side and terminating in solid vertical

Great Helmeted Hornbill *Buceros vigil* 193

plate of hornbill 'ivory' midway along bill, dark red from preen gland, except for yellow where worn off tip and front of casque. Bare face and neck skin blood-red. Eyes dark red or red-brown, legs and feet red-brown.

ADULT ♀: same overall coloration as ♂ but smaller; bill with black speckles at tip. Bare face and neck skin pale lilac, with extensive light turquoise under throat veined with pale blue; eyes red-brown.

IMMATURE: same overall coloration as ad ♀, but central rectrices still growing, casque undeveloped; bill yellowish green. Bare head and neck skin pale greenish blue, eyes pale red-brown.

SUBADULT: the layering of horn in the casque probably results from periodic growth, and may therefore allow ageing (Manger Cats-Kuenen 1961).

MOULT
Reported to drop central pair of tail feathers one at a time (Wetmore 1915), each apparently once a year starting Dec–Mar, but may not be so irregular since all pairs of tail feathers apparently shed asymmetrically (Stresemann and Stresemann 1966), though unlike typical tail moult of other hornbills (Friedmann 1930). In the chick, only one central tail feather appears to emerge initially (Wetmore 1915).

MEASUREMENTS AND WEIGHTS
SIZE: wing, ♂ (n = 40) 465–540 (494), ♀ (n = 17) 420–475 (443); tail, ♂ (n = 32) 730–982 (831), ♀ (n = 10) 615–790 (720); bill, ♂ (n = 25) 175–220 (196), ♀ (n = 9) 158–185 (171); tarsus, ♂ (n = 20) 60–78 (69), ♀ (n = 8) 62–67 (65). WEIGHT: ♂ (n = 1) 3060; ♀ (n = 2) 2610, 2840. One ♂ 2495, very thin.

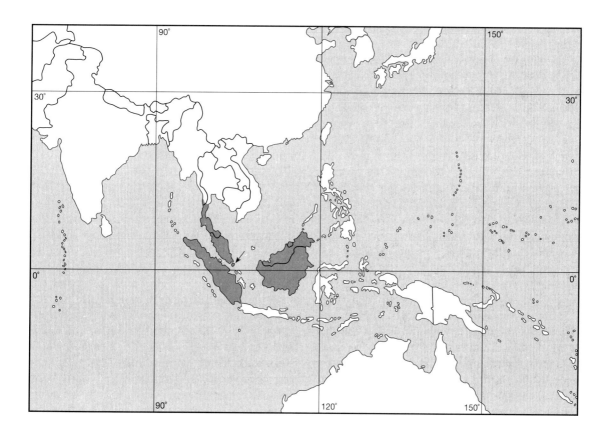

Great Helmeted Hornbill *Buceros vigil*

Field characters

Length 110–120 cm, including 20 cm of extended central tail feathers. A very large, brown and white hornbill, the relatively short, stubby bill, white tail with black tip, and elongated central tail feathers distinctive. Bare skin of head and neck impressive, either red (♂) or turquoise (♀). Sympatric with very large Great Rhinoceros Hornbill *B. rhinoceros* over all of range, and with Great Pied Hornbill *B. bicornis* on Sumatra and Malay Peninsula: these two spp. have long bills with prominent casques, and the latter has extensive areas of white in the wings and on the head and neck. The hooting and cackling call of *B. vigil* is loud and unique.

Voice

Loud call of ♂ audible to 2 km, one of loudest and most distinctive forest sounds (Haimhoff 1987). Described as a series of prolonged deep hooting *Hou* notes, rising and quickening into double notes until the crescendo is an unbroken series of shrill cackles *Hee-Hee-Hee*, before dying away to mocking *Ha* notes (Banks 1935). When alarmed, utters a quite distinct series of shrill, high startling cackles.

Three phases to loud call studied in detail (Haimhoff 1987): the repeated long, monotonal single *oo* notes, leading into the two-note *te-oo*, and finally the 10–20 short notes of the cackle phase. General pattern is of uniform acceleration in call rate, changing midway from *oo* to *te-oo* on course to the final cackle. May abort call sequence at any time during *oo* phase but not once into *te-oo* notes. Cackle notes formed like *oo* notes on sonograms, but louder, with more harmonics, and stress on harmonics 2–4, not on fundamental frequency as in *oo* notes; the resultant sound is quite different. One bout of 27 *oo* hoots lasted 2.75 min. Captive and wild birds fly upwards towards the end of the cackle phase (Haimhoff 1987), to repeat similar loud cackling and braying calls in flight (Kemp 1988*b*). Uttered alone or by pairs in syncopated duet, a series lasting up to 4 mins, and bouts of calling continuing for over 1 hr (Leighton 1982). ♀ also calls in higher pitch, but may begin and end the sequence before ♂ (Holmes and Nash 1990).

The loud call is ideal for forest transmission, given from high up, with low-middle frequency range, energy confined to a few harmonics, and consisting of pure notes. Separation into two parts may act to warn neighbours in large widespread territories before uttering the individually distinctive cackle climax. Convergent in design with characteristic loud forest calls of primates such as gibbons and call of the Great Argus Pheasant *Argusianus argus*. Tape recordings available.

Range, habitat, and status

Myanmar, S from Bankachon in S Tenasserim; SW Thailand; Peninsula Malaysia; Singapore, where extinct by 1950; Indonesia, on Sumatra and Kalimantan; Sarawak; Sabah; Brunei (Sanft 1960).

Inhabits moist evergreen forest at up to 1100 m, but absent from coastal peat-swamp forests (Smythies 1953). Least common on plains, and most abundant among foothills in the tops of large trees (Banks 1935). Extends into selectively logged but not secondary forests. Similar habitat requirements to but less common than *B. rhinoceros* (Smythies 1953; Gore 1968).

Not uncommon on Sumatra where forest undisturbed (van Marle and Voous 1988). Endangered in Thailand (Worth *et al.* 1994). Densities from census counts in Sabah of 1.8 km^2/pair in logged forest, up to 0.9 km^2/pair and 0.5 km^2/pair in 6-year logged and small isolates respectively (Johns 1988*a*); in Sarawak at densities of up to 0.12 km^2/pair in primary foothill forest (Kemp and Kemp 1974); in Kalimantan 7–8 km^2/pair (Leighton 1986).

Long subject to loss of prime habitat through shifting agriculture, and to being killed for carving and medicinal use of ivory. The tail feathers were also valued in Sarawak for the war helmets of Kayan, Kenyah, and Kelamantan men, worn only by those who had killed an enemy and removed by young men before dancing (Hose 1893; Banks 1935). The ivory was most often used locally as ear drops and plugs for old men (Harrison 1951), and casques and plumes fetched $4 each (Banks 1935). Casques had been exported to China since the Ming Dynasty, for use in *ho-ting* carving and making of belt buckles and bangles (Camman 1951), and to India as love charms fetching 50 rupees (Baker 1927). Skulls and tail feathers were still on sale more recently in Sabah (Thompson 1966), Sarawak (Kemp and Kemp 1974), and Singapore (pers. obs. 1991). A carved casque was used as a container to smuggle cocaine into the USA (B. Gilroy *in litt.* 1989). The Great Helmeted Hornbill is the emblem of W Kalimantan Province, Indonesia.

A Malay legend uses the unique calls to name it 'Kill your mother-in-law bird', based on the story of a youth who cut down the bamboo supports to his mother-in-law's hut (the repeated hooting notes) and was delighted when she died (the maniacal cackling). The penalty was to be changed into a hornbill and be condemned for ever to relive the act (Chhapgan 1977).

Feeding and general habits

Resident as territorial pairs, apart from small nomadic flocks of subad birds (Leighton and Leighton 1983). Mostly recorded singly or in pairs, sometimes with a single young or with 3–4 meeting at territory boundaries. Small flocks of 8, comprised of juvs, subads, and non-breeding ads, may briefly visit an area, moving and foraging steadily through the canopy (Kemp and Kemp 1974; Leighton 1986). Favours the canopy of larger trees, including emergents and clumps of trees standing in selectively logged forest. Not recorded low down or on the ground. Flight direct compared with other *Buceros* spp., with relatively fast flapping and the long tail waving astern, but also noisy.

Calls territorially at any time between 08.00 and 17.00 hrs, from 30–60 m up in the middle to upper canopy, usually with the ♀ mate perched 50–100 m away (Leighton 1986; Haimhoff 1987). Both sexes call and defend the territory (Kemp and Kemp 1974; Leighton 1986; D. Wells *in litt.* 1992), although not against nomadic non-breeding flocks. Pairs interact several times per day with neighbours and are very responsive to their calls, also reacting strongly to imitation or playback of the loud call, as has long been known to the Bornean Penan (Banks 1935; Smythies 1960; Kemp and Kemp 1974).

Forages by visiting fruit trees, between bouts of hunting small vertebrate and large arthropod animals, often prising off bark with the bill to uncover these small animals (Leighton 1986). Pair-members usually forage far apart during long and systematic bouts of hunting animals, keeping in contact by occasional calling (Leighton 1982). One ♀, and later ♂, visited an active nest of Sunda Pied Hornbill *Anthracoceros albirostris convexus* and hammered at the back of the nest, the resident ♀ screaming for 15 min but not helped by its mate (Bartels and Bartels 1937). Even the large *B. rhinoceros* shows intense aggression to a *B. vigil* passing near its nest, attacking it and calling loudly (Hetharia 1941), suggesting that *B. vigil* may be a serious predator.

The diet includes fruit, especially *Ficus* and *Urostigma* figs, and *Parkia speciosa* beans, whose odour affects the flesh. Takes at least 10 spp. of figs regularly in Malaysia (Lambert 1989), and in Kalimantan sugar-rich figs were the only fruits in the diet (Leighton 1986), suggesting that it is somewhat of a fig specialist. About half the time was spent foraging for animal foods ($n = 14$; first contacts: Leighton 1986), which are known to include the squirrel *Sciurus tennuis*, and birds.

The unique ivory-laden casque of the ad causes the skull and bill to form about 10% of the body weight (Manger Cats-Kuenen 1961). The bill is also shorter, straighter, and the casque less protruding than in other *Buceros* spp. Its function has long been debated: local Borneans suggested that it was for tamping down the nest sealing (Sharpe 1890); ♂♂ sometimes butt casques in flight for as long as 2 hr (Schneider 1945; Hubback in Madoc 1947; Reddish 1992b) suggesting an agonistic role; it has been observed hammering branches of large trees and knocking off bark like a woodpecker (Forbes in Manger Cats-

Kuenen 1961; Leighton 1982); and use of the bill as a weighted chisel for excavation of small animals from tree holes and under bark has been proposed (Kemp 1979; Burton 1984).

The design of the casque and skull (Manger Cats-Kuenen 1961) would support the hammering, digging, and butting functions. Skull of the juv is like that of other hornbills in mass, articulation, and form, but in ad is apparently designed in several ways to resist stress on the tip of bill and especially the rostral part of casque. Bony trabeculae aligned to resist stress almost fill the casque against the skull, which is itself reinforced laterally and centrally with the massive inter-orbital septum and thick neurocranium. There is only limited kinesis at the *fissura cranio-facialis* and the rod-like trabeculae are only in line to resist stress and receive pressure when the bill is closed. There are other kineses with ligaments that help to hold the bill closed without much muscular force, opening of the lower jaw being possible only with relaxation of these kineses. The skull both is denser and has the centre of gravity further forward and higher than in any other hornbills, owing to the amount of reinforcing bone and solid 'ivory' anterior to the casque; the neck and skull muscles, as indicated by the marks of their attachment, developed accordingly. The basi-occipital condyle, by which the head articulates on the neck, and the accessory supra-occipital condyle, which arrests the head when jerked backwards, are both especially well developed in this species (Bock and Andors 1992). In summary, *B. vigil* appears to have strong jaw-closing mechanisms, for squeezing prey or holding branches to climb, especially given the straight bill, and these, together with the bill–casque structure, might allow considerable force to be applied for knocking down fruit, fighting, or chipping away bark and soft wood.

Displays and breeding behaviour
Breed in a permanent territory as monogamous pairs without helpers. During territorial calling, the bill is pointed skywards as the notes of the loud call accelerate, and then the final cackle phrase of the loud call is uttered with the bill lowered and the mouth wide open (Wells 1974; Haimhoff 1987). A threat posture of a captive imm consisted of crouching with body horizontal, head retracted, bill open, wings drooped half open, and tail raised almost to the vertical.

Breeding and life cycle
Little studied. ESTIMATED LAYING DATES: Sumatra, Jan–Mar, May, Nov; Sarawak, Feb, Nov, with no obvious seasonality. c/1–2, but only one chick ever recorded with adults. Egg size and form unrecorded.

One nest in Sumatra was high up in a side branch of a large *Koompasia parviflora*, the entrance about 90 cm high × 50 cm wide but the cavity only 18 cm deep. The entrance was sealed with mud, fig remains, and nest debris, and the floor was lined with wood chips (Schneider 1945). Others were in isolated forest trees (van Marle and Voous 1988). The chick's skin remains pink except in areas where black plumage will develop, and the ♀ emerges sometime before the chick fledges (Schneider 1945). During two years in Kalimantan, pairs bred once during a period of high fruit availability but not the next year. A chick that fledged was supplied with all its food by the parents for the first few months, and was still receiving some food 6 months after fledging (Leighton 1986). Extremely difficult to maintain in captivity, but a captive pair has recently sealed in and laid eggs that failed to hatch (L. Kuah *in litt.* 1992).

Genus *Penelopides* Reichenbach, 1849

Small to medium-sized hornbills. Two outer primaries (9 and 10) emarginated, tail slightly graduated, bill with transverse or longitudinal ridging. Bare circumorbital and small gular skin patches. Ad ♀ usually distinctive. Juv of both sexes resembling ad ♂ in plumage, except in one sp. where sexes similar (*P. mindorensis*) and another in which juvs resemble their respective sex (*P. manillae*). Distinct ♀ plumage allies them with *Anorrhinus* and especially *Aceros* spp.

Biology very little known. Utter soft squeaking calls, providing onomatopoeic common name of tarictic hornbills. Some live in groups and probably breed co-operatively. Breeding biology little known, mainly from captivity. ♀ sealed nest alone, emerged with chick at end of 102 day nesting cycle, and fed at nest by ♂ through regurgitation. ♀ moults flight feathers simultaneously while breeding. Chicks retain pink skin throughout development.

At least 5 spp., restricted to the Wallacean biogeographical zone of Sulawesi and the Philippines. One sp. on Sulawesi. The other 4, previously considered as one sp., occur on various islands in the Philippines and surely form a supersp.; several of them have distinctive subspp. (Kemp 1988*b*). These 4 spp. are separated on aspects of plumage ontogeny, sexual dichromatism, and soft-part colours which suggest that they could not all be included in the single sp. *P. panini sensu lato*. Further detailed field and molecular studies are warranted.

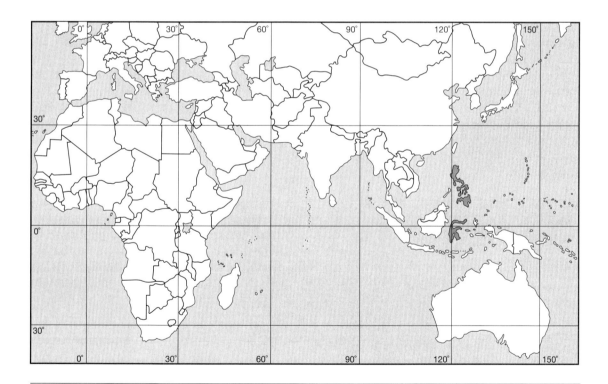

Sulawesi Tarictic Hornbill *Penelopides exarhatus*

Buceros exarhatus Temminck, 1823. *Planches colorees* **36,** plate 211. Celebes, restricted to Tondano, NE Sulawesi.

PLATES 9, 14

Other English name: Sulawesi Dwarf Hornbill, Sulawesi Hornbill, Celebes Hornbill, Celebes Tarictic Hornbill

Description
P. e. exarhatus (Temminck), 1823. North Sulawesi Tarictic Hornbill.

ADULT ♂: black, with metallic green sheen on upperparts, but for white face and throat. Bill, with low casque ridge grooved along length, the casque terminating abruptly just before bill tip, pale yellow, with brown casque and black patch at base of lower mandible. Bare circumorbital and throat-skin patches flesh-coloured. Eyes red, legs and feet black.

ADULT ♀: all black without white face, and smaller than ♂. Casque merging into bill midway along length, and black patch at base of lower mandible less developed or absent. Bare circumorbital skin black with yellow stripe below eye; throat pale yellow to flesh-coloured.

IMMATURE: plumage like ad ♂ in both sexes, but bill and casque less developed. Bill colour pale yellow, bare facial skin pale yellow, eyes dark brown.

SUBADULT: during Oct–Mar, when about one year old, imm ♀ moults the white face and the facial skin darkens to ad colours.

SUBSPECIES
P. e. sanfordi (Stresemann), 1932. South Sulawesi Tarictic Hornbill. Like nominate subsp. in all respects, but in ad ♂ the black area at the base of the lower mandible is barred with yellow and the casque is a redder brown. Imm apparently with cream-coloured superciliary stripes in some specimens (van Bemmel and Voous 1951).

MOULT
No obvious moulting season (Sanft 1960).

MEASUREMENTS AND WEIGHTS
P. e. exarhatus SIZE: wing, ♂ ($n = 13$) 226–245 (235), ♀ ($n = 7$) 215–231 (223); tail ♂ ($n = 10$) 180–205 (195), ♀ ($n = 6$) 176–191 (181); bill, ♂ ($n = 10$) 92–104 (98), ♀ ($n = 6$) 82–88 (86); tarsus, ♂ ($n = 10$) 39–43 (41), ♀ ($n = 4$) 39–42 (40). WEIGHT: ♀ 370.

P. e. sanfordi SIZE: wing, ♂ ($n = 20$) 223–252 (237), ♀ ($n = 17$) 208–235 (222); tail, ♂ ($n = 8$) 195–210 (199), ♀ ($n = 8$) 176–197 (185); bill, ♂ ($n = 15$) 89–107 (97), ♀ ($n = 15$) 82–97 (91); tarsus, ♂ ($n = 5$) 41–42 (41) ♀ ($n = 4$) 38–41 (40). WEIGHT: unrecorded.

Field characters
Length 45 cm. The smaller of two hornbill spp. on Sulawesi, all black with a dark bill and with some white around the face only in ad ♂ and imm of both sexes. Much larger Sulawesi Wrinkled Hornbill *Aceros cassidix* has a white tail, and the bill, with its high casque, and extensive bare facial skin are brightly coloured.

Voice
A loud, piercing *kerrekerre ... kerrekerre*, the 4 syllables repeated at short intervals. Also a chicken-like cackling (Stresemann 1940). Also described as a staccato *clack*, similar to Bushy-crested Hornbill *Anorrhinus galeritus* (O'Brien and Kinnaird in press). Generally a garrulous sp. (Watling 1983). Tape recordings available.

Range, habitat, and status
Indonesia. *P. e. exarhatus*: Sulawesi Is, S to Koelavi and Rano Lindoe as well as the island of Lebeh. *P. e. sanfordi*: Sulawesi Is, S from Tonkean and on islands of Muna and Buton (Sanft 1960; van Bemmel and Voous 1951).

Primary evergreen forest at up to 1100 m, especially that known as 'rimbu' (Watling

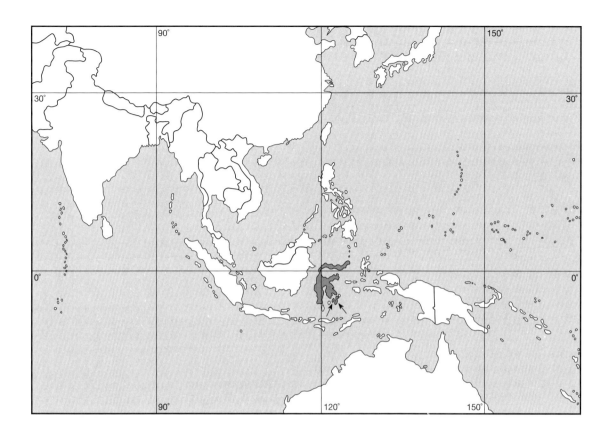

1983). Common in lowland and lower montane forests, despite regular hunting for food and pets (Mackinnon 1979), populations in S more divided by habitat destruction (Worth *et al.* 1994). Most common below 650 m in Tangkoko-Dua Saudara Nature Reserve, N Sulawesi, at density of 1.4 km²/group and each group with home range of 1.5–2 km² (n = 2, O'Brien and Kinnaird in press). Due to rapid loss of lowland forest, the species should be considered vulnerable (O'Brien and Kinnaird in press).

Feeding and general habits
Usually in small family groups of 4–8, moving mainly below the canopy and searching slowly through the foliage and from tree to tree (Stresemann 1940; Watling 1983; Whitten *et al.* 1987). This suggests that it may spend much time hunting small animal prey. Recorded chasing only other sympatric hornbill, *Aceros cassidix* (Holmes 1979).

Displays and breeding behaviour
Always in small groups of 2–10 (mean 4), even when breeding, suggesting that it may be group-territorial and breed co-operatively (Whitten *et al.* 1987).

Breeding and life cycle
Little known. ♂ collected on 10 Oct from Mt Tentolomatinan had well-developed testes. EGG SIZES: (n = 1) 47.3 × 34.8 (Schönwetter 1967). Reported as c/2 (Sanft 1960). Egg rough-shelled and dirty white.

Visayan Tarictic Hornbill *Penelopides panini*

Buceros panini Boddaert, 1783. *Table des Planches Enluminées d'Histoire Naturelle, de M. D'Aubenton.*, p. 48 (after Daubenton, 1783, *Planches Enluminées*, plate 781). Panay, restricted to San José de Buonavista, Panay I., Philippines.

PLATES 9, 14

Other English name: Tarictic Hornbill, Panay Tarictic Hornbill

Description
P. p. panini (Boddaert), 1783. Panay Tarictic Hornbill.

ADULT ♂: head, neck, and upper breast yellowish white. Ear-coverts and throat black. Lower breast, abdomen, thighs, and uppertail- and undertail-coverts rufous. Upperparts and wings black with olive-green metallic sheen. Tail reddish yellow with broad black tip. Bill, with casque a low ridge terminating midway along bill, red, with yellow ridges across upper and lower mandibles. Bare circumorbital skin white to flesh-coloured and throat-skin patches black. Eyes red, legs and feet dark brown.

ADULT ♀: like ad ♂, but head, neck, underparts, and uppertail-coverts black; bare facial skin pale blue, eyes red, legs and feet black.

IMMATURE: plumage like ad ♂, irrespective of sex, but uppertail-coverts and tip of tail brown. Bill less developed and olive-brown. Bare facial skin white, tinged blue; eyes brown, legs and feet dark greenish grey.

SUBADULT: imm ♀ with underparts moulting into black feathers of ad in Jan, probably about 6 months old, and bill colour also changing.

SUBSPECIES
P. p. ticaensis Hachisuka, 1930. Ticao Tarictic Hornbill. Larger than nominate subsp., and with darker rufous uppertail-coverts, tail white, not pale rufous, and underparts with rufous wash (du Pont 1972). Upperparts glossed with blue-green, not olive-green, and the ♀ darker.

MOULT
Occurs at beginning of rains (Sanft 1960).

MEASUREMENTS AND WEIGHTS
P. p. panini SIZE: wing, ♂ ($n = 15$) 248–275 (263), ♀ ($n = 14$) 236–265 (249); tail, ♂ ($n = 10$) 207–235 (220), ♀ ($n = 9$) 190–223 (207); bill, ♂ ($n = 8$) 100–120 (108), ♀ ($n = 6$) 85–102 (94); tarsus, ♂ ($n = 9$) 42–48 (46), ♀ ($n = 8$) 43–45 (44). WEIGHT: unrecorded.

P. p. ticaensis SIZE: wing, ♂ ($n = 1$) 292, ♀ ($n = 1$) 274; tail ♂ ($n = 1$) 280; bill, ♂ ($n = 1$) 123; tarsus, ♂ ($n = 1$) 51. WEIGHT: unrecorded.

Field characters
Length 45 cm. The only small hornbill on these W islands of the Philippines. All black, but for whitish head, neck, and underparts of ad ♂ and imm of both sexes. Red bill and ochre tail with broad black tip are distinctive. The much larger Visayan Wrinkled Hornbill *Aceros waldeni* has an extensive white band across the black tail, bright red bill and casque, loud calls, and noisy flight.

Voice
Calls *te.rik-tik-tik-tik*, the last notes rapidly following the first one (Ripley and Rabor 1956). Onomatopoeic for common collective name of genus, 'tarictic'. Noisy, keeping up incessant *Ta-ric-tic, ta-ric, ta-ric-tic* (Rabor 1977). A nasal, high-pitched trumpeting that does not carry far (Brooks *et al.* 1992). Also makes swishing wing noise in flight, but can fly silently when wary. No tape recordings located.

Range, habitat, and status
Philippines, *P. p. panini*: Panay, Masbate, Guimaras, Negros, and recently Pan de Azucar

Visayan Tarictic Hornbill *Penelopides panini*

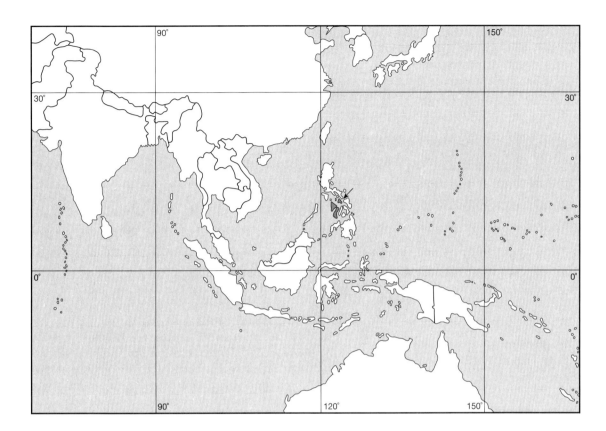

and Sicogon Is. (Sanft 1960; Dickinson *et al.* 1991). *P. p. ticaensis*: Ticao I.

Occupies primary evergreen dipterocarp forest at up to 1050 m, sometimes moving up into mid-montane forests or wandering into secondary forest or to lone fruiting trees adjacent to forest (Rabor 1977).

Was very common on Panay, Guimaras, Negros, and Masbate but absent from Cebu (McGregor 1909; Brooks *et al.* 1992). Masbate, Guimaras, and Ticao (with distinct subsp.) now largely deforested (Worth *et al.* 1994), and forest reduced to 10–30% on other islands. Recorded recently at three locations on Negros, but under extreme threat at least on Panay and Negros (Brooks *et al.* 1992).

Feeding and general habits

Recorded in small groups, with up to 12 gathering at a fruiting tree. Found low in the canopy, often by forest edge (Brooks *et al.* 1992). Fruit and beetles reported in diet (McGregor 1909). Known to hawk flying ants (Gonzales and Rees 1988). Stomachs of 22 specimens collected on Negros contained mainly fruit with a few insect remains, mainly beetles, and several pieces of earthworm (Rabor 1977).

Displays and breeding behaviour

Undescribed.

Breeding and life cycle

Little known (Sanft 1960). NESTING CYCLE: about 95 days. ESTIMATED LAYING DATES: Ticao, Mar–Apr. c/3. EGG SIZES: *P. p. ticaensis* (n = 3) 45.7–48.5 × 32.5–33.7 (47.0 × 33.1) (McGregor 1905). Eggs matt white.

Two nests recorded, 12 and 16 m up, the first in a tree by a small stream, the other in a large tree with DNH of 138 cm (McGregor 1909).

The entrance to the first was 11 × 9 cm, and to the second 10 × 15 cm. The nest-sealing material was mainly wood flakes and food remains (seeds, beetle elytra, feathers).

The first nest contained 2 small chicks and an addled egg, the second 3 eggs. Judging from the chicks' development, incubation started from the first egg. ♀ moulted when breeding, to become flightless. The first nest was reoccupied within 22 days of removing the contents, the ♂ feeding the ♀ by regurgitation and a soft call being uttered during feeding.

Luzon Tarictic Hornbill   *Penelopides manillae*

Buceros manillae Boddaert, 1783. *Table des Planches Enluminées d'Histoire Naturelle, de M. D'Aubenton.*, p 54 (after Daubenton *Planches Enluminées*, plate 891). Manila, Luzon I., Philippines.

PLATES 9, 14

Other English name: Tarictic Hornbill

TAXONOMY: spelling *manilloe* unsupported by careful examination of original description, rather *manillae* (Dickinson *et al.* 1991; Browning 1992).

Description

P. m. manillae (Boddaert), 1783. Luzon Tarictic Hornbill.

ADULT ♂: head, neck, and underparts pale yellowish white. Ear-coverts and throat black. Upperparts and wings dark brown, with slight metallic sheen on upperparts. Tail dark brown, with small white tips to outer vane of feathers and 1–4-cm-wide band of white or rufous across centre. Bill, with casque a low ridge terminating midway along bill, brown, with horn-coloured casque and pale yellow or pink ridges across upper and lower mandibles. Bare circumorbital and throat skin pink. Eyes dark red, legs and feet dark brown.

ADULT ♀: like ad ♂, but head, neck and underparts dark brown, although paler than upperparts; band across tail more brown. Circumorbital skin blue; throat pink, edged blue; eyes red-brown or orange.

IMMATURE: plumage like ad of respective sex, but for red-brown tips to rectrices and secondaries (unlike other *Penelopides* spp., where imm ♀ resembles ad ♂); sexes of chicks can be distinguished at fledging (Azua and Azua 1989). Bill and casque less developed, horn-coloured; bare facial skin and eye colours undescribed.

SUBSPECIES
P. m. subnigra McGregor, 1910. Polillo Tarictic Hornbill. Larger than nominate subsp. Underparts and wings black and with well-developed metallic sheen (not dark brown with slight sheen), and broader pale band across tail.

MEASUREMENTS AND WEIGHTS
P. m. manillae SIZE: wing, ♂ ($n = 19$) 226–248 (233), ♀ ($n = 9$) 207–230 (220); tail, ♂ ($n = 15$) 185–210 (198), ♀ ($n = 8$) 166–198 (184); bill, ♂ ($n = 14$) 83–100 (91), ♀ ($n = 6$) 62–80 (73); tarsus, ♂ ($n = 16$) 40–44 (42), ♀ ($n = 8$) 38–41 (40). WEIGHT: ♂ ($n = 9$) 400–479 (450); ♀ ($n = 2$) 470, 475.

P. m. subnigra Size: wing, ♂ ($n = 2$) 260, 262, ♀ ($n = 1$) 259; bill, ♂ ($n = 1$) 106, ♀ ($n = 1$) 90. WEIGHT: unrecorded.

Field characters

Length 45 cm. The only small hornbill on these N islands of the Philippines. All dark brown or black, but for cream underparts, head, and neck of ad and imm ♂. Pink gular skin, dark bill with pale transverse stripes, and white-tipped tail with pale band across centre are distinctive (Frith and Douglas 1978). The much larger *Buceros hydrocorax* has the tail

Luzon Tarictic Hornbill *Penelopides manillae*

white, bill and prominent casque bright red, loud calls, and noisy flight.

Voice
Loud, clear short squeak like child's squeeze toy (Frith and Douglas 1978), or trumpeting *toot-toot, tut-tut* (Delacour and Mayr 1946). Tape recording available.

Range, habitat, and status
Philippines. *P. m. manillae*: Luzon, Marinduque, and Catanduanes Is, plus some small adjacent islets (Sanft 1960; Dickinson *et al.* 1991). *P. m. subnigra*: Polillo Is.

Occurs along riparian forest or around forest edge, also moving out to lone fruiting trees in grassland. Abundant on Luzon in 1909 (McGregor 1909), uncommon at 450–900 m on Mt Isarog (Goodman and Gonzales 1990). There is much forest remaining on Luzon but little on other islands, including Polillo with its endemic subsp. (Worth *et al.* 1994).

Feeding and general habits
Recorded in groups of up to 14, feeding in the middle levels of the forest and using slender perches, vines, and tangles, often moving to the outermost twigs to reach berries (Frith and Douglas 1978). Several groups included imms, ♀♀, and at least 2 ♂♂ in their numbers. One flock of 10–14 called excitedly for 20 mins, pairs chasing one another or perching together within the noisy active group.

Recorded feeding on *Urostigma* fig and *Dysoxylum* fruit (McGregor 1909).

Displays and breeding behaviour
May breed co-operatively in the wild, based on group-living and unusual imm plumage differentiation. Probably monogamous and group-territorial, based on aggression to other hornbills in captivity.

A ♂ was seen twice to leap up and down off perch while still holding perch with the bill, possibly in threat to another ♂. The head and

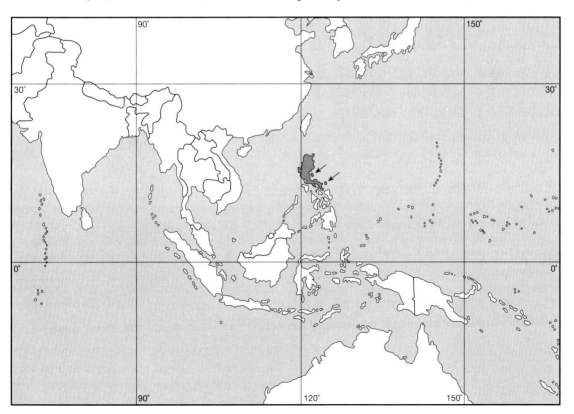

neck were also lowered and the bill pointed at the opponent in typical hornbill threat (Frith and Douglas 1978).

Breeding and life cycle
Recorded only in captivity (Jennings and Rundel 1976; Lieras 1983; Azua and Azua 1989). NESTING CYCLE: 80–102 days (mean 96, $n = 4$): 28–31 days incubation period, 50–65 days nestling period (mean 62, $n = 4$). ESTIMATED LAYING DATES: specimens in breeding condition late Mar to early Apr, a ♀ with a well-developed egg in oviduct (Goodman and Gonzales 1990). Captives: California, Apr (2). One captive pair re-laid 14 days after fledging a chick. c/3–5. Eggs about 32 mm long, details undescribed.

One nest supplied was a 76-cm-long palm log of 51 cm diameter with a 11-cm-square entrance hole, and erected 2 m above ground. ♀ sealed with soil, nest debris, and droppings, with little assistance from ♂ other than delivery of food, some of which used in sealing. Much banging around in nestbox prior to sealing in, which may be an important stimulus to breeding.

Eggs laid 1–5 days apart, c/5 over period of 9 days. Little lining added, and only once laying was under way. ♀ preened and also scooped in scattered eggs with the wings during incubation, escaping to high ledge when disturbed. ♂ regurgitated up to 21 items per visit to the nest. Animal food preferred by ♂ for feeding to the nest during the nestling period, although fruit also taken by the ♀ and passed to chicks until they could feed themselves. ♂ also offered leaf tips at nest.

Chicks have pink skin at hatching, and retain this thereafter. Within a week the chicks do not sleep under the ♀. By 17 days old their eyes are open and they start preening; by 22 days they make the first attempts at flapping. ♀ broke down sealing with almost no help from chicks, emerging a day before one chick fledged, the same day as two others, although in this brood the third chick emerged 10 days later. One pair raised 6 ♂♂, one of which started a second captive generation.

A captive ♂, probably of this sp., which is the commonest hornbill of the genus in aviculture, survived at least 14.8 years (K. Brouwer *in litt.* 1992).

Mindanao Tarictic Hornbill *Penelopides affinis*

Penelopides affinis Tweeddale, 1877. *Annals and Magazine of Natural History, London,* **(4)20**, 534, Butuan, Mindanao I., Philippines.

PLATES 9, 14

Other English name: Tarictic Hornbill

TAXONOMY: details of soft-part colours require confirmation, and field study necessary to confirm specific or subspecific status of each form.

Description
P. a. affinis Tweeddale, 1877. Mindanao Tarictic Hornbill.

ADULT ♂: head, neck, and underparts yellowish white. Ear-coverts and throat black. Upperparts and wings black with metallic green sheen. Tail with black base and tip bordering yellow-brown to red-brown patch in centre. Bill, with casque a low ridge terminating midway along bill, horn-coloured with black patch at base of upper mandible; casque and ridges across base of lower mandible dark brown. Bare circumorbital and throat-skin patches white with blue wash (especially around eyes). Eyes red, legs and feet dark greenish grey.

ADULT ♀: like ad ♂, but smaller, with head, neck, and underparts black (not yellowish), and facial skin darker blue; eyes red-brown to orange, even crimson.

IMMATURE: plumage like ad ♂ for both sexes, but uppertail-coverts chestnut, and rufous tips to tail feathers. Bill grey-brown, with greenish base to lower mandible; eyes brown to grey-brown, legs and feet dark greenish grey.

SUBSPECIES
P. a. samarensis Steere, 1890. Samar Tarictic Hornbill. Slightly larger than nominate subsp., with uppertail-coverts pale rufous in ad ♂ (not black in both sexes), and both sexes with more extensive yellowish to pale area on tail.

P. a. basilanica Steere, 1890. Basilan Tarictic Hornbill. Similar in size to nominate subsp., but rufous area on tail even more extensive than in most ♂ *P. a. samarensis*, so that black base to tail absent. Bill colour differs in ad and juv, with upper mandible and casque horn-coloured and with flesh (not black) patch at base (Parkes 1971); eyes in both sexes red.

MOULT
P. a. samarensis: 2 ♀♀ with flight-feather moult in Apr–May (Rand and Rabor 1960). Specimens indicate moult throughout the year (Sanft 1960).

MEASUREMENTS AND WEIGHTS
P. a. affinis SIZE: wing, ♂ (n = 18) 220–240 (233), ♀ (n = 12) 210–230 (218); tail, ♂ (n = 15) 180–197 (191), ♀ (n = 10) 170–187 (178); bill, ♂ (n = 14) 91–106 (97), ♀ (n = 11) 79–88 (84); tarsus, ♂ (n = 16) 40–44 (42), ♀ (n = 11) 39–42 (41). WEIGHT: ♂ 456 (Rand and Rabor 1960).

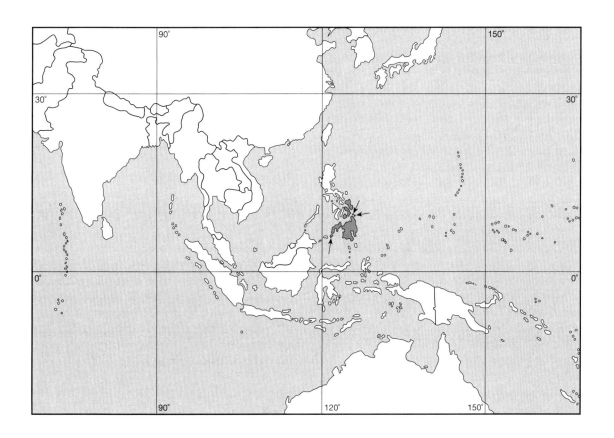

P. a. samarensis SIZE: wing, ♂ (*n* = 14) 230–247 (240), ♀ (*n* = 12) 210–230 (218); tail, ♂ (*n* = 14) 190–207 (199), ♀ (*n* = 3) 177–183 (179); bill, ♂ (*n* = 13) 93–105 (101), ♀ (*n* = 3) 80–84 (81); tarsus, ♂ (*n* = 14) 41–44 (43), ♀ (*n* = 3) 40–42 (41). WEIGHT: ♂ (*n* = 11) 453–584 (509), ♀ (*n* = 7) 335–506 (422) (Samar); ♂ (*n* = 4) 497–554 (523), ♀ (*n* = 6) 394–514 (450) (Leyte) (Rand and Rabor 1960).

P. a. basilanica SIZE: wing, ♂ (*n* = 5) 225–252 (236), ♀ (*n* = 5) 221–231 (227); tail, ♂ (*n* = 5) 188–205 (193), ♀ (*n* = 5) 184–192 (187); bill, ♂ (*n* = 5) 90–110 (95), ♀ (*n* = 5) 81–84 (83); tarsus, ♂ (*n* = 5) 40–46 (43), ♀ (*n* = 5) 41–42 (41). WEIGHT: unrecorded.

Field characters
Length 45 cm. The only small hornbill on these SE islands of the Philippines. All black, but for whitish underparts, head, and neck of ad ♂ and imms of both sexes. Dark bill and black and rufous tail are distinctive. The much larger *Buceros hydrocorax* and *Aceros leucocephalus* have extensive white in the tail, bright red on the bills and large casques, loud calls, and noisy flight.

Voice
Undescribed, and no tape recordings located.

Range, habitat, and status
Philippines. *P. a. affinis*: Mindanao, Dinagat, and recently Siargao I. (Sanft 1960; Dickinson *et al.* 1991). *P. a. samarensis*: Samar, Leyte, Bohol, and recently Calicoan I. (Dickinson *et al.* 1991). *P. a. basilanica:* Basilan I.

P. a. affinis: Primary evergreen dipterocarp forest remnants, as around clearings and secondary forest; absent from dense forest on Mt Malindang above about 900 m. Quite common on isolated forested hill between Salvacion and Buena-Suerte (Rand and Rabor 1960). Mindanao still has relatively much forest remaining, but Dinagat is small with little forest surviving (Worth *et al.* 1994).

P. a. samarensis: Originally primary forest and clearings, even lone fruit trees near forest. Estimated forest remaining: Samar 30%, and still logging; Leyte 20%; and Bohol 10%, with logging suspended (Worth *et al.* 1994).

P. a. basilanica: Basilan has very little forest remaining, and the one protected area is being settled.

Four specimens seen being prepared for a holiday feast (Lint and Stott 1948).

Feeding and general habits
Solitary or in pairs, sometimes in small groups of up to 12 at fruiting trees (Witmer 1989), especially figs (duPont and Rabor 1973*a*). Occupies central and lower levels of forest, not joining larger, more mobile spp. in high canopy (Stott 1947).

Diet little known. Stomachs of four collected specimens contained fruit, seeds, and a beetle, and one was seen to catch a lizard (Sanft 1960).

Displays and breeding behaviour
Undescribed.

Breeding and life cycle
Little known, except from captivity for undetermined subsp. (Buay 1991). ESTIMATED NESTING CYCLE: incubation period no more than 25 days, chick flying by about 47–54 days old. ESTIMATED LAYING DATES: *P. a. samarensis*, 2♀♀ with enlarged gonads in Apr–May (Rand and Rabor 1960). Captive: Singapore, Mar. c/3. Eggs undescribed.

Both sexes took part in sealing in captivity (Buay 1991), using a 46 × 41 × 56-cm-high nestbox of 2.5-cm-thick planks with a 13-cm diameter entrance. ♂ preferred to feed ♀ items of animal food before fruits. ♀ emerged prematurely at 37 days, and surviving chick, handreared, was probably ♀ at 365 g. Chick retained pale pink skin throughout development.

Mindoro Tarictic Hornbill *Penelopides mindorensis*

Penelopides Mindorensis Steere, 1890. *Listing of Birds and Mammals from the Steere Expedition*, p 13. Designated as Calapan, Mindoro I., Philippines.

PLATES 9, 14

Other English name: Tarictic Hornbill

Description

ADULT ♂: head, neck, and underparts yellowish white. Ear-coverts and throat black. Upperparts and wings black with metallic green sheen. Tail brick-red, with black tips to feathers increasing in width towards outer edge and with black outermost vane framing the rufous centre. Bill, with casque a low ridge terminating midway along bill, black, with yellow tip and yellow stripes across upper mandible. Bare circumorbital and throat-skin patches flesh-coloured. Eyes red-brown, legs and feet dark brown.

ADULT ♀: like ad ♂ in overall coloration, but smaller, casque less developed; facial skin dark blue, and eyes light brown. No extensive sexual plumage dichromatism as in other *Penelopides* spp., only differences in facial skin colour.

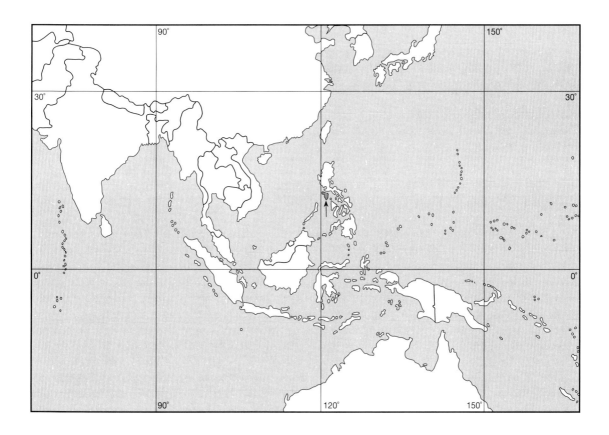

IMMATURE: plumage like ad, one specimen with brown on upperwing-coverts. Bill grey with yellow wash towards tip; facial skin and eye colours undescribed.

MEASUREMENTS AND WEIGHTS
SIZE: wing, ♂ ($n = 10$) 233–246 (239), ♀ ($n = 7$) 218–240 (230); tail, ♂ ($n = 10$) 191–210 (203), ♀ ($n = 7$) 185–200 (194); bill, ♂ ($n = 7$) 92–110 (100), ♀ ($n = 7$) 84–93 (90); tarsus, ♂ ($n = 7$) 40–43 (41), ♀ ($n = 7$) 39–42 (40).
WEIGHT: unrecorded.

Field characters
Length 45 cm. Only hornbill on Mindoro I. in the Philippines. Black upperparts and wings, white underparts (at all ages and in both sexes), dark bill, and red-brown tail rimmed with black are all distinctive.

Voice
Much deeper and less trumpet-like than Visayan Tarictic Hornbill *P. panini* (Dutson *et al.* 1992). No tape recordings located.

Range, habitat, and status
Philippines, on island of Mindoro (Sanft 1960; Dickinson *et al.* 1991).

Exceedingly abundant in 1909 (McGregor 1909), and still fairly common in primary lowland evergreen forest and around the edges of clearings in 1958 (Ripley and Rabor 1958) and more recently (Dutson *et al.* 1992). Extends into secondary forest and occasionally moves to fruiting trees out in the open, ranging up to just over 1000 m. The island may have only 20% of forest remaining but includes two national parks, and no plans for further timber-extraction (Worth *et al.* 1994). However, the extent of suitable habitat is so small that the species must be considered endangered, or at least vulnerable (Dutson *et al.* 1992).

Feeding and general habits
Most sightings in pairs, less often in groups of four, possibly families (Dutson *et al.* 1992). Feeds mainly on forest edges and in small patches of forest.

Displays and breeding behaviour
Undescribed.

Breeding and life cycle
Undescribed. ♀ with enlarged gonads in May (Ripley and Rabor 1958).

Genus *Aceros* J. E. Gray, 1844

Medium-sized to large hornbills. Plumage mainly black, but with tail rounded, relatively short in all but 2 spp., and largely or entirely white in all but one sp. (*A. everetti*). Ad ♂♂ and juvs of both sexes have more or less of the head, neck, and breast brown or white, only imm ♀♀ moulting these areas to all black when ad. Extensive naked throat area, inflatable in several species, is brightly coloured, as are bill and casque. Coloration of bill, casque, and pale plumage areas with orange, red, or yellow preen-gland oil from well-tufted uropygial gland in all 5 spp. in the subgenus *Aceros* and in the largest 3 spp. in the subgenus *Rhyticeros* (*A. plicatus*, *A. subruficollis*, *A. undulatus*). Outermost two primaries (9 and 10) have tips emarginated in all but one sp. (*A. comatus*). Most have strong powers of flight, with noisy wingbeats, and congregate from over long distances at fruiting trees or communal roosts, often responding to loud contact-calls of repeated barking notes.

Most spp. breed as pairs without helpers, using tree holes into which the ♀ seals herself and remains with the chicks throughout the nesting cycle. Only *A. comatus* breeds co-operatively. Breeding ♀ moults rectrices and remiges simultaneously while incubating and is temporarily flightless. The ♂ delivers food to the nest, mainly fruits carried in the gullet and regurgitated one at a time to pass into the nest,

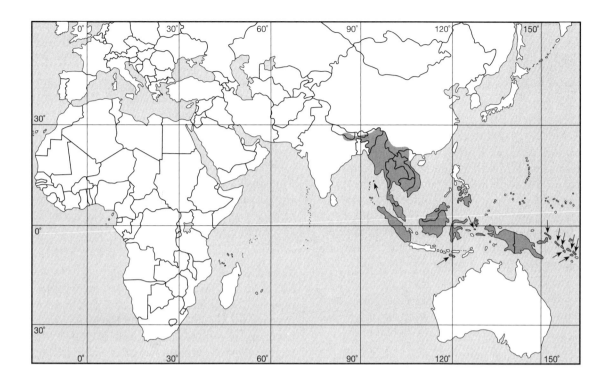

but including some small animals. Lay up to c/3 but usually only single chick reared, especially in larger species. Incubation 5–6 weeks and nestling period 12–13 weeks. Chick's skin turns from pink to black within a few days of hatching.

Total of 11 spp. in 3 subgenera, extending from NE India to New Guinea and Solomon Is. Each subgenus previously considered as a separate full genus, but, in two cases, with different composition of spp. (Sanft 1960; Kemp 1988b). Species in subgenus *Aceros* (5 spp.) have high, wrinkled casques and include *A. nipalensis* with its long tail and distinctive display, while those in the subgenus *Rhyticeros* (5 spp.) have low, wreathed casques. The single sp. in the subgenus *Berenicornis*, with its slight casque and shaggy crest, shows some affinities to spp. in the genus *Anorrhinus*. However, is little known, including in terms of its parasitic lice, and seems best included adjacent to these larger species with similar sexual plumage dichromatism.

Subgenus *Berenicornis* Bonaparte, 1822

A single sp., different in many respects from the rest of the genus and previously placed in the monotypic genus *Berenicornis*, but shares ad sexual plumage dichromatism and large size of other *Aceros* spp. Feather-lice not yet collected, but may support quite different alternative relationship with crested, co-operative-breeding *Anorrhinus* spp. Outer two primaries (9 and 10) not emarginated, preen-gland tuft little developed, tail long and graduated, crest prominent, white, and forward-pointing, and bare facial skin not extensive nor throat inflatable. Juv quite differently coloured from ads. Utters soft hooting calls. Breeds co-operatively. Confined to landmasses of Sunda Shelf.

White-crowned Hornbill   *Aceros (Berenicornis) comatus*

Buceros comatus Raffles, 1822. *Transactions of the Linnean Society, London* **13**, 339. Sumatra, restricted to Benkulen, Sumatra, Indonesia.

PLATES 10, 14
FIGURE 3.8

Other English names: White-crested Hornbill, Asian White-crested Hornbill, Long-crested Hornbill

Description

ADULT ♂: head, neck, breast, tail, and broad tips of remiges white. Back, wings, thighs, abdomen, and undertail-coverts black, upperparts with metallic green sheen. Crown feathers raised in forward-pointing crest; tail feathers graduated. Bill and low casque ridge black with greenish-yellow base. Bare circumorbital and throat skin matt blue. Eyes pale yellow, legs and feet black.

ADULT ♀: similar to ♂, but smaller, only crest white, often with black shaft streaks, but cheeks, neck, and underparts all black.

IMMATURE: unlike ads of either sex. Head, neck, and underparts black, with white feather tips giving mottled appearance; tail feathers black with broad white tips; bill yellow; bare facial skin grey with turquoise tinge; eyes greenish-yellow; legs and feet grey.

SUBADULT: possible sexual difference: imm ♀ with pure white crest, and probable imm ♂ with black shaft streaks in crest and more extensive white in tail (Frith and Douglas 1978).

MOULT

Three imms moulting Dec, Tenasserim. No moult season obvious from museum specimens (Sanft 1960).

MEASUREMENTS AND WEIGHTS

SIZE: wing, ♂ (n = 16) 365–405 (385), ♀ (n = 16) 332–380 (355); tail, ♂ (n = 9) 416–455 (426), ♀ (n = 9) 385–443 (417); bill, ♂ (n = 12) 150–185 (165), ♀ (n = 16) 140–165 (148); tarsus, ♂ (n = 14) 57–63 (61), ♀ (n = 9) 53–60 (56). WEIGHT: ♂ subad 1360–1250; ♀ 1470.

Field characters

Length 75–80 cm. Unmistakable, with black upperparts, black or white underparts, black bill, long, white, graduated tail, broad white wing tips, and erect white crest. Unobtrusive, and usually remains beneath canopy, so most often heard, or seen flying across gaps in forest as a small group.

Voice

Series of mellow double or triple cooing notes, repeated, *ho-ho ho-hoo-hoo...*, the first note of each phrase higher-pitched and the series slowly lowering in pitch (Kemp and Kemp 1974; Kemp 1988*b*), suggestive of a large pigeon (Medway and Wells 1976) or cuckoo (Smythies 1960). Call has considerable volume when heard from nearby. Utters single soft *hoo* alarm-note or at times a vast prolonged uproar (Smythies 1960). Large chick kept up continuous fast, mellow *koe-koe-koe-koe* in the wild (Leighton 1982) and in captivity (Dunselman 1937). Imm under 4 months old repeatedly gave typical hooting call, and this noted as being similar to that of several barbet spp. (Frith and Douglas 1978). Tape recordings available.

Range, habitat, and status

Myanmar, S from Mt Nwalabo in S Tenasserim; SW Thailand; Peninsula Malaysia; Indonesia, Sumatra and Kalimantan; Sarawak; Sabah; Brunei (Sanft 1960). Reports of specimens killed by herbicides in Vietnam require confirmation (Nguyen Cu pers. comm. 1992).

Prefers primary evergreen forest at up to 900 m, rarely up to 1680 m. Within lowland forest, favours extensive areas of thick tangled growth, as along rivers, edges of forest patches, selectively logged areas, bases of hills, and hence vicinity of limestone caves. Also

White-crowned Hornbill *Aceros (Berenicornis) comatus*

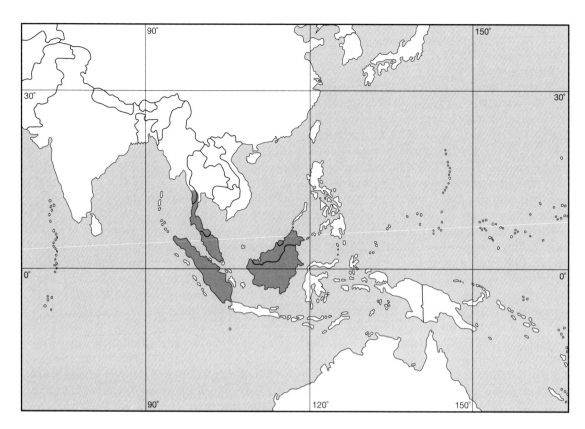

extends into oil-palm plantations adjacent to forest (Johns 1987) and cocoa plantations.

Uncommon to rare in much lowland forest, but easily overlooked if not calling; prefers dense undergrowth and often found close to the ground. Even survives in small conserved forest patches, from which it ranges into surrounding secondary forest. Recorded in Sabah at density of only one group per 10 km^2 in an isolated forest patch (Johns 1988*a*); in Sarawak at 1.6 km^2 per group (Kemp and Kemp 1974); in Kalimantan at slightly more than 1 km^2 per group (Leighton 1986).

Endangered in Thailand, and second-rarest hornbill species in Peninsula Malaysia (Worth *et al.* 1994). Used to be important omen bird to Kayans in Sarawak (Hose 1893).

Feeding and general habits

Resident and territorial (Leighton and Leighton 1983). Recorded in groups of 3–8, usually 4–6. Records of 12, 14, and 20 together may be groups combining during territorial clashes or gathering at fruiting trees. Most groups consist of a pair with 1–3 helpers and some juvs. Pairs call at intervals during day, and both sexes, and helpers, call and defend the territory (Leighton 1986). Shows distinct threat posture by facing opponent with wings open, tail spread, and bill lowered (Fig. 3.8, p. 29), similar to basic sunbathing posture.

Flies short distances with rather floppy flight, from one clump to another, then hops and climbs through foliage looking out for food. Flight almost silent, unlike that of other large hornbill spp. Groups move around together with low clucking contact-call, much like domestic chickens, and two groups moved only 300 m in 30 mins. Often makes fast jerky movements of head when looking from side to side (Dunselman 1937), and sometimes sunbathes with wings and tail outstretched (Frith and Douglas 1978). Appears to return to fixed roost sites, which are selected among isolated trees on the tips of twigs (Kemp and Kemp 1974; Leighton 1982).

Forages at all levels in foliage, from tops of largest emergents in primary forest to ground level in selectively logged areas (Sharpe 1890; Thompson 1966; Kemp and Kemp 1974; Johns 1987), even descending to dustbathe. Most often forages low down in vegetation or on the ground; 70% of first sightings under low canopy within primary forest (Leighton 1986). Digs among epiphytes, prises off bark, and splits husks of capsules, the latter especially important when fruiting poor. Caught weak-flying cave swiftlets on the wing (Smythies 1960).

Probably mainly carnivorous (Johns 1987), but also eats many fruits, including of cocoa and nibong palm *Onchosperma horridum*. Stomach contents of 7 specimens included fruit, a bird, and 2 lizards: a high proportion of animal foods. Stomach volume estimated at 144 ml (Leighton 1982). In Kalimantan, lipid-rich capsules and drupes were most important in the diet, sugar-rich figs being taken much less, and visits to fruit trees being interspersed with bouts of hunting animals. Groups larger during fruit-rich times, helpers splitting off for all or part of time when little fruit; may be driven from fruit by gangs of smaller Bushy-crested Hornbills *Anorrhinus galeritus* (Leighton 1982, 1986).

Also feeds on abundant arthropods and small vertebrates, including snakes, lizards, and small birds found in ground litter of oil-palm plantations, or on wood-boring insects and larvae excavated among logging debris (Johns 1987).

Displays and breeding behaviour

Monogamous as dominant pair, usually with helpers in a co-operative breeding system. Tail slightly raised and lowered while calling (Frith and Douglas 1978), or pumped back and forth under body.

Breeding and life cycle

Little known. ESTIMATED LAYING DATES: Malaysia, nesting June; Sarawak, Mar, Oct; Indonesia (Kalimantan), Dec, Jan. Usually only single chick recorded in nest (Smythies 1960) or with group, but two possibly raised together at times (Frith and Douglas 1978). Eggs unrecorded.

Found breeding in natural holes in trees in logged and unlogged forest (Johns 1987). In Kalimantan, bred and fledged chicks during Jan–May when there was a peak in fruit abundance, but then did not breed for remainder of two-year study (Leighton and Leighton 1983). The single chick raised per group was fed to at least 6 months after fledging by all older members of both sexes, who supplied all its food requirements for the first few months. Only one dominant ♀ bred, later chasing off all other ♀♀ in the group, especially when fruiting was poor (Leighton 1986).

Subgenus *Aceros* J. E. Gray, 1844

Casque slightly or extensively developed into vertical blade, usually of pleated or wrinkled form. Tail black and white in all but one sp., where all white (*A. cassidix*). Utter either soft hooting, short barking, or loud roaring calls, without any particular postures except for the elaborate display of *A. nipalensis*.

All 5 spp. allopatric, ranging from NE India to Sulawesi. *A. nipalensis* was previously separated in the monotypic genus *Aceros* on account of its low casque and the bill having ridges along side of upper mandible. The remaining spp. were joined with the 5 spp. in the subgenus *Rhyticeros* in a broader genus of the same name, although their close association with one another had long been recognized (Stressemann 1940) and may warrant supersp. recognition.

Rufous-necked Hornbill *Aceros (Aceros) nipalensis*

Buceros Nipalensis Hodgson, 1829. *Asiatic Research, Calcutta*, **18(1)**, 178. Designated as Katmandu, Nepal.

PLATES 10, 14

Description

ADULT ♂: head, neck, and underparts rufous, with voluminous nape feathers, and becoming darker on breast and abdomen. Wings and back black with metallic green sheen, five outer primaries with white tips. Tail black, with outer half white, feathers slightly graduated to produce rounded tip. Bill, and low casque ridge along its basal half, yellow, with up to eight black stripes across base of upper mandible. Bare circumorbital skin pale blue. Extensive inflatable bare throat skin scarlet, with deep violet-blue bases to lower mandible connecting as narrow band under throat and with blue flecks around perimeter. Eyes deep red-orange, legs and feet black.

ADULT ♀: smaller than ♂. Head, neck, and underparts black, not rufous; circumorbital skin duller blue; throat skin scarlet, with more extensive violet-blue forming rim around edge and band from angles of jaw across throat; eyes deep red.

IMMATURE: plumage like ad ♂ for both sexes, but bill smaller, unridged, and unmarked. Eyes greenish white, but other soft-part colours undescribed.

SUBADULT: imm ♀ with only one groove on bill still shows rufous feathers on head and neck (Deignan 1945). Imm ♀ moulting head, neck, and underparts to black ad colours in Dec–Apr, when about one year old, although birds with two stripes across bill may not be fully ad.

MOULT

Ads moult during rains, Apr–Sept (Deignan 1945; Sanft 1960).

MEASUREMENTS AND WEIGHTS

SIZE: wing, ♂ ($n = 14$) 415–455 (434), ♀ ($n = 16$) 392–440 (420); tail, ♂ ($n = 9$) 395–430 (419), ♀ ($n = 13$) 375–415 (394); bill, ♂ ($n = 9$) 195–240 (222), ♀ ($n = 14$) 170–205 (185); tarsus, ♂ ($n = 10$) 67–75 (71), ♀ ($n = 11$) 60–70 (66). Wingspan 136 cm (Bangs and van Tyne 1931). WEIGHT: ♂ 2500; ♀ 2270 (Ali and Ripley 1970).

Field characters

Length 90–100 cm. ♂ and imm with rufous head, neck, and underparts and glossy black upperparts. Tips of outer primaries and outer half of tail white. Heavy yellow bill with dark ridges at base of upper mandible, imm with smaller unridged bill. Facial skin scarlet, with blue patch around eyes. ♀ all black but for white on primaries and tail. Most like white-tailed Bar-pouched Wreathed Hornbill *Aceros undulatus* in habits, flying high above forest, but sound of wingbeats higher-pitched and less regular than in Great Pied Hornbill *Buceros bicornis* (Baker 1927).

Voice

Loud croaks, roars, and cackles uttered together with extravagant display. Also short bark, ♀ higher-pitched than ♂, uttered repeatedly in contact (Ali and Ripley 1970); usually monosyllabic, compared with di- or trisyllabic in *A. undulatus* (Tickell 1864), but may also be disyllabic (Baker 1927). No tape recordings located.

Range, habitat, and status

Nepal, in foothills of Himalayas S from Katmandu; India, from Assam to Mishmi Hills, N Cachar, Manipur; S Bhutan (Clements 1992); E Myanmar, S to Taok Plateau, Tenasserim; Thailand, in N and W at

214 Rufous-necked Hornbill *Aceros (Aceros) nipalensis*

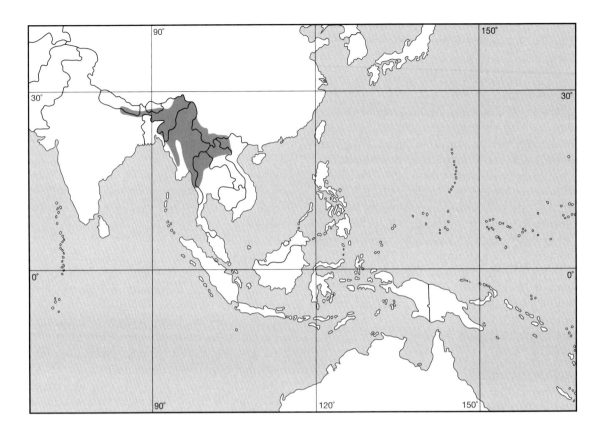

Um Prang, Huai Kha Khaeng, and Chieng Mai; China, in E Medoq of Xisang region and S Xishuangbanna in Yunnan province; N Laos, S to Xieng-Khouang; N Vietnam, at Tranninh (Sanft 1960).

Favours dense evergreen and deciduous forest at 700–2000 m. Locally common in tall evergreen forest, but secretive in heaviest evergreen forest on some higher peaks, as of N Thailand (Deignan 1945).

Habitat now lost in many areas. Extinct in Nepal, where was presumably resident in lower hills (Inskipp and Inskipp 1985). Used not to be rare in N Thailand (Deignan 1945) but now possibly extirpated in NW (Round 1985), and, although still in SW, there threatened since most of range coincides with that of hill-tribe shifting cultivators and found only in Thung Yai and Huai Kha Khaeng Sanctuaries. Very rare in China (Tso-Hsin 1987) and Vietnam (where hunted for food: Worth *et al.* 1994). Was fairly common in Laos (Beaulieu 1944). Unconfirmed in Cambodia (Collar and Andrew 1988).

Feeding and general habits

Usually in pairs or groups of 3–4, rarely 7–8 (Deignan 1945). Arboreal and frugivorous, rarely descending to ground to feed on fallen fruits, where moves with ungainly shuffling hops. Flies across valleys with 3–4 deep, noisy wingbeats followed by glide of 3–4 sec, at times making swooping dive with closed wings like woodpecker (Ali and Ripley 1970). Diet when breeding included *Dysoxylum* fruit (Gammie 1875), taken from the nest tree and others nearby.

Displays and breeding behaviour

Monogamous, apparently without helpers. ♂ displays while uttering the loud call and perched on a prominent bough with 'his head thrown back, red hair on end, his gular skin inflated and his bill erect and wide open. His wings hang loosely and slightly quivering, whilst every now and then his long tail is jerked up until it almost touches his head' (Baker 1927).

Breeding and life cycle
NESTING CYCLE: about 130 days. ESTIMATED LAYING DATES: India, Feb–Apr, laying at beginning of rains, May–June for replacement clutch; Thailand, ♂ with well-developed testes Jan. c/1–2. EGG SIZES: (n = 19) 53.3–68.0 × 39.9–46.5 (59.0 × 43.2) (Baker 1927). Egg a glossless, pitted, broad oval, slightly thinner at one end.

Nests in holes in trees 10–30 m above ground. One, near a stream in India, was 15 m up at the top of the trunk of a huge 25-m *Dysoxylum*, the entrance hole, 43 × 12 cm, just below the emergence of the first branches; the 43-cm-diameter cavity, with the floor level with the entrance, was just large enough to accommodate the ♀, and there was a high funk-hole above the nest (Gammie 1875). Other nests were in large trees in evergreen (but not necessarily dense) forest at 900–1700 m, even in scattered oaks on grassland (Baker 1927).

♀ seals nest with droppings and fruit pulp, possibly supplemented by mud delivered by ♂ (Ali and Ripley 1970). Same cavities used in successive years, and even to lay replacement clutches after chicks collected by local Nagar people in India. ♀ spends over 3 months in the nest, leaving only when the chick fledges. Incubates from first egg, and recorded with two chicks of different sizes in one nest (Gammie 1875).

Sulawesi Wrinkled Hornbill   *Aceros (Aceros) cassidix*

Buceros cassidix Temminck, 1823. *Planches colorées*, **36**, plate 210. Celebes, restricted to Tondano, NE Sulawesi.

PLATES 10, 14

Other English names: Red-knobbed Hornbill, Celebes Hornbill, Knobbed Hornbill, Greater Sulawesi Hornbill, Buton Hornbill

Description
ADULT ♂: crown and back of head rufous-brown, face and neck pale rufous to cream-coloured. Body and wings black, upperparts with metallic green sheen. Tail white. Bill yellow with orange-brown ridges across base of both mandibles, and with high, red-brown, wrinkled, helmet-like casque ridge above base. Bare circumorbital skin pale blue, with dark blue eyelids; extensive bare throat skin dark blue, with black band through lower edge and pale turquoise skin below band. Eyes orange to red, legs and feet black.

ADULT ♀: similar to ♂, but smaller, with less developed casque. Head and neck all black, not rufous and cream-coloured; throat skin with smaller black band; eyes brown to orange.

IMMATURE: plumage like ad ♂ for both sexes, but casque undeveloped; bill pale yellow, with red wash at base. Facial skin paler version of ad ♀; eyes dark brown with yellow rim.

SUBADULT: at about 10 months of age, the casque starts to develop and juv ♀ begins to moult into ad head and neck colours.

SUBSPECIES
Specimens from the islands of Muna and Buton are smaller, with shorter bills: wing, ♂ (n = 16) 397–440 (414), ♀ (n = 6) 368–395 (379); bill, ♂ (n = 6) 215–248 (232), ♀ (n = 2) 176, 180. They have been separated as *A. c. brevirostris* van Bemmel and Voous, 1951, but probably represent the southern end of a cline down Sulawesi (Kemp 1988*b*).

Sulawesi Wrinkled Hornbill *Aceros (Aceros) cassidix*

MOULT
No obvious season from museum specimens (Sanft 1960).

MEASUREMENTS AND WEIGHTS
SIZE: wing, ♂ (*n* = 37) 390–480 (441), ♀ (*n* = 37) 354–420 (396); tail, ♂ (*n* = 12) 250–308 (288), ♀ (*n* = 19) 234–283 (262); bill, ♂ (*n* = 7) 200–275 (264), ♀ (*n* = 8) 193–210 (197); tarsus, ♂ (*n* = 10) 60–67 (64), ♀ (*n* = 8) 55–60 (58). WEIGHT: ♂ 2360.

Field characters
Length 70–80 cm. The largest, and one of only two hornbill spp. on Sulawesi. The high red-brown casque, the colourful and extensive blue facial skin, and the white tail easily distinguish this sp. from the much smaller all-black Sulawesi Tarictic Hornbill *Penelopides exarhatus*. ♂ and juv have the head rufous and the neck cream-coloured, while in the ad ♀ these areas are black.

Voice
A single gruff bark, *grrok*, repeated 2–3 times (Stresemann 1940; Watling 1983), similar to a sympatric monkey species (Holmes 1979). Chick utters high wheezing call (Stresemann 1940). Tape recordings available.

Range, habitat, and status
Indonesia, on Sulawesi and adjacent islands of Lembeh, Togian, Buton, and Muna (Sanft 1960).

Mainly in tall evergreen forests, at up to 1800 m in central mountains, but emerging from forest to feed at figs and plantations (van Bemmel and Voous 1951).

Common in lowland and lower montane forests, despite regular hunting for food and

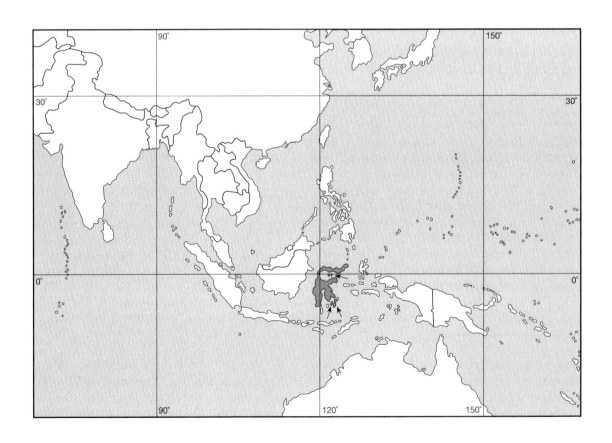

pets (MacKinnon 1979); only occasional in upper montane evergreen forests. Also common on Muna and Buton, where found in woodland patches with many banyan figs. Attain high densities where figs and other fleshy fruits abundant (Kinnaird and O'Brien 1993).

Skull and casque used in some areas for decorative purposes, such as as head-dresses (White and Bruce 1986). Official bird of S Sulawesi Province, Indonesia.

Feeding and general habits
Encountered mainly as pairs (Stresemann 1940, van Bemmel and Voous 1951, Holmes 1979), occupying upper canopy of forest, usually above the only sympatric hornbill *Penelopides exarhatus*. May gather in flocks of over 50 during the non-breeding season (Watling 1983), including both ads and juvs (M. Kinnaird *in litt.* 1994). Flight especially buoyant suggesting capable of extensive movements.

Diet primarily various fruits (Sanft 1960), including figs and drupes taken from within capsules, some even plucked off in flight. Consumes primarily figs in non-breeding season, especially those over 3 g and red in colour, and fruit crop related to flock size (S. Suryadi in M. Kinnaird *in litt.* 1994). Figs also principal food during breeding, comprising 69% of items delivered to nests (range 33–94%) (Kinnaird and O'Brien 1993), making this one of the more frugivorous hornbill spp. Capacity at least 263 ml of fruit (Leighton 1982). Fruit genera delivered to nests include *Koordersiodendron* (1 sp.), *Polyalthia* (3 spp.), *Canaga* (1 sp.), *Canarium* (3 spp.), *Garuga* (1 sp.), *Cryptocarya* (1 sp.), *Aglaia* (3 spp.), *Artocarpus* (1 sp.), *Ficus* (9 spp.), *Gymnocranthera* (2 spp.), *Knema* (1 sp.), *Syzygium* (3 spp.), *Livistona* (1 sp.), *Vitex* (1 sp.), and 23 unidentified spp. (Kinnaird and O'Brien 1993).

Also suspected of excavating and robbing communal nests of *Scissirostrum dubium* starlings (D. Bishop *in litt.* 1985) and insect remains often found in faecal plastering material of breeding ♀ (Kinnaird and O'Brien 1993). Tame in some areas and comes down among human habitation to feed (Holmes 1979). Captive ♀ roosted inside nest-barrel (P. Shannon *in litt.* 1990).

Displays and breeding behaviour
♀ inspects holes, especially in tall lone trees, well before breeding. Courtship-feeding and allopreening of ♀ by ♂ regular (Kinnaird and O'Brien 1993). Seen to chase off juv Sulawesi Hawk-eagle *Spizaetus lanceolatus* near an active nest (D. Bishop *in litt.* 1985). Not apparently territorial, nesting density 9.5 prs/km^2 ($n = 38$) in favourable habitat in N Sulawesi (M. Kinnaird *in litt.* 1994).

Breeding and life cycle
NESTING CYCLE: 133 days in one season ($n = 9$, range 58–120 days), mean 140 days in the next (Kinnaird and O'Brien 1993, M. Kinnaird *in litt.* 1994). ESTIMATED LAYING DATES: N Sulawesi late June to early Sept, also breeding in Mar (Rozendaal and Dekker 1987); S and central Sulawesi Sept–Nov; apparently synchronized within seasons but not regular between seasons (Watling 1983). Paired birds seen end of Aug, ♂ and ♀ on Buton in Sept–Oct with well-developed gonads, and nestlings and juvs of various ages recorded from Maros Mountains, SW Sulawesi, Nov (Stresemann 1940, Sanft 1960). Eggs unrecorded.

Sixteen nests were 13–42 m above ground in natural cavities in large trees of 115 cm mean DBH ($n = 13$, range 71–210). Tree spp. used include *Palaqium* sp. (6), *Dracontomelum dao* (2) and one each of *Diospyros rumphii*, *Alstonia ranvolfia*, *Pterospermum javanicum*, *Canaga odorata*, and *Ficus variegata*. Interior dimensions of nest cavities ($n = 5$) were 81 cm high, 33 cm wide and 50 cm deep, the entrance not facing in any special direction (Kinnaird and O'Brien 1993). Same site usually re-used in subsequent year (15/16), often by same individuals (M. Kinnaird *in litt.* 1994).

♀ seals nest-entrance to a narrow slit with own faeces, sometimes only partially whereafter ♀ emerges early, after only 58 days in nest (but chick fledged successfully). ♂ delivers food as multiple items in gullet and regurgitates one at a time to nest inmates, fruit comprising 59–98% of items, insects 1–4% and the rest unidentifiable (Kinnaird and O'Brien 1993). No pattern in number of items delivered per visit was evident, since size of food items was main factor in provisioning efficiency. Chick visible at entrance 42 days after ♀ sealed in, later develops dark purple skin. Breeding ♀ emerged on average 24 or 32 days before the chick fledged (two successive seasons, M. Kinnaird *in litt.* 1994) and assisted ♂ with provisioning of chick, which is unusual for the genus where ♀ unusually emerges with chick at fledging.

Ninety-four per cent of nests ($n = 16$) fledged a chick in one season (Kinnaird and O'Brien 1993), and 84% in the next ($n = 38$) (M. Kinnaird *in litt.* 1994). Some nests that failed were abandoned in mid-season, at others chicks close to fledging were found dead below, but predation and intraspecific aggression were minimal. Pattern of failures, early and variable emergence timing of ♀, and incomplete sealing at some nests are possible effects of shortages in food quality or quantity, such as the low numbers of oily Lauraceae fruits available (Kinnaird and O'Brien 1993). Lay replacement clutches in captivity after death of nestlings (Shannon 1993).

Sunda Wrinkled Hornbill *Aceros (Aceros) corrugatus*

Buceros corrugatus Temminck, 1832. *Planches colorées*, **89/90**, plate 531. W coast of Borneo, restricted to Pontianak, Kalimantan, Indonesia.

PLATES 10, 14

Other English name: Wrinkled Hornbill

Description

ADULT ♂: crown, back of neck, body, and wings black, upperparts with metallic green sheen. Face and front of neck white. Tail, with broad black base, white when fresh but soon stained yellow-brown to chestnut with preen-gland oils (Duckett 1985). Bill, with high wrinkled casque ridge above basal half, yellow, with red-brown base and ridges across lower mandible and with casque deep red from preen-gland oils. Bare circumorbital skin blue; extensive inflatable bare throat skin pale yellow. Eyes deep red, legs and feet black.

ADULT ♀: smaller than ♂, with casque a low ridge. Face and foreneck black, not white; bill paler yellow, with brown base; circumorbital and throat skin blue; eyes grey-brown, legs and feet greenish grey.

IMMATURE: plumage like ad ♂ for both sexes, but casque undeveloped and bill unridged. Bill pale yellow with wash of orange at base, and with black patch on base of upper mandible in some individuals of either sex. Facial skin pale yellow; eyes yellow with brown tinge, legs and feet blue-grey.

SUBADULT: imm ♀ moults into ad face and neck colours in Borneo when about a year old (in May–June), or at about 9 months old in captivity; even birds with two ridges across base of bill not fully ad. Casque of ♂ takes several years to develop fully, and extent of wrinkling may indicate age (Medway and Wells 1976).

SUBSPECIES

Specimens from Sumatra and the Malay Peninsula are larger than birds from Borneo: wing, ♂ ($n = 15$) 360–402 (373), ♀ ($n = 9$) 317–350 (338). They have been separated as *A.c. rugosus* (Begbie), 1834, but probably represent a cline of decreasing size to the SE, equivalent to several other hornbill spp. with similar ranges (Kemp 1988*b*).

MOULT

Dec–Apr, including some imm ♀♀.

Sunda Wrinkled Hornbill *Aceros (Aceros) corrugatus*

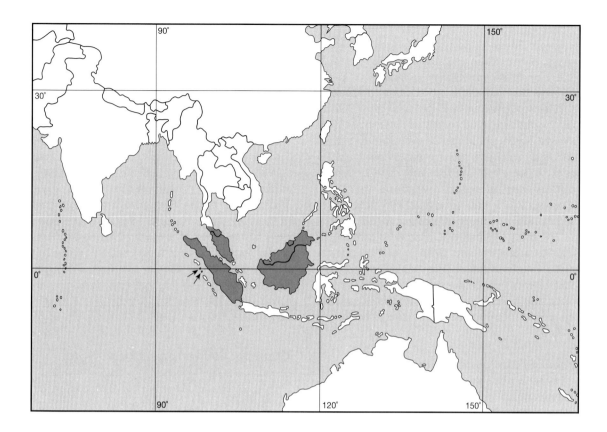

MEASUREMENTS AND WEIGHTS
SIZE: wing, ♂ (*n* = 16) 381–425 (402), ♀ (*n* = 5) 355–381 (366); tail, ♂ (*n* = 16) 258–289 (271), ♀ (*n* = 5) 220–251 (233); bill, ♂ (*n* = 13) 168–205 (181), ♀ (*n* = 4) 144–150 (147); tarsus, ♂ (*n* = 12) 52–60 (55), ♀ (*n* = 4) 48–52 (51). WEIGHT: ♂ 1590 (Ripley 1944).

Field characters
Length 65–70 cm. The high casque, yellow and red bill, blue and yellow (♂) or all-blue (♀) facial skin, and broad white end to the tail are distinctive against the predominantly black plumage. ♂ and imm have white faces and throats. Most similar to sympatric Bar-pouched Wreathed Hornbill *A. undulatus*, which differs most obviously in being larger, with a flat wrinkled casque, all-white tail, and red and yellow facial skin in ♂.

Voice
Main call 1–3 coughing notes repeated at intervals, *sok sok sok* or *kowwow*, rather monkey-like and ♀ at higher pitch (Smythies 1960; Kemp and Kemp 1974); a softer and more musical version in flight, *wakowwakowkow*, Also utters very harsh *pukekkek*. Captive ♂ said to utter high-pitched guttural call when feeding ♀ (Sigler and Myers 1992), but possibly ♀ acceptance-call. Wingbeats produce a loud rushing sound. Tape recordings available.

Range, habitat, and status
Thailand, S of Trang; Peninsula Malaysia; Singapore, last specimen taken 1941 (Medway and Wells 1976); Indonesia, on Sumatra and neighbouring islands of Rupat, Payong, and Batu, Kalimantan; Sarawak; Sabah; Brunei (Sanft 1960).

Inhabits lowland primary evergreen forest, especially swamp forest near the coast; also selectively logged but not secondary forest, rarely above 30 m (Kemp and Kemp 1974) but at up to 250 m (Holmes 1969).

Generally uncommon, but fairly common in N Sarawak, Brunei, and S Sumatra (Holmes 1969; Duckett 1985; Nash and Nash 1985a). Rare in Peninsula Malaysia and close to extinction in S Thailand; large numbers collected for trade in late 1980s (Worth *et al.* 1994). Highest density during census work in Peninsula Malaysia of 3.3 km^2/pair (Johns 1987); in Sabah 20 km^2/pair in unlogged primary forest, and up to 2.9 km^2/pair in forest logged selectively 12 years previously (Johns 1988a); and in Sarawak about 3.2 km^2/pair (Kemp and Kemp 1974). Suspected to cover large areas, so density can easily be overestimated, but usually seen more frequently than heard and not conspicuous.

Feeding and general habits

A highly mobile and nomadic species, usually in pairs or trios (Pan 1987a), but forms flocks of up to 30. Flocks of mixed sex with no hostility, but pairs usually remain close together and may leave one flock to join another in the course of a day (Leighton 1986). Imms usually in separate flocks, in which courtship-feeding and pair-association often seen. Loud calls used to elicit assembly into cohesive groups, which fly to and from roosts and fruiting trees for up to 10 km, high above the forest.

Forages mainly in the canopy of the largest emergent trees, flying from one to the next and so rapidly covering large areas: a pair covered 1 km in 15 mins and visited only four trees (Kemp and Kemp 1974). Hops through branches, searching for prey and sometimes plucking food from foliage in flight, using the broad wings and relatively long tail for dexterity. Capacity at least 156 ml of fruit (Leighton 1982). Leaves an area when fruit stocks drop, but returns even when only one tree species again becomes productive, suggesting good sampling abilities (Leighton and Leighton 1983). Seen to swoop at and chase off a Changeable Hawk Eagle *Spizaetus cirrhatus* (Nash and Nash 1985b).

Intersperses visits to fruit trees with bouts of hunting animals (Leighton 1986). Lipid-rich drupaceous fruits most important in diet, followed by lipid-rich capsules, then sugar-rich figs (20% of diet by numbers), and then animal food. Drupes (especially Lauraceae, Burseraceae) are more seasonally available than other fruits. Uses the bill to prise off bark and to split husks of capsules, the latter becoming most important when fruiting poor. The species is less dominant at fruiting trees than other spp. that are territorial residents, even when they are smaller. Not so attracted to figs as some other hornbills, taking only 6 spp. regularly at Kuala Lompat, Malaysia (Lambert 1989). ♂ in captivity regularly captured small birds to feed to nesting ♀ (Sigler and Myers 1992).

Displays and breeding behaviour

Breeds in logged and unlogged forests (Johns 1987), and pairs disperse when breeding but are not territorial (Leighton 1986). ♂ feeds and often chases ♀ during courtship in captivity (Sigler and Myers 1992). Pair was vocal or silent during copulation, which lasted 5–10 s and during which both birds spread their wings.

Breeding and life cycle

Little known, except from captivity (Sigler and Myers 1992; P. Shannon *in litt.* 1992). ESTIMATED NESTING CYCLE: 103–108 days: pre-laying period 4–6 days, incubation period 29 days, nestling period 65–73 days. ESTIMATED LAYING DATES: Sumatra, Mar; Kalimantan, bred during Jan–May peak in fruit abundance and fledged chicks, but not in subsequent year when fruit crop much lower (Leighton and Leighton 1983). Captives: Louisiana, Mar–May. c/3 ($n = 5$, same captive ♀), but third egg often failed to hatch or produced weak chick. EGG SIZES: ($n = 4$, same captive ♀) 52.5–54.5 × 39.0–41.0 (53.4 × 39.5). Eggs undescribed.

A wild-caught captive pair nested over five successive seasons in a 63.5-cm-diameter × 90.2-cm-high wooden barrel, with a 17.8-cm-square entrance edged with rope and located 52 cm above the base (Sigler and Myers 1992). ♀ did all sealing, using droppings and nest debris, although ♂ later assisted in breaking down sealing.

Eggs were laid at 48-hr (second egg) to 72-hr (third) intervals, one third egg laid 14 days after sealing in. One hatch occurred 45 days after ♀ sealed in. ♂ feeds ♀ and chicks, preferring animal food after hatching. Chicks hatch asynchronously, their skin remaining pink for at least 10 days before turning blue, and then blackish purple when pin feathers emerge after 12 days, except for yellowish face and throat pouch; their eyes open by age of 19 days.

No defecation out through nest slit observed, this possibly due to deep nest-chamber. Some flight-feather moult by ♀ during breeding, but apparently not simultaneous. ♀ and chicks leave the nest together at the end of the nesting cycle. Lay replacement clutch in captivity after death of nestlings (Shannon 1993). Wild birds probably fed chicks up to at least 6 months after fledging, family groups remaining together in feeding flocks until juvs become independent and move to own imm flocks (Leighton 1986). Captive ♂ later became aggressive to imm left in enclosure from previous breeding season (Sigler and Myers 1992).

Mindanao Wrinkled Hornbill *Aceros (Aceros) leucocephalus*

Buceros leucocephalus Vieillot, 1816. *Nouveau Dictionnaire d'Histoire Naturelle*, 4, 592–600. Moluccas, designated as Davao, Mindanao I., Philippines.

PLATES 10, 14

Other English names: Writhe-billed Hornbill, White-headed Hornbill, Wrinkled Hornbill

Description

ADULT ♂: crown and back of neck dark brown, body and wings black, upperparts with metallic green sheen. Face, front of neck, and upper breast white, later stained cream-coloured with preen oils. Tail white with black tip. Bill, with high wrinkled casque ridge above basal half, red, with blue and black grooves between ridges across base of lower mandible. Bare circumorbital skin and extensive inflatable bare throat skin bright deep orange, especially on throat, with yellow eyelids forming ring around eye. Eyes red, legs and feet dull black.

ADULT ♀: similar to ♂, but smaller, with casque a high straight ridge; head and neck all black, not brown and white. Bill slightly paler, and ridges across base of lower mandible brown.

IMMATURE: plumage like ad ♂ for both sexes, but with brown wash around the eyes and casque undeveloped. Bill coloured slightly paler than ad ♀ and without clear ridges or colours at base; facial skin bright yellow, but paler than ad; eyes blue-grey, legs and feet grey.

MOULT

Captive imm ♀ moulted to ad and started casque development in Sept, head and neck moult taking at least 2 months but casque development much longer (Sanft 1960). Another completed the same moult in 3 weeks (Lim Chong Keat *pers. comm.* 1991).

MEASUREMENTS AND WEIGHTS

SIZE: wing, ♂ ($n = 18$) 328–360 (344), ♀ ($n = 8$) 290–320 (305); tail, ♂ ($n = 10$) 225–245 (236), ♀ ($n = 6$) 215–235 (222); bill, ♂ ($n = 11$) 141–151 (145), ♀ ($n = 6$) 109–118 (113); tarsus, ♂ ($n = 12$) 47–55 (51), ♀ ($n = 7$) 44–49 (46). WEIGHT: ♂ ($n = 3$) 1012–1140 (1086) (Rand and Rabor 1960).

Field characters

Middle-sized of three hornbill spp. on the islands of Mindanao, Dinagat, and Siargao in

Mindanao Wrinkled Hornbill *Aceros (Aceros) leucocephalus*

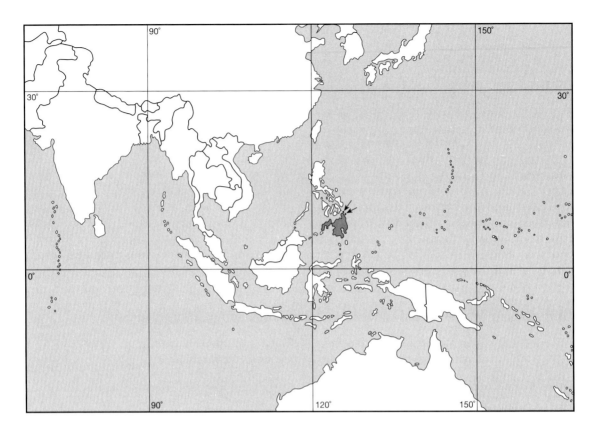

the Philippines. The predominantly black plumage, with white head and neck in ad ♂ and imm, and black tip to the white tail are distinctive. The high red casque and long bill are similar to those of the much larger Great Philippine Hornbill *Buceros hydrocorax*, but the latter has a flatter casque, the tip of the bill yellow, an all-white tail, and predominantly brown plumage. The much smaller Mindanao Tarictic Hornbill *Penelopides affinis* is mainly black, with a dark tail and bill, and with pale underparts in ad ♂ and imm.

Voice

Utters low *Ung–ngeek—ngeek—* when perched, but generally rather silent (duPont and Rabor 1973*a*). Also utters low hissing *phist* (Lim Chong Keat *pers. comm.* 1991). Tape recording available (from Mindanao, by Ben King; see Dickinson *et al.* 1991).

Range, habitat, and status

Philippines, on islands of Mindanao and Camiguin (Sanft 1960), also on Dinagat (Dickinson *et al.* 1991), and reported by locals to visit Siargao (duPont and Rabor 1973*a*).

Inhabits evergreen dipterocarp forest; most common in primary forest below 1000 m (Rand and Rabor 1960), rarely above 500–600 m (duPont and Rabor 1973*a*), 800 m (Dickinson *et al.* 1991), or 1100 m (Brooks *et al.* 1992).

Used to be common on Mindanao, but not found on Basilan (McGregor 1909). Now rare on Mt Apo, where historically was common, and the few isolated forests remaining probably preclude easy recolonization (Witmer 1989) and would be too small to support such a nomadic species (M. Witmer *in litt.* 1992). Still much forest on Mindanao relative to rest of Philippines, but little remains on smaller islands of Dinagat, Siargao, or Camiguin (Worth *et al.* 1994).

Feeding and general habits
Usually found singly, in pairs, or in small groups of 4–6. Up to 37 may gather at fruiting trees, from where they depart again in pairs (duPont and Rabor 1973a). Used to occur in flocks of 20–40 birds, often joined by larger *Buceros hydrocorax* for foraging, passage, and roosting (Stott 1947). Both species favour tops of larger forest trees and range over large areas, often flying high above the canopy (Witmer 1989). Most fruit taken from the canopy of the highest trees (McGregor 1909).

Displays and breeding behaviour
Apparently pairs associate within flocks and breed monogamously (Witmer 1989).

Breeding and life cycle
Little known. ♂ reported feeding ♀ in sealed nest in Mar (Dickinson *et al*. 1991). Eggs unrecorded. Captive pair in Florida laid in Mar, hatched two chicks on 20 April, and these fledged after a 92-day nestling period, at the same time as the ♀ emerged (Behler 1990).

Visayan Wrinkled Hornbill   *Aceros (Aceros) waldeni*

Cranorrhinus [sic] *waldeni* J. Sharpe, 1877. *Journal of the Linnean Society, London*, **13**, 156. Mountains W of Ilo-Ilo, Panay I., Philippines.

PLATES 10, 14

Other English names: Writhed-billed Hornbill, Panay Wrinkled Hornbill.

TAXONOMY: was considered a subsp. of *Aceros leucocephalus*, but differs in head, neck, tail, bill, and facial skin colours and has discrete range (Kemp 1988b). Comparative studies are necessary for confirmation of full specific status.

Description
ADULT ♂: head, neck, and upper breast rufous, body and wings black, upperparts with metallic green sheen. Tail black, with broad white band across centre that becomes stained rufous with preen oils. Bill, with high wrinkled casque ridge above basal half, red, including ridges across base of lower mandible. Bare circumorbital skin and extensive inflatable bare throat skin yellow to orange. Eyes red, legs and feet black.

ADULT ♀: smaller than ♂, with casque a high straight ridge. Head and neck all black, not rufous; circumorbital skin black and throat skin pale yellow; eyes red-brown.

IMMATURE: plumage like ad ♂ for both sexes, but casque undeveloped. Bill reddish pink, facial skin yellow, eyes pale yellow-brown.

SUBADULT: imm ♀, with few brown feathers remaining on head and neck, and ridges on base of bill just beginning to form, collected Dec (Sanft 1960).

MEASUREMENTS AND WEIGHTS
SIZE: wing, ♂ ($n = 9$) 342–371 (353), ♀ ($n = 6$) 300–330 (314); tail, ♂ ($n = 8$) 235–254 (246), ♀ ($n = 6$) 210–230 (219); bill, ♂ ($n = 8$) 140–158 (146), ♀ ($n = 5$) 114–125 (121); tarsus, ♂ ($n = 8$) 51–57 (53), ♀ ($n = 6$) 48–50 (49). WEIGHT: unrecorded.

Field characters
Length 60–65 cm. The largest of only two hornbill species on the islands of Panay, Guimaras, and Negros in the Philippines. The predominantly black plumage, with rufous head and neck in ♂ and imm, the broad white band across the tail, and the red bill and high casque are distinctive. The much smaller Visayan Tarictic Hornbill *Penelopides panini* is also black with pale underparts in ♂ and imm plumages, and the bill is red, but the latter has yellow stripes across the base and lacks the high casque, and the tail is pale rufous with a broad black tip.

Voice
Single mellow barks, repeated (Kemp 1988*b*). A loud nasal lamb-like bleating *wa-ha-ha* (Brooks *et al.* 1992). No tape recording located.

Range, habitat, and status
Philippines, on islands of Panay, Guimaras, and Negros (Sanft 1960).

Inhabits evergreen forest, up to the highest ridges W of Ilo-Ilo (Sharpe 1877), recently at 950 m (Brooks *et al.* 1992).

The forest on Panay was already much degraded when the species was first collected (Steere in Sharpe 1877). There is now almost no forest remaining on Guimaras, and it is reduced to 10–30% on Panay and Negros (Worth *et al.* 1994). A group of four hornbills has been recorded recently on S Negros, the first sightings for 80 years, but the species must be considered in imminent danger of extinction (Brooks *et al.* 1992).

Feeding and general habits
Undescribed.

Displays and breeding behaviour
Undescribed.

Breeding and life cycle
Undescribed.

Subgenus *Rhyticeros* Reichenbach, 1849

Casque developed into series of low transverse ridges or 'wreaths' across the base of the bill. Most spp. with short white tail. Main call of all spp. a single or short series of deep barking notes, in several spp. uttered with the bill jerked skywards with each note.

Five spp., all but one allopatric, including 2 island endemics with very restricted ranges, one differing from rest of genus in long black tail (*A. everetti*). Several species known in Dutch as 'jaarvogel' (year-bird), on the supposition that a new wreath is added annually to the casque.

Papuan Wreathed Hornbill *Aceros (Rhyticeros) plicatus*

Buceros plicatus J. R. Forster, 1781. *Indische Zoologie*, p. 40 PLATES 11, 14
(after W. Dampier, 1699, *Continental Voyage to New Holland*, 3, 164, 165, 170).
Designated as NW coast of Ceram I., Indonesia.

Other English names: Plicated Hornbill, New Guinea Hornbill, Blyth's Hornbill, Blyth's Wreathed Hornbill, Kokomo, New Guinea Wreathed Hornbill

Description
ADULT ♂: head and neck rufous, varying geographically in intensity from deep rufous to honey-coloured. Body and wings black, upperparts with metallic green sheen. Tail white. Bill, with casque a series of low wreaths across base, pale yellow, with dark red-brown base and casque. Bare circumorbital skin pale blue, with flesh-coloured eyelids forming ring around eye; extensive inflated bare throat skin white with blue tinge. Eyes red to red-brown, legs and feet black.

ADULT ♀: similar to ♂, but smaller; head and neck black, not rufous; eyes brown.

IMMATURE: plumage like ad ♂ for both sexes, but casque undeveloped; bill pale yellow with brown wash across base; facial skin pale blue; eyes grey-brown.

SUBADULT: imm ♀ moulted into adult plumage before even one wreath developed on casque (Sanft 1960).

SUBSPECIES

The total range extends to several smaller islands west and east of its core area on the island of New Guinea. Populations differ in body size, and in intensity of the rufous head coloration of ♂, but not in soft-part colours or calls so far as can be determined. Head colour grades from dark rufous in the west to golden yellow in the east, and size from largest in eastern New Guinea to smallest on islands to both west and east (see also Diamond 1972; Mees 1982). Various subspecies have been named (Sanft 1960), their range, size, and coloration being listed from west to east, although overlap in size exists and they probably do not warrant recognition (Kemp 1988b).

A. p. plicatus (J. R. Forster), 1781. Islands of Seram and Ambon. Head and neck dark rufous. Wing, ♂ ($n = 11$) 405–433 (422), ♀ ($n = 10$) 371–420 (392); weight, ♂ 1830, 1950, ♀ 1550 (Stresemann 1914).

A. p. ruficollis (Vieillot), 1816. W New Guinea and islands of Morotai, Halmahera, Bacan, Kasiruta, Obi, Misool, Salawati, Batanta, Gam, and Waigeo. Head and neck golden rufous. Size given below and description above.

A. p. jungei (Mayr), 1937. E New Guinea and straggler to island of Fergusson in d'Entrecasteaux Archipelago. Head and neck golden rufous. Wing, ♂ ($n = 40$) 416–465 (441), ♀ ($n = 21$) 386–430 (408); weight, ♂ 1190 (captive: Mayr and Gillard 1954), ♂ 1827 (Schodde and Hitchcock 1968), juv ♂ 1400 (Mees 1982).

A. p. dampieri (Mayr), 1934. Islands of New Hannover, New Ireland, and New Britain in Bismarck Archipelago. Head and neck yellowish. Wing, ♂ ($n = 8$) 389–415 (406), ♀ ($n = 5$) 360–380 (372).

A. p. harterti (Mayr), 1934. W Solomon Islands of Buka, Bougainville, Fauro, and Shortland. Head and neck golden yellow. Wing, ♂ ($n = 8$) 400–428 (415), ♀ ($n = 2$) 392, 394.

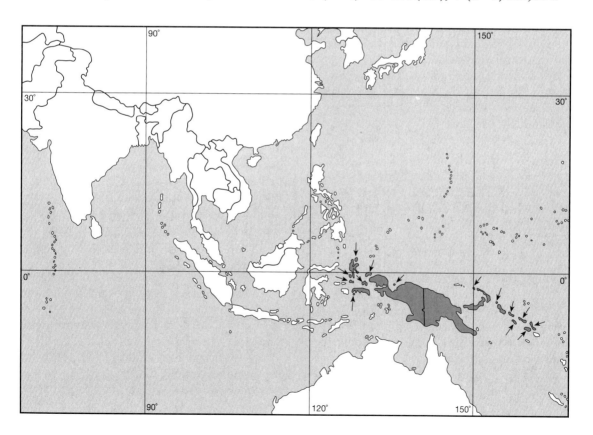

A. p. mendanae (Hartert), 1924. S Solomon Islands of Choiseul, Isabel, Vangunu, Malaita, and Guadalcanal. Head and neck golden yellow. Wing, ♂ (n = 23) 377–407 (393), ♀ (n = 11) 359–382 (373).

MOULT

No obvious moulting season from specimens (Sanft 1960). Moulting primaries, W New Guinea, Sept; moulting tail, Solomon Is, July.

MEASUREMENTS AND WEIGHTS

Details refer to *ruficollis* population of W New Guinea and adjacent small islands; see subspp. above and Sanft (1960) for other populations. SIZE: wing, ♂ (n = 56) 398–440 (417), ♀ (n = 41) 357–410 (385); tail, ♂ (n = 52) 222–256 (239), ♀ (n = 30) 200–236 (219); bill, ♂ (n = 35) 183–240 (209), ♀ (n = 25) 150–168 (169); tarsus, ♂ (n = 19) 52–59 (56), ♀ (n = 16) 49–55 (52). WEIGHT: ♂ (n = 2) 1500, 2000; ♀ (n = 3) 1500–2000 (1667).

Field characters

Length 60–65 cm. The only hornbill sp. on all these Australasian islands. The predominantly black plumage, white tail, and pale yellow bill with its flat wreathed casque are distinctive; ad ♂ and imm of both sexes with rufous to golden head and neck.

Voice

Loud raucous grunts while perched (Coupe 1967), or honk continuously in flight *ka-ka-ka-ka* (Gillard and LeCroy 1967). Short succession of grunts in alarm (Schodde and Hitchcock 1968). Tape recordings available.

Range, habitat, and status

E Indonesia; Papua New Guinea; Solomon Is. Occurs on 24 islands from the Moluccas in the W to the Bismarck Archipelago and Solomon Is in the E, the exact range being described under each putative subsp. above.

Occupies primary and secondary evergreen forest, extending to floodplain deciduous woodland and swamp forest, rarely as high as 1500 m.

Generally reported as common throughout its range e.g. in New Britain (Gillard and LeCroy 1967), New Guinea (Gillard and LeCroy 1966; Diamond 1972), Seram (Stresemann 1914), and islands of W Irian Jaya (D. Bishop *in litt.* 1984). Recently, 250 counted along 20 km of river in Irian Jaya (Worth *et al.* 1994). Apparently some decline on Waigeo (Ripley 1964), and in past two decades around Port Moresby, New Guinea (Peckover and Filewood 1976; Mackay 1970), where was once common (Ramsay 1878). Considered most vulnerable in eastern parts of range (Worth *et al.* 1994), heavily hunted in Papua New Guinea (D. Bishop *in litt.* 1984), and probably extinct on deforested Ambon.

The Iatmuls of the Sepik River, New Guinea, use wooden carvings of head and bill as important totems in house altars or on spires above. Respected by hunters for reputedly picking off colourful heads of fruit pigeons, in addition to fruit diet (Gillard and LeCroy 1966). Hunted for food and as trophies, the skull worn as an adornment (Rauzon 1988), and killing a hornbill was considered equivalent to taking a human life as part of attaining manhood (Peckover and Filewood 1976).

Feeding and general habits

Generally found singly, as pairs, or in small flocks, but hundreds roost together and as many as 45 leave roosts together at dawn (Gillard and LeCroy 1967; LeCroy and Peckover 1983). Flies straight and evenly, with steady wingbeats 'sounding like the hiss of escaping steam', and usually the ♂ leading each pair (Gillard and LeCroy 1967). The sound is apparently enhanced by the tip of the leading primary, which is emarginated, and bends outwards during the downstroke. Flies at 1500 m over Lake LoLoru on Bougainville I., from one side of the island to the other. Noted to freeze temporarily after landing in a tree (Gillard and LeCroy 1966).

Largely frugivorous (Sanft 1960), with the diet including orange figs, *Varingia grossularia*, *Arenga saccarifera*, *Myristica fatua*, *Pometia pinnata*, *Chisocheton* (Meliaceae), and other large purple fruits. Also recorded catching crabs on a beach. Observed regurgitating and reswallowing fruit (Gillard and LeCroy 1966).

Displays and breeding behaviour
Bounces on perch while calls with loud, raucous grunts (Coupe 1967).

Breeding and life cycle
Little known. ESTIMATED LAYING DATES: W New Guinea, inspecting nest-holes Sept–Oct, ♂ with enlarged gonads Jan (Ripley 1964); E New Guinea, Aug–Oct, large gonads in specimens Oct, ♂ feeding at nest 22 Sept and 26 Dec, half-grown nestling Nov, newly fledged chick 13 Jan; Bismarck Archipelago, ♀ with well-developed ovaries 31 Jan; Solomon Is, well-developed gonads in ♂ Apr–May. EGG SIZES: New Guinea 50.8–61.4 × 36.8–42.1 (Schönwetter 1967); Seram ($n = 1$) 59 × 42 (Bernstein 1861); Bismarck Archipelago ($n = 1$) 58 × 41.5, shell 3.93 g (Reichenow 1877); overall mean 57.3 × 40.6. Egg a long pointed oval with rough pitted shell.

One nest on Seram was 15 m up in a fig tree (Bernstein 1861). Another on Solomon Is was 30 m up in an ironwood (Gyldenstolpe 1955), the ♂ feeding at the nest by regurgitation. A captive pair in California was stimulated to pre-breeding activity by a palm-stump nest-cavity in which they could make some modifications (Jennings and Rundel 1976). Other captives became aggressive to keepers during summer (Coupe 1967), this possibly due to breeding territoriality.

Narcondam Wreathed Hornbill *Aceros (Rhyticeros) narcondami*

Rhyticeros narcondami Hume, 1873. *Stray Feathers, Calcutta*, 1, 411. PLATES 11, 14
Narcondam I., India.

Description
ADULT ♂: head and neck rufous. Body and wings black, upperparts with metallic green sheen. Tail white. Bill, with casque a series of low wreaths across base, white with dark crimson base. Bare circumorbital skin dark blue, with red eyelids forming ring around eye; extensive inflated bare throat skin white with blue tinge. Eyes deep orange-brown with yellow inner rim, legs and feet black.

ADULT ♀: similar to ♂, but smaller, head and neck all black, not rufous; eyes olive-brown with yellow inner rim. One ♀ with single black and pied tail feathers (Cory 1902).

IMMATURE: plumage like ad ♂ for both sexes, but casque undeveloped. Bill waxy yellow with base tinged red; facial skin like ad ♀ but paler; eyes pale grey, legs and feet dark grey.

SUBADULT: casque raised but not wreathed by about 9 months old; first wreath by about 18 months old (Hussain 1984), four wreaths by just under 4 years old (Frith and Douglas 1978). Captive ♂'s iris changed from grey to reddish at 4 years 9 months, and at same time ♀ showed interest in nestbox (Hussain 1984).

MOULT
♀ moults to ad plumage at first moult when about 10–14 months old (Hussain 1984). Adult moult evident at start of rains in Mar (Cory 1902).

MEASUREMENTS AND WEIGHTS
SIZE: ♂ ($n = 5$) 303–310 (307), ♀ ($n = 4$) 280–287 (283); tail, ♂ ($n = 5$) 185–198 (193), ♀ ($n = 4$) 172–182 (179); bill, ♂ ($n = 5$) 121–128 (124), ♀ ($n = 2$) 103, 106; tarsus, ♂ ($n = 3$) 40–43 (42), ♀ ($n = 2$) 39, 42. WEIGHT: ♂ ($n = 4$) 700–750; ♀ ($n = 3$) 600–750 (Ali and Ripley 1970).

Field characters
Length 45–50 cm. Only hornbill sp. on the island of Narcondam. Glossy black plumage with white tail; ♀ all black, ♂ and imm with rufous head and neck, imm without casque and smaller bill. Flight appears heavy, with slow and rather deep wingbeats.

Narcondam Wreathed Hornbill *Aceros (Rhyticeros) narcondami*

Voice
Loud, harsh *ka-ka-ka-ka-ka* or *kok-kok-kok kokok* followed by a cackle like a domestic fowl (Abdulali 1971), often 3–4 calling from same tree. Both sexes call *ka .. ka .. ka* in flight, together with laboured wheezy sound of wingbeats. ♂, when alarmed at nest, utters *ko .. kokokoko..ko..kok..ko kok kok kok*. ♀ utters single screech, *krwak*, when accepting food in nest; in alarm, calls *kraawk kok kok*. Chicks utter feeble *chew .. chew .. chew* when fed (Hussain 1984). Tape recordings available.

Range, habitat, and status
India: restricted to Narcondam I., about 500 km NW of Mergui Archipelago, 300 km SW of Gulf of Martaban in Myanmar, and 125 km E of N Andaman I. in Andaman and Nicobar Archipelago in the Bay of Bengal.

Occupies mainly evergreen forests, from the central peak at 750 m down to the coast, the forest originally dense or with open undergrowth under palms and huge figs (Ali and Ripley 1970). Most common on lower slopes (Osmaston 1905).

Narcondam I. is a recently inactive volcano rising steeply out of the sea, Sanskrit for 'hell island' (Hussain 1984). Total area of the island only 6.82 km^2; main summit in the W, and small bouldery bay on S the only flat area where landing is possible during Mar–May. A small spring 25 m up and near the bay is the only fresh water. Vegetation is moist forest, with average annual rainfall of 3056 mm, divided into littoral, mixed deciduous–evergreen, and evergreen areas.

Visited seven times by biologists between 1873, when first specimens collected, and 1969, each visit for periods of less than 5 days. Pigs, goats, and chickens released in the last century as food for shipwrecked sailors fortunately did not survive (Abdulali 1971). Total hornbill population estimated at about 200, of which 10 collected on each of three visits (St John 1898; Osmaston 1905; Baker 1927), and 'Five times that number might easily have

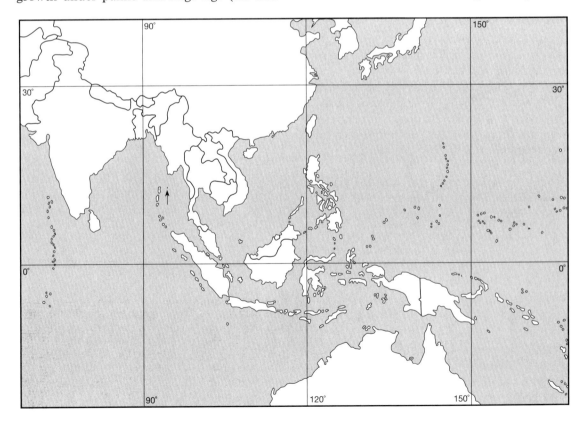

been shot, but I refrained from killing more owing to the rarity of the species' (St John 1898). Still plentiful during 1969 (Hussain 1984), and previous estimates probably low, since earlier visits either too short or during breeding in Apr when ♀ in nest. Max count over 2 km of trail during mid-Mar to mid-Apr 1972 of 72♂♂ and 28♀♀; male:female ratio recorded on the island at this time, during the breeding season, was 4.8:1 ($n = 877$).

Possible nest-site limitations owing to effects of cyclones on larger trees (Osmaston 1905). Coconut, papaya, and banana have since been introduced, and the island is currently occupied by a small police force with goats that are eliminating the undergrowth and seedlings of large trees in many areas (S. Hussain pers. comm. 1991). Detailed conservation measures have been proposed (Hussain 1984, in press).

Feeding and general habits

As many as 50 noisy and fearless birds congregate at fruiting fig trees, *Ficus ? religiosa*, the principal food (Mason and Lefroy 1912). Stomach contents included figs and three other species of fruit, and the birds, all in breeding condition, contained considerable fat ($n = 8$) (Abdulali 1971).

Seeds taken from a nest indicate that the breeding diet included fruits of *Anamirta cocculus*, *Capparis sepiaria*, *C. tenera* var. *latifolia*, *Garuga pinnata*, *Amoora rohituka*, *Bassia longifolia*, *Terminalia catappa*, and *Ixora brunniscens* (Hussain 1984).

Hornbills mobbed a White-bellied Sea Eagle *Haliaeetus leucogaster* at its perch and chased a Koel *Eudynamys scolopacea*. Regurgitates and reswallows food, and regurgitates indigestible seeds.

Displays and breeding behaviour

Nests as pairs without helpers and is not territorial, even towards other birds in the nest tree. Ad and imm make typical jerky head movements when uneasy (Hussain 1984). On 22 Mar, 4 ♂♂ and 3 ♀♀ calling together, the ♂ of the most active pair courtship-feeding the soliciting ♀ (Cory 1902). ♂ makes fast twisting descent in flight to land by ♀, possibly a form of display (S. Hussain pers. comm. 1991).

Breeding and life cycle

NESTING CYCLE: undescribed. ESTIMATED LAYING DATES: Feb–Apr. c/2. EGG SIZES: ($n = 1$) 45×33 mm (Hussain 1984).

Main study by Hussain (Mar–Apr 1972), when 73 active nests were found (Abdulali 1974), all in natural holes at 2.4–15.2 m above ground in *Sideroxylon*, *Sterculia*, and *Tetrameles* trees, discovered through noting debris accumulated below or through watching ♂♂ (Hussain 1984). Two nests at 2.50 and 2.74 m, and only 22.8 m apart, were studied in detail. Nest-entrances were 25–30 cm wide and 149–180 cm deep, one cavity sloping and tapering to the back and the other horizontal. The debris from one nest weighed 1360 g, including 8 seed types and moulted feathers.

♀ sealed nest-entrance, even after chicks hatched, and she and chicks held their tails vertically in nest. ♀ threw out debris and squirted out droppings, and later the chicks did the same. ♂ started foraging just before sunrise, delivering food at 10–30-min intervals depending on the distance to the food tree. Food was regurgitated at the nest, large fruit in pieces and small fruits whole, with 10–93 items being delivered per visit. Chicks called almost continuously.

Breeding ♀ in one nest was already growing in 3–4 remiges and a pair of tail feathers before starting the simultaneous flight-feather moult at laying, suggesting delayed conditions for breeding. This ♀ weighed 680 g and was flightless. Chicks at one nest were judged to have hatched 10 days apart, although the younger subsequently developed into a smaller ♀ in captivity. Estimated that fledging would have occurred in mid-May.

An ad ♂ refused to eat in captivity; an ad ♀ ate, but did not feed its chicks. Sibling chicks were raised to live together for 6 years in captivity, but at maturity at 5 years old the ♀ became very aggressive towards the ♂, even in the breeding season, and eventually injured it fatally (Hussain 1984).

Plain-pouched Wreathed Hornbill   *Aceros (Rhyticeros) subruficollis*

Buceros subruficollis Blyth, 1843. *Journal of the Asiatic Society of Bengal*, **12**, 177. Tenasserim, Myanmar.

PLATES 11, 14

Other English names: Blyth's Wreathed Hornbill, Blyth's Hornbill, Burmese Hornbill, Plain-pouched Hornbill

TAXONOMY: considered a good sp. for almost a century, until at least 1938 (Riley 1938), then more often considered conspecific with *A. plicatus*, until after 1953 when united with *A. undulatus* (Sanft 1953, 1960; type relabelled *A. u. ticehursti* Deignan, 1941). However, it has since been shown to differ from both in overall size, bill form, facial skin colours, and feather-lice (Elbel 1969a, 1977a; Hussain 1984; Kemp 1988b). It most resembles *A. undulatus* in form, colour, and lice, and is sympatric with that species on the Asian mainland. Its status in Peninsula Malaysia and on Sumatra, where *A. undulatus* is widespread, is uncertain. The confused taxonomic status of this hornbill has meant that many details of its soft-part colours and biology have been conflated with those of forms with which it was made conspecific. Sanft (1953) does report it as having a different wing shape, with the primaries projecting 8 cm or more past the secondaries, but being of equal length in *A. undulatus*.

Description

ADULT ♂: crown and nape deep rufous-brown; face, foreneck, and upper breast cream-coloured. Body and wings black, upperparts with metallic green sheen. Tail white. Bill, with casque a series of low wreaths across base and no ridges across base of mandibles, pale yellow, with red-brown base and narrow black base to lower mandible. Bare circumorbital skin reddish purple, extensive inflated bare throat skin plain yellow. Eyes red to red-orange, legs and feet black in front and dull grey behind.

ADULT ♀: similar to ♂, but smaller. Head and neck black, not rufous and cream-coloured; circumorbital skin dull pink, throat skin blue; eyes dark orange-brown.

IMMATURE: plumage like ad ♂ for both sexes, but casque undeveloped and bill pale yellow. Facial skin and eye colour undescribed.

MEASUREMENTS AND WEIGHTS

SIZE: wing, ♂ ($n = 10$) 395–450 (419), ♀ ($n = 9$) 363–400 (381); tail, ♂ ($n = 10$) 237–267 (247), ♀ ($n = 9$) 190–266 (221); bill, ♂ ($n = 10$) 138–177 (169), ♀ ($n = 9$) 131–143 (137); tarsus, ♂ ($n = 10$) 48–57 (53), ♀ ($n = 10$) 46–57 (51) (Kemp 1988b). WEIGHT: ♂ 2270, 2040, 1815 (Riley 1938).

Field characters

Length 65–70 cm. Extremely similar to *Aceros undulatus*, with which it is widely sympatric. Shares black body plumage, white tail, and ad ♂ and imm with rufous and white head and neck. Differs only in being obviously smaller, having proportionately longer and narrower wings, having a plain throat pouch without dark lines down the sides, lacking ridges across the base of the bill, and having a slightly different honking call. The throat skin is yellow (♂) or blue (♀) in both spp. Juv *A. undulatus* also lacks ridges on the sides of the bill and the bars on the throat are only dark traces. Both spp. differ from other larger and smaller sympatric species in having a flat wreathed casque and short all-white tail.

Voice

A harsh barking call (Anderson 1887). Call more like that of Papuan Wreathed Hornbill *A. plicatus* than Bar-pouched Wreathed Hornbill *A. undulatus* (H. Bartels *in litt.* 1982) The three-note flight call *ehk-ehk-ehk* is distinct from the normal two-note call of *A. undulatus* (Lim 1993). No tape recordings located.

Plain-pouched Wreathed Hornbill *Aceros (Rhyticeros) subruficollis*

Range, habitat, and status

Not well documented owing to earlier confusion with *A. undulatus*. Known as specimens from India, at Magherita and Upper Assam; Myanmar, at Kyaikkami, Lower Pegu by Sittang R., Atatan R., Three Pagodas Pass, Myitbyma, Amherst District, Payagalay on Toungoo Road, Myawadi, upper Thoungyin R., Taro in Dalu Valley of Chidwin District, Arakan, E Pegu Yomas, Karen Hills, Karrenin, Tenasserim, Tavoy R., Moulmein, Fort Hertz, Kadam Kyumand and Letsok-aw Kyun in Mergui Archipelago, Telok Krang and Telok Basar; Thailand, in Phet Buri and Karnchanaburi Provinces and at Ban Don, Huai Kha Khaeng, Mae Lem R., Ranong, and as a straggler to Isthmus of Kra (Davison 1883)

Reported from N Peninsula Malaysia (Sungai Tekam: Johns 1987; N Perak State: Lim 1993), which is likely but requires confirmation. Specimens and sightings from Sumatra (Medan, Boelse-Deli District, Gunung Leuser National Park, Tebingtinga, Pematang Siantar on Medan–Lake Toba Road: Kemp 1988*b*), but rejected for that island by van Marle and Voous (1988). The lower Malay states and Borneo have also been included in the range, the species being suggested as a migrant to the coastal lowlands of the latter (Allen 1958), but all specimens examined are of the smaller forms of *A. undulatus* with reduced ridges on the base of the mandibles (pers. obs.; D. Wells *in litt.* 1991).

Exact habitat requirement uncertain, but appears to favour tall evergreen forest in broken country. Used to be relatively common in Myanmar (Bingham 1879), especially on N islands of Mergui Archipelago (Anderson 1887). Already absent from N Thailand by 1945 (Deignan 1945), despite Gyldenstolpe's sight records and report of 'obtaining it at the

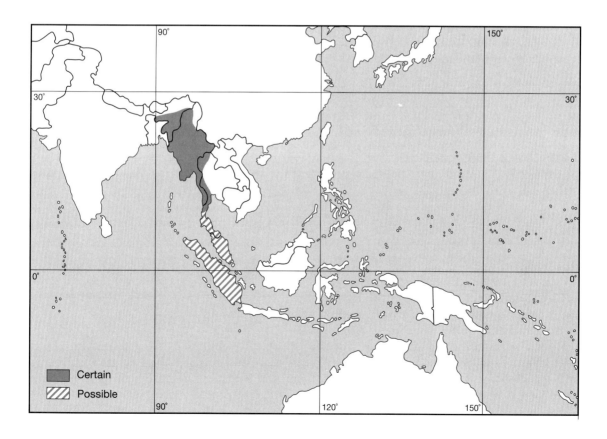

Meh Lem River in northern Siam'; now found only in mozaic of deciduous and evergreen forests in remote river valleys and lower hills of SW Thailand (Round 1985).

Feeding and general habits
Usually in flocks of 6–20, but up to 33 over town of Moulmein, Myanmar (Smythies 1953), and over 50 in lower hill forests of Karnchanaburi and Uthani Thani in Thailand (P. Poonswad *in litt.* 1992). Large flocks of several hundred were seen morning and evening flying long distances to feed (Oates 1883), even crossing the sea from island to island off Tenasserim. Often a wary species that remains in the tops of forest trees, strong on the wing with loud wingbeats and harsh calls (Anderson 1887; Johns 1987). Shared an immense roost with *A. undulatus* in bamboos in the headwaters of the Tavoy R., Myanmar (Oates 1883). Recent unconfirmed report of flocks totalling up to 764 birds, probably of this species, roosting by a lake in upper Perak State, Malaysia, and said to visit area seasonally during fruiting of *Parkia* trees in Aug–Sept (D. Wells *in litt.* 1992; Lim 1993).

Said to be entirely frugivorous (Mason and Lefroy 1912), but surely includes animal food in diet and is known to catch and eat small birds in captivity (Anderson 1876).

Displays and breeding behaviour
Undescribed.

Breeding and life cycle
Little known. ESTIMATED LAYING DATES: Myanmar, Feb–Mar (Bingham 1879; Oates 1883; Anderson 1887); Thailand, Jan–Feb (P. Poonswad *in litt.* 1992), active nest Apr (Kongtong 1989). c/1–3, usually 2 (Hume and Oates 1890). EGG SIZES: (n = 9) 52.9–60.3 × 38.5–47.0 (57.3 × 47.8) (Baker 1927).

Several nests found sealed in holes in tree trunks, notably much higher and in larger trees than sympatric *A. undulatus* or five other hornbill spp. nesting in area (Bingham 1897). One nest 18 m up in *Homalium tomentosum*, another 21 m up in a wood-oil tree (Hume and Oates 1890). ♂ regurgitates fruits to ♀ in nest (P. Poonswad *in litt.* 1992). Hundreds seen flying in the mornings across the Sittang R. to the Pegu plain in Myanmar, spending much time hopping around on the ground; when later shot returning to nests, their throats found to be full of snail shells and earth (Oates 1883).

Bar-pouched Wreathed Hornbill *Aceros (Rhyticeros) undulatus*

Buceros undulatus Shaw, 1811. *General Zoology or systematic natural history, by George Shaw*, **8**, 26. Java, restricted to environs of Batavia, Java, Indonesia.

PLATES 11, 14
FIGURES 1.1, 3.15

Other English names: Bar-throated Wreathed Hornbill, Wreathed Hornbill, Northern Waved Hornbill

Description
ADULT ♂: crown and nape rufous, face and foreneck white to cream-coloured. Body and wings black, upperparts with metallic green sheen. Tail white. Bill, with casque a series of low wreaths across base and obvious ridges across base of mandibles, pale yellow with dark orange-brown base and ridges. Bare circumorbital skin red, with pink eyelids; extensive inflated bare throat skin yellow with interrupted blue-black band across centre. Eyes dark red with narrow yellow inner rim, legs and feet dark olive-grey.

ADULT ♀: similar to ♂, but smaller. Head and neck black, not rufous and white; circumorbital skin flesh-coloured, throat skin blue with interrupted blue-black band across centre; eyes dark brown with narrow blue inner rim.

IMMATURE: plumage like ad ♂ for both sexes, but casque undeveloped. Bill white to pale yellow; facial skin like ad ♂, with lighter bands down sides of throat; eyes pale blue

SUBADULT: imm ♀ begins moult to ad plumage within 7–8 months of fledging (Sanft 1960). About 6 months after hatching, a paler casque area develops along the top of the basal half of the bill but is not yet raised. At a year old the basal third of the casque begins to fold into the first wreath, taking 2 months to spread across the bill and form complete fold. Thereafter, new wreaths grow from the base as the bill grows, approximately one each year, but showing the age only in a limited and general way until bill growth is complete and wear and replacement take over (Bartels and Bartels 1937; Frith and Douglas 1978). Ads attain 9 and rarely 11 wreaths on bill (Tirion 1933).

SUBSPECIES
The N Indian population has been separated as *A. u. ticehursti* (Deignan), 1941 on its large size (wing, ♂ ($n = 1$) 528, ♀ ($n = 3$) 504–510), and the Bornean population as *A. u. aequabilis* (Sanft), 1960 on its smaller size (wing, ♂ ($n = 24$) 425–460 (439), ♀ ($n = 10$) 388–415 (402)) and reduced development of ridges across the base of the mandibles. There appears to be a cline in these characters, with considerable overlap, but overall form, coloration, calls, and displays do not appear to differ from those of other populations (Kemp 1988*b*).

MOULT
Thailand, ♂ in moult in June (Deignan 1945).

MEASUREMENTS AND WEIGHTS
SIZE: wing, ♂ ($n = 27$) 472–527 (497), ♀ ($n = 15$) 418–490 (454); tail, ♂ ($n = 25$) 270–333 (298), ♀ ($n = 14$) 245–288 (267); bill, ♂ ($n = 27$) 180–225 (202), ♀ ($n = 19$) 144–190 (162); tarsus, ♂ ($n = 10$) 60–79 (68), ♀ ($n = 10$) 55–66 (59) (Kemp 1988*b*). WEIGHT: ♂ ($n = 28$) 1680–3650 (2515); ♀ ($n = 10$) 1360–2685 (1950).

Field characters
Length 75–85 cm. The black plumage and white tail, set against the pale yellow bill, are distinctive; and ♂ and imm with rufous and white head and neck and yellow gular skin, ad ♀ with head and neck black and blue gular skin. These characters, however, are shared with the Plain-pouched Wreathed Hornbill *A. subruficollis* where the two are sympatric in India, Myanmar, and Thailand (and maybe Malaysia and Sumatra). The present species is larger, with dark blue-black bands down the sides of the bare gular skin and dark ridges across the base of the mandibles. Both differ from other, larger and smaller, sympatric species in having the flat wreathed casque and short all-white tail.

Voice
Loud call uttered in phrases of 2–3 short raucous grunts *uk, oek-uk,* or *oek-uk-uk,* the first note softer and the whole call repeated in series of 3–4 phrases (Bartels and Bartels 1937; Kemp 1988*b*). Call given regularly by both sexes when ad, either as soft contact-call or as loud roar, but may be uttered occasionally by imm, even at the time of fledging (Bartels 1956). Small chicks make soft *eeeeh* call when moving to nest-entrance; larger chicks make sawing begging-call, a loud and continuous *egge-egge-egge-egge.* ♂ utters soft, fast *kok-kok-kok-kok-kok-kok* when ♀ slow to take food at nest-entrance. ♀ screeches at intruders at nest. Wingbeats audible to at least 1 km, but much quieter when feathers wet after rain (Bartels and Bartels 1937; McClure 1970). Tape recordings available.

Range, habitat, and status
Confusion on Asian mainland and Sumatra with range of *A. subruficollis*. Total range, including *A. subruficollis* (Sanft 1960): India, N to Buxa in Bengal, Dejoo and Rungagora in Assam; S Bhutan (Clements 1992); Myanmar,

234 Bar-pouched Wreathed Hornbill *Aceros (Rhyticeros) undulatus*

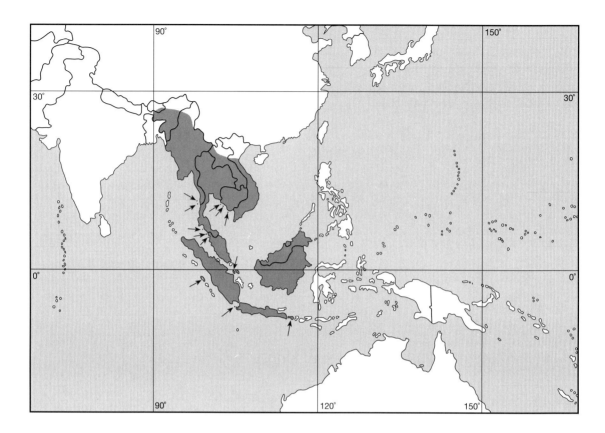

N to Dalu, and including islands of Kadan Kyun, Letsok-aw Kyun, and Kanmaw Kyun of Mergui Archipelago; Thailand, N to Chiang Saen, and including islands of Ko Kut and Ko Chang in Gulf of Thailand and of Junk Ceylon, Lanta and Tarutao along Strait of Malacca; Cambodia, including island of Quan Phu Quoc; Vietnam, N to Khuang Nam and Vinh Linh, and in S (Wildash 1968); Laos, at Tranninh (Beaulieu 1944); Peninsula Malaysia, including islands of Langkawi and Pinang; Indonesia, on Sumatra and islands of Musala, Lingga Archipelago, and Batu island of Telo (van Marle and Voous 1988), Java and W island of Panaitan, Bali, and Kalimantan; Sarawak; Sabah; Brunei.

Favours extensive primary evergreen forest, especially in foothills and on mountains, at up to 1675 m, rarely as high as 2560 m (Mt Trus Madio, Myanmar: Smythies 1953). Also uses and breeds in selectively logged forest, and visits mangrove forests.

Widespread but generally uncommon, except for a few localities such as in Myanmar (Smythies 1986), Gunong Mulu and Pagan Priok in Borneo (Kemp and Kemp 1974), and Khao Yai National Park, Thailand (McClure 1970; Poonswad *et al.* 1983). Not widespread in Java (Bartels 1931). Density, from groups, of about 1.3 km^2/bird in W Malaysia (Johns 1987), and in Sabah up to 0.2 km^2/bird in recently logged forest and 0.15 km^2/bird in 12-year logged forest (Johns 1988*a*). Nomadism makes assessment of density difficult, and use of communal day and night roosts exaggerates abundance. Reported as commoner than sympatric *A. subruficollis* (Riley 1938).

Nests often felled during logging (Kemp and Kemp 1974). Regular pet in longhouses (Dunselman 1937), but, if taken into captivity when ad, can be very aggressive and refuse to feed (Benson 1968). In Assam, up to 80 were shot from one tree in a single season and sold for medicinal purposes for 2–3 rupees (Baker 1927).

Feeding and general habits
A wide-ranging, non-territorial, and nomadic species, usually seen in pairs or groups of 4–14, less commonly as many as 20. May range over at least 100 km^2 daily, often making direct flights, high (300 m plus; Harrison 1974) above the canopy, of 3–10 km (Kemp and Kemp 1974; Leighton 1986). Even flies to offshore islands out of sight of mainland (Tickell 1864), crosses 30 km of non-forested country and commutes daily between Java and Bali. Flies with less gliding than other large spp., one of the first and last birds abroad in daylight (Bartels and Bartels 1937). Can be very shy, especially at nests, and will even fly in moonlight on hearing human voices (Bartels and Bartels 1937).

Pairs may leave one flock to join another during the day, but, although both sexes come close together in flocks with no hostility, they usually stay closest to their mate. After breeding, usually in trios with chick of the year (McClure 1970). Imm birds occur in separate flocks in which courtship-feeding and pair-association often seen (Leighton 1986). Not recorded to adopt spread-eagled sunning posture shown by so many other hornbills, merely gapes with 'wrists' held apart when hot (Frith and Douglas 1978).

Sleeps in communal roosts in a few trees, as many as 264 in one tree and over 400 at such a roost (Khao Yai, Thailand: McClure 1970), sometimes up to 1000 birds (Tsuji *et al.* 1987). Leaves roost in families, either sex leading, after a few squawks at first light, but at dusk forms into large flocks as the birds return, flying high above the canopy or working their way across the top of the forest. Usually uses the same routes to and from roost sites, often over a long period. Same roosts may or may not be used in successive years, and total number of birds present in area may vary: e.g. in 1981, 700 birds split between roosts I and II; in 1982, several smaller roosts existed but most birds (about 370) used roosts II and III; in 1983, most of 700 used roost IV, with none at I or II; in 1984, 1000 birds at II and III, with very few at IV; in 1985, only 470, split between roosts I and V (Tsuji *et al.* 1987). Larger flocks tend to stay for less time and to break up into smaller flocks more readily.

As many as 40 will congregate at fruiting trees or around midday rest sites. Large groups may form in an area for a few days and then not be seen there again for months. The loud calls, noisy wingbeats, and habit of perching in treetops and flicking the head while calling, to flash the bright gular-skin colours, all serve to maintain contact between perched and passing birds. Departs when fruit stocks drop but returns even when only one tree sp. again becomes productive, suggesting good sampling abilities (Leighton and Leighton 1983). Radio-tracked birds left roost to feed in valleys within area of 6–8 km^2; one feeding area, with more birds than other, with about 200 in 2.5 km^2 and only slightly fewer in 1 km^2 respectively (Tsuji *et al.* 1987). Mean home range of ♂♂ was 10 km^2 when breeding and 28 km^2 in non-breeding season (Poonswad and Tsuji 1994).

Omnivorous and generalized feeder, capable of exploiting wide range of areas and foods. Forages anywhere, from the upper canopy, where it bounds from branch to branch, down to the ground, where it hops clumsily about. Uses bill to prise off bark and split husks of lipid-rich capsules, the latter especially important when few fruits available. Plum-sized red fruits of *Dysoxylum* have small bright green kernel that is expelled by defecation 2 hr after a feed; seeds larger than a pea are regurgitated.

Alternates visits to fruit trees with bouts of hunting animals. Hoogerwerf (1950) noted one in low tree over a coral reef after crabs, and another that landed on the reef to take a crab but was disturbed. Often rests near a stream, moving below the canopy to the ground or to stand in the water in the heat of the day or at dusk (Tickell 1864): may then take animal foods such as molluscs and worms, but also comes down below fruiting trees for fallen fruit or small animals attracted to them. Solitary ♂ seen during breeding to move along stream beds, land, then move on, probably in search of animals, but incidence of solo foraging easily overlooked (Kemp and Kemp 1974).

Eats mainly fruit but also animal food, especially when breeding. Stomachs of 9 specimens all contained fruit and 6 also contained animal foods (Sody 1953). Capacity at least 288 ml of fruit (Leighton 1982). Lipid-rich drupes (Lauraceae, Burseraceae) form the major food in Kalimantan, but are more seasonally available than capsules. Sugar-rich figs form only 10% of diet by number, and animal items even less. The species is also less dominant at trees than territorial residents, even than smaller spp. (Leighton 1986). Diet includes several species of fig and other forest fruit, such as *Eusideroxylon*, *Dysoxylum*, *Canarium*, *Polyalthia*, *Beilschmiedia*, *Eugenia*, *Amomum* spp., some with large seeds, but was unable to break into fruits from 5 spp. of dipterocarp (Gould and Andau 1989).

Animals in diet include nestling birds, bird eggs, tree frogs, bats, snakes, lizards, snails, insects such as large beetles (Dynastidae, Lucanidae), spiders, millipedes, centipedes, and crabs (Baker 1927; Bartels and Bartels 1937; Sody 1953). Captives ignore water for drinking (Deignan 1945; Frith and Douglas 1978).

When breeding in Thailand, diet consisted of 57% figs, 29% non-fig fruits, and 5% animals by weight (Poonswad *et al.* 1983, 1986, 1987, 1988). Only fruits delivered at beginning of nesting cycle, animals appearing and increasing in diet after hatching, but not non-fig fruits. Fruits included 26 spp. in 12 families and the 19 genera *Polyalthia*, *Ovaria*, *Canarium*, *Elaegnus*, *Elaeocarpus*, *Cinnamomum*, *Litsea*, *Aphanamixis*, *Chrisocheton*, *Aglaia*, *Ficus*, *Horsfieldia*, *Knema*, *Eugenia*, *Connarus*, *Piper*, *Strombosia*, *Artocarpus*, and *Rourea*.

Animals included beetles (stag, passalid), cicadas (green and brown), psychid bug, leaf Orthoptera, tettigoniid, cockroaches, caterpillars, centipede, millipedes (long, broad, and flat spp.), crabs, land snails, lizards, and bird eggs and chicks (Pycnonotidae, Columbidae).

Displays and breeding behaviour

Monogamous without helpers, and non-territorial but for small area around the nest defended by calls and attacks. Raises bill vertically with each note of loud call, a flicking motion that exposes the bright gular skin, especially the yellow of ♂ and imm, and is done in rhythm with wagging the white tail below (Deignan 1945; Kemp and Kemp 1974; Frith and Douglas 1978). Gular pouch deflated when asleep and in cool mornings (Fig. 3.15, p. 32). Performs aggressive display by bouncing up and down on the perch (Frith and Douglas 1978).

Pairs often allofeed and allopreen. Nest visited only in weeks immediately prior to laying, with much calling in area (Bartels 1956). ♂ courtship-feeds ♀ with regurgitated fruit, making soft, fast bursts of calling, interrupted by soft clucks at nest while ♀ nearby. After chasing off woodpecker, and while ♀ trying to enter, ♂ calls *eu-eu-eu-eu-uh, uh, uh .... eu-eu-eu-eu—uh, uh, uh*. ♀ squeezes into and out of hole before laying, the wings trailing behind each time, and later spends much time chipping away at cavity wall from inside and rim of entrance from outside.

One nest-floor, initially flooded, contained seeds and old feathers. The ♀ sealed for about 1.5 hr on each morning visit, using mud from the floor and then dry leaves, the ♂ waiting outside. Returned briefly for evening visit, one evening sealing from outside. No help from ♂ but for some pattering of bill on outside. Took a week of visits before finally sealed into nest. ♀ rarely threw out seeds at first, so raising and levelling the damp uneven floor. ♀ always sat facing entrance slit, calling to ♂ from inside when saw him, roaring when taking food, and caught and ate a few ants and flies in nest (Poonswad *et al.* 1983, 1987).

Breeding and life cycle

NESTING CYCLE: 111–137 days (means 127, $n = 4$): pre-laying period 13–14 days, incubation period about 40 days, nestling period 90 days; ♀ and chick emerge together. ESTIMATED LAYING DATES: India (Assam), Apr–June in cooler season; Myanmar (Tenasserim), Feb–Mar, one clutch from Nagar locals in June, possibly after re-laying (Baker 1927); Thailand, Jan–Feb; Sumatra, Jan–Feb; Java, Jan, July–Sept; Borneo, Jan, June, pair with regressed gonads Oct (Thompson 1966), and 4 ♂♂ in May (Sanft 1960). Bred in Kalimantan during Jan–May peak in fruit abundance and not again during two years (Leighton and Leighton 1983). Captives: New York, Feb. c/2, rarely 1 or 3. Eggs laid several days apart and incubated from first, but only one chick ever recorded fledged. EGG SIZES: India ($n = 24$) 49.5–72.1 × 38.0–47.1 (62.0 × 43.2) (Baker 1927): Hume and Oates (1890) suggest that the eggs are inseparable from those of *A. subruficollis*, although they are apparently larger, as expected. Myanmar ($n = 3$) 50.8–57.4 × 38.1–41.9 (55.0 × 40.6) (Bingham 1879; Tickell 1864; possibly *A. subruficollis*) Java ($n = 1$) 64 × 43 (Bernstein 1861); Sumatra ($n = 1$) 75 × 48.7 (Coenraad-Uhlig 1930). Egg elongate oval, with slight gloss and rough, finely pitted shell, usually rounder and glossier than *Buceros* eggs.

Nests in natural holes in forest trees, often in large emergent specimens. Nests in Java and Sumatra were 18–26 m above ground, one 20 m up and only half-way up the trunk of a single live *Altingia excelsa* tree projecting above canopy (Bartels, E. 1931; Bartels and Bartels 1937; Bartels, H. 1956), another at same height in *Liquidambar altingiana* (Bernstein 1861). Entrance usually just large enough to admit ♀: two nests with entrance 15–33 cm high × 11–20 cm wide; another with DNH 3 m, entrance hole 13 cm diameter, sealed to 3.5–4-cm-wide slit. Two cavities were 60–85 cm deep, 45–47 cm wide, the floor 17.5 cm below entrance lip, and at least 40–75 cm high with funk-holes extending even further. A nest in Sarawak was in a softwood tree of DBH 1 m, the entrance 20 cm in diameter (Kemp and Kemp 1974).

Nest trees in Thailand ($n = 13$) included *Eugenia* (2), *Dipterocarpus* (5), *Sloanea* (1), *Pterospermum* (1), Ulmaceae (1), *Holoptelea integrifolia* (1), *Nephelium hypoleucum* (1), unknown (1). Most nests in trunks, two in large branches. Tree height 22–28 m (25.8); nest height 5–25 m (12.7); DBH 40–127 cm (76); DNH 40–95 cm (65). Interior dimensions ($n = 3$) 45–65 cm diameter, 35–46 cm width, 94–200 cm high, floor 3–14.5 cm below entrance lip. Favours rather round entrance holes. Nearest-neighbour distance between nests averaged 1422 m ($n = 12$), eight nests in forest, three in open forest, and one on the forest edge, at density in forest of 2.6 km^2/nest (Poonswad *et al.* 1983, 1986, 1987, 1988).

Openings faced in no special direction, and one nest had a funk-hole of at least 2 m. Two taken over by Great Pied Hornbill *Buceros bicornis* and one by Austen's Brown Hornbill *Anorrhinus austeni* in subsequent seasons. Not same pair at nest each year, and intense nest competition: e.g. a foreign ♀ invaded a sealed nest, broke the sealing, attacked the breeding ♀, defended itself from the ♂, drove off the ♂, and then left the area; the breeding ♀ resealed, but then left after a few days when deserted by ♂ (Poonswad *et al.* 1983). Cavities unlined but for rotten wood, seeds, and flight feathers of ♀.

Sealing only by ♀, assisted later by chick, using droppings, especially when sticky after eating animal foods, nest debris, and even a few feathers (Bartels and Bartels 1937). Captive ♀ sealed with materials brought by ♂ (Bell and Brunning 1979). Chick hatches with pink skin that turns dark purple-black within a week, the skin tight and shiny with an air sac up to 1 cm deep and always inflated; throat and underside of bill blue before feathered (Coenraad-Uhlig 1930). Gular pouch, neck, and back inflated in older chicks except in cool early mornings, suggesting heat-loss function (Frith and Douglas 1978). ♀ and chicks sit cramped, with tails above backs. Chick lowers tail when sleeps, droops wings, and rests head forward on bill tip. Chick tries indirect scratching and wing-and-leg stretch like ad before fledging. Defecation a tactile response, being done in dark or light, with maximum reach of anus 11.5 cm from floor to sill for young chick (Frith and Douglas 1978). Defecates forcefully out of slit, but throws out seeds and chitin that have been regurgitated. Chick in nest large once fig seedlings at base have 4–5 leaves.

♂ delivers food to nest by regurgitation, at 1.5–4-hr intervals from dawn to about 30 min before sunset, and ♀ continues to pass food to chick even when latter is quite large. Throat of ♂ obviously distended with fruit when flying to nest (Hume and Oates 1890). Main food is large seeds and arthropods, 50–60 cherry-sized fruits a typical load, but 120 raisin-sized fruits might be fed and the load repeated 30 min later. Feeding rate about 28 g/hr (Poonswad et al. 1983, 1987). Fruits regurgitated singly, rarely doubly, and large arthropods such as beetles also regurgitated. When ♀ satiated, refuses food and pecks at mate's bill; ♂ may then regurgitate remaining fruit on ground before flying off. ♂ often very shy at nest once disturbed, otherwise may dive directly in from high up with loud rushing sound of wings. ♂ roosts away from nest and may roost at different site each night while breeding (Tsuji et al. 1987).

♀ may or may not moult during breeding (Bernstein 1861; Bartels and Bartels 1937; Bell and Brunning 1979): one ♀ flightless when young hatched, with 4 primaries and 8 secondaries remaining on each wing and rest growing in; two others with small to half-grown chicks were fully flighted; no signs of moult in captive. Only single chick ever recorded being raised, and dried remains of second once found in nest indicating early death from starvation (Bartels and Bartels 1937). ♀ and chick emerge together, both chipping at plaster for about 3 hr, the chick rather ineffectually. Captive ♀ emerged 2 days after chick (Bell and Brunning 1979).

♂ fed ♀ and chick for few days after emergence while remained in nest area, even though ♀ flying perfectly (Bartels 1956). Radio-tracked chicks also left nest area within a few days of fledging; a month later one was 10 km away, then left area for a month, before returning to area 5 km from nest for a month until observations ended (Tsuji et al. 1987). Breeding ♂ foraged up to 5 km from nest area (mean 2.4 km) and thereafter followed same movements as chick, covering about 15 km^2 daily, of which 2.7 km^2 was unused open grassland. ♂ and chick did not join three communal roosts in the area. Pair probably fed chicks to at least 6 months after fledging while family remained together within larger flocks, juvs later forming their own flocks (Leighton 1986). Captive chicks required hand-feeding for 7 months (Benson 1968).

One nest used over at least 47 years (H. Bartels *in litt.* 1982), even though chick and ♀ removed on at least three occasions and hole occupied once by flying squirrel (Bartels, E., 1931; Bartels, H., 1956). In Thailand, 8 of 15 sites reused over four seasons. Three nestings successful one year, but all three failed the next when ♂ stopped feeding at nest; one ♀ deserted eggs within a few days after ♂ continually disturbed by other *A. undulatus* and *B. bicornis* feeding at fruiting fig 20 m away, and another deserted a chick after pair harassed by foreign ♀ (Poonswad et al. 1983, 1986, 1988). Overall success, Thailand, 14 out of 20 sealed nests fledged chicks over five seasons.

A captive ♀ survived at least 15.5 years and a pair 14.1 years (K. Brouwer *in litt.* 1992, 1993).

Sumba Wreathed Hornbill *Aceros (Rhyticeros) everetti*

Rhytidoceros everetti Rothschild, 1897. *Journal für Ornitologie*, 45, 513. Monjeli, Sumba I., Indonesia.

PLATES 11, 14

Description
ADULT ♂: head and neck dark brown. Body, wings, and tail black, upperparts with metallic green sheen. Bill, with casque a series of up to 6 low wreaths across the base, pale yellow, with red-brown patch across centre of both mandibles. Bare circumorbital skin dark blue, with pink eyelids forming ring around eye, and extensive throat skin pale blue with dark blue patch under throat. Eyes red-brown, legs and feet black.

ADULT ♀: similar to ♂ in overall coloration, but smaller; head and neck all black, not dark brown.

IMMATURE: plumage like ad ♂ for both sexes, but casque undeveloped; bill pale yellow, eyes grey-brown.

MOULT
Reported for rainy season, Dec–Mar (Sanft 1960).

MEASUREMENTS AND WEIGHTS
SIZE: wing, ♂ ($n = 5$) 345–350 (347), ♀ ($n = 3$) 315–325 (320); tail, ♂ ($n = 5$) 265–280 (270), ♀ ($n = 3$) 252–267 (257); bill, ♂ ($n = 3$) 144–153 (146), ♀ ($n = 1$) 140; tarsus, ♂ ($n = 3$) 50–52 (51), ♀ ($n = 1$) 45. WEIGHT: unrecorded.

Field characters
Length 55 cm. The only hornbill sp. on the island of Sumba. Predominantly black, with dark brown head and neck in ad ♂ and imm; blue facial skin, and red patch on mandibles contrasting with the otherwise pale yellow bill.

Voice
Harsh clucking notes, repeated (E. Sutter *in litt.* 1984). Short 2-note contact call *erm-err*, repeated clucking notes in alarm *kok-kok-kok-* (N. Yaacob in *in litt.* to P. Poonswad 1993.) Audible up to 500 m (Juhaeni 1993). Tape recordings available.

Range, habitat, and status
Indonesia, on island of Sumba in Lesser Sunda chain (Pulau Sumba in Nusa Tenggara division of Indonesia).

Dependent on primary deciduous forest, where nest sites probably exist and where majority of sightings made. Enters secondary forest and open parkland to feed at fruiting trees, but usually passes over grasslands only on passage between forest patches.

About 16% of island now still suitable forested habitat, restricted to patches in steep-sided high valleys and on hilltops. The island, with an area of 13 720 km^2, had little evergreen forest even in 1931 (Rensch in White and Bruce 1986), being covered mainly in deciduous woodland, bushy savanna, and 'alang alang' grass. Large fig trees were common in the interior (Doherty in White and Bruce 1986), but have since been much reduced by subsistance agriculture for the population of about 0.4 million people. A dry seasonal monsoon climate with mean annual rainfall of about 500 mm at coast and 2000 mm on interior hills mainly in Oct–Feb.

Scarce to not uncommon in few remaining areas of forest that were visited (White and Bruce 1986). Commonly seen during fieldwork in Pengaduhahar and Tabundung areas (Jones and Banjaransani 1990). Most frequent in undisturbed forest (Jepson 1993), and also on forest edge and in clumps of trees near villages. A preliminary population estimate of 4000 birds, greatly exceeds previous figures of below 200 (Worth *et al.* 1994) and thankfully does not suggest a serious decline or the need for captive breeding.

Trapped by liming at fig trees and sold live or used for food (D. Yong *in litt.* 1987) but

240 Sumba Wreathed Hornbill *Aceros (Rhyticeros) everetti*

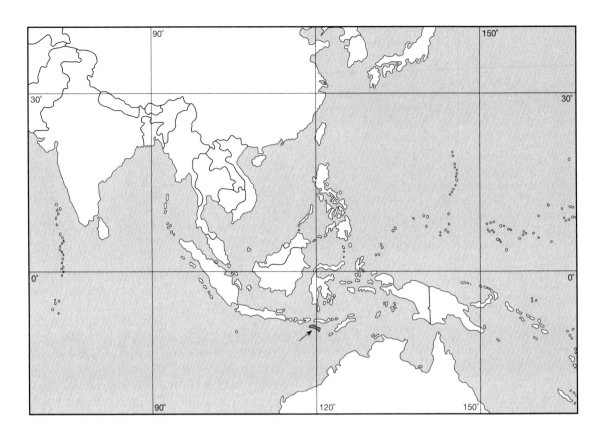

probably not major threat (Juhaeni 1993). All primary and secondary forest in SE of island (Gunung Wanggamet, Tanjung Ngunju, Tagaludon, and Manupeu) requires total protection. Control of fires used to enhance grazing most obvious management requirement (Juhaeni 1993).

Feeding and general habits
Usually in pairs with or without offspring or solitary juvs, but up to 70 may gather at roost sites. Pairs separate off 1–2 months before the breeding season, but juvs remain in small flocks and later only single ♂ recorded once ♀ sealed into nest (Juhaeni 1993). Fifteen is the largest group recorded (Jepson 1993), and 5 flying over Pringkarsha in Tabundung area (Jones and Banjaransani 1990). Usually seen as pairs with no territoriality but keeping well separated (N. Yaacob *in litt.* to P. Poonswad 1993). May displace pigeons from their perches, and move between several fruiting trees during the day (Jepson 1993).

Only fruit recorded thus far in diet, from at least seven species of tree (Jepson 1993). Older individuals, with 5 wreaths on their casques, dominate others with fewer wreaths. Make long flights above canopy when returning to roosts. Defend fruiting trees when breeding, calling at and attacking intruders (Juhaeni 1993).

Displays and breeding behaviour
Copulation only recorded after a suitable nest tree located, otherwise pairs just remain together during the breeding season (Juhaeni 1993).

Breeding and life cycle
Little known, but subject of study by Deddy Juhaeni, National University, Jakarta. (Jepson 1993; Juhaeni 1993). Copulation observed in Sept, and 9/10 nest-holes in live trees.

Genus *Ceratogymna* Bonaparte, 1859

Medium-sized to large hornbills with boldly marked black and white plumage. Crown and nape formed of large, broad feathers. ♂ with well-developed, often massive, casque, ♀ with much smaller casque. Imm with more or less brown facial feathers.

Five smaller spp., previously in genus *Bycanistes* Cabanis and Heine, 1860, have white rump, no bare skin on throat, and emarginated outer 2 primaries. Two largest spp., previously alone in genus *Ceratogymna*, have extensive bare facial and throat skin, the latter formed into pendulous wattles, and ♀♀, as well as imms of both sexes, have brown head and neck feathers.

Largest African forest hornbills and primarily frugivorous. Calls very loud and resonant, possibly as a result of the large hollow casque in ♂. Both sexes take part in sealing the nest. ♂ has unique form of transporting sealing material, swallowing lumps of soil and forming them into mud pellets in his gullet before delivering them to ♀ by regurgitation. ♀ remains in nest until chicks fledge, and chicks' skin changes from pink to purplish black within a few days. All share similar feather-lice (Elbel 1967, 1976).

Total of 7 spp., endemic to Africa. Two larger wattled spp. (*C. atrata, C. elata*), two smallest white-rumped spp. (*C. fistulator, C. bucinator*), and three grey- or brown-faced spp. (*C. cylindricus, C. subcylindricus, C. brevis*) form separate clades, placed respectively in the subgenera *Ceratogymna, Bycanistes,* and *Baryrhynchodes* (Kemp and Crowe 1985).

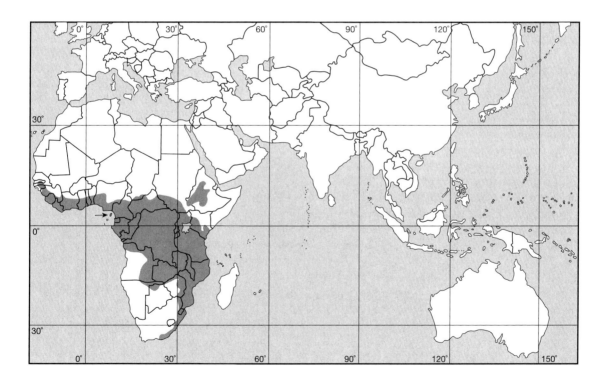

Subgenus *Bycanistes* Cabanis and Heine, 1860

Smallest species in the genus, previously united in the genus *Bycanistes* with spp. in the following subgenus *Baryrhynchodes*. Both subgenera have white of varying extent on rump and underparts, emarginated outer primaries, no bare throat skin, and share the same species of *Chapinia* and *Buceronirmus* feather-lice. The subgenus *Bycanistes*, however, is the only one with *Bucerocophorus pachycnemis* lice (Elbel 1967, 1976). Also the only subgenus in which the ♀ regularly (but not always) undergoes simultaneous rectrix and remex moult while breeding.

Two spp. in African lowland forests and woodlands, one with bill heavily ridged in both sexes, the other with the casque much enlarged, especially in ad ♂. Ranges of each are largely allopatric, at least when breeding, but the two seem too different to be joined as a supersp.

Piping Hornbill *Ceratogymna (Bycanistes) fistulator*

Buceros Fistulator Cassin, 1850. *Proceedings of the Academy of Natural Sciences of Philadelphia*, **5**, 68. W Africa, restricted to St Paul's River, Liberia.

PLATES 12, 13

Other English names: Laughing Hornbill, White-tailed Hornbill, Whistling Hornbill

Description

C. f. fistulator (Cassin), 1850. Upper Guinea Piping Hornbill.

ADULT ♂: head, neck, and back glossy black, undertail- and uppertail-coverts white. Tail black, with broad white tips to all but central pair of feathers. Upper breast black, rest of underparts white. Wings black, except for broad white tips to central secondaries. Bill dusky brown with cream base and tip, casque a low prominence at base with high ridges extending down across bill. Circumorbital skin dark blue. Eyes red-brown, legs and feet sooty brown.

ADULT ♀: like ♂ in overall coloration, but smaller, bill lighter, and slight casque even more ridged. Circumorbital skin greenish blue, eyes brown.

IMMATURE: like ad ♂, but bill smaller and darker, without casque. Brown feathers around base of bill; eyes grey; circumorbital skin pale green.

SUBSPECIES

C. f. sharpii (Elliot), 1873. Sharpe's Piping Hornbill. Intergrades with *C. f. fistulator* along Niger R. (Elgood 1982), but larger. Inner primaries, central secondaries, and outer rectrices almost all white. Bill, unridged with raised casque ridge running to tip, cream, with dusky-brown patch in centre. Circumorbital skin black.

C. f. duboisi (W. Sclater), 1884. Dubois's Piping Hornbill. Intergrades with *C. f. sharpii* in E Cameroon (Louette 1981), but even larger. White on secondaries, outer rectrices, and inner primaries as in *C. f. sharpii*, but outer primaries also with broad white tips. High casque ridge extends whole length of bill in ♂, cream with dusky-brown patch in centre. Circumorbital skin black (♂) or grey-green (♀); feet blue-grey (van Someren and van Someren 1949).

MEASUREMENTS AND WEIGHTS

C. f. fistulator SIZE: wing, ♂ ($n = 15$) 233–261 (241), ♀ ($n = 11$) 217–236 (226); tail, ♂ ($n = 14$) 170–200 (185), ♀ ($n = 8$) 160–180 (171);

Piping Hornbill   *Ceratogymna (Bycanistes) fistulator*

bill, ♂ (n = 14) 85–105 (97), ♀ (n = 11) 76–90 (82); tarsus, ♂ (n = 14) 36–42 (39), ♀ (n = 8) 33–37 (35). WEIGHT: unrecorded.

C. f. sharpii SIZE: wing, ♂ (n = 38) 235–285 (259), ♀ (n = 17) 223–255 (240); tail, ♂ (n = 27) 174–207 (192), ♀ (n = 12) 162–183 (176); bill, ♂ (n = 29) 95–118 (105), ♀ (n = 14) 79–93 (86); tarsus, ♂ (n = 26) 36–42 (40), ♀ (n = 13) 35–39 (37). WEIGHT: ♂ 490, 710; ♀ 500; unsexed 420–555 (3 ♂♂, 4 ♀♀: da Rosa Pinto 1983).

C. f. duboisi SIZE: wing, ♂ (n = 16) 242–283 (267), ♀ (n = 6) 236–258 (246); tail, ♂ (n = 12) 186–210 (200), ♀ (n = 3) 182–185 (184); bill, ♂ (n = 16) 100–122 (112), ♀ (n = 4) 82–87 (86); tarsus, ♂ (n = 13) 40–44 (42), ♀ (n = 2) 38, 40. WEIGHT: ♂ 463; ♀ 413, 485.

Field characters

Length 45 cm. The smallest *Ceratogymna* in African lowland evergreen forests, with the smallest bill and the most reduced casque. Shrill laughing calls also distinctive. Distinguished from larger Brown-cheeked Hornbill *C. cylindricus* and Grey-cheeked Hornbill *C. subcylindricus* by more extensive white underparts, less white on rump, and different tail pattern, latter being either black with white tip (*C. f. fistulator*) or white with black central feathers (other subspp.). *C. f. duboisi* overlaps in SE Zaïre with similar-sized Trumpeter Hornbill *C. bucinator*, which has dark bill and casque and less white in wings and tail. Separable from African Pied Hornbill *Tockus fasciatus* by white on secondaries and more direct flap-and-glide flight.

Voice

Noisy, uttering a raucous cackling laugh *kah-k-k-k-k* or shrill piping *peep-peep-peep*, both increasing and decreasing in volume and tempo. Contact-note while feeding is a soft bark. Tape recordings available.

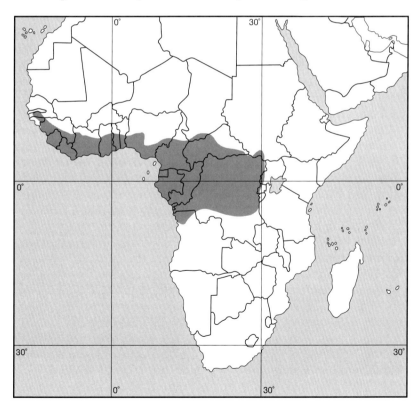

Range, habitat, and status

C. f. fistulator: Senegal, S of The Gambia; Guinea-Bissau; NE and S Guinea; Sierra Leone; Liberia; Côte d'Ivoire; Ghana; Togo; Benin; Nigeria, W of the Niger R. *C. f. sharpii*: Nigeria, E of Niger R.; Cameroon; S Chad (Blancou 1939); Rio Muni; Gabon; Congo; W Zaïre; Angola, N of Cuango R. *C. f. duboisi*: Zaïre, in Congo R. basin; Central African Republic; S Sudan, at Bengengai; W Uganda, at Bwamba and Budongo forests.

Inhabits edges of lowland primary forest, secondary forest, oil-palm plantations, forest patches within savanna, and mangrove swamps. Also enters dense primary forest during the fruiting season of favoured trees. Found at up to 600 m on Mt Nimba (Colston and Curry-Lindahl 1986). Common throughout its range.

Feeding and general habits

Prefers forest canopy and crowns of emergent trees. May land high in forest, 23–50 m up, calling to passing birds and so attracting them to fruiting trees. Encountered singly, more usually in pairs or family trios, sometimes in small parties of 12–20 (Brosset and Erard 1986; Colston and Curry-Lindahl 1986), rarely 40–50 together during the rains (Marchant 1953). Families interchange between groups in the course of a day, but reassemble at roosts. Usually roosts communally, sometimes in a tree on a hillside, dispersing during the day to wander widely in search of fruiting trees. Flight fast, direct, undulating, with fast flapping alternating with gliding, and without notably noisy wingbeats. Often covers distances exceeding 6 km in follow-my-leader fashion, high above forest canopy, and may forage over hundreds of ha of forest (Brosset and Erard 1986).

Usually lands near fruiting trees before moving cautiously in to feed, 2–3 together staying for up to 30 min in a fig vine and visiting daily (Breitwisch 1983). Takes 2–3 fruits/min and selects ripe red fruit. Moves from tree to tree once feeding, resulting in many fruits being dropped to forest floor. Catches winged termites in the air, and once seen sallying for insects disturbed by army ants (Willis 1983).

Largely frugivorous. In one sample ($n = 9$), 3 stomachs containing insects as well as fruit, including beetles, mantids, large winged ants, and wasps (Chapin 1939). In another ($n = 27$), unidentified fruits were found in 23, *Musanga* fruits in 3, termite alates in 2, Rhynchophoridae beetles in 2, and unidentified insects in 2 (Germain *et al.* 1973). Over 30 plant genera recorded in diet, including *Ficus*, *Dacryodes*, *Tricalysia*, *Polyalthia*, *Xylopia*, *Trichoscypha*, *Dracaena*, *Scorodophloeus*, *Trichilia*, *Coelocaryon*, *Pycanthus*, *Staudia*, *Uapaca*, *Cissus*, *Pseudospondias*, *Macaranga*, *Beilschmiedia*, *Lovoa*, *Griffonia*, *Strombosia*, *Strombosiopsis*, *Pachypodantium*, *Musanga*, *Heisteria*, *Cassia*, *Croton*, *Viscum*, *Calamus*, *Piptadeniastrum*, and *Morinda* spp. (Gabon: Brosset and Erard 1986). Also eats maize and oil-palm fruit, flowers, buds, and young leaves.

Displays and breeding behaviour

Undescribed.

Breeding and life cycle

Little known. ESTIMATED LAYING DATES: Gabon, Sept–Dec in short dry season; Zaïre, Jan–Feb, July; Nigeria, active nest July; Ghana, ♂ in breeding condition Jan; Cameroon, gonads well developed in Apr, Oct–Dec (Germain *et al.* 1973); Uganda, ♂ with enlarged gonads Apr, ♀ on point of laying Aug (van Someren and van Someren 1949). c/1–2 (but one ♀ with 3 large-yolked ova when collected), but usually only one chick seen with ads. EGG SIZES: about 50 × 35. Eggs white with pitted shells.

Nests in natural holes in trees, 8–15 m above ground, the entrance sealed to a narrow slit. Feathers of some breeding ♀♀ indicate that simultaneous rectrix and remex moult does occur, but other specimens suggest that this may not be invariable.

There are two records of Black Sparrowhawks *Accipiter melanoleucus* killing an adult and two recently fledged chicks (Brosset and Erard 1986). Captive lived at least 8.7 years (K. Brouwer *in litt.* 1992).

Trumpeter Hornbill *Ceratogymna (Bycanistes) bucinator* PLATES 12, 13

Buceros bucinator Temminck, 1824. *Planches colorees*, **48**, plate 284.
Cape of Good Hope, restricted to Knysna, Cape Province, South Africa.

Description

ADULT ♂: head, neck, upper breast, and back glossy black, but for a few grey flecks on face. Uppertail- and undertail-coverts white; tail black, with white tips to all but central pair of feathers. Underparts white. Wings black, except for white tips to secondaries and inner primaries. Bill and casque black; casque large, cylindrical, almost as long as and projecting above bill tip. Rear of casque becomes red (probably vascularized) in breeding ♂. Circumorbital skin dark purple, red, or pink, depending on breeding condition. Eyes red-brown, legs and feet black.

ADULT ♀: like ♂ in overall coloration, but smaller. Bill black with pale yellow tip and dusky-yellow base, casque slighter and terminating midway along bill; eyes brown.

IMMATURE: like ad, but casque undeveloped. Bill dusky black like ♀; brown-tipped feathers around face and base of bill. Eyes grey, later becoming brown; circumorbital skin dark grey. Juv ♂ distinguishable by larger more vascular casque area, even at fledging (G. Ranger unpublished notes).

MEASUREMENTS AND WEIGHTS
SIZE: wing, ♂ (n = 33) 273–302 (288), ♀ (n = 37) 252–280 (263); tail, ♂ (n = 19) 200–230 (216), ♀ (n = 25) 188–209 (198); bill, ♂ (n = 25) 121–154 (136), ♀ (n = 35) 94–125 (112); tarsus, ♂ (n = 19) 41–47 (44), ♀ (n = 23) 37–44 (40). WEIGHT: ♂ (n = 9) 607–941 (721); ♀ (n = 10) 452–670 (567).

Field characters

Length 50–55 cm. White underparts and tips to wing and tail feathers, dark purple-pink skin around eye, dark bill, large casque, and loud trumpeting calls all aid identification. Overlaps in several areas with larger Silvery-cheeked Hornbill *C. brevis*, which has cream-coloured casque, blue circumorbital skin, lacks white in wings, and has less white on underparts and more white on back. Overlaps in SE Zaïre with smaller Piping Hornbill *C. fistulator*, which has smaller, bicoloured bill and casque, more white in wings and tail, plus a very different voice.

Voice

Loud, high, prolonged nasal wailing *nhaa nhaa ha ha ha ha*, like a crying child, breaking into loud braying calls in flight. Nasal, cackling croak in alarm. Contact-call while feeding a quiet guttural croak or subdued *ke-er-er* note, which may be uttered by either sex. ♀ makes panting sound in threat when nest examined. Utters high screeching bray when attacked by large falcon. Young chicks make a high cheeping call, and fledged juvs utter low, bugling, rapidly repeated *uh-uh-uh-uh* begging-call (G. Ranger unpublished notes). Tape recordings available.

Range, habitat, and status

S Kenya, N and W to Witu, Meru, and Nairobi; Tanzania, N to Kibondo in W; SE Zaïre; N Angola, W to Chitau; Zambia; Malawi; Mozambique; Zimbabwe, excluding central plateau and more arid W; South Africa, in E Transvaal, Natal, and SE Cape Provinces S to Alexandria and Knysna forests. Distribution rather patchy, but ranges widely between seasons. Exact status in area of contact with *C. fistulator* uncertain.

Inhabits interior and edge of montane, riparian, and costal evergreen forests, also moist deciduous woodlands and mangroves. Often wanders far into or across surrounding savanna, especially along riverine forest. Resident from sea-level to 2200 m, but nomadic or even a breeding migrant in some areas (Dowsett-Lemaire 1989), congregating into the largest groups during the dry season. Generally uncommon, as in Angola (Dean *et al.* 1988), but may be locally abundant at

246 Trumpeter Hornbill *Ceratogymna (Bycanistes) bucinator*

patches of suitable habitat or where many trees are fruiting. In Malawi, 3 pairs in 110 ha of montane forest (Dowsett-Lemaire 1989); 4 pairs bred in 240 ha of riverine forest in E Cape, S Africa (M. du Plessis *in litt.* 1992).

Feeding and general habits
Most often found as pairs or family groups of 3–5, but regularly gathers in small flocks of up to 48 birds, although more than 100 may feed and up to 200 may roost in same general area. Pairs seem to remain together throughout the year, foraging and roosting together. Sometimes roosts just as families, but often communally, when birds may travel up to 15 km to reach a roost. Roosts in remote stands of large trees, along watercourses, in the hills near major rivers, or sometimes in dry savanna far from riverine forest. Same site may be used for many years, and numbers in attendance drop during the breeding season (M. du Plessis *in litt.* 1992). Radio-tagged ♀ moved between adjacent communal roost sites about 5 km apart within one week. Often very active in chasing and play during hour before going to roost. At sunrise streams out in pairs or small groups, heading off in different directions to fruiting trees in surrounding area.

Two radio-tracked ♀♀ spent each day around a single fruiting source, although visiting other sources *en route* to and from the day's destination. Daily feeding sites may be at least 15 km apart and involve quite different spp. of fruiting trees. Did not usually return to the same source on successive days, as also found by direct observation (G. Ranger unpublished notes). Between bouts of feeding rests in dense tangles near food source, during which regurgitates many seeds that later germinate there.

Feeds mainly on small fruits, especially figs, also berries and drupes, including in N Transvaal *Trichilia emetica, Ekebergia capensis, Diospyros mesptiliformes, Pseudocadia zambesiaca, Afzelia quanzensis, Pterocarpus angolensis, Ficus*

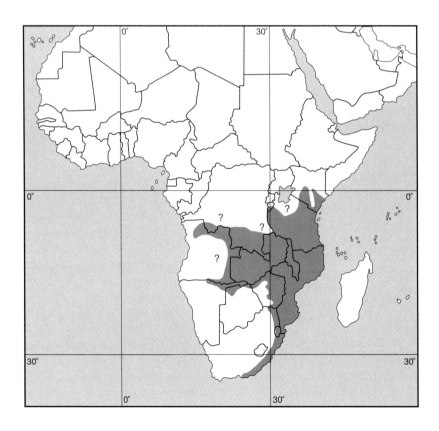

sycomorus, Strychnos potatorum; in Malawi *Drypetes gerrardii, Ekebergia capensis, Ficus capensis, F. kirkii, F. lutea, F. polita, F. sansibarica, F. thonningii, F. sycomorus, Polyscias fulva, Rauvolfia caffra, Syzygium cordatum, S. quineense, Landolphia buchananii* (Dowsett-Lemaire 1988); in E Cape *Ekebergia capensis, Ficus capensis, F. craterostoma, Dovyalis tristis* (G. Ranger unpublished notes) in Natal *Ficus natalensis, Drypetes gerradii, Rauvolfia caffra, Syzygyium cordatum*, and *Chrysophyllum viridifolium* (H. Chittenden *in litt.* 1994) and in Zimbabwe *Strychnos potatorum* (Vernon 1987). Also takes exotic fruits such as papaw/papaya, mangoes, litchis and syringa *Melia azederach*, and eats some nectar-filled flowers such as *Schotia brachypetala*. Where sympatric with Silvery-cheeked Hornbill, appears to take somewhat different and smaller fruits (Dowsett-Lemaire 1988).

Takes some animal food, such as caterpillars, beetles, spiders, bird nestlings and eggs, wasp nests, woodlice, millipedes, and a crab, both from vegetation and from the ground, and hawks termites on the wing. Often mobbed by small birds, suggesting that it preys regularly on nests; seen carrying a *Camaroptera brachyura* warbler (pers. obs.) and poking about in and raiding sunbird nests (Onderstall 1989; Vernon 1991). Deliveries to one nest ($n = 1295$) comprised 11% animal food items (mainly woodlice, millipedes, and wasp nests) and 89% fruits (mainly figs, *Drypetes* and *Rauvolfia*) (H. Chittenden *in litt.* 1994).

Displays and breeding behaviour

Monogamous, but probably not territorial except for area immediately around nest; active nests sometimes only 200 m apart. Co-operative breeding recently reported (N. Myburg *in litt.* 1992; du Plessis 1994), but nests also visited at times by other non-breeding birds (H. Chittenden *in litt.* 1994), which even try to feed the inmates. At one nest, 2 ♂♂ fed ♀ sealed in nest (du Plessis 1994). Loud wailing calls of ♂ may serve for mate-attraction, interspersed with circuits of short energetic display flights with full braying call.

♂ courtship-feeds ♀ before nesting, and some preliminary sealing of nest-entrance and allopreening both precede breeding and occur among family-members. Allopreening, billing, chasing, food passing, fighting, and play often seen among members of larger groups (Vernon 1987). Head and white rump feathers usually raised during allopreening. ♂ often hops along branch to ♀ and jumps over her back several times in initial stages of courtship, a possible prelude to copulation. ♂ and ♀ may visit nest-cavity 2 months before laying, ♀ enter, and both show sealing movements without using any materials.

Breeding and life cycle

NESTING CYCLE: at least 94 days: pre-laying period 10–15 days, incubation 28 days ($n = 3$), nestling period at least 50 days (G. Ranger unpublished notes). ESTIMATED LAYING DATES: Kenya, nest-preparation Oct; Angola, Oct–Dec; Zambia, Oct; Zimbabwe, Sept–Dec; South Africa (Transvaal), Oct –Nov, (Natal) Oct–Jan, (E Cape) Oct –Dec. c/2–4. EGG SIZES: ($n = 13$) 45.0–50.2 × 33.3–36.8 (48.0 × 34.5) (McLachlan and Liversidge 1957; G. Ranger unpublished notes). Eggs white with pitted shells.

Nests in natural holes in tree as low as 2–3 m (Pomeroy and Tengecho 1980), up to 13 m, or high on rock faces as in the gorges below Victoria Falls. One entrance hole was 14 cm wide × 11 cm high and sealed to a 9 × 2-cm slit (G. Ranger unpublished notes). The nest may be sufficiently small to be used later by African Crowned Hornbills *Tockus alboterminatus*. Tree spp. used include live *Adansonia digitata, Ficus sycamorus, Acacia albida, Harpephyllum caffrum, Erythrina caffra, Schotia latifolia*, and dead *Calodendron capense*. Often nests some distance from main food source, as in savanna away from riparian fruiting trees or in wooded gorges far from coastal forest. One ♂ known to travel up to 3 km to gather food for nest-inmates (G. Ranger unpublished notes), but others must cover at least 8 km.

In wild and captivity, entrance sealed to narrow vertical slit by ♀ with mud pellets brought by ♂ in gullet and regurgitated to her, ♂ sometimes bringing lumps of mud in the bill (Millar 1921; Stonor 1936). ♀ uses some food remains and droppings for sealing from within, especially during final closure of nest, and millipedes brought at this time are used for sealing and for food. Sealing at one nest weighed 1.47 kg (G. Ranger unpublished notes) and often the area sealed up is very extensive. The ♂ also brings many pieces of bark, wood chips, fern fronds and small sticks for lining, both before and during nesting. ♀ and chicks squirt their brown droppings out of the nest, forming a distinctive faecal shadow in the relatively dry forest, and the ♂ may also remove droppings and debris accumulated on the nest rim.

The eggs are laid at intervals of 2–3 days, and incubation begins with the first egg. Captive ♂ attacked and flew around cage screaming when nest inspected (Stonor 1936). Brood-members hatch over several days, and in one brood of four the youngest was much smaller than the rest (Swynnerton 1908). ♀ undergoes rectrix and remex moult while breeding, sometimes simultaneously soon after laying, at others sequentially throughout the nesting cycle (G. Ranger unpublished notes). ♂ feeds ♀ by regurgitation of fruit brought about hourly (Millar 1921) or every 20–47 min ($n = 8$) (G. Ranger unpublished notes); 2–38 fruits are regurgitated per visit (mean 14, $n = 12$), a full load ranging from 4 large to 38 small figs. At another nest, 148 visits were made over 78 hr of observation, averaging a visit every 32 min, 9 items per visit and about 17 food items per hr (H. Chittenden *in litt.* 1994). Animal food items may also be regurgitated, but are often delivered to the nest carried in the bill tip. Up to two additional ♂♂ have been recorded feeding at nests during the incubation and nestling periods (N. Myburg *in litt.* 1992; du Plessis 1994).

Chicks hatch asynchronously, and smallest of brood may die of starvation even after 2–3 weeks, resulting in final brood sizes of 1–3. Their eyes open by age of 12 days. The ♀ emerges with the chicks at the end of the nesting cycle. The ♂ is notably quiet during the breeding cycle, but begins calling loudly again around fledging. Once the chicks leave the nest they can fly only weakly for the first 2–3 days, remaining in the vicinity of the nest for 5–7 days, by which time they fly strongly and join the parents on foraging forays (G. Ranger unpublished notes).

At one nest, a genet broke open the nest and ate the ♀ and brood. At another, a chick was dug out and eaten from behind the ♀ by a genet and the ♀ deserted a week later. ♀ in captivity remained sealed in nest for 94 days, but emerged without any offspring. Reported to seal up nest-hole between breeding attempts (Millar 1921), but this certainly not done in most cases (although extensive areas of sealing may remain in place between seasons).

Captive lived at least 21.7 years (K. Brouwer *in litt.* 1992).

Subgenus *Baryrhynchodes* Strand, 1926

Larger white-rumped spp., in which ads develop brown or grey colouring of the facial feathers, much more extensive than the normal brown colouring shown by juvs. All have marked sexual dimorphism, with the casque very large in ♂, and the two outer primaries are emarginated, as in the subgenus *Bycanistes*. Show extensive areas of white on back, wings, and tail, the bill and casque also pied in one or other sex. Share *Chapinia* and *Buceronirmus* feather-lice with smaller spp. in subgenus *Bycanistes*, but have *Bucerocophorus* lice in common with larger spp. in following subgenus *Ceratogymna* (Elbel 1967, 1976).

Three spp. in African moist lowland and montane forests, the ranges of two of these overlapping considerably and suggesting no supersp. groups.

Brown-cheeked Hornbill *Ceratogymna (Baryrhynchodes) cylindricus*

Buceros cylindricus Temminck, 1831. *Planches colorées*, **84**, plate 521, figure 2, Cape Coast, W Africa.

PLATES 12, 13

Description
C. c. cylindricus (Temminck), 1831. Upper Guinea Brown-cheeked Hornbill.

ADULT ♂: head, neck, back, and underparts mainly glossy black, but feathers of cheeks and throat tipped red-brown. Rump, uppertail- and undertail-coverts, and belly white. Tail white with broad black band across centre. Outer half of primaries and secondaries white. Bill and casque cream, base of lower mandible grooved, the casque high and tubular with grooves along sides and wrinkles at base. Circumorbital skin red. Eyes carmine, legs and feet black.

ADULT ♀: like ♂ in overall coloration, but smaller. Bill dusky with blackish-brown base and blue-grey tip; casque a low, heavily folded projection at base of bill. Circumorbital skin pale pink to cream; eyes dark brown, legs and feet black (van Someren and van Someren 1949).

IMMATURE: like ad, but bill small and pale yellow, lacking casque. Brown feathers around base of bill; eyes grey.

SUBSPECIES
C. c. albotibialis (Cabanis and Reichenow), 1877. White-thighed Hornbill (Lower Guinea Brown-cheeked Hornbill). A very distinct subsp. which should probably be considered as separate sp., the White-thighed Hornbill (Kemp and Crowe 1985). Same size, but thighs as well as belly white, and no brown on cheeks or throat. Casque of ♂ more elongate, prominent, and flattened laterally, with discrete projection at tip and only slight grooving; casque cream-coloured, but bill dark brown, with only base and tip cream. Circumorbital skin pale yellow, eyes brown. ♀ with cream tip to dark brown bill, eye grey-brown, circumorbital skin undescribed.

MOULT
♂ replacing restrices and remiges Feb (Liberia: Rand 1951).

MEASUREMENTS AND WEIGHTS
C. c. cylindricus SIZE: wing, ♂ ($n = 10$) 311–330 (322), ♀ ($n = 8$) 276–314 (291); tail, ♂ ($n = 9$) 235–260 (244), ♀ ($n = 5$) 225–240 (233); bill, ♂ ($n = 8$) 145–165 (162), ♀ ($n = 6$) 125–138 (133); tarsus, ♂ ($n = 9$) 47–52 (51), ♀ ($n = 5$) 45–50 (47). WEIGHT: ♀ 921.

C. c. albotibialis SIZE: wing, ♂ ($n = 37$) 310–342 (327), ♀ ($n = 32$) 268–315 (298); tail, ♂ ($n = 30$) 225–257 (240), ♀ ($n = 24$) 198–240 (222); bill, ♂ ($n = 33$) 143–170 (159), ♀ ($n = 26$) 121–143 (136); tarsus, ♂ ($n = 28$) 46–53 (50), ♀ ($n = 26$) 42–50 (46). WEIGHT: ♂ ($n = 3$) 1200–1411 (1286); ♀ 908, 1043; juv 820.

Field characters
Length 60–70 cm. Colour and form of bill and casque, together with distribution of white in plumage, distinctive. Raucous calls also notable. Sympatric with similar-sized Grey-cheeked Hornbill *C. subcylindricus*, but has cream on bill, differently shaped and much whiter casque, and white tail with black band across centre. In flight shows white tips to all flight feathers, whereas outer primaries black in *C. subcylindricus*. Sympatric Piping Hornbill *C. fistulator* is smaller, with small bill and insignificant casque, and has different tail pattern (nominate subsp. black with white tips, other subspp. with central tail feathers black and rest white).

250 Brown-cheeked Hornbill *Ceratogymna (Baryrhynchodes) cylindricus*

Voice

Loud call, of harsh, coughing notes descending in tone, *rack kack kak-kak-kak*, sometimes given singly as contact-call (Chapin 1939; van Someren and van Someren 1949). Less often, a more clamorous series of raucous notes running into one another. Low-pitched piping *ooh ooh* uttered when birds perched together, and high piping note also reported. No differences between subspp. described. Whooshing wingbeats clearly audible. Tape recordings available.

Range, habitat, and status

C. c. cylindricus: Sierra Leone; S Guinea; Liberia; Côte d' Ivoire; Ghana; Togo. *C. c. albotibialis*: Benin; Nigeria, although no recent records for N; Cameroon; Rio Muni; Gabon; Congo, N and S; N Angola, in Cabinda and Zombo provinces; Zaïre, in Congo R. basin; S Sudan (Lambert 1987); W Uganda, in Bwindi (Impenetrable), Maramagombo, Bwamba, Bugoma, and Budongo forests. Occurs as scattered populations in W Africa, with main contiguous range through Lower Guinea forests.

Resident in lowland evergreen forests, with some local movements. Uncommon in W Africa, but frequent to locally common elsewhere. Less common in Ghana than 30 years ago (Grimes 1987). The W African subsp., if confirmed as a separate sp., may be the most threatened hornbill sp. on the continent. Inhabits extensive stands of primary forest, chiefly in lowlands, but once wandered as high as 4054 m in Cameroon, where a skeleton was found (Boulton and Rand 1952). Two pairs occupied 300 ha of forest (Gabon: Brosset and Erard 1986). Prefers higher, more mature forest than Grey-cheeked Hornbill.

This and other large, pied forest hornbills, and pied colobus monkeys, are not eaten by the Wabali and Bandaka tribes of the Ituri forest in Zaïre because of a local taboo (Chapin 1939). A casque was used as an ornament in E Zaïre (Prigogine 1971).

Feeding and general habits
Usually found in pairs or family trios, less often in groups of up to 6, rarely in assemblages of 90 at fruiting trees (Thiollay 1985). Flight buoyant, with deep, noisy wingbeats interspersed with short glides. Roosts, after much preliminary moving around near sunset, at the end of a hanging dead limb or high in a dead tree (Good 1952).

Forages in foliage, most often in the canopy and crown of emergent trees, or concentrates at fruiting trees and sometimes descends to secondary growth or isolated trees in abandoned fields. When feeding on fruit, wrenches it from stem, and often knocks off many other fruits (Maclatchy 1937). Usually 2–3 visit together, stay up to 1 hr 15 min at a fig vine, and eat 2–3 fruits/min (Breitwisch 1983). At fruiting trees dominant over Piping Hornbill, but not over larger wattled hornbills or monkeys (Brosset and Erard 1986). Sometimes flies out to hawk insects such as winged termites.

Primarily frugivorous, but also takes such small animals as it encounters. Total of 18 stomachs contained mainly fruit, but 4 held insects (Chapin 1939). In another sample ($n = 17$), unidentified fruits were found in 13, *Musanga* fruits in 5, and four types of insects each in 1 (Germain *et al.* 1973). Tiny figs and oil-palm fruit in one stomach (Eisentraut 1963), only figs in another (Friedmann 1966). Eats fruits from 17 plant genera in Gabon: *Dacryodes, Guibourtia, Trichoscypha, Polyalthia, Xylopia, Dracaena, Trichilia, Pycnanthus, Ficus, Musanga, Heisteria, Strombosia, Strombosiopsis, Piptadeniastrum, Cassia, Beilschmiedia* and *Morinda* (Brosset and Erard 1986).

Insects taken include damselflies, dragonflies, mantids, winged ants, grasshoppers, long-horned grasshoppers, elaterid beetles, and vespid wasps. Seen to eat eggs and chicks of weavers (*Ploceus cucullatus, Malimbus nigerrimus*), either tearing a hole in the side of the nest or inserting the bill into the nest-entrance and then tipping up the nest to bring the contents within reach (van Someren and van Someren 1949).

Displays and breeding behaviour
Undescribed.

Breeding and life cycle
Little known. ESTIMATED LAYING DATES: Liberia, ♀ in breeding condition Jan, Aug (Rand 1951; Eisentraut 1963); Ghana, ♂ feeding to nest Jan; Nigeria, birds in breeding condition July; Cameroon, birds in breeding condition Nov–Apr, July (Eisentraut 1963; Germain *et al.* 1973); Gabon, July–Aug, Nov; Zaïre, Apr, June, Aug, Oct, Dec; Uganda, ♀ in breeding condition Apr (Friedmann 1966). No obvious breeding season. Clutch size and eggs undescribed.

Nests are in natural holes 20–25 m above ground in trunks of large forest trees, such as *Petersia africana*, the entrance sealed to a narrow vertical slit. Two nests were 2.4 km apart. A nest was found to contain flightless, possibly coprophagous cockroaches and a few lepidopterous larvae (Chapin in Moreau 1937). ♂ feeds ♀ and chick in nest with loads of up to 12 fruits carried in the gullet, sometimes with a single fruit held in the bill (Chapin 1931); fruits usually regurgitated and passed one at a time to ♀, and 14–18 visits made by ♂ per day. ♀ moults flight feathers gradually during enclosure, not simultaneously. One record of 2 nestlings (Brosset and Erard 1986), but usually only a single chick found in nests and seen flying with parents after fledging.

Grey-cheeked Hornbill *Ceratogymna (Baryrhynchodes) subcylindricus*

Buceros subcylindricus P. Sclater, 1870. *Proceedings of the Zoological Society of London*, **1870**, 668, plate 39. W Africa, restricted to Ashanti, Ghana.

PLATES 12, 13

Other English names: Black-and-white Casqued Hornbill

Description

C. s. subcylindricus (P. Sclater), 1870. Upper Guinea Grey-cheeked Hornbill.

ADULT ♂: head, neck, and back glossy black, feathers of cheeks and base of bill edged grey. Rump, uppertail- and undertail-coverts, abdomen, and thighs white; breast black. Central pair of tail feathers all black, rest white with broad black band across centre. Secondaries and inner primaries white with black bases, greater wing-coverts tipped white, rest of wing black. Bill and casque dark brown, but for cream-coloured knob at rear of casque; casque a high, curved ridge, broad at base, then flattened laterally to end in point projecting over the bill tip, and wrinkled on sides. Circumorbital skin dusky flesh-coloured. Eyes red, legs and feet black. Reported to colour plumage with yellow preen-gland oils (Lowe 1937).

ADULT ♀: like ♂ in overall coloration, but smaller, with low casque only a rounded, wrinkled projection at the base of the bill. Eyelids and facial skin pale pink, swelling and flushing red in breeding season; eyes brown.

IMMATURE: like ad, but bill small and without casque. Brown feathers around base of bill, especially on forehead; eyes grey.

SUBADULT: young ♂ has markedly larger bill than ♀ on leaving nest, and vascularization evident in area of future casque. Facial feathers change from brown to grey by 10th month; casque of ♂ well developed but without projection at 12 months (Kilham 1956).

SUBSPECIES
C. s. subquadratus (Cabanis), 1881. Lower Guinea Grey-cheeked Hornbill. Larger size, rear half of casque cream-coloured, and more white on underparts. Casque development assists ageing and coloration aids individual recognition (Kalina 1988). Extent of pale area on casque varies considerably among individuals, even from the same area (Kalina 1988), but seems an inadequate character for subspecific separation. Calls reported as different (Chappuis 1974), but further comparative field studies required.

MEASUREMENTS AND WEIGHTS
C. s. subcylindricus SIZE: wing, ♂ ($n = 4$) 315–335 (326), ♀ ($n = 3$) 284–305 (298); tail, ♂ ($n = 3$) 238–250 (243), ♀ ($n = 2$) 230, 234; bill, ♂ ($n = 4$) 140–166 (151), ♀ ($n = 3$) 100–133 (121); tarsus, ♂ ($n = 2$) 49, 51, ♀ ($n = 2$) 45, 47. WEIGHT: unrecorded.

C. s. subquadratus SIZE: wing, ♂ ($n = 25$) 320–378 (358), ♀ ($n = 18$) 306–352 (325); tail, ♂ ($n = 20$) 258–290 (276), ♀ ($n = 11$) 222–263 (247); bill, ♂ ($n = 20$) 160–192 (171), ♀ ($n = 12$) 136–152 (142); tarsus, ♂ ($n = 18$) 49–55 (53), ♀ ($n = 10$) 44–52 (47). WEIGHT: ♂ ($n = 9$) 1078–1390 (1311), ♀ ($n = 4$) 1000–1200 (1090); captive ♂ 1525, juv ♂ 1240–1450, ♀ 1250, juv ♀ 1160 (Bourne and Chessell 1982).

Field characters

Length 60–70 cm. Distinguished from similar-sized Brown-cheeked Hornbill *C. cylindricus* by black bill, and by largely black casque with cream area confined to rear portion of casque in ♂. In flight, tail appears white with black band at base and black central feathers. Outer primaries black and rest of wing feathers largely white. Larger than Piping Hornbill *C. fistulator*, which has slighter bill with more cream areas; can also be distinguished by black band near base of outer tail feathers, which in Piping Hornbill are either all white or black with white tips. Voice distinctive.

Voice

Loud calls of each subsp. may differ (Bates 1930; Chappuis 1974), but deserve further study. *C. s. subcylindricus* makes loud, mournful, slowly repeated, hooting notes, like toots on a cornet, or sometimes a hoarser version given at a faster tempo. *C. s. subquadratus* makes very loud, but less raucous, quacking notes, uttered at higher pitch and faster, *kakaka kawack kawack* or *kaaak kaaak kaaak*, and audible at up to 2 km. Other calls include a hoarse *ark* contact-note, a screech of distress at intruders near the nest, a low croak by the ♂ arriving at a nest with food, a low chuckle given when feeding the ♀, and low guttural begging groans given by the ♀ (Kilham 1956; Kalina 1988). ♂ has deeper voice than ♀. Young chicks chirp softly. Food-begging call of older chick a harsh *brick-a-brick-a-brick* ...; once fledged, juv makes squeaky version of ad call. Ad ♂ often makes a loud noise by rubbing its bill on a branch, especially in agonistic encounters (Kalina 1988). Tape recordings available.

Range, habitat, and status

C. s. subcylindricus: Côte d' Ivoire; Ghana; Togo; Benin; Nigeria, W of the Dahomey Gap and only a vagrant to E of Niger R. *C. s. subquadratus*: Cameroon; Central African Republic; S Sudan; S Chad (Blancou 1939); NE Zaïre; Uganda, in S and W; W Kenya, N to Cherenganis; Tanzania, in N and W and including W Serengeti and Kibondo; Rwanda; Burundi; NE Angola, at Malanje and Lunda. Isolated populations in W Africa and NE Angola.

Inhabits evergreen forest, especially secondary forest, plantations, and forest edge on the periphery of the main African forest blocks, also extending from forest into adjacent deciduous savanna and woodland. Ranges up to 2600 m, but allopatric from Silvery-cheeked Hornbill *C. brevis* in E Africa. Local and least common in W and S isolates, but common and widespread within main range, especially along Congo R. Now much less common in Ghana than in early 1960s (Grimes 1987). Densities ranged from 49 birds/km^2 in primary forest to 7/km^2 in logged forests in Uganda (mean 12.9/km^2 in primary forest), with highest densities and most transient birds in primary forests (Kalina 1988).

Feeding and general habits

Pairs remain together for years (Kalina 1988), but gather in groups of 20–50 at feeding sites or to mob predators. Subad ♂♂ (12–36 months old) and ♀♀ of unknown age often seen travelling together in larger groups, swapping partners, preening one another, and both sexes skirmishing together (Kalina 1988). Roosts in pairs on outer branches of trees, using same site regularly, sometimes congregating in large loose flocks, as at Kakamega, W Kenya. May spend up to 12 hours at roost.

Feeds mainly in the canopy, picking fruits, rarely plucking them in flight, or breaking open capsules prior to dehiscence. Tears off loose bark or searches in old capsules for small animals; also snaps them up in passing, as in the case of insects and fruit bats (Kalina 1988). Usually in pairs, taking 2–3 fruits/min at a fig vine (Breitwisch 1983). Also clings to creepers, scans ground from low perch, or descends to ground for items such as lizards and millipedes. May travel over 6 km in search of fruit, especially during the dry season. Sometimes hawks passing insects, such as termite alates (Allen 1978). Crashes through foliage while chasing birds and roosting fruit bats (Pitman 1974). Searches for birds' nests to rob (Pitman 1928). More flexible in foraging techniques than congeners, and most similar to Silvery-cheeked Hornbill, including intensive group-foraging drives after animal prey (Pitman 1928).

Twelve stomachs contained only fruit (Chapin 1939; Blancou 1939; Germain *et al.* 1973), and most some grit (Moreau and Moreau 1937). Fruit comprises 91% of the diet by volume in Uganda ($n = 13\,235$), with *Ficus* figs making up 57% of all fruits. Eats a wide variety of husked, capsular, and drupaceous fruits, 5–60 mm in diameter but most pea- to olive-sized. Total of 67 spp. in 26 families and 38 genera recorded in Uganda: *Antiaris, Beilschmiedia, Bequaetiodendron, Bersame, Blighia, Bosqueia, Bridelia, Canarium, Carica, Celtis, Clorophora, Cola, Cordia, Diospyros, Dracaena, Eugenia, Fagaropsis, Ficus, Harrisonia, Lychnodiscus, Maesopsis, Mimusops, Monodora, Morus, Musanga, Olea, Phytolacca, Premna, Pseudospondias, Pycanthus, Rauvolfia, Sapium, Sterculia, Teclea, Trichilia, Uvariopsis, Vangueria* and *Voacanga* (Kalina 1988); and 4 genera in Gabon (*Dacryodes, Pycanthus, Coelocaryon, Elaeis*) (Brosset and Erard 1986). Also feeds on introduced figs and papaw/papaya.

Varied diet includes wide selection of animal foods, mainly insects such as winged termites, beetles (8 families), fly pupae, moths, bugs, bees, caterpillars, cockroaches, mantids and their oothecae, and crickets, but also nestlings, small birds, lesser galagos, bats, lizards and their eggs, chameleons, snails, and millipedes. Mosses, fungi, and lichens also eaten (Kalina 1988).

Displays and breeding behaviour

Monogamous, with no obvious territoriality (3 active nests recorded only 40 m apart), but defends nest tree against conspecifics (Kalina 1988). Groups sometimes gather, and then pairs call loudly together with bill somewhat raised (Granvik 1923). Pair-members also utter loud call regularly from nest tree and chase intruders to advertise nest territory.

Nest-site inspection, allopreening, especially of ♀ on throat, and later courtship-feeding precede breeding. The ♂ may break off a piece of bark, flip it about in its bill, then present it to the ♀. ♂ and ♀ alternate at screaming and groaning into the nest-hole, producing a distinctive hollow sound, and may vibrate their bills against the sides of the en-

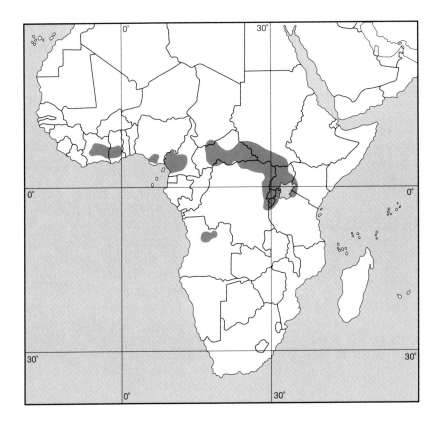

trance as though plastering, clacking the tips rhythmically (Kilham 1956; Kalina 1988). Head-shaking also part of courtship, with head feathers raised, and of aggression towards intruders, with head feathers flattened.

In captivity, copulation was recorded from 17–55 days before the ♀ sealed into the nest (Bourne and Chessell 1982). Nest display and nest-preparation may occur for entire six months of breeding season without further progress, and display may even continue in area where nest tree has since fallen.

Both sexes undertake preliminary nest-sealing, the ♀ applying material brought by the ♂, but the ♂ also applies sealing around the outside. ♂ swallows soil, sometimes from termite mounds, holding it in gullet for up to 30 mins, regurgitating it to ♀ as pellets of sticky mud, bringing up to 10 pellets per load. ♂ sometimes brings mud in the bill, and ♀ sometimes swallows mud or pellets presented by ♂. Most active at nest in morning, when ♀ often enters nest to seal, patting material into place with rapid sideways vibrations of the bill and sometimes using her droppings if no mud available. Some plant and insect food remains included in sealing material.

Breeding and life cycle

NESTING CYCLE: 115–142 days ($n = 56$): incubation period less than 42 days, nestling period 70–79 days (Kalina 1988); 112–124 days in captivity (Harvey 1973; Porritt and Riley 1976; Bourne and Chessell 1982). ESTIMATED LAYING DATES: Zaïre, Jan–Mar; E Africa (rainfall region A: Brown and Britton 1980), Feb, (region B) Aug, Oct–Feb; Uganda, Sept–Mar, ♂ with enlarged gonads Apr–May (Friedmann 1966); Kenya, Apr–May. Laying usually starts with onset of rains. Captives: England, May (Harvey 1973; Porritt and Riley 1976); Canada, Mar–May (Bourne and Chessell 1982). c/2. EGG SIZES: captive *C. s. subquadratus* ($n = 1$) 49.3 × 37.4 (Bourne and Chessell 1982). Eggs white with pitted shells.

Nests are placed in a natural hole 9–30 m above the ground in a large tree of DBH over 48 cm ($n = 45$: Kalina 1988), within (or in lone tree outside) the forest (Kilham 1956). Nest densities highest where many large trees occur in primary forest (11 pairs/km^2) and lowest in heavily logged forest (3/km^2) (Uganda: Kalina 1988). Favours cavities greater than 25 cm wide, with entrance 10–66 cm high × 10–26 wide (Kalina 1988). Will also use a hole in a rock face, as on Mt Elgon (Granvik 1923; North 1942), with an entrance as large as 60 × 46 cm to be sealed. The entrance is sealed to form a narrow slit about 30–50 mm wide.

Nestboxes in captivity were 1.5–7.6 m above ground, at least 46 cm square, and 61–91 cm high, with entrance holes 15–23 cm in their longest dimensions (Bourne and Chessell 1982). The floor was filled with wood shavings to level with the entrance, but was lowered by the hornbills to about 13 cm below. Natural nest lining is formed from bark and small sticks brought by the ♂, together with pieces of wood chipped from the nest-interior by the ♀.

Incubation begins with the first egg, since the chicks hatch asynchronously. ♀ does all incubation, fed by the ♂, mainly on fruit. Food is delivered at intervals of about 1 hr (range: a few mins to 3 hr). Over 200 pea-sized fruits (Kalina 1988), or 2–17 olive-sized fruits and some locusts, may be carried in the gullet on each visit, regurgitated, and passed one or more at a time to the ♀. *Ficus* and *Diospyros* comprise 80% of fruit delivered to nests in Uganda, both probably relatively high in protein (Kalina 1988). Also increases carnivorous component of diet while breeding, at least in captivity (Harvey 1973; Porritt and Riley 1976; Bourne and Chessell 1982).

Captive pair consumed about 560 g of food daily, 70% fruit and 30% animal supplement (Bourne and Chessell 1982). During incubation, the daily animal component of the diet doubled to 320 g while the fruit remained constant at 420 g; but when feeding nestlings the total consumption doubled, with fruit 830 g and the carnivorous component 620 g (plus 6 newborn mice, 30 mealworms, and 25 crickets), dropping to the original amount and composition before fledging.

The ♀ probably does not invariably undergo simultaneous rectrix and remex moult, since some desert well into the nesting cycle, but others moult (Bates 1930; Moreau and Moreau 1937; Harvey 1973; Bourne and Chessell 1982). One chick hatched 42 days after ♀ sealed into nest (Harvey 1973), but incubation period probably less. Another ♀ refused food for 3 days at hatching (Bourne and Chessell 1982).

Eggshells are thrown out of the nest at hatching. The skin of newly hatched chicks is pink, soon turning dark purple-grey, and by 20 days old the eyes are open and all the feather-quills out. ♀ and chicks usually emerge together in the morning, but the chick may sometimes seal itself in again to emerge a day or two later. The chick may also be deserted by the parents at this late stage and die of starvation in the nest (Kalina 1988).

Usually only the older chick survives, the younger chick, hatching a few days later, usually dying of starvation. Two chicks have been recorded in the wild (North 1942) and have been reared by pairs breeding in captivity (Harvey 1973; Anonymous 1975; Porritt and Riley 1976; Bourne and Chessell 1982). The chick may fly well on leaving the nest, or may just flutter to the ground, spending hours trying to reach a low branch on which to roost. Fledgings usually remain near the nest for a few days, then join their parents on foraging trips.

Some ads lead their young from the nest area immediately after fledging, and there is high mortality of chicks at this time (Kalina 1988). Crowned Eagles *Stephanoaetus coronatus* and African Gymnogenes *Polyboroides typus* are suspected predators. Captive chicks refused food from the ♂ after fledging, and were fed items by the ♀ (Anonymous 1975). Captive chicks first fed themselves 40–72 days after fledging (Porritt and Riley 1976; Bourne and Chessell 1982), and after 101–139 days began being harassed by the ad ♂.

Many pairs court, copulate, and defend a nest site with no further activity. Of pairs where ♀ started to seal herself into the nest, at least 36–66% fledged a chick anually ($n = 56$) but only one reared both chicks (Uganda, 1981–84: Kalina 1988). More pairs were successful in a year of high rainfall, and even large chicks starved during a drought. Crowned Eagle killed ♂ at nest, but flooding, visits by monkeys, and even occupation of nest-cavity by squirrels may limit success.

Apart from seasonal availability of food, breeding success is most affected by attacks from conspecific ♀♀ on the breeding ♀ or on newly fledged chicks, and their effects on ♂ foraging (20–80% of annual failures). Interference at nests by paired ♀♀ suggests a shortage of nest-holes limiting the population, or competition with the adoption of infanticide as an alternative reproductive strategy (Kalina 1988, 1989).

Even in captivity, ♀ may seal into the nest for 104–114 days, leaving behind eggs with dead embryos or no contents (Bourne and Chessell 1982). First breeding in captivity at 3 years old (Porritt and Riley 1976). One captive survived for at least 31.8 years (K. Brouwer *in litt*. 1992).

Silvery-cheeked Hornbill *Ceratogymna (Baryrhynchodes) brevis*

Bycanistes cristatus brevis Friedmann, 1929. *Proceedings of the New England Zoological Club*, **11**, 32. Mt Lutindi, Usambara Mts, Tanzania.

PLATES 12, 13

Description

ADULT ♂: mainly glossy black, facial feathers tipped with pale silvery grey. Lower back, rump, uppertail- and undertail-coverts, and belly white. Tail black, with white tips to all but central pair of feathers; wings all black. Bill dark brown with yellow band across base; large, curved casque cream-coloured in

Silvery-cheeked Hornbill *Ceratogymna (Baryrhynchodes) brevis*

contrast, cylindrical at base, and tapering to blade-like projection overhanging bill tip. Circumorbital skin blue. Eyes brown, legs and feet black.

ADULT ♀: like ♂ in overall coloration, but smaller. Bill and casque dark brown with yellow band across base, the casque only a raised eminence on base of bill.

IMMATURE: like ad, but feathers around base of bill and on sides of face black with brown edges. Bill small, without casque. Circumorbital skin undescribed; eye pale cream-coloured.

MOULT
Mar–May, Tanzania, after breeding (Moreau 1935).

MEASUREMENTS AND WEIGHTS
SIZE: wing, ♂ ($n = 72$) 345–395 (369), ♀ ($n = 37$) 321–360 (340); tail, ♂ ($n = 50$) 256–295 (276), ♀ ($n = 21$) 245–280 (260); bill, ♂ ($n = 43$) 153–195 (171), ♀ ($n = 21$) 138–155 (149); tarsus, ♂ ($n = 35$) 50–60 (55), ♀ ($n = 17$) 47–52 (50). WEIGHT: ♂ ($n = 5$) 1265–1400 (1308); ♀ ($n = 5$) 1050–1450 (1162).

Field characters
Length 60–70 cm. Distinguished from all other arboreal hornbills in its range by large size, white back and rump, and very prominent cream-coloured casque that contrasts with brown bill. Only sympatric congener is Trumpeter Hornbill *C. bucinator*, which is smaller, with black casque, purple-pink facial skin, mainly black cheek feathers, white trailing edge to wing, and different voice.

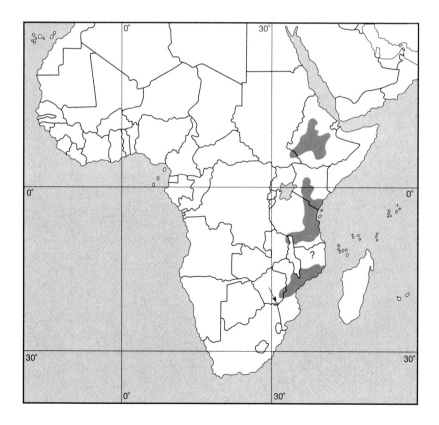

Voice
Loud braying and growling *quark quark quark* notes uttered with head thrust forward. Softer grunting, hooting, and quacking contact-calls given while feeding. May call in middle of moonlit nights (Duckworth *et al.* 1992). Wingbeats clearly audible as loud whooshing sound. Tape recordings available.

Range, habitat, and status
Ethiopia, throughout the highlands; SE Sudan, in the Boma hills; Kenya, patchy distribution E of Rift Valley in Ndotos, Meru, Mt Kenya, and Abedare forests but more widespread in S lowland forests, including Arabuko-Sokoke; Tanzania, extending W to Loita Hills, Maranga Forest, Mt Rungwe, and Umalila; Malawi, in extreme N and in S lowlands; NE Zambia; Mozambique, widespread N of Zambezi R., marginal in SW; E Zimbabwe; NE South Africa, a rare vagrant to Pafuri. Patchy distribution N of the Equator, more widespread in E lowlands.

Inhabits both forest interior and edge, mainly of evergreen coastal and montane forests up to 2600 m, but also extends to dense deciduous woodland and riparian gallery forest. The patchiness of such forests results in the disjunct distribution of the sp. Locally common in some areas but generally uncommon, wandering widely in dry non-breeding season in search of fruit (e.g. Malawi: Dowsett-Lemaire 1988, 1989).

Feeding and general habits
Usually found in pairs throughout the year, or as family trios with an imm present. At times young birds form separate small flocks, and flocks of up to 100 congregate at fruiting trees or other concentrations of food such as locusts. Often roosts communally, with up to 200 birds together in adjacent large trees. Disperses from the roost singly, in pairs, or in small flocks to feed, leaving about half hour after sunrise and returning again before sunset (Trump 1982). Basks in sun, with head lolling about, before departing from roost. Also roosts singly or in pairs, and the breeding ♂ roosts alone away from nest tree. Roost sites may be regular or vary from night to night.

Ranges widely in daily and seasonal search for food, visiting some areas regularly (Trump 1982) and others erratically. May return to same fruiting tree daily for 3 months (Allen 1978), or be found far from its normal forest haunts during a drought or when exceptional rains make marginal habitat more suitable. Often associates with Trumpeter Hornbill at roost and feeding sites. Most food is picked from the canopy and branches but will descend to the ground for insects, or fruits dropped by monkeys, as well as to dustbathe. Also hawks flying insects (Durfey and Durfey 1978) and hovers to tear at nests (Baker and Allen 1978).

Largely frugivorous, eating many figs and favouring cherry-sized, stoned fruits as well as small, hard nuts. Favoured fruits in Tanzania are *Sersalisia usumbarensis*, *Polyathia oliveri*, and *Ficus*, also taking *Myriantus arboreus*, *Harongana madagascariensis*, *Chrysophyllum msolo*, *Parinarium goetzeanum*, *Rauvolfia obliquinerva*, *Sapium ellipticum*, *Odyendea zimmermannii*, and *Eugenia*, *Strychnos*, *Heterophylla*, and *Canthium* spp. plus exotics such as *Maesopsis eminii*, *Passiflora edulis*, *Dichopsis gutta*, *Hovenia*, and *Psidium* (Moreau 1935; Moreau and Moreau 1941). In Malawi, eats *Blighia unijugata*, *Bridelia brideliifolia*, *B. micrantha*, *Ekebergia capensis*, *Ficus capensis*, *F. kirkii*, *F. lutea*, *F. polita*, *F. sansibarica*, *F. thonningii*, *Rauvolfia caffra*, *Syzygium cordatum*, *S. quineense*, *Englerina inaequilatera*, and *Uapaca kirkiana* (Dowsett-Lemaire 1988). Large seeds may be rapidly cast within 50 m of the tree, and habit probably important in seed dispersal. Smaller seeds are expelled by defecation, but sticky mistletoe seeds are wiped off on the perch. Also eats fruits of *Acokanthera*, and flowers of *Conopharyngia holstii* are delivered to nests. Fruits delivered to nests differ markedly between seasons (Moreau and Moreau 1941).

May be more carnivorous at times than is generally believed, eating bird eggs, nestlings, skinks, chameleons, caterpillars, locusts, long-horned grasshoppers, mantids, termites, spiders, and centipedes, sometimes in large quantities (Durfey and Durfey 1978), and including prey as large as an African Green Pigeon *Treron calva* (Brass 1983). Pair-members may forage some distance apart when hunting animal food, keeping up low contact-calls, and moving steadily through foliage with short flights every few minutes (Duckworth *et al*. 1992). At times aggressively chases small animals, bouncing up and down on branches to disturb prey, often in small parties that attack roosting fruit bats, or tearing up birds' nests (*Lonchura, Ploceus*) in search of edible contents (Kingdon 1973; Baker and Allen 1978; Allen 1978). This behaviour may occur only when protein foods are in demand.

Displays and breeding behaviour

Monogamous, without obvious territoriality other than defence of immediate nest area. Both sexes take part in sealing the nest-entrance (Moreau 1935). The ♂ swallows lumps of soil, forming them into pellets in its gullet, and then regurgitates them to the ♀, who applies them to the sides of the entrance hole. The diameter of the pellets is 1.5–2.5 cm and 3–48 pellets are regurgitated per load, with up to 15 loads (235 pellets) being brought per day during nest-preparation periods of up to 5 hr. Sealing may continue for months, even several seasons, before conditions become suitable and nesting is attempted. The ability of the ♂ to produce sufficient saliva to form pellets may be a limiting factor in the short term.

The ♂ courtship-feeds the ♀ for some time before she finally enters the nest, the ♀ uttering a grunting acceptance-call. Copulation occurs with no obvious ritual, being recorded at least 10 days before final sealing of the entrance by the ♀, often when she emerges from the nest after a bout of sealing. The ♀ enters and leaves the nest by first inserting one wing, then the body, then the other wing. The ♂ is very defensive of the nest area, especially against conspecific ♂♂.

Breeding and life cycle

NESTING CYCLE: 107–138 days; incubation period about 40 days, nestling period 77–80 days. ESTIMATED LAYING DATES: Ethiopia, Feb–July; E Africa (rainfall region D: Brown and Britton 1980), Oct–Nov; Kenya, Aug, Oct; Tanzania, Aug–Sept, Usambara Mts Sept–Nov in dry season; Malawi, excavating hole Oct, Feb; Zimbabwe, Apr, Sept–Oct. Laying takes place within a 3-week window around 5 Nov in Usambara Mts, despite lack of obvious day length and climatic cues (Moreau 1935; Moreau and Moreau 1941). Breeding further S occurs during the early part of the rainy season, chicks fledging at the end of the rains. c/1–2. EGG SIZES: ($n = 3$) 51.9–53.6 × 36.4–38.2 (53.1 × 37.1). Eggs white with pitted shells.

Nests in natural tree holes 7–25 m above ground, with the entrance sealed to a narrow vertical slit. Recorded nesting in a mahogany tree, *Ocotea usambarensis, Polyscias kikuyuensis, Parinarium*, and also in rock holes (Roberts 1940). One 32 × 20-cm entrance hole was sealed to a 25 × 2-cm slit, another 25 × 37.5 cm was closed to 12.5 × 5 cm (Start and Start 1978).

♂ regularly brings bark flakes and small sticks throughout the nesting cycle, especially in early stages, probably as lining, also small stones (Start and Start 1978). Much bark lining is also brought during period of chick growth, probably to assist nest sanitation. A nest with a chick contained larvae or nymphs of bugs, beetles, moths, cockroaches, and flies (Britton 1940); these probably helped to keep the hole clean, and were quite different from insects found a year later when the hornbills were not resident.

Only the ♀ incubates, beginning with the laying of the first egg. The ♂ feeds the ♀ and chick throughout the breeding cycle, regurgitating small fruits one at a time from gullet, or occasionally bringing larger fruit or other food items carried in the bill. The loaded gullet may hold up to 69 fruits, which show no signs of digestion even after being retained for 35 min. Loads delivered number 10–14 per day during incubation, an average of 21 (max 24) per day during chick growth, declining to 16–19 per day towards the end of the nesting cycle. Some 100–300 fruits (about 360 g) brought daily, with an estimated 24 000 fruits delivered during 1600 visits to the nest over a complete breeding cycle.

♀ remains in the nest with the chick until it fledges, usually early in the morning. The sealing is broken on emergence by both ♀ and chick, which fly well, and neither returns to the nest until its next breeding attempt. Usually only one chick is fledged, rarely two. ♀ does not appear to undergo simultaneous rectrix and remex moult while breeding, being able to desert the nest midway through the cycle. ♀ may moult only when the breeding cycle is completed.

Eight breeding attempts produced at least 4 and possibly 6 young (Moreau 1935; Moreau and Moreau 1941). Breeding was often sporadic and irregular. In some years, the numbers of ad ♀♀ at roosts may show no obvious decline during the breeding season and even regular nest holes may not be occupied. Breeding may be abandoned at any stage during the cycle, from sealing of the nest to after the chicks have hatched, although desertions were usually during the early stages. Crowned Eagles *Stephanoaetus coronatus* are likely predators (Start and Start 1978).

Subgenus *Ceratogymna* Bonaparte, 1859

Large, predominantly black hornbills with extensive bare facial skin, unique within the order for their pendulous throat wattles, on account of which they were previously separated into their own genus *Ceratogymna*. Head feathers form loose shaggy nape; adult ♀♀ and imms of both sexes notable for their rufous head and neck plumage. Lack the white rump of smaller spp. in the genus, but share with them spp. of *Chapinia* and *Buceronirmus* feather-lice (Elbel 1967). Also share with them the method of transporting sealing materials and, with larger spp., the lack of simultaneous flight-feather moult in breeding ♀.

Two spp. in W and central African moist evergreen forests, differing in extent of white in the outer tail feathers and, in adult ♂, in casque and head form and colour. Otherwise ♀ and imm of both spp. very similar. Ranges overlap completely in W African forests, to which one sp. is totally confined (an unusual phenomenon among African forest spp.).

Black-casqued Wattled Hornbill *Ceratogymna (Ceratogymna) atrata*

Buceros atratus Temminck, 1835. *Planches colorées*, **94**, plate 558. Ashanti, Ghana.

PLATES 12, 13

Other English names: Black-casqued Hornbill, Wattled Hornbill, Black-wattled Hornbill

Description
ADULT ♂: plumage all black, with a metallic sheen on the upperparts, except for white tips to all but central pair of tail feathers. Bill and casque black, the casque a massive cylindrical structure dwarfing even the heavy bill. Circumorbital skin, extensive inflatable bare throat, and limp wattles cobalt-blue. Eyes red, legs and feet black.

ADULT ♀: similar to ♂, but smaller; head and neck feathers red-brown, foreneck cream. Bill and casque horn-coloured, with casque a smaller cylinder ending abruptly half-way along bill. Circumorbital skin pale blue with pink patch below the eye; eyes brown.

IMMATURE: both sexes like ad ♀, but brown feathers of head and neck darker. Bill light green

Black-casqued Wattled Hornbill *Ceratogymna (Ceratogymna) atrata*

without any casque; no loose throat wattles, but throat inflatable, and skin greyish violet.

SUBADULT: juv ♂ moults directly to black ad plumage at end of first year.

MOULT
A pair moulting Apr (Uganda: Friedmann 1966).

MEASUREMENTS AND WEIGHTS
SIZE: wing, ♂ ($n = 27$) 368–440 (395), ♀ ($n = 34$) 330–375 (354); tail, ♂ ($n = 23$) 275–350 (306), ♀ ($n = 30$) 253–290 (270); bill, ♂ ($n = 22$) 170–200 (186), ♀ ($n = 30$) 130–155 (145); tarsus, ♂ ($n = 22$) 50–58 (53), ♀ ($n = 34$) 45–52 (47). WEIGHT: ♂ ($n = 4$) 1069–1600 (1348), ♀ ($n = 3$) 907–1182 (1059).

Field characters
Length 60–70 cm. One of two very large forest spp. with extensive bare blue facial skin and throat wattles, lacking white on body or wings of smaller spp. Black bill and casque, shaggy appearance of head, and plain black head and neck of ad ♂ distinctive; ♀ and imm with brown head and neck. Overlaps in W Africa with Yellow-casqued Wattled Hornbill *C. elata*, which has yellow casque, ad ♂ with scaly appearance to head and neck, and tail white with black centre (rather than black with white tips) in both sexes and at all ages. Noisy, with loud calls and wingbeats, and massive casque of ♂ obvious even in flight.

Voice
Main loud call a clamorous braying with resonant nasal quality, *wha-o wha-o wha-a-a-aw whaaaw*, pitch and tempo variable and audible at up to 2 km. ♂ calls louder and deeper than ♀. ♂ also gives loud, resonant *squark* and long fluting *toot*, while ♀ has single crying note. Alarm-note a soft chuckling. Tape recordings available.

Range, habitat, and status
S. Guinea; Liberia; Côte d'Ivoire; Ghana; Togo; Benin; Nigeria, with no recent records; Cameroon; Rio Muni; Gabon; S Congo; Zaïre, in Congo R. basin with one S record from Shaba province; S Central African Republic; S Sudan, on Aloma Plateau and at Bengengai; W Uganda, in Bwamba forest; N Angola, in Cabinda and Lunda provinces with further isolate at Cuanza Norte. Scattered populations in W Africa, with main range in Lower Guinea forests. Only African hornbill resident on island of Bioko.

Inhabits moist primary lowland evergreen forest, extending to gallery forest along rivers and into coffee, cocoa, and oil-palm plantations. Ascends to 1500 m in Cameroon. Only locally common in Upper Guinea forests (Liberia, Côte d'Ivoire), possibly extinct in Nigeria, but common in main part of range in Congo R. basin. Hunted locally to extinction in some areas such as in Côte d'Ivoire, Cameroon and Congo, in addition to rapid disappearance through habitat loss from fragmented forests.

Feeding and general habits
Usually occurs in pairs or family parties of 3–5, probably with juvs from successive breeding attempts since usually only one chick reared per season. Sometimes seen singly, probably breeding ♂♂, but groups of up to 40 may gather at fruiting trees. Pairs appear to maintain a core area to their home-range, where they also roost, but then travel far to feed. Larger groups of mainly ♂♂, possibly unpaired, or of several families may be temporarily associated (Brosset and Erard 1986). Often feeds in same trees as *C. elata* and Brown-cheeked Hornbill *C. cylindricus*. Local movements probably depend on patterns of fruiting trees, such as *Dacryodes* spp. in Congo (Dowsett-Lemaire and Dowsett 1991).

Lives mainly in forest canopy, feeding discreetly and dominating all but monkeys at fruiting trees. Descends to the ground to feed on seeds of certain fruits which are available only after the pods burst and scatter the contents. Roosts on large branches high in forest trees, choosing similar sites to pass the hot

Black-casqued Wattled Hornbill *Ceratogymna (Ceratogymna) atrata*

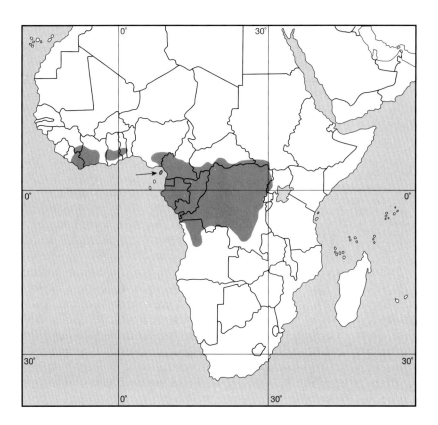

period of the day. The flight is direct and strong, interspersed with gliding, and accompanied by braying calls. The whooshing wing-beats are audible for a considerable distance. Imitation of the calls or wing noise by hunters attracts the birds, especially ad ♂♂, which suggests that flight noise is important in communication and that the calls are territorial (Brosset and Erard 1986).

Feeds mainly on fruit and berries, especially oil-palm nuts. Total of 19 plant genera recorded from diet in Gabon (*Coelocaryon*, *Heisteria*, *Dacryodes*, *Guibourtia*, *Polyalthia*, *Trichilia*, *Pycnanthus*, *Ficus*, *Staudia*, *Strombosia*, *Strombosiopsis*, *Grewia*, *Castanola*, *Canarium*, *Beilschmiedia*, *Xylopia*), as well as *Calamus*, *Raphia*, and *Elaeis* palm fruits (Brosset and Erard 1986). Diet on Bioko includes oil-palm and calabo *Pycnanthus kombo* fruits (Basilio 1963). Seeds of 8 different fruits were found in a nest, mainly *Ricinodendron heudeloti*, together with remains of a crab (Chapin 1939).

Of 16 stomachs, 6 contained insects such as winged ants, beetles, and caterpillars, as well as fruit (Chapin 1939). Another five stomachs contained only fruit, two from oil palms. Probably takes any animal food available, and recorded raiding nests of colonial *Ploceus cucullatus* weavers (Heinrich 1958) and eating insect larvae and termites. Observed to drink water in captivity and in the wild.

Displays and breeding behaviour

Allopreening between mates recorded in captivity (Poulsen 1970). Defends the area around the nest, which includes the regular roost sites. Once two ♂♂ were caught on the ground when engrossed in combat (Prigogine 1971). A second ♂ assisted in feeding the occupants at one nest (Brosset and Erard

1986), and this, together with the extended family groups and indications of territoriality, suggests a social organization of group-territorial and co-operatively breeding groups.

Breeding and life cycle

Little known. ESTIMATED LAYING DATES: Ghana, nestling in Nov; Cameroon, ♀ on point of laying July (Germain *et al.* 1973); Bioko, birds in breeding condition Sept, lone ♂♂ carrying food Jan–Feb; Gabon, Oct–Dec; Zaïre, Apr, June, chick on point of fledging Apr (Prigogine 1971); Uganda, Oct, ♂ in breeding condition July (Friedmann and Williams 1971). Captive, Denmark, sealed into nest Mar, May (Poulsen 1970). c/1–2. Three chicks captured just after fledging may have come from the same nest, but usually only one fledged chick is seen with the parents (Brosset and Erard 1986). EGG SIZES: (*n* = 2) 58.7 × 37.1, 59.0 × 37.8 (Schönwetter 1967).

Eggs white with pitted shells, rather more elongated than those of other hornbills.

Nests in natural holes in trees 10–23 m above ground, the nest-entrance being sealed to a narrow vertical slit. One nest was 10–12 m up in a tree in a clearing on an island in a river (Brosset and Erard 1986), another 21 m up in a 44-m *Macrolobium dewerei*. Both sexes take part in sealing, in captivity and in the wild (Bates 1930; Poulsen 1970). The ♂ forms pellets of mud, droppings, and sticky food in its gullet, regurgitates them one at a time to ♀, who uses them to seal the nest. ♂ brings food to ♀ when in the nest, mainly seeds and fruit regurgitated from the gullet. Undigested seeds may raise level of the nest-floor significantly (Chapin 1939).

A Crowned Eagle *Stephanoaetus coronatus* successfully removed a half-grown chick from a nest and was pursued by the ♂ hornbill (Keith 1969). A captive specimen lived at least 19.5 years (K. Brouwer *in litt.* 1993).

Yellow-casqued Wattled Hornbill *Ceratogymna (Ceratogymna) elata*

Buceros elatus Temminck, 1831. *Planches colorées*, **88**, plate 521, figure 1. Designated as Ghana.

PLATES 12, 13
FIGURE 1.1

Other English names: Yellow-casqued Hornbill

Description

ADULT ♂: mainly black, with a metallic sheen on upperparts. Neck feathers with white bases and brown tips, producing a scaly effect. Tail white but for black central pair of feathers. Bill and underside of casque dark grey, rest of casque pale yellow, grooves along bill running along base of casque; casque a high, curved tube arising over the skull well behind the base of the bill and ending abruptly, as though cut off, above the bill. Circumorbital skin, extensive inflatable bare throat, and limp wattles cobalt-blue, but for a strip of brown feathers up the centre of the throat. Eyes red, legs and feet black.

ADULT ♀: similar to ♂, but smaller; head and neck red-brown, with pale bases to neck feathers and cream throat. Bill and small casque pale yellow; eyes brown. One ♀ with breast white (Eisentraut 1963).

IMMATURE: both sexes like ad ♀, but neck darker brown, bill and casque smaller.

SUBADULT: imms begin body moult about 6 months after they fledge; this evident in juv ♂, which resembles ad ♀ until begins moult. A year-old ♂ had the mature black head and neck.

264 Yellow-casqued Wattled Hornbill *Ceratogymna (Ceratogymna) elata*

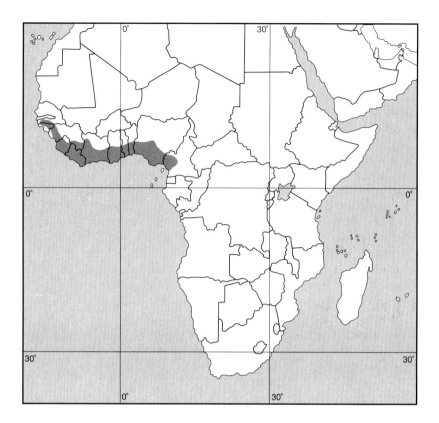

MOULT
Adult pair in moult Mar (Liberia: Hald-Mortensen 1971).

MEASUREMENTS AND WEIGHTS
SIZE: wing, ♂ ($n = 22$) 380–415 (402), ♀ ($n = 22$) 344–377 (362); tail, ♂ ($n = 19$) 285–325 (309), ♀ ($n = 21$) 248–288 (272); bill, ♂ ($n = 18$) 200–230 (212), ♀ ($n = 17$) 143–176 (165); tarsus, ♂ ($n = 20$) 52–58 (56), ♀ ($n = 18$) 47–53 (50). WEIGHT: ♂ 2100; ♀ 1500, 2000.

Field characters
Length 60–70 cm. One of two very large African forest hornbills, lacking any of the white on wings or body of smaller spp. Sympatric with Black-casqued Wattled Hornbill *C. atrata*, from which it differs in having smaller, cut-off, pale yellow casque (♂) or pale yellow bill (♀), white outer tail feathers, and scaly neck pattern.

Voice
Loud call a resonant, drawn-out, braying *aa-a aa-a*. Similar to call of Black-casqued Wattled Hornbill, but less raucous, more prolonged, and at slower tempo. Also utters a hoarse croak in alarm and as a contact-call. Wingbeats audible over a considerable distance. Tape recordings available.

Range, habitat, and status
Senegal, in extreme SW at Basse-Casamance; The Gambia, no recent records; Guinea-Bissau; Guinea, in N and S; Sierra Leone; Liberia; Côte d'Ivoire; Ghana; Togo; Benin; Nigeria; Cameroon, E to Lake Tissongo. Distribution fragmented throughout.

Inhabits moist lowland primary evergreen forest, but extends into secondary forest, riparian forest, patches of forest in savanna, and oil-palm plantations, at up to 1000 m in Liberia. Uses more secondary habitats than

C. atrata. Uncommon to locally frequent, common in Liberia (Colston and Curry-Lindahl 1986), but numbers greatly reduced in Nigeria through forest destruction.

Feeding and general habits
Usually in pairs, sometimes with single imm, occasionally in groups of 6–12. Communal roost of about 50 birds recorded, perched low down in a swamp (Büttikofer in Bates 1930). Heard flying about on moonlit nights (Young 1946).

Forages mainly in the forest canopy, but also descends to the ground to feed on fallen fruit or emerging winged ants and termites (Bouet 1961). Usually visits as a pair, feeding in a fig vine for up to 15 min and taking 2–3 fruits/min (Breitwisch 1983).

Feeds largely on fruit, and especially those of oil-palms, with which its distribution is closely associated. Also takes *Calamus* and *Raphia* palm nuts, as well as Myristicaceae fruit. Includes some small animals in the diet, such as beetles, winged ants and termites, and millipedes. Regurgitates indigestible fibre from oil-palm fruits as a pellet of about 2-cm diameter (Young 1946).

Displays and breeding behaviour
Descends to ground to obtain mud for sealing of nest. Allopreening between members of a pair (Breitwisch 1979) may be part of courtship. Fighting between ♂♂ recorded, and another ♂ recorded to feed accompanying ♀ and juv (Young 1946).

Breeding and life cycle
Little known. ESTIMATED BREEDING DATES: Ghana, ♂♂ feeding at two nests Jan, pair with fully fledged chick Apr (Grimes 1987); Liberia, pair with newly fledged imm Aug, possibly from egg laid Mar, ♀ just finishing moult Mar, possibly after breeding (Hald-Mortensen 1971); Cameroon, ♀♀ with developed eggs in their oviducts Jan, ♂ seen sealing nest-hole in July, imm usually appears with ads after the rains (suggesting laying Jan–Feb for emergence to coincide with ripening of oil-palm fruit Feb–May).

♂ seen sealing a natural hole high in a tall forest tree and a pair with an imm seen examining a tree hole, suggesting that this is the normal nest site. Two active nests were in *Terminalia superba* trees. A captive bird lived at least 22.2 years (Reinhard and Blaskiewitz 1986).

Glossary

Afrotropical Region a biogeographical term for the sub-Saharan area of Africa with its characteristic fauna and flora, also known as the Ethiopian Region, which abuts the Palaearctic Region in the north but does not include Madagascar or the Mascarene Islands

agonistic a term for animal behaviour with a primarily aggressive and combative motivation

alate the mature, winged, reproductive stage of an insect, especially ants (Formicidae) and termites (Isoptera), which often emerge *en masse* and attract many small animals to feed on them

allofeed to pass food from one individual animal to another of the same species, other than when provisioning young but sometimes forming part of courtship-feeding

allometry the differential growth of one part of the body in relation to the whole or to other parts

allopatric a group of two or more species, normally close relatives, whose ranges do not overlap and who are usually mutually exclusive (see sympatric, parapatric)

allopreen to preen the feathers of another individual of the same species

allospecies a group of two or more species, shown to be each other's closest relatives, with adjacent but non-overlapping ranges (see parapatric, sympatric), often termed superspecies

altricial a collective term for the chicks of birds which are helpless at hatching, opposite to precocial

aposematic a feature having a protective role, such as a colour or behaviour that warns a predator of distasteful prey

aril an extra seed-covering (often coloured, hairy, or fleshy) on a seed, such as nutmeg or cashew

aspect ratio a measure of wing proportions, wing span squared divided by wing area, useful in describing aerodynamic performance

basi-occipital the skull bone (or region of the skull when fused) that forms the floor of the opening at which the spinal cord enters the skull and which includes a single protrusion (condyle) on which the skull articulates with the neck

biogeography the description and explanation of patterns of geographical distribution of living or extinct animals and plants, also termed zoogeography when applied only to animals

carnivorous meat-eating

casque an enlargement of the upper surface of the bill, as in hornbills, or on top of the head, as in cassowaries

clade an inclusive group of species, all of which are hypothesized to have arisen from a common ancestor and therefore to be each other's closest relatives

cladistics a theoretical approach to determining relationships between species, based on recognition of monophyletic clades, also known as phylogenetic systematics

cline a geographical gradient between individuals of a species in some feature, such as size or colour

condyle a protruding articulation surface on a bone

co-operative breeding the phenomenon where non-breeding birds, often themselves sexually mature, assist others in various aspects of their reproductive cycle, such as territory defense, nest building and chick rearing

coprophagous dung- or faeces-eating

dehiscent bursting open, applied to seed capsules

diastataxis a condition of the wing where the fifth from outer secondary remex is absent, but its position is revealed by the presence of its greater wing covert; opposite of eutaxis

diploid number the number of paired chromosomes in the nucleus of non-reproductive cells of an organism, usually written $2n$

drupe a succulent fruit that contains a stony seed within

egg-tooth a small sharp protrusion on the tip of the upper mandible of the bill of a young chick, used to cut its way out of the egg shell

emargination a term applied to the margin of a feather, usually a flight feather, where the vane narrows suddenly near the tip

epigamic a term applied to features or behaviour that promote reproductive behaviour

eutaxis a condition of the wing where the fifth from outer secondary remex is present and its base covered by a greater wing covert; opposite of diastataxis

Fitch tree analysis a mathematical procedure, using the Fitch algorithm, to estimate the minimum number of times that characters have changed during evolution and to link species, according to the distribution of such characters, into an evolutionary tree of possible relationships

fledging the point in development at which a young bird leaves its nest or has attained its first true feathers, which is often coincident

fledgling the stage in development of a young bird from when it leaves its nest to when it becomes independent of its parents

frugivorous fruit-eating

Helmholtz resonance a formula for the frequency of resonance of air in a sphere of known volume connected to a neck of known volume and cross-sectional area, derived empirically by Helmholtz in 1875

Indomalayan Region a biogeographical term for the area of the Indian and Asian mainland, approximately in a line south of the Himalaya Mountains, and for the islands stretching east to the Philippines, Sulawesi, and the Lesser Sundas, with their characteristic fauna and flora, also known as the Oriental or Indian Region. Abuts the Australasian Region in the east and the area where these two regions overlap is sometimes separated as the Wallacean Region

incubation the behaviour and process by which the necessary heat is supplied to eggs for embryo development This heat is normally supplied by the parent birds

karyotype a description of the number and gross structure of chromosomes in the cell nucleus of an organism

keystone species an ecological term for species that play a fundamental or 'keystone' role in ecosystem functioning, such as dispersal of seeds from the main types of trees in a habitat, or exertion of a major pressure on a habitat through predation

macrosome a relatively large chromosome within the cell nucleus

Malesia a geographical region describing the peninsular and island chain stretching from the Isthmus of Kra, between Thailand and Malaysia, to New Guinea and the Solomon Islands

microsome a relatively small chromosome within the cell nucleus

mobbing a term for the behaviour of birds, often in numbers and of several species, when they collect around, call at, and fly close to a potential predator, such as an owl or cat

monophyletic a group of species sharing features which support the hypothesis that they have all arisen from a common ancestor, are therefore each other's closest relatives, and have evolved as a single lineage

mutualism a relationship between organisms of different species, from which both species benefit and for which special features, such as behaviour, are developed to initiate and maintain the relationship

nestling the stage of development of a young bird from hatching to when it leaves the nest, sometimes also termed the fledging period

omnivorous eats a mixed diet

oocyte a cell in the ovary of a female bird which develops, or has the potential to develop, into an egg or ovum

ootheca egg case of an arthropod

Oriental, or Indian Region *see* Indomalayan Region

parapatric a group of two or more species determined to be each other's closest relatives and whose ranges abut or just overlap (see allopatric, sympatric)

pericarp the structure formed around a seed during its development, derived from the ovary wall which is either hard and compact or variously swollen

phylogeny the evolutionary history of a species, analogous to the ontogeny of an individual organism from birth to maturity, often used as an adjective to describe aspects of the study of this history, such as phylogenetic systematics

post-fledging the period in the development of a young bird from immediately after it leaves the nest until it starts to move with agility and feed itself, the first part of the fledgling period

remex, plural remiges the main primary (attached to the hand and wrist bones) and secondary (attached to the ulna or forearm bone) wing feather(s)

retrix, plural rectrices the main tail feather(s)

ritual display the use, elaboration, and incorporation of normal behavioural patterns into ritualized displays that serve functions different to the original role of the behaviour, such as incorporation of a preening movement into a courtship display

selective logging cutting and extraction only of those forest trees suitable for commercial exploitation, rather than clear felling of the entire forest

shifting agriculture growing of crops in one area for only a few seasons, until the soil becomes too infertile, and then moving on to a new area

solifugid sun spider, member of the order Solifugae in the Class Arachnida

sonogram a visual representation of sound, usually with time on the horizontal axis, frequency on the vertical axis and relative intensity represented by the density of the trace

supra-occipital the skull bone (or region of the skull when fused) that forms the roof of the opening at which the spinal cord enters the skull and which includes, in hornbills, an accessory condyle which helps support the skull in its articulation with the neck

superspecies see allopatry

swidden agriculture the cutting and burning of dense vegetation, such as forest, to supply nutrients for growing crops, leading inevitably to rotational farming (*see* shifting agriculture)

syndactyl the condition of having the centre and outer toes (III and IV) fused at the base for part of their length

sympatric a group of two or more species, applied mainly to close relatives, whose ranges overlap more or less completely (see allopatric, parapatric)

systematics that discipline of biology which studies, describes, and orders the classificatory units of biology (taxonomy) and which arranges them according to some evolutionary system (see cladistics, phylogeny)

wing loading a measure of the body weight supported per unit area of wing, useful in describing aerodynamic performance

References

Abdulali, H. (1942). The nesting of the Malabar Grey Hornbill. *Journal of the Bombay Natural History Society*, **43**, 102–3.

Abdulali, H. (1951). Some notes on the Malabar Grey Hornbill *Tockus griseus* (Bath.). *Journal of the Bombay Natural History Society*, **50**, 403–4.

Abdulali, H. (1971). Narcondam Island and notes on some birds from the Andaman Islands. *Journal of the Bombay Natural History Society*, **68**, 385–411.

Abdulali, H. (1974). The fauna of Narcondam Island. Part 1. Birds. *Journal of the Bombay Natural History Society*, **71**, 496–505.

Alexander, G. (1991). A study of the possible acoustic function of hornbill casques with reference to environmental effects. Unpublished project, University of Glasgow.

Alexander, G. D., Houston, D. C., and Campbell, M. (1994). A possible acoustic function for the casque structure in hornbills (Bucerotidae). *Journal of Zoology* (London), **233**, 57–67.

Ali, S. (1936). The ornithology of Travancore and Cochin. *Journal of the Bombay Natural History Society*, **39**, 3–35.

Ali, S. (1976). Why Hornbill House? *Hornbill*, **1**, 2–3.

Ali, S. and Ripley, S. D. (1970). *Handbook of the birds of India and Pakistan*, Vol. 4. Oxford University Press, Oxford.

Ali, S. and Ripley, S. D. (1987). *Handbook of the birds of India and Pakistan*, (2nd edn), Vol. 4. Oxford University Press, Oxford.

Allen, F. G. H. (1958). Wreathed Hornbills. *Malayan Nature Journal*, **66**, 214–15.

Allen, P. M. (1978). Carnivorous habits of hornbills. *East African Natural History Society Bulletin*, **1978**, 84–5.

Alston, M. (1937). *Wanderings of a bird-lover in Africa*. H. F. and G. Witherby, London.

Anderson, J. (1876). On the habits of hornbills. *Journal of the Proceedings of the Linnaean Society of London*, **13**, 156–7.

Anderson, J. (1887). List of birds, chiefly from the Mergui Archipelago, collected for the trustees of the Indian Museum, Calcutta. *Proceedings of the Linnaean Society of London*, **21**, 136–53.

Andrew, P. (1985). An annotated checklist of the birds of the Cibodas-Gunung Nature Reserve. *Kukila*, **2**, 10–28.

Anonymous. (1974). Where's Charlie. *Zoonooz*, **47**, 12–16.

Anonymous. (1975). Report of zoo breeding of *Bycanistes subcylindricus*, Birdworld, Farnham. *Animals, the International Wildlife Magazine* **17**, 219.

Anonymous. (1982a). Masai Mara National Reserve. *Swara*, **5**, 20.

Anonymous. (1982b). Crowned Hornbills irrupt again. *Bee-eater*, **33**, 24.

Anonymous. (1983). Surprise visitors call on PE gardens. *Eastern Province Herald*, **June 16 1983**, 1.

Anonymous. (1984). Environment — wound in the world. *Asiaweek* **July 13 1984**, 35–41.

Anonymous. (1990). Rhinoceros hornbills. *Zoonooz*, **58**, 14.

Archer, G. and Godman, E. M. (1961). *The birds of British Somaliland and the Gulf of Aden*, Vol. 3. Oliver and Boyd, Edinburgh.

Aschoff, J. and Pohl, H. (1970). Der Ruheumsatz von Vögeln als Funktion der Tagezeit und der Körpergrösse. *Journal für Ornitologie*, **111**, 38–47.

Aurelian, Fr (1957). *Les oiseaux du Ruanda-Urundi*, Vols 1 and 2. Frères de la charité, Group Scolaire, Astrilda.

Azua, J. and Azua, C. (1989). A review of hornbill breeding at the Zoological Society of San Diego. Notes from American Association of Zoos, Parks and Aquaria workshop on hornbill propagation and management.

Babich, K. (1983). Ground Hornbills (R430). *Witwatersrand Bird Club News*, **123**, 15.

Baker, E. C. St (1927). *The fauna of British India, Birds*, Vol. 4. Taylor and Francis, London.

Baker, E. C. St (1934). *The nidification of the birds of the Indian Empire*, Vol. 3. Taylor and Francis, London.

Baker, W. J. and Allen, P. B. (1978). Silvery-cheeked hornbills. *East African Natural History Society Bulletin*, **1978**, 56–7.

Bangs, O. and van Tyne, J. (1931). Birds of the Kelly-Roosevelt Expedition to French Indo-China. *Fieldiana: Zoology*, **18**, 33–119.

Banks, E. (1935). Notes on birds in Sarawak, with a list of native names. *Sarawak Museum Journal*, **4**, 267–326.

Bannerman, D. A. (1910). On a collection of birds made in northern Somaliland by Mr. G. W. Bury. *Ibis*, **(4)9**, 291–327.

Barnes, R. F. W. (1990). Deforestation trends in tropical Africa. *African Journal of Ecology*, **28**, 161–73.

Bartels, E. (1931). Vogels van Kole Beres [Birds of Kole Beres]. *Natuurkundig Tijdschrift voor Nederlandsche Indië*, **91**, 308–48.

Bartels, H. (1956). Waarnemingen bij een broedhol van de Jaarvogel, *Aceros u. undulatus*, (Shaw) op Sumatra [Observations at the nest hole of the Wreathed Hornbill *Aceros u. undulatus* on Sumatra]. *Limosa*, **29**, 1–18.

Bartels, M. and Bartels, H. (1937). Uit het level der Neushoornvogels, I, II, III. [From the life of hornbills] *Tropische Natuur*, **26**, 117–27, 140–7, 166–72.

Bartlett, A. D. (1869). Remarks on the habits of the hornbills (*Buceros*). *Proceedings of the Zoological Society of London*, **1869**, 142–6.

Basilio, A. J. (1963). *Aves de la Isla de Fernando Po* [Birds of the island of Fernando Po]. Editorial Coculsa, Madrid.

Bates, G. L. (1930). *Handbook of the birds of West Africa*. John Bale, Sons and Danielson, London.

Beasley, A. J. (1985). Grey Hornbill attempting to rob weaver nests. *Honeyguide*, **31**, 57.

Beaulieu, D. (1944). *Les oiseaux du Tranninh*. Publications de l'ecole supérieure des Sciences, Université Indochinoise, Hanoi.

Becker, P. and Wong, M. (1985). Seed dispersal, seed predation, and juvenile mortality of *Aglaia* sp. (Meliaceae) in lowland dipterocarp rainforest. *Biotropica*, **17**, 230–7.

Beehler, B. (1983). Frugivory and polygamy in birds of paradise. *Auk*, **100**, 1–12.

Behr, M. G. (1970). Another cliff nest. *Lammergeyer*, **11**, 83–4.

Behler, D. (1990). Hornbill chicks leave nest in Miami ... finally. *Wildlife Conservation*, **Jan/Feb 1990**, 21.

Beintema, A. J. (1991). Penguins shed stomach linings. *Nature*, **352**, 481–2.

Bell, J. and Brunning, D. F. (1979). Breeding the Malayan wreath-billed hornbill *Aceros undulatus undulatus* at the New York Zoological Park. *International Zoo Yearbook*, **19**, 145–7.

Bell, K. J. and Seibels, R. (1990). A visit to some Indonesian zoos and bird markets. *Avicultural Magazine*, **96**, 50–4.

Belterman, R. H. R. and De Boer, L. E. M. (1984). A karyological study of 55 species of birds, including karyotypes of 39 species new to cytology. *Genetica*, **65**, 39–82.

Belterman, R. H. R. and De Boer, L. E. M. (1990). A miscellaneous collection of bird karyotypes. *Genetica*, **83**, 17–30.

Bennett, E. L. (1988). Proboscis monkeys and their swamp forests in Sarawak. *Oryx*, **22**, 69–74.

Bennett, G. F., Earlé, R. A., du Toit, H., and Huchzermeyer, F. W. (1992). A host-parasite catalogue of the haematozoa of sub-Saharan birds. *Onderstepoort Journal of Veterinary Research*, **59**, 1–73.

Bennett, G. F., Earlé, R. E., Peirce, M. A., and Nandi, N. C. (1993). The Leucocytozoidae of South

African birds. The Coliiformes and Coraciiformes. *South African Journal of Zoology*, **28**, 74–80

Benson, C. W. and Benson, F. M. (1977). *The birds of Malawi*. Privately published, Malawi.

Benson, W. (1968). Behavioural differences among captive Wreathed Hornbills. *Avicultural Magazine*. **68**, 98–9.

Berlioz, J. and Roche, J. (1960). Etude d'une collection d'oiseaux de Guinée. *Bulletin de la Museum Nationale d'Histoire Naturelle*, **32**, 272–83.

Bernstein, H. A. (1861). Ueber Nester und Eier javascher Vögel. *Journal für Ornitologie*, **9**, 113–28.

Bigalke, R. (1951). A note on the drinking habits of hornbills. *Fauna and Flora*, **2**, 35–6.

Binggeli, P. (1989). The ecology of *Maesopsis* invasion and dynamics of the evergreen forest of the East Usambaras, and their implications for forest conservation and forestry practices. In *Forests of the East Usambaras: the resources and their conservation*, (ed. A. C. Hamilton and R. Bensted-Smith), pp. 265–6. International Union for the Conservation of Nature, Nairobi.

Bingham, C. T. (1879). Notes on the nidification of some hornbills. *Stray Feathers*, **8**, 459–63.

Bingham, C. T. (1897). The Great Indian Hornbill in the wild state. *Journal of the Bombay Natural History Society*, **11**, 308–10.

Biswas, B. (1961). The birds of Nepal. *Journal of the Bombay Natural History Society*, **58**, 100–34.

Blancou, F. (1939). Contribution à l'étude des oiseaux de l'Oubangui-chari occidental. *L'Oiseau, et la Revue française d'ornithologie*, **47**, 1–162.

Blankespoor, G. W. (1991). Slash-and-burn shifting agriculture and bird communities in Liberia, West Africa. *Biological Conservation*, **57**, 41–71.

Blasius, W. (1882). Neuer Beitrag zur Kenntniss der Vögelfauna von Borneo. *Journal für Ornitologie*, **30**, 241–55.

Bock, W. J. and Andors, A. V. (1992). Accessory occipital condyle in hornbills (Aves: Bucerotidae), *Zoologischer Jahrbücher Anatomie und Ontogonie der Tiere*, **122**, 161–6.

Bodmer, R. E., Mather, R. J., and Chivers, D. J. (1991). Rain forests of central Borneo — threatened by modern development. *Oryx*, **25**, 21–6.

Bohmke, B. W. (1987). Breeding the Great Indian Hornbill *Buceros bicornis* at the St. Louis Zoological Parks, USA. *Avicultural Magazine*, **93**, 159–61.

Bouet, G. (1961). Faune Tropical XVII. *Oiseaux de l'Afrique Tropicale*, Vol. 2. Office de la recherche scientifique et technique outre-mer, Paris.

Boulton, R. and Rand, A. L. (1952). A collection of birds from Mount Cameron. *Fieldiana: Zoology*, **34**, 35–64.

Bourne, D. and Chessell, D. (1982). Breeding of the Black-and-white Casqued Hornbill *Bycanistes subcylindricus* at the Metro Toronto Zoo. *Avicultural Magazine*, **88**, 15–25.

Brain, C. K. and Sillen, A. (1988). Evidence from the Swartkrans cave for the earliest use of fire. *Nature*, **336**, 464–6.

Brass, D. (1983). Bendera birds. *East African Natural History Society Bulletin*, **1983**, 74.

Breitwisch, R. (1979). Allopreening by the Yellow-casqued Hornbill *Ceratogymna elata*. *Bulletin of the British Ornithologists' Club*, **99**, 114.

Breitwisch, R. (1983). Frugivores at a fruiting *Ficus* vine in a southern Cameroon tropical wet forest. *Biotropica*, **15**, 125–8.

Briggs, J. C. (1989). The historic biogeography of India: Isolation or contact? *Systematic Zoology*, **38**, 322–32.

Britton E. B. (1940). The insect fauna of a nest of the Silvery-cheeked Hornbill, including the description of *Oecornis nidicola* G. et Sp.N. (Col., Carabidae) from Tanganyika. *Entomologist's Monthly Magazine*, **912**, 108–12.

Brooke, R. K. (1956). Speed of S.A. Grey Hornbill *Lophoceros nasutus*. *Ostrich*, **27**, 184.

Brooke, R. K. and Kemp, A. C. (1973). Specimen data on *Bucorvus leadbeateri*. *Bulletin of the British Ornithologists' Club*, **93**, 89–92.

Brooks, T. M., Evans, T. D., Dutson, G. C. L., Anderson, G. Q. A., Asane, D. C., Timmins, R. J., and Toledo, A. G. (1992). The conservation status of the birds of Negros, Philippines. *Bird Conservation International*, **2**, 273–302.

Brosset, A. (1973). Evolution des *Accipiters* forestiers de l'est du Gabon. *Alauda*, **41**, 185–201.

Brosset, A. and Dragesco, J. (1967). Oiseaux collectés et observés dans le Haut-Ivindo. *Biologica Gabonica*, **3**, 59–88.

Brosset, A. and Erard, C. (1986). *Les oiseaux des régions forestières du nord-est du Gabon*, Vol. 1.

Ecologie et comportement des espèces. Société Nationale de Protection de la Nature, Paris.

Brouwer, K. (1990). *European hornbill survey*. National Foundation for Research in Zoological Gardens, Amsterdam.

Brouwer, K. (1991). *European hornbill survey: 1991 update*. National Foundation for Research in Zoological Gardens, Amsterdam.

Brouwer, K. (1993). *Great Hornbill Buceros bicornis European studbook*. Vol. 1. National Foundation for Research in Zoological Gardens, Amsterdam.

Brown, A. (1940). Nesting of a grey hornbill. *Nigerian Field*, 9, 80–4.

Brown, L. H. (1976). Breeding records of Hemprich's Hornbill, *Tockus hemprichii* in Tigai, Ethiopia. *Bulletin of the British Ornithologists' Club*, 96, 81–2.

Brown, L. H. and Britton, P. B. (1980). *The breeding seasons of East African birds*. East African Natural History Society, Nairobi.

Browning, M. R. (1992). Comments on the nomenclature and dates of publication of some taxa in Bucerotidae. *Bulletin of the British Ornithologists' Club*, 112, 22–7.

Buay, J. (1991). Breeding and hand-rearing the Tarictic hornbill *Penelopides panini* at Jurong Bird Park, Singapore. *International Zoo Yearbook*, 30, 180–7.

Burton, P. J. K. (1984). Anatomy and evolution of feeding apparatus in the avian orders Coraciiformes and Piciformes. *Bulletin of the British Museum of Natural History, Zoology*, 47, 331–443.

Butler, A. L. (1905). A contribution to the ornithology of the Egyptian Sudan. *Ibis*, (8)5, 301–401.

Büttiker, W. (1960). Artificial nesting devices in southern Africa. *Ostrich*, 31, 39–48.

Büttikofer, J. (1885). Zoological researches in Liberia. A list of birds collected by J. Büttikofer and C. F. Sala in Western Liberia, with biological observations. *Notes from the Museum*, 7, 129–255.

Büttikofer, J. (1889). On Birds from S.W. Africa. *Notes from the Leyden Museum*, 11, 67–8.

Calder, W. A. (1984). *Size, function and life history*. Harvard University Press, Cambridge, Massachusetts.

Camman, S. (1951). Chinese carvings in hornbill ivory. *Sarawak Museum Journal*, 5, 393–9.

Carr, A. and Carr, P. (1984). Ground Hornbill chasing fishing owl. *Laniarius*, 22, 4.

Cawkell, E. M. and Moreau, R. E. (1963). Notes on birds in the Gambia. *Ibis*, 105, 156–78.

Chadwick, P. (1984). Bradfield's Hornbill in Bulawayo. *Honeyguide*, 30, 125.

Chapin, J. P. (1931). Day by day at Lukolela. *Natural History*, 31, 600–14.

Chapin, J. P. (1932). Birds of the Belgian Congo I. *Bulletin of the American Museum of Natural History*, 65, 1–756.

Chapin, J. P. (1939). Birds of the Belgian Congo II. *Bulletin of the American Museum of Natural History*, 75, 338–70.

Chappuis, C. (1974). Illustrations sonores de problèmes bioacoustique posés par les oiseaux de la zone Ethiopienne. *Alauda*, 42, 88–491 (inc. supplement sonore).

Cheesman, R. E. and Sclater, W. L. (1935). On a collection of birds from north-western Abyssinia. *Ibis*, (13)5, 151–91, 297–329, 594–622.

Chhapgan, B. F. (1977). A hornbill story. *Hornbill*, **Jan–Mar** 1977, 2.

Chin, L. (1971). Protected animals in Sarawak. *Sarawak Museum Journal*, 19, 359–61.

Choy, P. K. (1978). Breeding the Great Indian Hornbills at Jurong Bird Park. *Avicultural Magazine*, 84, 181–3.

Choy, P. K. (1980). Breeding the Great Indian Hornbill *Buceros bicornis* at Jurong Bird Park. *International Zoo Yearbook*, 20, 204–6.

Claasen, J. (1992). Gekroonde neushornvoëls val Karoo binne. *Mirafra*, 9, 60.

Clancey, P. A. (1959). Miscellaneous taxonomic notes on African birds XIV. Notes on variation in the South African populations of the Yellowbilled Hornbill *Tockus flavirostris* (Rüppell), with the characters of a new race. *Durban Museum Novitates*, 5, 238–42.

Clancey, P. A. (1964). Miscellaneous taxonomic notes on African birds XXI. A new subspecies of Redbilled Hornbill *Tockus erythrorhynchus* (Temminck), from the south-eastern lowlands of Africa. *Durban Museum Novitates*, 7, 130–2.

Clay, T. (1957). *First symposium on host specificity among parasites of vertebrates*. Paul Attinger S. A., Neuchâtel.

Clements, F. A. (1992). Recent records of birds from Bhutan. *Forktail*, **7**, 57–73.

Coates-Estrada, R. and Estrada, A. (1986). Fruiting and frugivores at a strangler fig in the tropical rain forest of Los Tuxtlas, Mexico. *Journal of Tropical Ecology*, **2**, 349–57.

Coenraad-Uhlig, V. (1930). Bemerkungen über einen jungen Jahrvögel. *Zoologische Garten*, **3**, 323–5.

Collar, N. J. and Andrew, P. (1988). Birds to watch. The ICBP world checklist of threathened birds. *ICBP Technical Publication*, **8**.

Collins, M. (1990). *The last rain forests*. Mitchell Beazley, London.

Collins, N. M., Sayer, J. A., and Whitmore, T. C. (1991). *The conservation atlas of tropical forests: Asia and the Pacific*. Macmillan Press, London.

Colston, P. R. and Curry-Lindahl, K. (1986). *The birds of Mount Nimba, Liberia*. British Museum (Natural History), London.

Cory, C. P. (1902). Some further notes on the Narcondam Hornbill (*Rhyticeros narcondami*). *Journal of the Bombay Natural History Society*, **14**, 372.

Coupe, M. F. (1967). Aggressive behaviour in Wreathed Hornbills at Chester Zoo. *Avicultural Magazine*, **67**, 170.

Crome, F. H. J. (1975). The ecology of fruit pigeons in tropical northern Queensland. *Australian Wildlife Research*, **2**, 155–85.

Crowe, T. M. and Kemp, A. C. (1988). African historical biogeography as reflected by galliform and hornbill evolution. *Acta XIX Congressus Internationalis Ornithologici*, **2**, 2510–18.

Cunningham-van Someren, G. R. (1977). Some observations on a captive Grey Hornbill *Tockus nasutus*. *Scopus*, **1**, 52–3.

Cunningham-van Someren, G. R. (1986). Avifauna of the Kora National Reserve. In *Kora: An ecological inventory of the Kora National Reserve, Kenya* (ed. M. Coe and N. M. Collins). Royal Geographical Society, London.

Curl, H. C. (1911). Notes on the digestive system of *Hydrocorax*. *Philippine Journal of Science, Manila*, **6d**, 31–6.

da Rosa Pinto, A. A. (1983). *Ornithologia de Angola* [Ornithology of Angola], Vol. 1 (non passeres). Instituto de Investigação Cientifica Tropical, Lisbon.

Davison, W. (1883). Notes on some birds collected in the Nilghiris and in parts of Wynad and Southern Mysore. *Stray Feathers*, **90**, 330–419.

Davidson, J. (1891). Notes on nidification in Kanara. *Journal of the Bombay Natural History Society*, **6**, 331–40.

Davidson, J. (1898). The birds of North Kanara. *Journal of the Bombay Natural History Society*, **12**, 43–72.

Dean, W. R. J. (1973). Notes on a Lanner with a malformed bill, and on hornbills feeding on oil palm fruits. *Bulletin of the British Ornithologists' Club*, **93**, 55.

Dean, W. R. J. (1988). The avifauna of Angolan miombo woodlands. *Tauraco*, **1**, 99–104.

Dean, W. R. J., Huntley, M. A., Huntley, B. J., and Vernon, C. J. (1988). Notes on some birds of Angola. *Durban Museum Novitates*, **14**, 63–4.

Deignan, H. G. (1945). The birds of northern Thailand. *Bulletin of the United States National Museum*, **186**.

Delacour, J. and Mayr, E. (1946). *Birds of the Philippines*. The Macmillan Company, New York.

de Silva, G. S. (1981). Some birds of the Kabili-Sepilok Forest Reserve. *Sarawak Museum Journal*, **29**, 151–66.

Desselberger, H. (1930). Bewegungsmöglichkeiten und Bewegungshemmungen im Hals der Bucerotidae. *Journal für Ornitologie*, **78**, 86–106.

Dev, U. N. (1992). Rearing of the Malabar Pied Hornbill (*Anthracoceros coronatus*) in Bihang Institute for Ornithology and Mass Education. *Bihang News Letter*, **1**, 1–6.

de Zylva, T. S. U. (1984). *Birds of Sri Lanka*. Trumpet Publishers, Sri Lanka.

Diamond, J. (1972). Avifauna of the eastern highlands of New Guinea. *Publication of the Nuttall Ornithologists' Club*, **12**.

Dickinson, E. C., Kennedy, R. S., and Parkes, K. C. (1991). Birds of the Philippines. *British Ornithologists' Union Check-list*, **199**.

Dingle, R. V., Siesser, W. G., and Newton, A. R. (1983). *Mesozoic and Tertiary Geology of Southern Africa*. Balkema, Rotterdam.

Dingle, R. V. and Rogers, J. (1972). Pleistocene palaeogeography of the Agulhas Bank. *Transactions of the Royal Society of South Africa*, **40**, 155–65.

Diop, M. S. (1993). Eco-éthologie du petit calao à bec rouge, *Tockus* (*Lophoceros*) *erythrorhynchus* (Temminck, 1823) en zone de savane. Unpublished thesis, Université Cheikh Anta Diop de Dakar, Senegal.

Dixon, J. E. W. (1968). Prey of large raptors. *Ostrich*, **39**, 203–4.

Dowsett-Lemaire, F. (1988). Fruit choice and seed dissemination by birds and mammals in the evergreen forests of upland Malawi. *Revue de l'Ecologie*, **43**, 251–85.

Dowsett-Lemaire, F. (1989). Ecological and biogeographical aspects of forest bird communities in Malawi. *Scopus*, **13**, 1–80.

Dowsett-Lemaire, F. and Dowsett, R. J. (1991). The avifauna of the Kouilou basin in Congo. *Tauraco Research Report*, **4**, 189–239.

Duckett, J. E. (1985). Some general bird notes from the Fourth Division of Sarawak. *Sarawak Museum Journal*, **34**, 145–59.

Duckett, J. E. (1986). A second set of general bird notes from the Fourth Division of Sarawak. *Sarawak Museum Journal*, **37**, 123–38.

Duckworth, J. W., Evans, M. I., Safford, R. J., Telfer, M. G., Timmins, R. J., and Zewide, C. (1992). A survey of Nechisar National Park, Ethiopia. *International Council for Bird Preservation Study Report*, **50**.

Dunselman, J. (1937). Iets over neushoornvogels in Borneo [Items about hornbills in Borneo]. *Tropische Natuur*, **26**, 16–19.

duPont, J. E. (1972). Notes on Philippine birds (No. 2). Birds of Ticao. *Nemouria*, **6**, 1–13.

du Plessis, M. A. (1994). Cooperative breeding in the Trumpeter Hornbill *Bycanistes bucinator*. *Ostrich*, **65**, 45–7.

duPont, J. E. and Rabor, D. S. (1973*a*). Birds of Dinagat and Siargao, Philippines. *Nemouria*, **10**, 1–111.

duPont, J. E. and Rabor, D. S. (1973*b*). South Sulu Archipelago birds. *Nemouria*, **9**, 43–4.

Durfey, F. D. and Durfey, F. D. (1978). Silverycheeked hornbills congregated in large numbers. *East African Natural History Society Bulletin*, **1978**, 85.

Dutson, G. C. L., Evans, T. D., Brooks, T. M., Asane, D. C., Timmins, R. J., and Toledo, A. (1992). Conservation status of birds on Mindoro, Philippines. *Bird Conservation International*, **2**, 303–25.

Eisentraut, M. (1963). *Wirbeltiere des Kamerungebirges*. Verlag Paul Parey, Hamburg.

Elbel, R. E. (1967). Amblyceran Mallophaga (biting lice) found on the Bucerotidae (hornbills). *Proceedings of the United States National Museum*, **120(3558)**, 1–76.

Elbel, R. E. (1969*a*). The taxonomic position of the hornbill *Rhyticeros plicatus subruficollis* (Blyth) as indicated by the Mallophaga. *Condor*, **71**, 434–5.

Elbel, R. E. (1969*b*). *Chapinia elbeli* Tendeiro, a synonym of *Chapinia fasciata* Elbel, (Mallophaga, Menoponidae). *Proceedings of the Biological Society of Washington*, **82**, 489–90.

Elbel, R. E. (1976). *Bucerocophorus*, a new genus of ischnoceran Mallophaga from African hornbills (Bucerotidae). *Proceedings of the Biological Society of Washington*, **89**, 313–24.

Elbel, R. E. (1977*a*). Two new *Buceronirmus* (Mallophaga: Philopteridae) from *Rhyticeros undulatus* and *R. plicatus* (hornbills). *Pacific Insects*, **17**, 413–18.

Elbel, R. E. (1977*b*). *Buceroemersonia* a new genus of ischnoceran Mallophaga found on the hornbill genus *Tockus* (Bucerotidae). *Proceedings of the Biological Society of Washington*, **90**, 798–807.

Elgood, J. H. (1982). The birds of Nigeria. An annotated check-list. *British Ornithologists' Union Check-list*, **4**.

Elliot, D. G. (1877–1882). *A monograph of the Bucerotidae, or family of the hornbills*. Taylor and Francis, London.

Elliott, C. C. H. (1972). An ornithological survey of the Kidepo National Park, northern Uganda. *Journal of the East African Natural History Society*, **129**, 1–31.

Ellis, M. (1993). Von der Decken's and Jackson's Hornbills: two species or one? *Avicultural Magazine*, **99**, 9–11.

Ellison, B. C. (1923). Notes on the habits of a young hornbill. *Journal of the Bombay Natural History Society*, **29**, 280–1.

Encke, W. (1970). Breeding the Red-billed Hornbill *Tockus erythrorhynchus* at Krefeld Zoo. *International Zoo Yearbook*, **10**, 101–2.

Fairfield, J. R. (1973). Observations on hatching North African Ground Hornbill at San Diego Wild Animal Park. *Avicultural Magazine*, **78**, 27–9.

Falzone, C. K. (1989). Maintaining and breeding the Abyssinian ground hornbill *Bucorvus abyssinicus* at the Dallas Zoo. *International Zoo Yearbook*, **28**, 246–9.

Finlay, J. D. (1928). The nesting habits of the northern grey hornbill (*Lophoceros birostris*). *Journal of the Bombay Natural History Society*, **33**, 444–5.

Fischthal, J. H. and Kuntz, R. E. (1973). Brachylairid and Dicrocoeliid trematodes of birds from Palawan Island, Philippines. *Proceedings of the Helminthological Society of Washington*, **40**, 11–12.

Fleming, R. L. (1968). *Buceros bicornis* Linnaeus in Nepal. *Pavo*, **6**, 59–61.

Flower, S. S. (1925). Contribution to our knowledge of the duration of life in vertebrate animals. IV. Birds. *Proceedings of the Zoological Society of London*, **1925**, 1365–422.

Flower, S. S. (1938). Further notes on the duration of life in animals. IV. Birds. *Proceedings of the Zoological Society of London*, **1938**, 95–235.

Friedmann, H. (1930). The caudal moult of certain Coraciiform, Coliform and Piciform birds. *Proceedings of the United States National Museum*, 77, 1–6.

Friedmann, H. (1966). A contribution to the ornithology of Uganda. Scientific results of the 1963 Knudsen-Machris expedition to Kenya and Uganda. *Bulletin of the Los Angeles County Museum of Natural History, Science*, **3**, 1–55.

Friedmann, H. (1978). Results of the Lathrop Central African Republic Expedition 1976, Ornithology. *Contributions in Science*, **287**, 1–22.

Friedmann, H. and Loveridge, A. (1937). Notes on the ornithology of tropical East Africa. *Bulletin of the Museum of Comparative Zoology, Cambridge, Massachusetts*, **81**, 1–413.

Friedmann, H. and Williams, J. G. (1971). The birds of the lowlands of Bwamba, Toro Province, Uganda. *Contributions in Science*, **221**, 1–70.

Frith, C. B. and Douglas, V. E. (1978). Notes on ten Asian hornbill species (Aves: Bucerotidae); with particular reference to growth and behaviour. *Natural History Bulletin of the Siam Society*, **27**, 35–82.

Frith, C. B. and Frith, D. W. (1978). Bill growth and development in the Northern Pied Hornbill *Anthracoceros malabaricus*. *Avicultural Magazine*, **84**, 20–31.

Frith, C. B. and Frith, D. W. (1983a). A systematic review of the hornbill genus *Anthracoceros* (Aves, Bucerotidae). *Zoological Journal of the Linnean Society*, **78**, 29–71.

Frith, C. B. and Frith, D. W. (1983b). Hornbill habits. *Wildlife*, **September 1983**, 328–32.

Fry, C. H., Keith, S., and Urban, E. K. (1988). *Birds of Africa*, Vol. 3. Academic Press, London.

Gammie, J. (1875). On the breeding of *Aceros nipalensis*. *Stray Feathers*, **3**, 209–11.

Garrod, A. H. (1876). On a peculiarity in the carotid arteries and other points in the anatomy of the ground-hornbill (*Bucorvus abyssinicus*). *Proceedings of the Zoological Society of London*, **1876**, 60–1.

Gautier-Hion, A. and Michaloud, G. (1989). Are figs always keystone resources for tropical frugivorous vertebrates? A test in Gabon. *Ecology*, **70**, 1826–33.

Gautier-Hion, A., Duplantier, J.-M., Quris, R., Feer, F., Sourd, C., Decoux, J.-P., et al. (1985). Fruit characters as a basis of fruit choice and seed dispersal in a tropical forest vertebrate community. *Oecologia (Berlin)*, **65**, 324–37.

Gee, E. P. (1933a). Note on the Indo-Burmese pied hornbill *Hydrocissa malabaricus leucogaster*. *Journal of the Bombay Natural History Society*, **36**, 505–7.

Gee, E. P. (1933b). Note on the development of the casque of the Indo-Burmese pied hornbill (*Anthracoceros albirostris*). *Journal of the Bombay Natural History Society*, **36**, 750–1.

Germain, M. Dragnesco, F. R., and Garcin, H. (1973). Contribution à l'ornithologie du sud-Cameroun. 1. Non-passeriformes. *L'Oiseau, et la Revue française d'ornithologie*, **43**, 119–82.

Gibson-Hill, C. A. (1950). A collection of birds' eggs from North Borneo. *Bulletin of the Raffles Museum*, **21**, 108–15.

Gillard, E. T. and LeCroy, M. (1966). Birds of the Middle Sepik Region, New Guinea. *Bulletin of the American Museum of Natural History*, **132**, 247–75.

Gillard, E. T. and LeCroy, M. (1967). Results of the 1958–1959 Gillard New Britain Expedition. 4. Annotated list of birds of the Whiteman Mountains, New Britain. *Bulletin of the American Museum of Natural History*, **135**, 173–216.

Godfrey, R. (1941). Bird-lore of the eastern Cape Province. *'Bantu Studies' Monograph Series*, **2**.

Golding, F. D. (1935). Nigerian pets. The Ground Hornbill. *Nigerian Field*, **4**, 123–6.

Golding, R. R. and Williams, N. G. (1986). Breeding the Great Indian Hornbill *Buceros bicornis*. *International Zoo Yearbook*, **24/25**, 248–52.

Gonzales, P. C. and Rees, C. P. (1988). *Birds of the Philippines*. Haribon Foundation for the Conservation of Natural Resources, Manila.

Good, A. I. (1952). The birds of French Cameroons. *Memoire de l'Institut Francais d'Afrique (Centre du Cameroun) Serie: Sciences Naturelles*, **2**.

Goodman, S. M. and Gonzales, P. C. (1990). The birds of Mt. Isarog National Park, Southern Luzon, Philippines, with particular reference to altitudinal distribution. *Fieldiana: Zoology*, **60**, 1–39.

Gore, M. E. J. (1968). A check-list of the birds of Sabah, Borneo. *Ibis*, **110**, 165–96.

Gore, M. E. J. (1981*a*). Grey hornbills accompanying Olive Baboons. *East African Natural History Society Bulletin*, **1981**, 115.

Gore, M. E. J. (1981*b*). Birds of the Gambia. *British Ornithologists' Union Check-list*, **3**.

Gore, M. E. J. (1990). Birds of the Gambia, (revised edn). *British Ornithologists' Union Check-list*, **3**.

Gould, K. and Andau, M. (1989). Selection and rejection of five species of dipterocarp fruits by wild and captive primates, squirrels and hornbills. *Malayan Nature Journal*, **42**, 245–50.

Granvik, H. (1923). Contributions to the knowledge of the East African Ornithology. Birds collected by the Swedish Mount Elgon expedition 1920. *Journal für Ornitologie*, **71**, supplement.

Greenberg, D. A. (1975). Some breeding records from Zambia. *Bulletin of The Zambian Ornithological Club*, 7, 7.

Grimes, L. G. (1987). The birds of Ghana. An annotated check-list. *British Ornithologists' Union Check-list*, **9**.

Guichard, K. M. (1947). Birds of the inundation zone of the River Niger, French Soudan. *Ibis*, **89**, 450–88.

Guichard, K. M. (1950). A summary of the birds of the Addis Ababa Region, Ethiopia. *Journal of the East African Natural History Society*, **19**, 154–81.

Gurney, J. H. (1861). On some additional species of birds received in collections from Natal. *Ibis*, **3(10)**, 128–36.

Gyldenstolpe, N. (1955). Birds collected by Dr Sten Bergman during his expedition to Dutch New Guinea 1948–1949. *Arkiv för Zoologi, Stockholm*, **(2)8**, 183–397.

Haagner, A. and Ivy, R. H. (1908). *Sketches of S. African bird life*. R. H. Porter, London.

Hachisuka, M. (1930). Contributions to the birds of the Philippines. *Tori*, **supplement 14**, 138–222.

Hachisuka, M. (1934). Notes on birds from the Philippine Islands and Borneo. *Tori*, **8**, 220–2.

Haimhoff, E. H. (1987). A spectrographic analysis of the loud calls of Helmeted Hornbills *Rhinoplax vigil*. *Ibis*, **129**, 319–26.

Hald-Mortensen, P. (1971). A collection of birds from Liberia and Guinea (Aves). *Streenstrupia*, **1**, 115–25.

Hall, E. F. (1918). Notes on the nidification of the common grey hornbill *(Lophoceros birostris)*. *Journal of the Bombay Natural History Society*, **25**, 503–5.

Hallé, F., Oldeman, R. A. A., and Tomlinson, P. B. (1978). *Tropical trees and forests*. Springer-Verlag, Berlin.

Happel, R. (1986). Observations of birds and other frugivores feeding at *Tetotchidium didymostemon*. *Malimbus*, **8**, 77–8.

Harrison, T. (1951). Humans and hornbills in Borneo. *Sarawak Museum Journal*, **5**, 400–13.

Harrison, T. (1974). The food of *Collocalia* swiftlets (Aves, Apodidae) at Niah Great Cave in Borneo. *Journal of the Bombay Natural History Society*, **71**, 376–93.

Harvey, P. M. (1973). Breeding the Casqued Hornbill at 'Birdworld'. *Avicultural Magazine*, **78**, 23–5.

Hauge, P., Terborgh, J., Winter, B., and Parkinson, J. (1986). Conservation priorities in the Philippine Archipelago. *Forktail*, **2**, 83–91.

Heaney, L. R. (1984). Mammalian species richness on islands on the Sunda Shelf, Southeast Asia. *Oecologia (Berlin)*, **61**, 11–7.

Hegner, R. E. (1980). Group dispersal in Hemprich's Hornbill *Tockus hemprichii*. *Scopus*, **4**, 67–8.

Heinrich, G. (1958). Zur Verbreitung und Lebensweise der Vögel von Angola. *Journal für Ornitologie*, **99**, 121–41, 322–62, 399–421.

Helton, D. (1991). Borneo ablaze again as New El Niño stirs. *BBC Wildlife*, **9**, 792.

Henry, G. M. (1978). *A guide to the birds of Ceylon*, (2nd edn). Oxford University Press, London.

Herholdt, J. J. (1989). Interesting bird observations in the Kalahari Gemsbok National Park (1988). *Mirafra*, **6**, 13–18.

Hetharia, J. C. (1941). Waarnemingen bij het nesthol van een neushornvogel (*Buceros rhinoceros rhinoceros* L.). [observations at the nest hole of a hornbill *Buceros rhinoceros rhinoceros* L.] *Tropische Natuur*, **30**, 185–7.

Hines, C. J. H. (1990). Leaves and flowers in the diet of Grey Louries and Yellowbilled Hornbills in Namibia. *Lanioturdus*, **25**, 53–4.

Hodgson, B. H. (1832). Characters and descriptions of new species of Mammalia and birds from Nepal. *Proceedings of the Zoological Society of London*, **1832**, 10–7.

Hoesch, W. (1933). Beiträge zur Naturgeschichte der Toko's. *Ornithologische Monatsberichte*, **41**, 97–106.

Hoesch, W. (1937). Brut und Mausebeobachtungen an verscheiden *Lophoceros*-Arten. *Ornithologische Monatsberichte*, **45**, 106–14.

Hofer, A. (1982). Some facts about the breeding of the Ground Hornbill. *East African Natural History Society Bulletin*, **1982**, 10–12.

Holgersen, H. (1956). On a collection of birds from N'Zerekore, French Guinea. *Sterna*, **25**, 3–19.

Holmes, D. A. (1969). Bird notes from Brunei: December 1967–September 1968. *Sarawak Museum Journal*, **17**, 399–402.

Holmes, D. and Nash, S. (1989). *The birds of Java and Bali*. Oxford University Press, Oxford.

Holmes, D. and Nash, S. (1990). *The birds of Sumatra and Kalimantan*. Oxford University Press, Oxford.

Holmes, P. (1979). The report of the ornithological expedition to Sulawesi, 1979. Unpublished manuscript.

Hoogerwerf, A. (1949). Bijdrage tot die de oologie van Java [Contribution to the oology of Java]. *Limosa*, **22**, 1–279.

Hoogerwerf, A. (1950). De avifauna van Tijbodas-Gn. Gede [The avifauna of Tijbodas-Gunong Gede]. *Limosa*, **23**, 1–158.

Horne, C. (1869). Notes on the common Grey Hornbill of India (*Meniceros bicornis*). *Proceedings of the Zoological Society of London*, **1869**, 241–3.

Hose, C. (1893). On the avifauna of Mount Dulit and the Baram District in the territory of Sarawak. *Ibis*, **(6)5**, 381–424.

Howe, H. F. (1977). Bird activity and seed dispersal of a tropical wet forest tree. *Ecology*, **58**, 539–50.

Howe, H. F. (1984). Implications of seed dispersal by animals for tropical reserve management. *Biological Conservation*, **30**, 261–81.

Hoyt, D. (1979). Practical methods of estimating volume and fresh weight of birds' eggs. *Auk*, **96**, 73–7.

Hutton, A. F. (1986). Mass courtship display by Great Pied Hornbill, *Buceros bicornis*. *Journal of the Bombay Natural History Society*, Centenary supplement, 209–10.

Hume, A. O. and Oates, E. W. (1890). *The nests and eggs of Indian birds* (2nd edn), Vol 3. R. H. Porter, London.

Hussain, S. A. (1984). Some aspects of the biology and ecology of Narcondam Hornbill (*Rhyticeros narcondami*). *Journal of the Bombay Natural History Society*, **81**, 1–18.

Hutchins, M. (1976). Breeding biology and behaviour of the Indian Pied Hornbill *Anthracoceros malabaricus malabaricus*. *International Zoo Yearbook*, **16**, 99–104.

Hutchison, C. (1989). *Geological evolution of Southeast Asia*. Clarendon Press, London.

Inskipp, C. and Inskipp, T. (1985). *A guide to the birds of Nepal*. Croom Helm, London.

Irwin, M. P. S. (1981). *The birds of Zimbabwe*. Quest Publishing, Salisbury.

Irwin, M. P. S. (1982). Seasonal range overlap between Bradfield's and Crowned Hornbills in the Hwange (Wankie) National Park. *Honeyguide*, **111/112**, 18–19.

Jackson, F. J. (1938). *Birds of the Kenya Colony and the Uganda Protectorate*, Vol. 2. Gurney and Jackson, London.

James, D. and Kannan, R. (1993). Great hornbills in SW India. *ICBP Hornbill Group Newsletter*, **2(1)**, 1.

Jansen, D. H. (1979). How to be a fig. *Annual Review of Ecology and Systematics*, **10**, 13–51.

Jennings, J. T. and Rundel, R. (1976). First captive breeding of the Tarictic Hornbill *Penelopides panini*. *International Zoo Yearbook*, **16**, 98–9.

Jepson, P. (1993). Sumba Island project. *Birdlife International Hornbill Group Newsletter*, **2(2)**, 1.

Johns, A. D. (1982). Observations on nesting behaviour in the Rhinoceros Hornbill, *Buceros rhinoceros*. *Malayan Nature Journal*, **35**, 173–7.

Johns, A. D. (1987). The use of primary and selectively logged rainforest by Malaysian hornbills (Bucerotidae) and implications for their conservation. *Biological Conservation*, **40**, 179–90.

Johns, A. D. (1988*a*). Long-term effects of selective logging operations on Malaysian Wildlife. II. Case studies in the Ulu Segama Forest Reserve, Danum Valley and Tabin conservation areas, Sabah, east

Malaysia. Final report to the Danum Valley Management Committee, Sabah Foundation, and the Socioeconomic Research Unit of the Prime Minister's Department. Unpublished manuscript.

Johns, A. D. (1988b). New observations on hornbills and logging in Malaysia. *Bulletin of the Oriental Bird Club*, **8**, 11–15.

Johns, A. D. (1988c). Effects of 'selective' timber extraction on rain forest structure and composition and some consequences for frugivores and folivores. *Biotropica*, **20**, 31–7.

Johns, A. D. (1989). Recovery of a Peninsular Malaysian rainforest avifauna following selective timber logging: the first twelve years. *Forktail*, **4**, 89 105.

Johnson, O. W. (1979). Urinary organs. In *Form and function in birds*, Vol. 1. (ed. A. S. King and J. McLelland), pp. 183–236. Academic Press, London.

Johnson, T. H. and Stattersfield, A. J. (1990). A global review of island endemic birds. *Ibis*, **132**, 167–80.

Jones, B. C. (1969). Cliff nesting. *Lammergeyer*, **10**, 103–4.

Jones, D. K. (1992). Red-billed Hornbill cover photograph. *Swara*, **15**, cover.

Jones, M. and Banjaransani, H. (1990). The ecology and conservation of the birds of Sumba and Buru. Unpublished preliminary report.

Jordano, P. (1983). Fig-seed predation and dispersal by birds. *Biotropica*, **15**, 38–41.

Juhaeni, D. (1993). Little-known bird. Sumba Hornbill *Aceros everetti*. *Oriental Bird Club Bulletin*, **18**, 19–20.

Kalina, J. (1988). Ecology and behaviour of the Black-and-white Casqued Hornbill (*Bycanistes subcylindricus subquadratus*) in Kibale Forest, Uganda. Unpublished Ph.D. Thesis, Michigan State University.

Kalina, J. (1989). Nest intruders, nest defence and foraging behaviour in the Black-and-white Casqued Hornbill *Bycanistes subcylindricus*. *Ibis*, **131**, 567–71.

Kannan, R. (1992). Breeding biology of the Great Indian Hornbill *Buceros bicornis* in the Anaaimalai Hills, Southern India. Unpublished manuscript.

Keith, A. R. (1969). Crowned Hawk-eagle raids hornbill nest. *Journal of the East African Natural History Society and National Museum*, **28**, 64.

Kemp, A. C. (1970). Some observations on the sealed-in nesting method of hornbills (Family: Bucerotidae). *Ostrich*, **supplement 8**, 149–55.

Kemp, A. C. (1973). Environmental factors affecting the onset of breeding in some southern African hornbills, *Tockus* spp. *Journal of Reproduction and Fertility*, **supplement 19**, 319–31.

Kemp, A. C. (1976a). A study of the ecology, behaviour and systematics of *Tockus* hornbills (Aves: Bucerotidae). *Transvaal Museum Memoir*, **20**.

Kemp, A. C. (1976b). Factors affecting the onset of breeding in African hornbills. In *Proceedings of the 16th International Ornithological Congress*, pp. 248–57. Australian Academy of Science, Canberra.

Kemp, A. C. (1979). A review of the hornbills: biology and radiation. *Living Bird*, **17**, 105–36.

Kemp, A. C. (1988a). The behavioural ecology of the southern ground hornbill: are competitive offspring at a premium? In *Current topics in avian biology*, pp. 267–71. Proceedings of the International-100 Deutscheornitologen-Gesellschaft Meeting, Bonn.

Kemp, A. C. (1988b). The systematics and zoogeography of Oriental and Australasian hornbills (Aves: Bucerotidae). *Bonner zoologische Beiträge*, **39**, 315–45.

Kemp, A. C. (1990). What is the status of the Damara Redbilled Hornbill? *Lanioturdus*, **25**, 51–2.

Kemp, A. C. and Crowe, T. M. (1985). The systematics and zoogeography of Afrotropical hornbills (Aves: Bucerotidae). In *Proceedings of the International Symposium on African Vertebrates* (ed. K. L. Schuschmann), pp. 279–324. Zoologische Forschungsinstitut und Museum Alexander Koenig, Bonn.

Kemp, A. C. and Kemp, M. I. (1972). A study of the biology of Monteiro's Hornbill. *Annals of the Transvaal Museum*, **27**, 255–68.

Kemp, A. C. and Kemp, M. I. (1974). Report on a study of hornbills in Sarawak, with comments on their conservation. *World Wildlife Fund Project Report 2/74*. Unpublished manuscript.

Kemp, A. C. and Kemp, M. I. (1980). The biology of the Southern Ground Hornbill *Bucorvus leadbeateri* (Vigors) (Aves: Bucerotidae). *Annals of the Transvaal Museum*, **32**, 65–100.

Kemp, A. C. and Kemp, M. I. (1991). Timing of egglaying by Southern Ground Hornbills *Bucorvus leadbeateri* in the central Kruger National Park. *Ostrich*, **62**, 80–2.

Kemp, A. C., Joubert, S. C. J., and Kemp, M. I. (1989). Distribution of southern ground hornbills in the Kruger National Park in relation to some environmental features. *South African Journal of Wildlife Research*, **19**, 93–8.

Kemp, E. (1978). Tertiary climatic evolution and vegetation history in the southeast Indian ocean region. *Palaeogeography, Palaeoeclimatology and Palaeoecology*, **24**, 169–209.

Kemp, M. I. and Kemp, A. C. (1978). *Bucorvus* and *Sagittarius*: two modes of terrestrial predation. In *Proceeding of the Symposium on African Predatory Birds* (ed. A. C. Kemp), pp. 13–16. Transvaal Museum, Pretoria.

Kilham, L. (1956). Breeding and other habits of Casqued Hornbills (*Bycanistes subcylindricus*). *Smithsonian Miscellaneous Collection*, **131**, 1–45.

Kingdon, J. (1973). Hornbills and bats. *East African Natural History Society Bulletin*, **December 1973**, 160.

Kinnaird, M. F. and O'Brien, T. G. (1993). Preliminary observation on the breeding biology of the endemic Sulawesi red-knobbed hornbills (*Rhyticeros cassidix*). *Tropical Biodiversity*, **1**, 107–12.

Knight, G. M. (1990). Status, distribution and foraging ecology of the Southern Ground Hornbill (*Bucorvus cafer*) in Natal. Unpublished M. Sc. Thesis, University of Natal, Durban.

Koenig, A. (1937). Bericht über ein in der Freiheit gesammeltes Ei von *Bucorvus abyssinicus*. *Beiträge zur Fortpflanzungsbiologie der Vögel mit Berucksichtigung der Oologie*, **13**, 113–15.

Komen, J. (1984). Grey Hornbills and Helmeted Guineafowl feed at vulture restaurant. *Vulture News*, **11**, 23.

Kongtong, W. (1989). Recent reports. *Bulletin of the Oriental Bird Club*, **8**, 32–6.

Kutter, F. (1883). Beitrag zur Ornis der Philippinen. *Journal für Ornitologie*, **31**, 291–317.

Lack, P. L. (1985). The ecology of the land-birds of Tsavo East National Park, Kenya (continued). *Scopus*, **9**, 57–96.

Lambert, F. (1987). New and unusual bird records from the Sudan. *Bulletin of the British Ornithologists' Club*, **107**, 17–19.

Lambert, F. (1989). Fig-eating by birds in a Malaysian lowland forest. *Journal of Tropical Ecology*, **5**, 401–12.

Lambert, F. (1991). The conservation of fig-eating birds in Malaysia. *Biological Conservation*, **58**, 31–40.

Lambert, F. R. and Marshall, A. G. (1991). Keystone characteristics of bird-dispersed *Ficus* in a Malaysian lowland rain forest. *Journal of Ecology*, **79**, 793–809.

LeCroy, M. and Peckover, W. S. (1983). The birds of the Kimbe Bay Area, west New Britain, Papua New Guinea. *Condor*, **85**, 297–304.

Legge, V. (1983). *A history of the birds of Ceylon*, Vol. 2. Tisara Prakasakayo, Dehiwala.

Leighton, M. (1982). Fruit resources and patterns of feeding, spacing and grouping among sympatric Bornean hornbills (Bucerotidae). Unpublished Ph.D. Dissertation, University of California, Davis.

Leighton, M. (1986). Hornbill social dispersion: variations on a monogamous theme. In *Ecological aspects of social evolution*, (ed. D. I. Rubenstein and R. W. Wrangham), pp. 108–30. Princeton University Press, Princeton.

Leighton, M. and Leighton, D. R. (1983). Vertebrate responses to fruiting seasonality within a Bornean rain forest. In *Tropical rain forest: ecology and management*, (ed. S. L. Sutton, T. C. Whitmore, and A. C. Chadwick), pp. 181–96. *British Ecological Society*, **special publication No. 2**. Blackwell Scientific Publications, Oxford.

Lewis, A. D. (1984). Grey hornbills accompanying baboons. *East African Natural History Society Bulletin*, **May/June 1984**, 58.

Lewis, A. D. and Pomeroy, D. E. (1989). *A bird atlas of Kenya*. A. A. Balkema, Rotterdam.

Lewis, D. (1992). The chopped continent. *BBC Wildlife*, **November 1992**, 66–8.

Lewis, R. E. (1992). Mt Apo and other national parks in the Philippines. *Oryx*, **22**, 100–9.

Lieras, M. (1983). A birds's eye view. *Zoonooz*, **66**, 4–10.

Lim, K. K. (1993). Spectacular movements of hornbills in Perak. *The Asian Hornbills*, **1(1)**, 11–12.

Lint, K. C. (1972). Those odd hornbills. *Zoonooz*, **50**, 5–17.

Lint, K. C. and Stott, K. (1948). Notes on birds of the Philippines. *Auk*, **65**, 41–6.

Lipscomb, C. G. (1938). Nesting of *Bucorvus abyssinicus*. *Ibis*, **(14)2**, 153–4.

Louette, M. (1981). The birds of Cameroon. An annotated check-list. *Verhandlingen van de Koninklijke Academie voor Wetenschappen, Letteren en Schone Kunsten van België*, **43(163)**, 1–295.

Lowe, W. P. (1937). Report on the Lowe-Waldron expedition to the Ashanti Forests and Northern Territories of the Gold Coast. Part IV. *Ibis*, **(14)1**, 635–62.

Lowther, E. H. N. (1942). Notes on some Indian birds VII – Hornbills. *Journal of the Bombay Natural History Society*, **43**, 389–401.

Maasman, D. and Klaasen, M. (1987). Energy expenditure during free flight in trained and free-living Eurasian kestrels. *Auk*, **104**, 603–16.

MacDonald, M. (1960). *Birds in my Indian garden*. Jonathan Cape, London.

Mace, M. E. (1992). Methods used to collect data on parentally raised African Grey Hornbills *Tockus nasutus*. *Avicultural Magazine*, **98**, 155–62.

Mackay, R. D. (1970). *The birds of Port Moresby and district*. Thomas Nelson, Melbourne.

MacKinnon, J. (1979). A glimmer of hope for Sulawesi. *Oryx*, **15**, 55–9.

Maclatchy, A.-R. (1937). Contribution à l'étude des oiseaux du Gabon méridional. *L'Oiseau et la Revue française d'ornithologie*, **7**, 311–64.

Madge, S. G. (1969). Notes on the breeding of the Bushy-crested Hornbill. *Malayan Nature Journal*, **23**, 1–6.

Madoc, G. C. (1947). An introduction to Malayan birds. *Malayan Nature Journal*, **2**, 1–123.

Manger Cats-Kuenen, C. S. W. (1961). Casque and bill of *Rhinoplax vigil* (Forst.) in connection with the architecture of the skull. *Verhandelingen der Koninklijke Nederlandsche Akademie van Wetenschappen, Afdeling Natuurkunde*, **53**, 1–51.

Manuel, M. F. (1969). Gametogenic development of *Leucocytozoon* sp. (*andrewsi* type) of chestnut mannikin (*Lonchura malaca jagori*) and Rufous Hornbills (*Buceros h. hydrocorax*). *Philippine Journal of Veterinary Medicine*, **8**, 20–4.

Marais, E. R. (1993). Captive breeding project for the Southern Ground Hornbill at National Zoological Gardens, Pretoria. *Birdlife International Hornbill Group Newsletter*, **2(2)**, 4.

Marchant, S. (1953). Notes on the birds of south-eastern Nigeria. *Ibis*, **95**, 38–69.

Marshall, B. (1984). Breeding the Yellow-billed Hornbill *Tockus flavirostris*. *Avicultural Magazine*, **90**, 36–40.

Mason, C. W. and Lefroy, H. M. (1912). The food of birds in India. *Memoir of the Department of Agriculture, India*, **3**.

Masterson, A. N. B. (1979). Of birds, the war and the future. *Honeyguide*, **100**, 7–11.

Maurer, D. R. (1984). *The appendicular myology and relationships of the avian order Coraciiformes*. University Microfilms International, Ann Arbor, Michigan.

Maurer, D. R. and Raikow, R. J. (1981). Appendicular myology, phylogeny and classification of the avian order Coraciiformes (including Trogoniformes). *Annals of the Carnegie Museum*, **50**, 417–34.

Mayr, E. and Gillard, E. Th. (1954). Birds of central New Guinea. *Bulletin of the American Museum of Natural History*, **103**, 311–74.

McClure, H. E. (1966). Flowering, fruiting and animals in the canopy of a tropical rain forest. *Malayan Forester*, **29**, 182–203.

McClure, H. E. (1970). Three notes on Thai Birds. *Natural History Bulletin of the Siam Society*, **23**, 331–43.

McGregor, R. C. I. (1905). Birds from the Islands of Romblon, Sibuyan and Crista de Gallo. II. *Further notes on birds from Ticao, Cuyo, Culion, Calayan, Lubang and Luzon*. Bureau of Government Laboratories in Manila, **25**, 1–34.

McGregor, R. C. (1909). *A manual of Philippine birds. Galliformes to Eurylaemiformes*. Bureau of Printing, Manila.

McLachlan, G. R. and Liversidge, R. (1957). *Roberts' Birds of South Africa*. Trustees of the John Voelcker Bird Book Fund, Johannesburg.

Medway, Lord. (1972). The Gunong Benon Expedition 1967. 6. The distribution and altitudinal zonation of birds and mammals on Gunong Benon. *Bulletin of the British Museum (Natural History), Zoology*, **23**, 105–54.

Medway, Lord and Wells, D. R. (1971). Diversity and density of birds and mammals at Kuala Lompat, Pahang. *Malayan Nature Journal*, **24**, 238–47.

Medway, Lord and Wells, D. R. (1976). *Birds opf the Malay Peninsula*. H. F. and G. Witherby and Penerbit Universiti Malaya, London.

Mees, G. F. (1982). Birds from the lowlands of Southern New Guinea (Meranke and Koembe). *Zoologische Verhandelingen*, **191**, 1–188.

Mendelsohn, J. (1990). Yellowbilled Hornbill feeds Grey Hornbill nestlings. *Lanioturdus*, **25**, 54–5.

Mendelsohn, J. M., Kemp, A. C., Biggs, H. C., Biggs, R., and Brown, C. J. (1989). Wing areas, wing loadings and wing spans of 66 species of African raptors. *Ostrich*, **60**, 35–42.

Millar, A. G. (1921). The nesting habits of the Trumpeter Hornbill *Bycanistes bucinator* (Temm.). *South African Journal of Natural History*, **3**, 217–19.

Miller, K. G. and Fairbanks, R. G. (1985). Caenozoic record of climate and sea level. *South African Journal of Science*, **81**, 248–9.

Mlikovsky, J. (1989). Brain size in birds: 3. Columbiformes through Piciformes. *Vestnik Ceskoslovenske Spolecnosti Zoologicke*, **53**, 252–64.

Mlingwa, C. (1990). Ground Hornbill *Bucorvus cafer* and Marabou Storks *Leptoptilos crumeniferus* feeding in association with mammals. *Scopus*, **14**, 23–4.

Modse, S. V. (1988). Some aspects of ecology and behaviour of hornbills with special reference to *Anthracoceros coronatus* (Boddaert) from North Kanara District of Western Ghats. Unpublished Ph.D. Thesis, Karnatak University, Dharwad.

Moreau, R. E. (1934). Breeding habits of hornbills. *Nature*, **134**, 899.

Moreau, R. E. (1935). The breeding biology of certain East African hornbills (Bucerotidae). *Journal of the East African and Uganda Natural History Society*, **13**, 1–28.

Moreau, R. E. (1937). The comparative breeding biology of the African hornbills (Bucerotidae). *Proceedings of the Zoological Society of London*, **107A**, 331–46.

Moreau, R. E. (1938). Nesting of the Red-billed Hornbill observed by Mr. J. Lawes in the Sudan. *Ibis*, **(14)2**, 533–6.

Moreau, R. E. (1966). *The bird faunas of Africa and its islands*. Academic Press, London.

Moreau, R. E. and Moreau, W. M. (1937). Biological and other notes on some East African birds. *Ibis*, **(14)1**, 152–74, 321–45.

Moreau, R. E. and Moreau, W. M. (1940). Hornbill studies. *Ibis*, **(14)4**, 639–56.

Moreau, R. E. and Moreau, W. M. (1941). Breeding biology of Silvery-cheeked Hornbills. *Auk*, **58**, 13–27.

Morel, G. J. and Morel, M.-Y. (1990). *Les Oiseaux de Sénégambie*. Editions de l'ORSTOM, Paris.

Moynihan, M. (1978). An 'ad hoc' association of hornbills, starlings, coucals and other birds. *La Terre et la Vie*, **32**, 557–76.

Nash, S. V. and Nash, A. D. (1985*a*). A checklist of the forest and forest edge birds of the Padang-Suhigan Wildlife Reserve, South Sumatra. *Kukila*, **2**, 51–9.

Nash, S. V. and Nash, A. D. (1985*b*). Breeding notes on some Padang-Suhigan birds. *Kukila*, **2**, 59–63.

Neunzig, R. (1930). Zur Brutbiologie des Rotschnabeltoko (*Lophoceros erythrorhynchus* (Temm.)). *Beiträge zur Fortpflanzungsbiologie der Vögel mit Berücksichtigung der Oologie*, **6**, 114–16.

Newnham, A. (1911). Hornbills devouring young paroquets. *Journal of the Bombay Natural History Society*, **21**, 263–4.

Newton, A. (1893). *A dictionary of birds*. A. and C. Black, London.

North, M. E. W. (1942). The nesting of some Kenya Colony hornbills. *Ibis*, **(14)6**, 499–508.

North, M. E. W. and McChesney, D. S. (1964). *More voices of African birds: supplementary notes*. Laboratory of Ornithology, Ithaca.

Nuttall, R. J. (1992). Further Grey Hornbill sightings. Central OFS. *Mirafra*, **9**, 44.

Oates, E. W. (1883). *A handbook of the birds of British Burma*, Vol. 2. R. H. Porter, London.

O'Brien, T. G. and Kinnaird, M. F. (in press). Notes on the density and distribution of the endemic Sulawesi Hornbill (*Penelopides exarhatus exarhatus*) in the Tangkoko-Dua Saudara Nature Reserve, North Sulawesi. *Tropical Biodiversity*.

Olson, S. L. (1985). The fossil record of birds. In *Avian Biology*, (ed. D. S. Farner, J. R. King, and K. C. Parkes), Vol. 8, pp. 136–8. Academic Press, London.

Onderstall, J. (1989). Trumpeter Hornbill antics. *Hornbill*, **19**, 22–3.

Osborne, W. (1904). Nesting of hornbills. *Journal of the Bombay Natural History Society*, **15**, 715–16.

Osborne, T. O. and Colebrooke-Robjent, J. F. R. (1988). The diurnal raptors of Lochinvar National Park, Zambia 1970–1980. *Proceedings of the Sixth Pan-African Ornithological Congress*, 223–31.

Osmaston, B. B. (1905). A visit to Narcondam. *Journal of the Bombay Natural History Society*, **16**, 620–1.

Osmaston, A. E. (1913). The birds of Gorakhpur. *Journal of the Bombay Natural History Society*, **22**, 532–49.

Pan, Khang Aun. (1987a). The study of species composition and behaviour of sympatric hornbills in Beshout, Perak. *Journal of Wildlife and Parks*, **6**, 84–96.

Pan, Khang Aun. (1987b). The breeding behaviour of the southern pied hornbill (*Anthracoceros coronatus*) in Peninsular Malaysia. *Journal of Wildlife and Parks*, **6**, 97–101.

Pannell, C. M. and Koziot, M. J. (1987). Ecological and phytochemical diversity of arillate seeds in *Aglaia* (Meliaceae): a study of vertebrate dispersal in tropical trees. *Philosophical Transactions of the Royal Society, London*, **B316**, 303–33.

Parker, V. (1984). Snake Eagle robs hornbill nest. *Witwatersrand Bird Club News*, **124**, 23.

Parkes, K. C. (1965). A small collection of birds from the island of Buad, Philippines. *Annals of the Carnegie Museum*, **38**, 49–67.

Parkes, K. C. (1971). Taxonomic and distributional notes on Philippine birds. *Nemouria*, **4**, 1–67.

Paterson, L. (1992). Breeding the von der Decken's hornbill at Leeds Castle, Maidstone, Kent in 1990. *Avicultural Magazine*, **98**, 44–7.

Pearce, F. (1990). Hit and run in Sarawak. *New Scientist*, **126(1716)**, 46–9.

Peckover, W. S. and Filewood, L. W. C. (1976). *Birds of New Guinea and tropical Australia*. A. H. and A. W. Reed, Sydney.

Penny, C. G. (1975). Breeding the Abyssinian ground hornbill *Bucorvus abyssinicus* at San Diego Wild Animal Park. *International Zoo Yearbook*, **15**, 111–15.

Penry, E. H. (1987). Bradfield's Hornbill in croton oil trees at Savuti. *Babbler*, **14**, 23–4.

Peters, J. L. (1945). *Check-list of birds of the world*, Vol. 5. Harvard University Press, Cambridge, Massachusetts.

Phillips, J. C. (1913). Notes on a collection of birds from the Sudan. *Bulletin of the Museum of Comparative Zoology*, **58**, 3–27.

Phillips, W. W. A. (1978). *Annotated checklist of the birds of Ceylon (Sri Lanka)*. Wildlife and Nature Protection Society of Sri Lanka in association with the Ceylon Bird Club, Colombo.

Phillips, W. W. A. (1979). Nests and eggs of Ceylon birds. X. Coraciiformes, Bucerotidae, Upupidae, Coraciidae, Meropidae and Alcedinidae. *Ceylon Journal of Science (Biological Science)*, **13**, 131–58.

Phipson, H. M. (1897). The Great Indian Hornbill in captivity. *Journal of the Bombay Natural History Society*, **11**, 307–8.

Pitman, C. R. S. (1928). The foods of hornbills (in Uganda). *Journal of the Bombay Natural History Society*, **33**, 206.

Pitman, C. (1974). Hornbills and bats. *East African Natural History Society Bulletin*, **April 1974**, 56–7.

Pomeroy, D. E. and Tengecho, B. (1980). Probable breeding of Trumpeter Hornbills in Tsavo West National Park. *East African Natural History Society Bulletin*, **1980**, 91.

Poonswad, P. and Kemp, A. C. (ed.) (1994). *Manual of the status and study of Asian hornbills*. Hornbill Project Thailand, Bangkok.

Poonswad, P. and Tsuji, A. (1994). Range of males of the Great Hornbill *Buceros bicornis*, Brown Hornbill *Ptilolaemus tickelli* and Wreathed Hornbill *Rhyticeros undulatus* in Khao Yai National Park, Thailand. *Ibis*, **136**, 79–86.

Poonswad, P., Tsuji, A., and Ngarmpongsai, C. (1983). A study of the breeding biology of hornbills (Bucerotidae) in Thailand. In *Proceedings of a Delacour/International Foundation for the Conservation of Birds Symposium on Breeding Birds in Captivity*, pp. 239–65. International Foundation for the Conservation of Birds, Los Angeles.

Poonswad, P., Tsuji, A., and Ngarmpongsai, C. (1986). Zur Brutbiologie der Nashornvögel. *Natur und Museum*, **116**, 129–44.

Poonswad, P., Tsuji, A., and Ngarmpongsai, C. (1987). A comparative study of breeding biology of sympatric hornbill species (Bucerotidae) in Thailand with implications for breeding in captivity. In *Proceedings of the Jean Delacour/International Foundation for the Conservation of Birds Symposium on Breeding Birds in Captivity*, pp. 250–77. International Foundation for the Conservation of Birds, Los Angeles.

Poonswad, P., Tsuji, A., and Ngarmpongsai, C. (1988). A comparative ecological study of four sympatric hornbills (Family Bucerotidae) in Thailand. *Acta XIX Congressus International Ornithologi* (ed. H. Ouellet), pp. 2781–91.

Porritt, R. and Riley, M. (1976). Breeding the Black-and-white Casqued Hornbill *Bycanistes subcylindricus* at Birdworld, Farnham. *International Zoo Yearbook*, **16**, 104–5.

Poulsen, H. (1970). Nesting behaviour of the Black-casqued Hornbill *Ceratogymna atrata* (Temm.) and the Great Hornbill *Buceros bicornis* L. *Ornis Scandinavica*, **1**, 11–15.

Prater, S. H. (1921). Notes on two young Indian hornbills. *Journal of the Bombay Natural History Society*, **28**, 550–2.

Preuss, M. and Preuss, B. (1973). Brut von Doppelhornvögeln (*Buceros bicornis* L. 1758) im Zoologischen Garten Rostock. *Zoologische Garten, Leipzig*, **43**, 65–73.

Prigogine, A. (1971). Les oiseaux de l'Itombwe et de son hinterland. *Annales Musée royale de l'Afrique Centrale*, **8(15)**, 1–299.

Primrose, A. M. (1921). Notes on the habits of *Anthracoceros albirostris*, the Indo-Burmese Pied Hornbill, in confinement. *Journal of the Bombay Natural History Society*, **27**, 950–1.

Prozesky, O. P. M. (1965). Development and behaviour of a yellow-billed hornbill (*Lophoceros flavirostris leucomelas*) chick. *Koedoe*, **8**, 109–28.

Punchihewa, G. (1968). The mystery of hornbill nesting. *Loris*, **2**, 156–8.

Rabor, D. S. (1977). *Philippine birds and mammals*. University of the Philippines Press, Quezon City.

Raich, J. W. and Christensen, N. L. (1989). Malaysian dipterocarp forest: tree seedling and sapling species composition and small-scale disturbance patterns. *National Geographic Research*, **5**, 348–63.

Ramsay, E. P. (1878). Contribution to the zoology of New Guinea. *Proceedings of the Linnean Society of New South Wales*, **3**, 241–305.

Rand, A. L. (1951). Birds from Liberia. *Fieldiana: Zoology*, **32**, 561–653.

Rand, A. L. and Rabor, D. S. (1960). Birds of the Philippine Islands: Siquijor, Mount Malindang, Bohol and Samar. *Fieldiana: Zoology*, **35**, 223–539.

Rand, A. L., Friedmann, H., and Traylor, M. A. (1952). Birds from Gabon and Moyen Congo. *Fieldiana: Zoology*, **41**, 223–411.

Ranger, G. (1931). The Ground Hornbill at home. *Blythswood Review*, **8**, 9–10, 21–2, 34.

Ranger, G. (1941). Observations on *Lophoceros melanoleucos melanoleucos* (Lichstenstein) in South Africa. *Ibis*, **(14)**5, 402–7.

Ranger, G. (1949). Life of the Crowned Hornbill. *Ostrich*, **20**, 45–65, 152–67.

Ranger, G. (1950). Life of the Crowned Hornbill. *Ostrich*, **21**, 2–13.

Ranger, G. (1951). Life of the Crowned Hornbill. *Ostrich*, **22**, 77–93.

Ranger, G. (1952). Life of the Crowned Hornbill. *Ostrich*, **23**, 26–36.

Rasa, O. E. A. (1981). Raptor recognition: an interspecific tradition? *Naturwissenschaften*, **68.S**, 151–2.

Rasa, O. E. A. (1983). Dwarf mongoose and hornbill mutualism in the Taru Desert, Kenya. *Behavioural Ecology and Sociobiology*, **12**, 181–90.

Rau, B. (1988). Hornraben im Hellabrun. *Der Zoofreund*, **Mar 1988**, 16–18.

Rauzon, M. (1988). The Cassowary. *Living Bird Quarterly*, **7(4)**, 36–7.

Reddy, K. R., and Rao, B. V. (1983). Nematode parasites of captive birds at Nehru Zoological Park at Hyderabad, Andhra Pradesh, India. *Current Science*, **52**, 316.

Reddish, P. (1992a). Bloodless coup. *BBC Wildlife*, **10(2)**, 10.

Reddish, P. (1992b). Good head for fights. *BBC Wildlife*, **10(12)**, 10.

Reichenow, A. (1877). Die Vögel der Bismarckinselen. *Mitteilungen aus dem zoologischen Museum in Berlin*, **1(3)**, 1–106.

Reilly, S. E. (1988). Breeding the Rhinoceros Hornbill *Buceros rhinoceros* at the Audubon Park and Zoological Garden. *International Zoo Yearbook*, **27**, 263–9.

Reilly, S. E. (1989). The captive management and breeding of the Rhinoceros Hornbill, *Buceros rhinoceros* (AVES: BUCEROTIDAE) at the Audubon Park and Zoological Garden. Unpublished manuscript, Regional Conference Proceedings 1988, American Association of Zoological Parks and Aquaria, Wheeling, West Virginia.

Reinhard, R. and Blaskiewitz, B. (1986). Zur Haltung von Nashornvögeln im Zoologischer Garten Berlin. *Gefiederte Welt*, **110**, 4–6.

Riekert, B. R. (1988). Some factors affecting productivity of three *Tockus* hornbills in central Nambia. In *Proceedings of the 6th Pan-African Ornithological Congress*, pp. 163–72.

Riekert, B. R. and Clinning, C. F. (1985). The use of artificial nest boxes by birds in the Daan Viljoen Game Park. *Bokmakierie*, **37**, 84–6.

Riley, T. H. (1938). Birds from Siam and the Malay Peninsula in the United States National Museum collected by Drs. Hugh M. Smith and William L. Abbot. *Bulletin of the United States National Museum*, **172**.

Ripley, S. D. (1944). The bird fauna of the West Sumatra Islands. *Bulletin of the Museum of Comparative Zoology*, **94**, 305–430.

Ripley, S. D. (1964). A systematic and ecological study of birds of New Guinea. *Bulletin of the Peabody Museum of Natural History*, **19**, 1–85.

Ripley, S. D. and Beehler, B. M. (1989). Ornithogeographic affinities of the Andaman and Nicobar islands. *Journal of Biogeography*, **16**, 323–32.

Ripley, S. D. and Beehler, B. M. (1990). Patterns of speciation in Indian birds. *Journal of Biogeography*, **17**, 639–48.

Ripley, S. D. and Rabor, D. S. (1956). Birds from Canlaon Volcano in the highlands of Negros Island in the Philippines. *Condor*, **58**, 283–91.

Ripley, S. D. and Rabor, D. S. (1958). Notes on a collection of birds from Mindoro Island, Philippines. *Bulletin of the Peabody Museum of Natural History*, **13**, 1–81.

Roberts, A. (1912). Notes on a collection of birds in the Transvaal Museum from Boro, Portuguese East Africa. *Journal of the South African Ornithologists' Union*, **8**, 22–61.

Roberts, A. (1940). *The birds of South Africa*. H. F. and G. Witherby, Johannesburg.

Roberts, T. J. (1991). *The birds of Pakistan*, Vol. 1. Oxford University Press, Karachi.

Roberts, W. W. (1924). Ground Hornbills. *Blythswood Review*, **1**, 106.

Roots, C. (1968). Breeding the red-billed hornbill at the Winged World. [*Tockus erythrorhynchus*]. *Avicultural Magazine*, **74**, 144–6.

Round, P. D. (1985). The current status of hornbills Bucerotidae in Thailand. *Oriental Bird Club Bulletin*, **2**, 6–9.

Rozendaal, F. G. and Dekker, R. W. R. J. (1987). An annotated checklist of the birds of the Dumoga-Bone National Park, North Sulawesi. *Kukila*, **4**, 85–109.

Ryall, C. (1984). More about hornbills and monkeys. *East African Natural History Society Bulletin*, **1984**, 105.

Sanft, K. (1953). On the status of the hornbill *Aceros subruficollis* (Blyth.). *Ibis*, **95**, 702–3.

Sanft, K. (1954). *Tockus nasutus dorsalis* subsp. nova. Zur Frage der Vallidität der Neushornvögels *Tockus jacksoni* (Ogilvie-Grant). *Journal für Ornithologie*, **95**, 416–18.

Sanft, K. (1960). Bucerotidae (Aves/Upupae). *Das Tierreich*, **76**, 1–176.

Santharam, V. (1990). Common Grey Hornbill *Tockus birostris* (Scopoli) dust bathing. *Journal of the Bombay Natural History Society*, **87**, 300–1.

Sayer, J. A., Harcourt, C., and Collins, N. M. (1992). *The conservation atlas of tropical forest: Africa*. Macmillan Press, London.

Schenker, A. (1990). Longevity records at Basle Zoo. *International Zoo News*, **27/28**, 9–28.

Schifter, H. (1986). Beiträge zur Ornithologie des nordhehen Senegal. *Annalen des Naturhistorische Museums in Wien*, **878**, 83–116.

Schneider, G. (1945). *Rhinoplax vigil* (Forst.) und sein Nestling. *Verhandlungen der Naturforschenden Gesellschaft im Basel*, **61**, 1–36.

Schodde, R. and Hitchcock, W. B. (1968). Contributions to Papuasian ornithology. 1. Report on the birds of the Lake Kutubu Area, Territory of Papua and New Guinea. *Division of Wildlife Research*, **13**, 1–73.

Scholtz, C. H. (1972). Grey hornbill (R424) feeding on peanuts. *Laniarius*, **3(2)**, 3.

Schönland, S. (1895). Nesting habits of *Tockus melanoleucos*, Licht. *Transaction of the South African Philosophical Society*, **9**, 1–7.

Schönwetter, M. (1967). *Handbuch der Oologie*. Akademie-Verlag, Berlin.

Schupp, E. W. (1988). Seed and early seedling predation in the forest understory and in treefall gaps. *Oikos*, **51**, 71–8.

Sclater, W. L. (1902). Some account of the ground hornbill or Brom-vögel *Bucorax cafer*. *Zoologist*, **1902**, 49–52.

Searle, R. (1993). Diet. *Witwatersrand Bird Club News*, **158**, 12.

Seibels, R. E. (1989). A brief history of the Pied Hornbill at Riverbanks Zoological Park. Unpublished manuscript, Regional Conference Proceedings 1988, American Association of Zoological Parks and Aquaria, Wheeling, West Virginia.

Seibt, U. and Wickler, W. (1977). Ein Stimmfuhlungs-duett beim Hornraben, *Bucorvus leadbeateri* (Vigors). *Journal für Ornithologie*, **118**, 195–8.

Siesser, W. G. and Dingle, R. V. (1981). Tertiary sea-level movements around southern Africa. *Journal of Geology*, **89**, 523–6.

Shannon, P. (1993). Double-clutching in captive Asian hornbills. *ICBP Hornbill Group Newsletter*, **1**, 4.

Sharp, C. (1985). Yellowbilled Hornbills eating rodents. *Honeyguide*, **32**, 19.

Sharpe, R. B. (1877). Description of a new hornbill from the Island of Panay (1877). *Proceedings of the Linnaean Society of London, Zoology*, **13**, 155–6.

Sharpe, R. B. (1890). On the ornithology of North Borneo, with notes by John Whitehead. *Ibis*, **(6)2**, 1–24.

Shelford, R. (1899). On some hornbill embryos and nestlings. With field notes by C. Hose. *Ibis*, **(7)5**, 538–49.

Short, L. L. (1973). Habits of some Asian woodpeckers (Aves, Picidae). *Bulletin of the American Museum of Natural History*, **152(5)**.

Short, L. L., Horne, J. F. M., and Muringo-Gichuki, C. (1990). Annotated check-list of the birds of East Africa. *Proceedings of the Western Foundation of Vertebrate Zoology*, **4(3)**.

Sibley, C. G. and Ahlquist, J. E. (1972). A comparative study of the egg white proteins of non-passerine birds. *Bulletin of the Peabody Museum of Natural History*, **39**, 1–276.

Sibley, C. G. and Ahlquist, J. E. (1985). The relationships of some groups of African birds, based on comparisons of the genetic material DNA. In *Proceedings of the International Symposium of African Vertebrates* (ed. K. L. Schuschmann), pp. 115–61. Zoologische Forschungsinstitut und Museum Alexander Koenig, Bonn.

Sibley, C. G. and Ahlquist, J. E. (1991). *Phylogeny and classification of birds: a study in molecular evolution*. Yale University Press, New Haven.

Sibley, C. G. and Monroe, B. (1990). *Taxonomy and distribution of birds of the world*. Yale University Press, New Haven.

Siemens, L. (1981). Crowned Hornbills nesting near Eldoret. *East African Natural History Society Bulletin*, **1981**, 65–6.

Sigler, E. U. and Myers, M. S. (1992). Breeding the Wrinkled Hornbill *Aceros corrugatus* at the Audubon Park and Zoological Garden. *International Zoo Yearbook*, **31**, 147–53.

Silsby, J. (1980). *Inland birds of Saudi Arabia*. Immel Publishing, London.

Silvius, M. J. and Verheught, W. J. M. (1985). The birds of Berbak Game Reserve, Jambi Province, Sumatra. *Kukila*, **2**, 76–85.

Simmons, R. (1988). Offspring quality and the evolution of cainism. *Ibis*, **130**, 339–57.

Smith, K. D. (1957). An annotated check list of the birds of Eritrea. *Ibis*, **99**, 1–26, 307–37.

Smythies, B. E. (1953). *The birds of Burma*. Oliver and Boyd, Edinburgh.

Smythies, B. E. (1960). *The birds of Borneo*. J. and J. Gray, Edinburgh.

Smythies, B. E. (1986). *The birds of Burma* (revised edn). Nimrod Press, Liss.

Snelling, J. C. (1969). A raptor study in the Kruger National Park. *Bokmakierie*, **21, supplement**, vii–xi.

Snelling, J. C. (1971). Some information obtained from marking large raptors in the Kruger National Park, Republic of South Africa. *Ostrich*, **supplement 8**, 415–28.

Snow, D. W. (ed.) (1978). *An atlas of speciation in African non-passerine birds*. Trustees British Museum (Natural History), London.

Snow, D. W. (1981). Tropical frugivorous birds and their food plants: a world survey. *Biotropica*, **13**, 1–14.

Sody, H. J. V. (1953). Vogels van het Javaanse Djatibos [Birds from the Javan Djatibos]. *M. I. A. I.*, **4-6**, 125–72.

Starck, D. (1940). Beobachtungen an der Trigeminus muskulatur der Nashornvögel. *Morphologisches Jahrbuch*, **84**, 585–623.

Stark, A. and Sclater, W. L. (1903). *The birds of South Africa*, Vol. 3. R. H. Porter, London.

Start, J. and Start, H. (1978). Nesting record of the Silvery-cheeked Hornbill. *East African Natural History Society Bulletin*, **1978**, 125–7.

Steyn, P. (1980). Breeding and food of the Bateleur in Zimbabwe (Rhodesia). *Ostrich*, **51**, 168–78.

Steyn, P. (1982). *Birds of prey of Southern Africa*. David Philip, Cape Town.

St John, J. H. (1898). Some notes on the Narcondam Hornbill (*Rhyticeros narcondami*). *Journal of the Bombay Natural History Society*, **12**, 211–14.

Stonor, C. R. (1936). On the attempted breeding of a pair of Trumpeter Hornbills (*Bycanistes bucinator*) in the gardens in 1936; together with some remarks on the physiology of the moult in the female. *Proceedings of the Zoological Society of London*, **152A**, 89–95.

Stott, K. (1947). Notes on the Philippine Brown Hornbill. *Condor*, **49**, 35.

Stott, K. (1951). A nesting record of hornbills in captivity. *Avicultural Magazine*, **57**, 113–18.

Stresemann, E. (1914). Die Vögel von Seran (Ceram). *Novitates Zoologicae*, **21**, 25–153.

Stresemann, E. (1940). Die Vögel von Celebes. Teil 3. Biologische Beiträge von Gerd Heinrich. *Journal für Ornithologie*, **88**, 389–487.

Stresemann, E. and Stresemann, V. (1966). Die Mauser der Vögel. *Journal für Ornithologie*, **107**, 1–445.

Swynnerton, C. M. F. (1908). Further notes on the birds of Gazaland. *Ibis*, **(9)2**, 391–443.

Tarboton, W. R. and Allan, D. G. (1984). The status and conservation of birds of prey in the Transvaal. *Transvaal Museum Monograph*, **3**, 1–115.

Tarboton, W. R., Kemp, M. I., and Kemp, A. C. (1987). *Birds of the Transvaal*. Transvaal Museum, Pretoria.

Technau, G. (1936). Die Nasendrüse der Vögel. *Journal für Ornithologie*, **84**, 511–617.

Terborgh, J. (1990). Mixed flocks and polyspecific associations: Cost and benefits of mixed groups to birds and monkeys. *American Journal of Primatology*, **21**, 87–100.

Thiollay, J.-M. (1970). L'exploitation par les oiseaux des essaimages de fourmis et termites dans une zone de contact savana-forêt en Côte d'Ivoire. *Alanda*, **38**, 255–73.

Thiollay, J.-M. (1976). Besoins alimentaires quantatifs de quelques oiseaux tropicaux. *La Terre et la Vie*, **30**, 229–45.

Thiollay, J. M. (1985). The birds of the Ivory Coast. *Malimbus*, **7**, 1–59.

Thompson, M. C. (1966). Birds from North Borneo. *University of Kansas Publications of the Museum of Natural History*, **17**, 402–3.

Thormahlen, M. P. and Healy, S. Y. (1989). Breeding Great Hornbill (*Buceros bicornis*) at Sacramento Zoo. Unpublished manuscript, Regional Conference Proceedings 1988, American Association of Zoological Parks and Aquariums, Wheeling, West Virginia.

Thornhill, W. (1986). An unwelcome intruder. *Witwatersrand Bird Club News*, **132**, 2.

Tickell, O. (1992*a*). Cross-country fire-fighting. *BBC Wildlife*, **February 1992**, 56.

Tickell, O. (1992*b*). Borneo eternal flames. *BBC Wildlife*, **August 1992**, 55.

Tickell, S. R. (1864). On the hornbills of India and Burmah. *Ibis*, **(1)6**, 173–82.

Tirion, C. J. (1933). Kleintjies van den Jaarvogel [Chicks of the Wreathed Hornbill]. *Tropische Natuur*, **22**, 46–9.

Toffic, G. S. (1985). Development of captive raised Abyssinian ground hornbills. *American Association of Zoological Parks and Aquaria 1985 Annual Proceedings*, 241–5.

Toschi, A. (1959). Contributo alla Ornitofauna d'Etiopia. 1. Uccelli raccolti ed osservati in Abissinia dal 1939 al 1942 [Contributions to the avifauna of Ethiopia. 1. Birds collected and observed in Abyssinia from 1939 to 1942]. *Supplemento alle Richerche di Zoologia Applicata alla Caccia, Bologna*, **2**, 301–412.

Tramontana, R. and Rider, P. (1988). Captive breeding program of the Southern Ground Hornbill (*Bucorvus leadbeateri*) at Jacksonville Zoological Park. *American Association of Zoological Parks and Aquaria 1988 Regional Proceedings*, 476–82.

Tree, A. J. (1963). Grey hornbill *Tockus nasutus* as prey of the Lanner Falcon *Falco biarmicus*. *Ostrich*, **34**, 179.

Tree, A. J. (1979). Grey hornbill *Tockus nasutus* preying on nestling weavers. *Honeyguide*, **100**, 6.

Trump, D. (1982). Notes on the Silvery-cheeked Hornbill. *East African Natural History Society Bulletin*, **1982**, 13.

Tso-Hsin, Cheng. (1973). *A distribution list of Chinese birds. Vol. 1. Non-passeriformes*. Joint Publications Research Service, Arlington, Virginia.

Tso-Hsin, Cheng. (1987). *A synopsis of the avifauna of China*. Science Press, Beijing.

Tsuji, A., Poonswad, P., Jirawatkari, N. (1987). Application of radio tracking to study ranging pattern of hornbills (Bucerotidae) in Thailand. In *Proceedings of the Jean Delacour/International Foundation for the Conservation of Birds Symposium on Breeding birds in Captivity*, pp. 316–51. International Foundation for the Conservation of Birds, Los Angeles.

Tweedale, A. (1877). Reports on the collection of birds made during the Voyage of H.M.S. 'Challenger' — No. 11. On the birds of the Philippine Islands. *Proceedings of the Zoological Society of London*, **1877**, 535–51.

Urban, E. K., Brown, L. H., Buer, C. E., and Plage, G. D. (1970). Four descriptions of nesting previously undescribed, from Ethiopia. *Bulletin of the British Ornithologists' Club*, **90**, 162–4.

van Bemmel, A. C. V. (1947). Two small collections of New Guinea birds. *Treubia*, **19**, 1–45.

van Bemmel, A. C. V. and Voous, K. H. (1951). On the birds of the islands of Muna and Buton, S. E. Celebes. *Treubia*, **21**, 27–104.

van Ee, C. A. (1963). Yellow-billed hornbill (*Lophoceros flavirostris*) breeding in captivity. *Ostrich*, **34**, 252.

van Marle, J. G. and Voous, K. H. (1988). The birds of Sumatra. *British Ornithologists' Union Check-list*, **10**.

van Someren, G. L. (1922). Notes on the birds of East Africa. *Novitates Zoologica*, **29**, 1–246.

van Someren, V. D. (1958). *A bird watcher in Kenya*. Oliver and Boyd, Edinburgh.

van Someren, V. G. L. and van Someren G. R. C. (1949). *The birds of Bwamba*. Uganda Journal Special Supplement, Uganda Society, Kampala.

Verheyen, R. (1953). Oiseaux. *Exploration du Parc National d'Upemba, Mission de Witte*, **19**, 1–687.

Vernon, C. (1982). Ground Hornbill. *Bee-eater*, **33**, 35.

Vernon, C. (1984). Ground Hornbills and the drought. *Bee-eater*, **35**, 32.

Vernon, C. J. (1987). Comments on some birds of Zimbabwe. *Honeyguide*, **33**, 54–7.

Vernon, C. (1988). Regional roundup — notes from the Border. *Bee-eater*, **39**, 15–16.

Vernon, C. (1991). Trumpeter Hornbill raids sunbird nest. *Bee-eater*, **42**, 28.

Vigors, N. A. (1825). Observations on the natural affinities that connect the orders and families of birds. *Transactions of the Linnean Society of London*, **14**, 395–517.

Voisin, J. C. (1953). Note sur la nidification en Côte d'Ivoire de Petit Calao à Bec Jaune (*Lophoceros semifasciatus*). *L'Oiseau, et la Revue française d'Ornithologie*, **23**, 148.

von Wieschke, R. (1928). Beobachtungen beim Brutgeschäft des Rotschnabeltoko (*Lophoceros erythrorhynchus* Temm.) im Zoologischen Garten Frankfurt a. M. *Ornithologische Monatsberichte*, **36**, 1–2.

Walton, P. B. and Watt, J. P. (1987). Observations on a bushy crested hornbill nest. Unpublished manuscript.

Wambuguh, O. J. G. (1987). Report on the breeding biology of the Redbilled Hornbill (*Tockus erythrorhynchus*) in Tsavo National Park (East). *Occasional Research Paper of the Wildlife Conservation and Management Department*, unpublished.

Wambuguh, O. (1988). A day spent watching a nest of red-billed hornbills in Tsavo National Park (East). *East African Natural History Society Bulletin*, **18**, 2–4.

Watling, R. J. (1983). Ornithological notes from Sulawesi. *Emu*, **83**, 247–61.

Wells, D. R. (1974). Bird report: 1970 and 1971. *Malay Nature Journal*, **27**, 30–49.

Wells, D. R. (1975). Bird report: 1972 and 1983. *Malay Nature Journal*, **28**, 186–213.

Western, D. and Ssemakula, J. (1982). Life history patterns of birds and mammals and their evolutionary interpretation. *Oecologia*, **54**, 281–90.

Wetmore, A. (1915). A peculiarity in the growth of the tail feathers of the giant hornbill (*Rhinoplax vigil*). *Proceedings of the United States National Museum*, **47**, 497–9.

Whistler, H. (1944). The avifaunal survey of Ceylon. *Spolia Zeylanica*, **23**, 119–322.

White, F. N., Kinney, J., Siegfried, W. R., and Kemp, A. C. (1975). Thermal and gaseous conditions of hornbill nests. Unpublished research report to National Geographic Society.

White, C. M. N. and Bruce, M. D. (1986). The birds of Wallacea. *British Ornithologists' Union Check-list*, **7**.

Whitehead, J. (1890). Notes on the birds of Palawan. *Ibis*, (2)**6**, 38–61.

Whitten, A. J., Musfafa, M., and Henderson, G. S. (1987). *The ecology of Sulawesi*. Gadjah Mada University Press, Bulaksumur.

Wildash, P. (1968). *Birds of S. Vietnam.* Charles E. Tuttle, Rutland and Tokyo.

Wiley, E. O. (1981). *Phylogenetics: the theory and practice of phylogenetic systematics.* John Wiley and Sons, New York.

Wilkinson, R. (1992). Any hope for hornbills? *Avicultural Magazine*, **98**, 119–23.

Wilkinson, R. and McLeod, W. (1990). Breeding the African Grey Hornbill *Tockus nasutus epirhinus* at Chester Zoo. *Avicultural Magazine*, **96**, 167–70.

Williams, A. A. E. (1978). Notes on *Tockus* hornbills breeding at Lake Baringo, Kenya. *Scopus*, **2**, 21–3.

Willis, E. O. (1983). Toucans (Ramphastidae) and hornbills (Bucerotidae) as ant followers. *Le Gerfaut*, **73**, 239–42.

Wilson, C. C. and Wilson, W. L. (1975). The influence of selective logging on primates and some other animals in East Kalimantan. *Folia Primatologica*, **23**, 245–74.

Wilson, R. T. (1982). Environmental changes in Western Dafur, Sudan, over half a century and their effects on selected bird species. *Malimbus*, **4**, 15–26.

Wilson, W. L. and Johns, A. D. (1982). Diversity and abundance of selected animal species in undisturbed forest and plantations in East Kalimantan, Indonesia. *Biological Conservation*, **24**, 205–18.

Witmer, M. C. (1989). Natural history and feeding habits of Philippine hornbills: implications for their conservation and captive breeding. Unpublished manuscript, Regional Conference Proceedings 1988, American Association of Zoological Parks and Aquaria, Wheeling, West Virginia.

Witmer, M. C. (1993). Cooperative breeding by Rufous Hornbills on Mindanao Island, Philippines. *Auk*, **110**, 933–6.

Wong, M. (1986). Trophic organization of understory birds in a Malaysian dipterocarp forest. *Auk*, **103**, 100–16.

Wood, W. S. (1927). Is the large hornbill (*Dichoceros bicornis*) carnivorous? *Journal of the Bombay Natural History Society*, **32**, 374.

Woodward, R. B. and Woodward, J. D. S. (1875). Notes on the natural history of South Africa. *Zoologist*, **1875**, 4389–98.

Worth, W., Sheppard, C., Kemp, A., Ellis, S., and Seale, U. (ed.) (1994). *Hornbill conservation and management plan.* First review draft. IUCN/SSC Captive Breeding Specialist Group, Apple Valley.

Young, C. G. (1946). Notes on some birds of the Cameroon Mountain district. *Ibis*, **88**, 348–82.

Zeller, H. G., Cornet, J. P., and Camicas, J. L. (1992). Experimental transmission of Crimean-Congo hemorrhagic fever virus from West-African ground-feeding birds to *Hyaloma marginatum rufipes* ticks. *American Journal of Tropical Medicine and Hygiene*, **47**, 243–4.

Index

Species accounts can be found on the page numbers set in bold type.

Hornbill species can be found by looking up either the English name (as a subentry under 'Hornbill') or the scientific name. Cross-references or page references can also be found under the Latin specific and subspecific names.

abbreviations xii
Aberia 113
abyssinicus, Bucorvus, see Bucorvus
Acacia 94, 113, 115, 126, 133–4, 139, 142, 144, 247
Acanthoplus 129
Accipiter melanoleucus 139, 244
 toussenelii 152
Aceros 5, 22, 23, 24, 28, 31, 34, 36, 43, 46, 65, 68–71, 73, 79, 84, 100, 191, 197, **208–9**, 212, Plates 10, 11, 14
 cassidix (Sulawesi Wrinkled Hornbill) 7, 12, 15, 18, 21, 39, 47, 64, 67, 79, 198–9, 212, **215–18,** Plates 10, 14
 comatus (White-crowned Hornbill) 7, 12, 15, 18, 21, 28, 29, 38, 40, 47, 49, 64, 67, 69–70, 73, 79, 100, 208–9, **210–12,** Plates 10, 14
 corrugatus (Sunda Wrinkled Hornbill) 7, 12, 15, 18, 21, 39, 47, 64, 67, 79, **218–21,** Plates 10, 14
 everetti (Sumba Wreathed Hornbill) 7, 12, 15, 18, 21, 39, 47, 64, 67, 78, 208, 224, **239–40,** Plates 11, 14
 leucocephalus (Mindanao Wrinkled Hornbill) 7, 12, 15, 18, 21, 39, 47, 64, 67, 79, 191–2, 206, **221–3** 223, Plates 10, 14
 narcondami (Narcondam Wreathed Hornbill) 7, 12, 15, 18, 21, 37, 39, 47, 64, 67, 78, **227–9,** Plates 11, 14
 nipalensis (Rufous-necked Hornbill) 7, 12, 15, 18, 21, 39, 47, 64, 67, 79, 209, 212, **213–15,** Plates 10, 14
 plicatus (Papuan Wreathed Hornbill) 7, 12, 15, 18, 21, 39, 47, 64, 67, 69, 73, 79, 208, **224–7,** 229, 230, Plates 11, 14
 subgenus 64, 67, 208–9, **212**, Plate 10
 subruficollis (Plain-pouched Wreathed Hornbill) 7, 12, 21, 39, 47, 64, 67, 79, 208, **230–2,** 233–4, 237, Plates 11, 14
 undulatus (Bar-pouched Wreathed Hornbill) 4, 7, 12, 15, 18, 21, 32, 37, 39, 44, 47, 64, 67, 70, 77, 79, 103, 169, 181, 183, 208, 213, 219, 230–2, **232–8,** Plates 11, 14
 waldeni (Visayan Wrinkled Hornbill) 7, 12, 21, 39, 47, 64, 67, 79, 200, **223–4,** Plates 10, 14
Achatina 97
Acokanthera 258
acutovulvata, Chapinia 67, 68, 160
Adansonia 94, 98, 113, 134, 247
aequabilis, Aceros undulatus **233**
affinis, Penelopides, see Penelopides
Africa, ecology 5, 14, 40, 42, 71–3, 76, 78–80, 90, 100, 260; *see also* geography, Afrotropical
africanus, Bucerophagus 66, 68
Afrotropical Region, *see* geography, Afrotropical
Afzelia 246
Agama 119
Agamidae 103
age, *see also* species accounts
 juvenile differences 23, 34–5, 62, 70
 longevity 16, 60
 at maturity 47–8, 57–8, 60
Aglaia 41, 45, 103, 167, 174, 181, 217, 236
agriculture, shifting *or* swidden 77
Ailanthes 159
air sac 56
Alangium 108
albescens, Buceronirmus 67, 68
albipes, Buceronirmus 66, 68
albirostris, Anthracoceros, see Anthracoceros
albirostris, Anthraceros albirostris (Asian Pied Hornbill) 11, 20, **165,** Plate 7
Albizzia 134, 139
albocristatus, Tockus, see Tockus
albocristatus, Tockus albocristatus Western Guinea Long-tailed Hornbill) 11, 20, **149,** Plates 6, 13
alboterminatus, Tockus, see Tockus
albotibialis, Ceratogymna cylindricus (White-thighed Hornbill) 12, 21, **249–50,** Plates 12, 13
Alcedinidae, *see* Coraciiformes
allofeeding 33
Allophylus 113
allopreening 33
allospecies 91
Alseodaphne 181–2
Alstonia 217
Altingia 103, 169, 183, 237
Amblycera, *see* Menoponidae
Amomum 236
Amoora 103, 167, 181, 229
Amorphophallus 174
Anamirta 229
anatomy 5–8, 89, 100
 brain size 16
 neck 26, 62, 89–90, 100
Andaman Islands 73, 78, 228
Angola 95, 112, 115, 118, 123, 125, 128, 133, 138, 146, 148, 150, 244–5, 250, 253, 261
Anomma nigricans 149
Anorrhinus 36, 46, 65, 68–70, 79, 84, **100,** 197, 209, Plates 3, 14
 austeni (Austen's Brown Hornbill) 6, 11, 20, 37, 38, 47, 49, 64, 66, **101–4,** 104–5, 169, 237, Plates 2, 3, 14

Anorrhinus cont.
  galeritus (Bushy-crested Hornbill) 6, 11, 15, 17, 20, 38, 41, 44, 47, 49, 64, 66, 69, 73, **106–9**, 173–4, 198, 212, Plates 3, 14
  tickelli (Tickell's Brown Hornbill) 6, 11, 15, 17, 20, 38, 47, 49, 64, 66, 101, **104–5**, Plates 3, 14
Anseriformes 13
Anthracoceros 22, 36, 38, 46, 65, 68–70, 79, 153, **159–60**, Plates 7, 14
  albirostris (Oriental Pied Hornbill) 4, 6, 11, 20, 27, 37, 38, 47, 59, 64, 67, 79, 103, 160–2, **164–70**, 171, 173, 195, Plates 7, 14
  coronatus (Indian Pied Hornbill) 6, 11, 15, 17, 20, 37, 38, 47, 64, 67, 69, 79, 102, 155, 160, **161–4**, 181, Plates 7, 14
  malayanus (Malay Black Hornbill) 6, 11, 15, 17, 20, 38, 39, 45, 47, 49, 64, 67, 79, 106–7, 160, 167, 170, **172–5**, Plates 7, 14
  marchei (Palawan Hornbill) 6, 11, 15, 17, 20, 38, 47, 64, 67, 79, 160, **170–2**, Plates 7, 14
  montani (Sulu Hornbill) 6, 11, 15, 17, 20, 38, 47, 64, 67, 79, 160, **175–6**, Plates 7, 14
Antiaris 254
Aphanamixis 103, 167, 181, 236
Apodiformes 13
Aptenodytes fosteri 13
Aquila gurneyi 14
  rapax 97, 136, 139
  verreauxii 97
  wahlbergi 127, 136, 139
Ardina 164
area cladogram *see* geography, biogeography
Arenga 227
Argusianus argus 194
artificial nest, *see* nest box
Artocarpus 103, 168, 217, 236
Ascaradia galli 182
aspect ratio 14–16
atrata, Ceratogymna, see Ceratogymna
austeni, Anorrhinus, see Anorrhinus
Australasia, *see* geography
australis, Tockus alboterminatus **111**
aviculture, *see* breeding, captive

Baikaea 115
Bali 73, 235
Bangladesh 157, 166, 180
Barbet, Red-and-Yellow 126

Baryrhynchodes 64, 67, 69, 241–2, **248**, Plate 12
basilanica, Penelopides affinis (Basilan Tarictic Hornbill) 12, 21, **205–6**, Plates 9, 14
Bassia 229
Bateleur 136, 139
bathing 26–8, 97
Bauhinia 158
bee-eater, *see* Coraciiformes
  Carmine 97
behaviour, *see also* communication; species accounts
  development of 56
  female emergence 47–8
  maintenance 25–8
  nesting 7, 46–60; *see also* nest
  non-breeding 24–35
  play 30, 57, 97
  preening 26
  pre-layingl 33–4, 52, 58
  roosting 31–2
  sealing 7, 8–9, 34, 42, 45, 50–1, 62; *see also* nest preparation
  stretching 26
Beilschmiedia 236, 244, 251, 254, 262
Benin 118, 125, 133, 146, 148, 150, 244, 250, 253, 261, 264
Bequaetiodendron 254
Berenicornis 64, 65, 67, 69–71, **209**, Plate 10
Bersame 254
Bhutan 166, 213
bicornis, Buceros, see Buceros
bill design 37
  length xiv, 17–18, 20–1; *see also* species accounts
  volume 17–18
biogeography, *see* geography
biology, non-breeding 24–35
birostris, Ocyceros, see Ocyceros
Bismarck Archipelago 225
blakei, Chapinia 67, 68
Blattidae 103, 181
Blighia 254, 258
body mass, *see* mass
body proportions, *see* proportions
body shape, *see* shape
body size, *see* size
Bombax 159
boonsongi, Chapinia 67, 68
Borassus 94
Borneo 3, 40, 43, 45, 72, 100; *see also* Malaysia; Indonesia
borneoensis, Buceros rhinoceros (Bornean Great Rhinocers Hornbill) 11, 21, **185–6**, Plate 8
Boscia 134–5, 138
Bosqueia 254
Botswana 95, 115, 125, 133, 138
Brachylaima fuscata 172
Brachypteracidae, *see* Coraciiformes

Brachystegia 72, 98, 123
bradfieldi, Tockus, see Tockus
brain size, *see* anatomy
breeding xv; *see also* species accounts
  biology 7, 46–60; *see also* behaviour
  captive 78–80, 82–5, 99
  co-operative 23, 48, 49, 63, 70, 84
  timing of 48–9, 58
brelihi, Buceroemersonia 66, 68, 127
brevirostris, Aceros cassidix **215**
brevis, Ceratogymna, see Ceratogymna
Bridelia 254, 258
Britain, *see* England
Brunei 106, 167, 173, 186, 194, 210, 219, 234
Bubo africanus 139
  bubo 13
  lacteus 139
Bucconidae, *see* Galbuliformes
Bucerocolpocephalum 66–8, 70, 100
Bucerocophorus 66–9, 110, 241, 248
Buceroemersonia 66–8, 109, 127, 152
Buceronirmus 66–8, 110, 241, 248, 260
Bucerophagus 66–8, 90
Buceros 8, 19, 22, 28, 34, 36, 43, 46, 68–70, 78, 79, 83–4, 90, 160, **177–8**, 195, 237, Plates 8, 14
  bicornis (Great Pied Hornbill) 6, 8, 11, 15, 17, 20, 25, 27, 37, 38, 47, 57, 64, 67, 70, 77, 79–80, 162, 178, **178–84**, 185–6, 194, 237–8, Plates 2, 8, 14
  hydrocorax (Great Philippine Hornbill) 4, 6, 12, 15, 17, 21, 38, 47, 49, 64, 67, 69, 79, 177–8, **189–92**, 202, 206, 222–3, Plates 8, 14
  rhinoceros (Great Rhinoceros Hornbill) 6, 8, 11–12, 15, 17, 21, 27, 37, 38, 47, 49, 64, 67, 69, 73, 77, 79, 178–80, **184–9**, 194–5, Plates 8, 14
  vigil (Great Helmeted Hornbill) 4, 6, 12, 13, 14, 15, 17, 21, 22, 28, 38, 47, 64, 67, 69, 79, 170, 177–8, 186, 189, **192–6**, Plates 1, 2, 8, 14
Bucerotidae 9, 69–70, 89, **100**
Bucerotiformes 4, 5, 9, 10, 13, 25, 40, 48, 63–5, **89**
Bucerotimorphae 89
bucerotis, Chapinia 67, 68
bucerus 8–9
Buchaniana 164
bucinator, Ceratogymna, see Ceratogymna
Bucorvellus 66–8, 90

Bucorvidae   7, 63, 69–70, 89, **90**, 91
Bucorvus   5, 19, 22, 24, 25, 28, 30, 32, 34, 36, 39, 41, 43, 46, 50, 52, 62, 63, 64, 67, 68–9, 71–2, 79, 89–90, **91**, 130, Plates 3, 13
   abyssinicus (Northern Ground Hornbill)   4, 6, 11, 15, 17, 20, 38, 47, 64, 67, 69–70, 90, **91–4**, Plates 3, 13
   leadbeateri (Southern Ground Hornbill)   6, 11, 13, 15, 17, 20, 25, 29, 33, 37, 38, 47, 49, 52, 55, 60, 62, 64, 67, 69–70, 79–80, 90, 92–3, **94–9**, Plates 1, 2, 3, 13
Buprestidae   113, 168
Burchelia   113
Burkina Faso   93, 125, 133
Burseraceae   41, 71, 220
Burundi   95, 112, 125, 253
Bustard, Great   13
Button-quail, Spotted   13
Bycanistes   64, 65, 67, 69, 241, **242**, 248, Plate 12

cafer, Bucorvus   94
Calamus   244, 262, 265
California   94, 126, 182, 204, 227
calls, *see* communication, vocal
Calodendron   113, 247
Calophyllum   182
Calotes   163–4
Camaroptera brachyura   247
Cambodia   166, 180, 214
Cameroon   93, 118, 125, 146, 148, 150, 244, 253, 261, 264
Campephilus imperialis   13
camurus, Tockus, see Tockus
Canaga   217
Canarium   103, 108, 167, 181, 217, 236, 254, 262
Canthium   113, 258
Capparis   229
Caprimulgus   181
Carallia   155
Carica   254
carotid artery   90, 100
carrion   93, 97
Caryota   164
casque form   5, 19–22, 62, 71, 89; *see also* species accounts
   function   19–22
   volume   18–22
Cassia   244, 251
cassidix, Aceros, see Aceros
Cassine   113
cassini, Tockus albocristatus (Lower Guinea Long-tailed Hornbill)   11, 20, **150**, Plates 6, 13
Castanola   262
Celtis   254

Central African Republic   93, 118, 125, 133, 146, 148, 150, 244, 253, 261
Centropus sinensis   13
cephalotes, Bucerornirrus   67, 68
Cerambycidae   113, 168, 174, 181
Ceratogymna   5, 19, 23, 31, 34, 36, 43, 46, 52, 65, 68–72, 78–9, 84, 117, **241**, 243, 260, Plates 12, 13
   atrata (Black-casqued Wattled Hornbill)   7, 12, 13, 16, 18, 21, 39, 47, 49, 64, 67, 79, 241, **260–3**, 264–5, Plates 12, 13
   brevis (Silvery-cheeked Hornbill)   7, 9, 12, 16, 18, 21, 37, 39, 45, 47, 64, 67, 79, 241, 245, 247, 253, **256–60**, Plates 12, 13
   bucinator (Trumpeter Hornbill)   7, 12, 16, 18, 21, 39, 45, 47, 49, 62, 63, 64, 67, 69–70, 79, 241, 243, **245–8**, 257–8, Plates 12, 13
   cylindricus (Brown-cheeked Hornbill)   7, 12, 16, 18, 21, 39, 47, 64, 67, 73, 79, 241, 243, **249–51**, 252, 261, Plates 12, 13
   elata (Yellow-casqued Wattled Hornbill)   4, 7, 12, 13, 16, 18, 21, 39, 47, 49, 64, 67, 73, 79, 241, 261, **263–5**, Plates 12, 13
   fistulator (Piping Hornbill)   7, 12, 16, 18, 21, 39, 47, 64, 67, 241, **242–4**, 245, 249, 251–3, Plates 12, 13
   subcylindricus (Grey-cheeked Hornbill)   7, 9, 16, 18, 21, 37, 39, 42, 47, 64, 67, 69–70, 79, 241, 243, 249, **252–6**, Plates 12, 13
   subgenus   64, 67, 241, 248, **260**, Plate 12
Cetoniidae   168, 181
Chad   93, 118, 125, 133, 244, 253
Chapinia   66–9, 110, 127, 152, 160, 241, 248, 260
chick, *see* nestling
China   102, 166, 180, 214
Chiroptera   103
Chisocheton   103, 167, 181, 227, 236
chromosome, *see* karyology
Chrysocolaptes lucidus   156
Chrysophyllum   247, 258
Cicadidae   168, 181
Ciconiiformes   13
Cinnamomum   103, 167, 181, 183, 236
Cissus   143, 244
cladistics, *see* systematics
clarkei, Buceroemersonia   66–7, 68
clayae, Chapinia   67, 68

climate, *see* geography
Clorophora   254
clutch size   47–8, 58; *see also* species accounts
Coelocaryon   118, 244, 254, 262
Cola   254
Coliiformes   13
Colius castanotus   13
Colophospermum   115, 126, 134, 139
Columbidae   168, 181, 236
Columbiformes   13
comatus, Aceros, see Aceros
Combretum   98, 126, 133–4, 139
Commiphora   113, 126, 134–5, 141–3
communication   28–31, 62; *see also* species accounts
   visual   28–31, 49–50
   vocal   xiv, 28–31, 61
Communication Library   xiv
Condor, California   13
Congo   118, 146, 148, 150, 244, 250, 261
Connaraceae   108
Connarus   41, 103, 108, 167, 181, 236
Conopharyngia   258
conservation   8–9, 73–4, 75–85; *see also* species accounts
convexus, Anthracoceros albirostris (Sunda Pied Hornbill)   11, 20, 164, **166–8**, 170, 195, Plate 7
Copsychus saularis   108
copulation   33, 34, 51
Coraciidae, *see* Coraciiformes
Coraciiformes   7, 13, 55, 56, 63, 65, 70, 89
Cordia   254
coronatus, Anthracoceros, see Anthracoceros
corrugatus, Aceros, see Aceros
Corvus corax   13
Coryaethaeola cristata   13
Corypha   167
Cossidae   103, 168
Côte d'Ivoire   93, 118, 133, 146, 148, 150, 244, 250, 253, 261, 264
courtship feeding *see* food
Croton   116, 244
Crow-pheasant, Common   13
Cryptocarya   217
cuckoo-roller, *see* Coraciiformes
Cuculiformes   13
Cullenia   182
Cyclophoridae   103
Cygnus cygnus   13
cylindricus, Ceratogymna, see Ceratogymna
cylindricus, Ceratogymna cylindricus (Upper Guinea Brown-cheeked Hornbill)   12, 21, **249**, Plates 12, 13

Dacryodes 108, 118, 151, 244, 251, 261–2
Dahlbergia 159
Dahomey Gap 72
damarensis, Tockus erythrorhynchus (Damaraland Red-billed Hornbill) 11, 20, **131, 133–4**, Plates 5, 13
dampieri, Aceros plicatus **225**
deckeni, Tockus, see Tockus
defecation, *see* nest sanitation
degens, Tockus erythrorhynchus **132**
Dehasia 108
deignani, Bucerocolpocephalum 67–8, 69
Delonix 134
deminutus, Anthracoceros malayanus **172**
demography 8, 58–60, 75–6
Denmark 263
density 46–8, 49–50
d'Entrecasteaux Archipelago 225
descriptions, of hornbills, *see* plumage
Dichopsis 258
Dicrostachys 143
Dicrurus parodiscus 181
Dictyoptera 135
diet, *see* food
digestion, *see* food, processing
dimorphism, *see also* species accounts
  coloration 20–1
  sexual 19–21, 62
Diospyros 98, 126, 139, 217, 246, 254–5
diploid number 70; *see also* karyology
Dipterocarpus 103, 169, 183, 236
display, *see* communication
distribution, *see* maps
Djibouti 120
DNA 63, 65, 69–71
docophorus, Bucorvellus 66, 68
Dodo 13
dorsalis, Tockus nasutus **124**
Dovyalis 247
Dracaena 244, 251, 254
Draco 107
Dracontomelum 217
drinking 25, 39–40; *see also* species accounts
Drongo, Greater Racquet-tailed 181
Drypetes 247
duboisi, Ceratogymna fistulator (Dubois's Piping Hornbill) 12, 21, **242–4**, Plates 12, 13
dust bathing, *see* bathing
Dynastidae 174, 236
Dysoxylum 203, 214–15, 235–6

Eagle, Ayres' 127
  Black 97

Crowned 14, 119, 256, 260, 263
  Gurney's 14
  Martial 94, 136, 139
  Monkey-eating 14
  Papuan Harpy 14
  Tawny 97, 136, 139
  Wahlberg's 127, 136, 139
White-bellied Sea 229
egg, form 52, 89
  hatching 55, 58; *see also* nestling survival
  size xv, 53; *see also* species accounts
  volume 47–8
egg-laying xv, 52; *see also* species accounts
Ehretia 134
Ekbergia 246, 258
Elaegnus 103, 167, 181, 236
Elaeis 254, 262
Elaeocarpus 103, 181, 236
elata, Ceratogymna, see Ceratogymna
Elateridae 113, 168, 181
elegans, Tockus leucomelas **137**
Elliot, Daniel Giraud 9
El Niño–Southern Ocean Oscillation 77
emersoni, Bucerocolpocephalum 66, 68
Encephalartos 113
England 126, 134, 143, 182, 255
Englerina 258
Entada 108
epirhinus, Tockus nasutus (South African Grey Hornbill) 11, 20, **124–6**, Plate 4
Erythrina 113, 121, 247
erythrorhynchus, Tockus, see Tockus
erythrorhynchus, Tockus erythrorhynchus (North African Red-billed Hornbill) 11, 20, **131**, Plates 5, 13
Ethiopia 72, 93, 111–12, 120, 125, 133, 141, 143, 258
Eudynamys scolopacea 229
Eugenia 103, 108, 169, 181–3, 236–7, 254, 258
Euphorbia 113, 120–1
Eusideroxylon 236
everetti, Aceros, see Aceros
Evodia 108
evolution, *see* systematics
exarhatus, Penelopides, see Penelopides
exarhatus, Penelopides exarhatus (North Sulawesi Tarictic Hornbill) 12, 21, **198**, Plate 9
exploitation, *see* conservation; species accounts
eyelashes 62, 89–90

faecal 'shadow' 55
faeces, *see* nest, sanitation

Fagaceae 108
Fagaropsis 254
Falco biarmicus 127, 139
Falcon, Lanner 127, 139
fasciati, Chapinia 66, 68
fasciatus, Tockus, see Tockus
fasciatus, Tockus fasciatus (Lower Guinea Pied Hornbill) 11, 20, **116**, Plates 4, 13
feather-lice, *see* parasites
feeding, *see* food
female, emergence, *see* nest, emergence from
moult, *see* plumage
Ficus 39, 41–5, 77, 98, 103, 108, 113, 118, 159, 162, 164, 167, 174, 180–1, 187, 217, 229, 236, 244, 246, 251, 253–5, 258, 262
field characters xiv; *see also* species accounts
fig, *see Ficus*
fistulator, Ceratogymna, see Ceratogymna
fistulator, Ceratogymna fistulator (Upper Guinea Piping Hornbill) 12, 21, **242**, Plates 12, 13
Fitch tree analysis 68
FitzPatrick Bird Communication Library xiv
'flagship' 81
flavirostris, Tockus, see Tockus
flooding 59
foliage bathing *see* bathing
food xiv; *see also* species accounts
  animal 41–2
  associations 40–1
  availability 36
  captives 82–3
  collection 24–5, 37–9, 42–4
  courtship and 49, 50
  ecology 36
  feeding rates 57
  fruit 41–2
  location 24–5, 42–4
  nesting and, *see* nest, provisioning
  processing 44–5
  selection 36, 40–2
footspan 17–18
forcipatus, Bucerocophagus 67, 68
forest types xv, 72–3; *see also* geography
forskalii, Tockus nasutus **124**
fossil 63, 65, 90
francolin 71

Gabon 118, 146, 148, 150, 244, 250, 261
Galbulidae, *see* Galbuliformes
Galbuliformes 13, 63
galeritus, Anorrhinus, see Anorrhinus
Galliformes 13

Index

Gambia, The 93, 118, 125, 133, 264
Garuga 217, 229
geloensis, Tockus alboterminatus **111**
general habits xv
geography
  biogeography 71–3
    Afrotropical 71, 89, 100
    Australasian 73, 89, 100
    Indomalayan 71, 89, 100
    Wallacean 197
  climate 61, 71
  geology 61
  vegetation 71, 76–8
    forest types xv, 72–3
geology, *see* geography
Germany 134, 182
Ghana 93, 118, 125, 133, 146, 148, 150, 250, 253, 261, 264
gingalensis, Tockus, see Tockus
Glaucidium radiatum 181
Goshawk, chanting 41
  Dark Chanting 97, 139
  Eastern Chanting 144
  Gabar 136
  Red-chested 152
Goura cristata 13
grandiceps, Buceronirmus 67, 68
granti, Tockus hartlaubi (Lower Guinea Dwarf Black Hornbill) 11, 20, **145–6**, Plate 6
Grevillea 113
Grewia 164, 262
Griffonia 244
griseus, Ocyceros, see Ocyceros
ground-rollers, *see* Coraciiformes
Gruiformes 13
Gryllidae 103, 181
Guibourtia 115–16, 118, 251, 262
Guinea 93, 118, 125, 133, 146, 148, 150, 244, 250, 261, 264
  –Bissau 118, 125, 133, 244, 264
  forests 72–3, 250, 261
gular fluttering 27, 28
Gymnocranthera 217
Gymnogene, African 33, 127, 139, 256
Gymnogyps californianus 13

habitat xiv–xv; *see also* species accounts
  conservation 75–8
habits, general xv; *see also* species accounts
haematozoa, *see* parasites
Haliaeetus leucogaster 229
handrearing, *see* breeding, captive
Harongana 258
Harpephyllum 113, 247
Harpyopsis novaeguineae 14
Harrisonia 254

hartlaubi, Tockus, see Tockus
hartlaubi, Tockus hartlaubi (Upper Guinea Dwarf Black Hornbill) 11, 20, **145**, Plate 6
hatching, *see* egg
Hawk-eagle 41
  African 136, 139
  Changeable 14, 220
  Sulawesi 217
Heisteria 118, 244, 251, 262
Helmholtz resonance 22
Helogale undulata 141, 143
Hemiprocne longipennis 187
hemprichii, Tockus, see Tockus
HENNIG86 64
Heterophylla 258
Hieraaetus ayresii 127
  spilogaster 136, 139
hirta, Chapinia 67, 68
Hirundapus celebensis 13
history, evolutionary, *see* systematics
Hodotermes 116
Holigarna 162
Holoptelea 237
Homalium 182, 232
homrai, Buceros bicornis **178**
hoopoe, *see* Upupiformes
hop, *see* progression
Hopea 105, 182
hoplai, Chapinia 67, 68
Hornbill
  Abyssinian Ground, *see* Hornbill, Northern Ground
  African Crowned (*Tockus alboterminatus*) 6, 9, 11, 15, 17, 20, 30, 32, 37, 38, 40, 47, 59, 64, 66, **110–14**, 115, 118–20, 125, 247, Plates 4, 13
  African Grey (*Tockus nasutus*) 6, 11, 15, 17, 20, 25, 30, 37, 38, 40, 48, 53, 58, 64, 66, 69, 72, 79, 112, 128, 115, 119, 122–3, **123–7**, 138–9, Plates 4, 13
  African Ground, *see* Hornbill, Southern Ground
  African Pied (*Tockus fasciatus*) 6, 11, 15, 17, 20, 30, 38, 47, 64, 66, 110, 112, **116–19**, 146, 243, Plates 4, 13
  African Red-billed (*Tockus erythrorhynchus*) 6, 8, 11, 15, 17, 20, 27, 31, 37, 38, 39, 40, 48, 54, 64, 66, 69, 72–3, 79, 113, 127–8, **131–6**, 137–8, 140–1, Plates 5, 13
  African White-crested, *see* Hornbill, Long-tailed
  Allied, *see* Hornbill, African Pied
  Asian Black, *see* Hornbill, Malay Black
  Asian Pied (*Anthracoceros albirostris albirostris*) 11, 20,

165, Plate 7; *see also* Hornbill, Oriental Pied
  Asian White-crested, *see* Hornbill, White-crowned
  Assam Brown-backed, *see* Hornbill, Tickell's Brown
  Austen's Brown (*Anorrhinus austeni*) 6, 11, 20, 37, 38, 47, 49, 64, 66, **101–4**, 104–5, 169, 237, Plates 2, 3, 14
  Bar-pouched Wreathed (*Aceros undulatus*) 4, 7, 12, 15, 18, 21, 32, 37, 39, 44, 47, 67, 70, 77, 79, 103, 168, 181, 183, 208, 213, 219, 230–2, **232–8**, Plates 11, 14
  Bar-throated Wreathed, *see* Hornbill, Bar-pouched Wreathed
  Basilan Tarictic (*Penelopides affinis basilanica*) 12, 21, **205–6**, Plates 9, 14; *see also* Hornbill, Mindanao Tarictic
  Black, *see* Hornbill, Malay Black
  Black-and-white Casqued, *see* Hornbill, Grey-cheeked
  Black-casqued, *see* Hornbill, Black-casqued Wattled
  Black-casqued Wattled (*Ceratogymna atrata*) 7, 12, 13, 16, 18, 21, 39, 47, 49, 64, 67, 79, 241, **260–3**, 264–5, Plates 12, 13
  Black-wattled, *see* Hornbill, Black-casqued Wattled
  Blyth's (Wreathed), *see* Hornbill, Papuan Wreathed; Hornbill, Plain-pouched Wreathed
  Bornean Great Rhinoceros (*Buceros rhinoceros borneoensis*) 11, 21, **185–6**, Plate 8; *see also* Hornbill, Great Rhinoceros
  Bradfield's (*Tockus bradfieldi*) 6, 11, 15, 17, 20, 30, 38, 48, 64, 66, 72, 110, 112, **114–16**, 125, Plates 4, 13
  Brown, *see* Hornbill, Austen's Brown; Hornbill, Tickell's Brown
  Brown-backed, *see* Hornbill, Austen's Brown
  Brown-cheeked (*Ceratogymna cylindricus*) 7, 12, 16, 18, 21, 39, 47, 64, 67, 73, 79, 241, 243, **249–51**, 252, 261, Plates 12, 13
  Burmese, *see* Hornbill, Plain-pouched Wreathed
  Bushy-crested (*Anorrhinus galeritus*) 6, 11, 15, 17, 20, 38, 41, 44, 47, 49, 64, 66, 69, 73, **106–9**, 173–4, 198, 212, Plates 3, 14

Hornbill *cont.*
  Buton, *see* Hornbill, Sulawesi Wrinkled
  Celebes, *see* Hornbill, Sulawesi Wrinkled
  Ceylon Grey, *see* Hornbill, Sri Lankan Grey
  Common Grey, *see* Hornbill, Indian Grey
  Concave-casqued, *see* Hornbill, Great Pied
  Crowned, *see* Hornbill, African Crowned
  Damaraland Red-billed (*Tockus erythrorhynchus damarensis*) 11, 20, **131**, **133–4**, Plates 5, 13; *see also* Hornbill, African Red-billed
  Dubois's Piping (*Ceratogymna fistulator duboisi*) 12, 21, **242–4**, Plates 12, 13; *see also* Hornbill, Piping
  Dwarf, *see* Hornbill, Dwarf Red-billed
  Dwarf Black (*Tockus hartlaubi*) 6, 11, 13, 15, 17, 20, 38, 48, 64, **145–7**, 148, Plates 6, 13
  Dwarf Red-billed (*Tockus camurus*) 4, 6, 11, 15, 17, 20, 38, 48, 49, 64, 66, 146, **147–9**, Plates 6, 13
  Eastern Guinea Long-tailed (*Tockus albocristatus macrourus*) 11, 20, **150**, Plates 6, 13; *see also* Hornbill, Long-tailed
  Eastern Pale-billed (*Tockus pallidirostris neumanni*) 11, 20, **121–3**, Plate 4; *see also* Hornbill, Pale-billed
  Eastern Yellow-billed (*Tockus flavirostris*) 6, 11, 20, 31, 38, 39, 40, 48, 64, 66, 72, 79, 127, 133, 136, **140–2**, 143, Plates 5, 13
  Great, *see* Hornbill, Great Pied
  Great Helmeted (*Buceros vigil*) 4, 6, 12, 13, 14, 15, 17, 21, 22, 28, 38, 47, 64, 67, 69, 79, 170, 177–8, 186, 189, **192–6**, Plates 1, 2, 8, 14
  Great Indian, *see* Hornbill, Great Pied
  Great Luzon (*Buceros hydrocorax hydrocorax*) 12, 21, **189**, Plate 8; *see also* Hornbill, Great Philippine
  Great Mindanao (*Buceros hydrocorax mindanensis*) 12, 21, 189, **190–1**, 192, Plate 8; *see also* Hornbill, Great Philippine
  Great Philippine (*Buceros hydrocorax*) 4, 6, 12, 15, 17, 21, 38, 47, 49, 64, 67, 69, 79, 177–8, **189–92**, 202, 206, 222–3, Plates 8, 14
  Great Pied (*Buceros bicornis*) 6, 8, 11, 15, 17, 20, 25, 27, 37, 38, 47, 57, 64, 67, 70, 77, 79–80, 162, 178, **178–84**, 185–6, 194, 237–8, Plates 2, 8, 14
  Great Rhinoceros (*Buceros rhinoceros*) 6, 8, 11–12, 15, 17, 21, 27, 37, 38, 47, 49, 64, 67, 69, 73, 77, 79, 178–80, **184–9**, 194–5, Plates 8, 14
  Great Samar (*Buceros hydrocorax semigaleatus*) 12, 21, 189, **190–1**, Plate 8; *see also* Hornbill, Great Philippine
  Great Sulawesi, *see* Hornbill, Sulawesi Wrinkled
  Grey, *see* Hornbill, African Grey; Hornbill, Indian Grey
  Grey-cheeked (*Ceratogymna subcylindricus*) 7, 9, 12, 16, 18, 21, 37, 39, 42, 47, 64, 67, 69–70, 79, 241, 243, 249, **252–6**, Plates 12, 13
  Ground, *see* Hornbill, Southern Ground
  Helmeted, *see* Hornbill, Great Helmeted
  Hemprich's (*Tockus hemprichii*) 6, 11, 15, 17, 20, 30, 38, 47, 64, 66, 72, 110, **119–21**, Plates 4, 13
  Homrai, *see* Hornbill, Great Pied
  Indian, *see* Hornbill, Great Pied
  Indian Concave-casqued, *see* Hornbill, Great Pied
  Indian Grey (*Ocyceros birostris*) 6, 11, 15, 17, 19, 20, 38, 47, 64, 67, 152–4, **157–9**, 161, Plates 6, 14
  Indian Pied (*Anthracoceros coronatus*) 6, 11, 15, 17, 20, 37, 38, 47, 64, 67, 69, 79, 102, 160, **161–4**, 181, Plates 7, 14; *see also* Hornbill, Oriental Pied
  Jackson's (*Tockus deckeni jacksoni*) 11, 20, 79, **142–4**, Plates 5, 13; *see also* Hornbill, Von der Decken's
  Javan Great Rhinoceros (*Buceros rhinoceros sylvestris*) 12, 21, **185–6**, Plate 8; *see also* Hornbill, Great Rhinoceros
  Knobbed, *see* Hornbill, Sulawesi Wrinkled
  Laughing, *see* Hornbill, Piping
  Long-crested, *see* Hornbill, White-crowned
  Long-tailed (*Tockus albocristatus*) 4, 6, 11, 15, 17, 20, 38, 47, 64, 66, 70, **149–52**, Plates 6, 13
  Lower Guinea Brown-cheeked, *see* Hornbill, White-thighed
  Lower Guinea Dwarf Black (*Tockus hartlaubi granti*) 11, 20, **145–6**, Plate 6; *see also* Hornbill, Dwarf Black
  Lower Guinea Grey-cheeked (*Ceratogymna subcylindricus subquadratus*) 12, 21, **252–5**, Plate 12; *see also* Hornbill, Grey-cheeked
  Lower Guinea Long-tailed (*Tockus albocristatus cassini*) 11, 20, **150**, Plates 6, 13; *see also* Hornbill, Long-tailed
  Lower Guinea Pied (*Tockus fasciatus fasciatus*) 11, 20, **116**, Plate 4; *see also* Hornbill, African Pied
  Luzon Tarictic (*Penelopides manillae*) 6, 12, 21, 38, 47, 49, 64, 67, 79, 197, **202–4**, Plates 9, 14
  Malabar Grey (*Ocyceros griseus*) 6, 11, 15, 17, 20, 38, 48, 64, 67, **153–4**, 155, 157, 159, 161, Plates 6, 14
  Malayan Great Rhinoceros (*Buceros rhinoceros rhinoceros*) 11, 21, **184**, Plate 8; *see also* Hornbill, Great Rhinoceros
  Malay Black (*Anthracoceros malayanus*) 6, 11, 15, 17, 20, 38, 39, 45, 47, 49, 64, 67, 79, 106–7, 160, 167, 170, **172–5**, Plates 7, 14
  Malaysian Black, *see* Hornbill, Malay Black
  Malaysian Pied, *see* Hornbill, Oriental Pied
  Mindanao Tarictic (*Penelopides affinis*) 6, 12, 21, 38, 47, 49, 64, 67, **204–6**, 222, Plates 9, 14
  Mindanao Wrinkled (*Aceros leucocephalus*) 7, 12, 15, 18, 21, 39, 47, 64, 67, 79, 191–2, 206, **221–3**, Plates 10, 14
  Mindoro Tarictic (*Penelopides mindorensis*) 6, 12, 21, 38, 47, 49, 64, 67, 197, **207–8**, Plates 9, 14
  Montano's, *see* Hornbill, Sulu
  Monteiro's (*Tockus monteiri*) 6, 11, 15, 17, 20, 31, 37, 38, 47, 54, 64, 66, 69, 72, 127, **128–31**, 133, 137, 138, Plates 1, 5, 13
  Narcondam Wreathed (*Aceros narcondami*) 7, 12, 15, 18, 21, 37, 39, 47, 64, 67, 78, **227–9**, Plates 11, 14
  New Guinea (Wreathed), *see* Hornbill, Papuan Wreathed

Index

Hornbill *cont.*
North African Grey (*Tockus nasutus nasutus*) 11, 20, **123**, Plate 4; *see also* Hornbill, African Grey
North African Red-billed (*Tockus erythrorhynchus erythrorhynchus*) 11, 20, **131**, Plates 5, 13; *see also* Hornbill, African Red-billed
Northern Ground (*Bucorvus abyssinicus*) 4, 6, 11, 15, 17, 20, 38, 47, 64, 67, 69–70, 79–80, 90, **91–4**, 95, Plates 3, 13
Northern Pied, *see* Hornbill, Oriental Pied
Northern Waved, *see* Hornbill, Bar-pouched Wreathed
North Sulawesi Tarictic (*Penelopides exarhatus exarhatus*) 12, 21, **198**, Plate 9; *see also* Hornbill, Sulawesi Tarictic
Oriental Pied (*Anthracoceros albirostris*) 4, 6, 11, 20, 27, 37, 38, 47, 59, 64, 67, 79, 103, 160–2, **164–70**, 171, 173, 195, Plates 7, 14
Palawan (*Anthracoceros marchei*) 6, 11, 15, 17, 20, 38, 47, 64, 67, 79, 160, **170–2**, Plates 7, 14
Pale-billed (*Tockus pallidirostris*) 6, 11, 15, 17, 20, 30, 38, 47, 64, 66, 69, 72, 112, **121–3**, 125, Plates 4, 13
Panay Tarictic (*Penelopides panini panini*) 12, 21, **200**; *see also* Hornbill, Visayan Tarictic
Panay Wrinkled, *see* Hornbill, Visayan Wrinkled
Papuan Wreathed (*Aceros plicatus*) 7, 12, 15, 18, 21, 39, 47, 64, 67, **224–7**, 241, 245, 249, 251–3, Plates 12, 13
Philippine Brown, *see* Hornbill, Great Philippine
Philippine Rufous, *see* Hornbill, Great Philippine
Piping (*Ceratogymna fistulator*) 7, 12, 16, 18, 21, 39, 47, 64, 67, 241, **242–4**, 245, 249, 251–3, Plates 12, 13
Plain-pouched, *see* Hornbill, Plain-pouched Wreathed
Plain-pouched Wreathed (*Aceros subruficollis*) 7, 12, 21, 39, 47, 64, 67, 79, 208, **230–2**, 233–4, 237, Plates 11, 14
Plicated, *see* Hornbill, Papuan Wreathed
Polillo Tarictic (*Penelopides manillae subnigra*) 12, 21, **202**, Plate 9; *see also* Hornbill, Luzon Tarictic
Red-billed, *see* Hornbill, African Red-billed
Red-knobbed, *see* Hornbill, Sulawesi Wrinkled
Rhinoceros, *see* Hornbill, Great Rhinoceros
Rufous, *see* Hornbill, Great Philippine
Rufous-necked (*Aceros nipalensis*) 7, 12, 15, 18, 21, 39, 47, 64, 67, 79, 209, 212, **213–15**, Plates 10, 14
Samar Tarictic (*Penelopides affinis samarensis*) 12, 21, **205–6**, Plates 9, 14; *see also* Hornbill, Mindanao Tarictic
Sharpe's Piping (*Ceratogymna fistulator sharpii*) 12, 21, **242–4**, Plates 12, 13; *see also* Hornbill, Piping
Silvery-cheeked (*Ceratogymna brevis*) 7, 9, 12, 16, 18, 21, 37, 39, 45, 47, 64, 68, 79, 241, 245, 247, 253, **256–60**, Plates 12, 13
Solid-billed, *see* Hornbill, Great Helmeted
South African Grey (*Tockus nasutus epirhinus*) 11, 20, **124–6**, Plate 4; *see also* Hornbill, African Grey
South African Red-billed (*Tockus erythrorhynchus rufirostris*) 11, 20, 79, **131–4**, 136, Plates 5, 13; *see also* Hornbill, African Red-billed
Southern Ground (*Bucorvus leadbeateri*) 6, 11, 13, 15, 17, 20, 25, 29, 33, 37, 38, 47, 49, 52, 55, 60, 62, 64, 67, 69–70, 79–80, 90, 92–3, **94–9**, Plates 1, 2, 3, 13
Southern Yellow-billed (*Tockus leucomelas*) 6, 11, 15, 17, 20, 31, 37, 38, 40, 42, 48, 53, 64, 66, 69, 128, 133, **136–40**, 140–2, Plates 5, 13
South Sulawesi Tarictic (*Penelopides exarhatus sanfordi*) 12, 21, **198**, Plate 9; *see also* Hornbill, Sulawesi Tarictic
Sri Lankan Grey (*Ocyceros gingalensis*) 4, 6, 11, 20, 38, 47, 49, 64, 67, 153, **155–6**, Plates 2, 6, 14
Sulawesi Tarictic (*Penelopides exarhatus*) 6, 12, 15, 18, 21, 38, 47, 49, 64, 67, **198–9**, 216–17, Plates 9, 14
Sulawesi Wrinkled (*Aceros cassidix*) 7, 12, 15, 18, 21, 39, 47, 64, 67, 79, 198–9, 212, **215–18**, Plates 10, 14
Sulu (*Anthracoceros montani*) 6, 11, 15, 17, 20, 38, 47, 64, 67, 79, 160, **175–6**, Plates 7, 14
Sumba Wreathed (*Aceros everetti*) 7, 12, 15, 18, 21, 39, 47, 64, 67, 78, 208, 224, **239–40**, Plates 11, 14
Sunda Pied (*Anthracoceros albirostris convexus*) 11, 20, 164, **166–8**, 170, 195, Plate 7; *see also* Hornbill, Oriental Pied
Sunda Wrinkled (*Aceros corrugatus*) 7, 12, 15, 18, 21, 39, 47, 64, 67, 79, **218–21**, Plates 10, 14
Tarictic, *see under* Hornbill: Luzon Tarictic; Mindanao Tarictic; Mindoro Tarictic; Ticao Tarictic; Visayan Tarictic
Ticao Tarictic (*Penelopides panini ticaensis*) 12, 21, **200–1**, Plate 9; *see also* Hornbill, Visayan Tarictic
Tickell's Brown (*Anorrhinus tickelli*) 6, 11, 15, 17, 20, 38, 47, 49, 64, 66, 101, **104–5**, Plates 3, 14
Trumpeter (*Ceratogymna bucinator*) 7, 12, 16, 18, 21, 39, 45, 47, 49, 62, 63, 64, 67, 69–70, 243, **245–8**, 257–8, Plates 12, 13
Upper Guinea Brown-cheeked (*Ceratogymna cylindricus cylindricus*) 12, 21, **249**, Plates 12, 13; *see also* Hornbill, Brown-cheeked
Upper Guinea Dwarf Black (*Tockus hartlaubi hartlaubi*) 11, 20, **145**, Plate 6; *see also* Hornbill, Dwarf Black
Upper Guinea Grey-cheeked (*Ceratogymna subcylindricus subcylindricus*) 12, 20, **252**, Plate 12; *see also* Hornbill, Grey-cheeked
Upper Guinea Pied (*Tockus fasciatus semifasciatus*) 11, 20, **117–19**, Plates 4, 13; *see also* Hornbill, African Pied
Upper Guinea Piping (*Ceratogymna fistulator fistulator*) 12, 21, **242**, Plates 12, 13; *see also* Hornbill, Piping
Visayan Tarictic (*Penelopides panini*) 6, 12, 15, 18, 21, 38, 47, 49, 64, 67, 79, 197, **200–2**, 223, Plates 9, 14
Visayan Wrinkled (*Aceros waldeni*) 7, 12, 21, 39, 47, 64, 67, 79, 200, **223–4**, Plates 10, 14

Hornbill cont.
  Von der Decken's (*Tockus deckeni*) 4, 6, 11, 15, 17, 20, 31, 38, 48, 64, 66, 69, 79, 127, 133, 140–1, **142–4**, Plates 5, 13
  Wattled, *see* Hornbill, Black-casqued Wattled
  Western Guinea Long-tailed (*Tockus albocristatus albocristatus*) 11, 20, **149**, Plates 6, 13; *see also* Hornbill, Long-tailed
  Western Pale-billed (*Tockus pallidirostris pallidirostris*) 11, 20, **121**, Plate 4; *see also* Hornbill, Pale-billed
  Whistling, *see* Hornbill, Piping
  White-crested, *see* Hornbill, Long-tailed; Hornbill, White-crowned
  White-crowned (*Aceros comatus*) 7, 12, 15, 18, 21, 28, 29, 38, 40, 47, 49, 64, 67, 69–70, 73, 79, 100, 208–9, **210–12**, Plates 10, 14
  White-headed, *see* Hornbill, Mindanao Wrinkled
  White-tailed, *see* Hornbill, Piping
  White-thighed (*Ceratogymna cylindricus albotibialis*) 12, 20, **249–50**, Plates 12, 13; *see also* Hornbill, Brown-cheeked
  White-throated (Brown-backed), *see* Hornbill, Austen's Brown
  Wreathed, *see* Hornbill, Bar-pouched Wreathed
  Wrinkled, *see* Hornbill, Sunda Wrinkled; Hornbill, Mindanao Wrinkled
  Writhe-billed, *see* Hornbill, Mindanao Wrinkled; Hornbill, Visayan Wrinkled
  Yellow-billed, *see* Hornbill, Eastern Yellow-billed; Hornbill, Southern Yellow-billed
  Yellow-casqued, *see* Hornbill, Yellow-casqued Wattled
  Yellow-casqued Wattled (*Ceratogymna elata*) 4, 7, 12, 13, 16, 18, 21, 39, 47, 49, 64, 67, 73, 79, 241, 261, **263–5**, Plates 12, 13
  Zande, *see* Hornbill, African Pied
Horsfieldia 103, 108, 167, 181, 236
ho-ting 195
Hovenia 258
Hummingbird, Giant 13
hydrocorax, *Buceros*, *see* Buceros
hydrocorax, *Buceros hydrocorax* (Great Luzon Hornbill) 12, 21, **189**, Plate 8
Hymenoptera 118

immature, *see* age
incertae sedis (subgenus *Tockus*) 64, **145**, Plate 6
incubation period 47–8, 53–4; *see also* species accounts
India 5, 9, 40, 72–3, 100, 102–3, 153, 157, 162, 166, 180, 209, 213, 228, 231, 233
Indo-China 40
Indomalayan Region, *see* geography
Indonesia 9, 73, 78, 106, 167, 173, 180, 186, 194, 198, 210, 216, 219, 225–6, 239
information centre, *see* roost, communal
Ischnocera, *see* Philopteridae
Ixora 154, 229

jacamar, *see* Galbuliformes
jacksoni, *Tockus deckeni* (Jackson's Hornbill) 11, 20, 79, **142–4**, Plates 5, 13
Jasminum 167
Java 9, 72, 76
javanicus, *Paroncophorus* 67, 68
Julidae 168, 181
jungei, *Aceros plicatus* **225**
juvenile, *see* age

Kakapo 13
Kalimantan 40, 42
karyology 63, 65, 70–1
Kenya 91, 93, 95, 112, 120, 123, 125, 133, 141, 143, 245, 253, 258
'keystone' species 45, 77
kidney structure 39–40, 62, 89
Kigelia 98, 139
Kingfisher, Banded 168
  Giant 13
  see also Coraciiformes
Kirkia 98
Knema 103, 167, 181, 217, 236
Koel 229
Kokomo, *see* Hornbill, Papuan Wreathed
Koompasia 168, 170, 196
Koordersiodendron 217

Lacedo pulchella 168
Lagerstroemia 168, 182
Landolphia 247
Laniarius atrococcineus 134
Lannea 98, 126, 134
Lanraecae 108
Laos 102, 166, 180, 214
latifrons, *Bucerocophorus* 66, 68
Lauraceae 40, 41, 71, 181, 183, 218, 220
laying dates, estimated xv; *see also* species accounts

leadbeateri, *Bucorvus*, *see* Bucorvus
Leguminosae 108
length, perched xiv; *see also* species accounts
Lepidoptera 103, 129, 181
Lepisanthes 167, 181
Leptosomatidae, *see* Coraciiformes
Lesser Sundas 73
leucocephalus, *Aceros*, *see* Aceros
Leucocytozoon 192
leucomelas, *Tockus*, *see* Tockus
Liberia 118, 146, 148, 150, 244, 250, 261, 264
life cycle xv; *see also* species accounts
  expectancy, *see* age
Liquidambar 237
Lithocarpus 108, 181
Litsea 103, 167, 181, 188, 236
Livistona 217
logging, *see* conservation
Lombok 73
Lonchocarpus 126, 134
Lonchura 259
longevity, *see* age
longicuneatus, *Buceronirmus* 67, 68
Lophoceros 64, 66, 69, 110, **121**, Plate 4
lophocerus, *Chapinia* 66, 68, 127
Louisiana 220
Lovoa 244
Lower Guinea, *see* Guinea, forests
Luangwa valley 72
Lucanidae 168, 181, 236
Lychnodiscus 254
lydae, *Chapinia* 67, 68

Mabuya 164
Macaranga 244
Machilus 163–4
Macrolobium 263
macrosome, *see* karyology
macrourus, *Tockus albocristatus* (Eastern Guinea Long-tailed Hornbill) 11, 20, **150**, Plates 6, 13
Madagascar 63
Maesopsis 45, 254, 258
major, *Paroncophorus* 67, 68
Makassar Straight 73
Malawi 92, 112, 123, 125, 133, 138, 245, 258
malayanus, *Anthracoceros*, *see* Anthracoceros
Malay Archipelago 73
malayensis, *Chapinia* 67, 68
Malay Peninsula 40, 72; *see also* Malaysia; Peninsula, Malaysia
Malaysia 43, 77–8
Malesia 40, 73
Mali 93, 125, 133
Malimbus nigerrimus 251

Mallophaga, *see* parasites
Mallotus 164
management, *see* conservation; breeding, captive
manillae, Penelopides, see Penelopides
manillae, Penelopides manillae (Luzon Tarictic Hornbill) 12, 21, **202**, Plate 9
map, distribution xiv–xv; *see also* species accounts
marchei, Anthracoceros, see Anthracoceros
mass, body xiv, 6–7, 10–14, 20–1; *see also* species accounts
maturity, *see* age
Mauritania 93, 125, 133
measurements xiv, 89; *see also* species accounts
Megaceryle maxima 13
Megalaima 181
Melanoxylon 167
Meleagris gallopavo 13
Melia 163–4, 247
Meliaceae 107, 187, 227
Melierax 41
    metabates 97, 136, 139
    poliopterus 144
mendanae, Aceros plicatus **226**
Meniceros 153
Menoponidae 66–8; *see also* parasites, Mallophaga
Meropidae, *see* Coraciiformes
Merops nubicoides 97
metabolic rate 14–16
Micronisus gabar 136
microsome, *see* karyology
Mimusops 254
mindanensis, Buceros hydrocorax (Great Mindanao Hornbill) 12, 21, 189, **190–1**, 192, Plate 8
mindorensis, Penelopides, see Penelopides
Miocene 63, 72, 90
Moluccas 73, 226
Momotidae, *see* Coraciiformes
Monasa nigrifrons 13
Mongoose, Dwarf 41, 141, 143
Monodora 254
monophyly 61, 62–3; *see also* systematics
montani, Anthracoceros, see Anthracoceros
monteiri, Tockus, see Tockus
Morinda 118, 244, 251
Morocco 90
mortality 58–60; *see also* predation
Morus 254
motmot, *see* Coraciiformes
moucheti, Buceronirmus 67, 68
moult, *see* plumage
Mousebird, Red-backed 13
Mozambique 95, 112, 123, 125, 133, 138, 245, 258
muesebecki, Chapinia 67, 68

Mulleripicus pulverulentus 103, 169, 175
Musanga 118, 149, 244, 251, 254
Musophagiformes 13
Myanmar 79, 100, 102, 105, 166, 180, 194, 210, 213, 231, 233
Myrianthus 258
Myristica 174, 181, 227
Myristicaceae 107, 265

Namib Desert 3, 128
Namibia 72, 78, 95, 112, 115, 125, 128, 133, 138
narcondami, Aceros, see Aceros
Narcondam Island 9, 73, 227–9
nasal aperture 22
nasutus, Tockus, see Tockus
nasutus, Tockus nasutus (North African Grey Hornbill) 11, 20, **123**, Plate 4
neck, *see* anatomy
Needletail, Purple 13
Nepal 157, 166, 180, 213
Nephelium 237
nest, *see also* species accounts
    box 78, 83–4
    emergence from 57
    fauna 55
    lining 51–2
    preparation 50, 83–4
    provisioning 54, 56–7
    sanitation 45, 51, 54–5
    sealing, *see* behaviour, sealing
    site 46, 50, 51, 78, 84
nesting cycle xv; *see also* species accounts
nestling, development 55–7, 89; *see also* species accounts
    emergence 57
    period 47–8, 56
    survival 58–9
neumanni, Tockus pallidirostris (Eastern Pale-billed Hornbill) 11, 20, **121–3**, Plate 4
New Guinea 5, 14, 73, 209, 225
New York 237
Nicobar Islands 73, 228
Niger 93, 125, 133
Nigeria 93, 118, 125, 133, 146, 148, 150, 244, 250, 253, 261, 264
nipalensis, Aceros, see Aceros
Nunbird, Black-fronted 13

Ocotea 41, 259
Ocyceros 46, 65, 68, 79, **152–3**, 160–1, Plates 6, 14
    birostris (Indian Grey Hornbill) 4, 6, 11, 15, 17, 19, 20, 38, 47, 64, 67, 152–4, **157–9**, 161, Plates 6, 14

gingalensis (Sri Lankan Grey Hornbill) 6, 11, 20, 38, 47, 49, 64, 67, 153, **155–6**, Plates 2, 6, 14
griseus (Malabar Grey Hornbill) 6, 11, 15, 17, 20, 38, 48, 64, 67, **153–4**, 155, 157, 159, 161, Plates 6, 14
Odyendea 258
oil gland 26, 90; *see also* behaviour, maintenance
oil-palm 78, 113, 118, 244, 251, 262, 265
Olea 254
Olinia 113
Onchosperma 212
organization, *see also* species accounts
    social 46–8
    spatial 46–8, 49–50
Oriolus chinensis 167
Orthoptera 103, 147, 168, 181, 236
Ostrich 13
Otis tarda 13
Otus bakkamoena 181
outgroup 61; *see also* systematics
Ovaria 236
Owl, Barn 98
    Collared Scops 181
    Eurasian Eagle 13
    Milky Eagle 139
    Pel's Fishing 98
    Spotted Eagle 139
Owlet, Barred Jungle 181

pachycnemis, Bucerocophorus 67, 68
Pachypodantium 244
Pakistan 157
Palaquium 182, 217
Palawan Island 73, 170–2
pallidirostris, Tockus, see Tockus
pallidirostris, Tockus pallidirostris (Western Pale-billed Hornbill) 11, 20, **121**, Plate 4
Palmae 71, 108
panini, Penelopides, see Penelopides
panini, Penelopides panini (Panay Tarictic Hornbill) 12, 21, **200**, Plate 9
panting 27, 28
Papua New Guinea 226
parasites 5, 61; *see also* species accounts
    haematozoa 59
    Mallophaga 59, 65, 66–9
Parinarium 258–9
Parkia 195, 232
Paroncophorus 66–9
Passalidae 168, 181
Passeriformes 13
Passiflora 258
Patagona gigas 13

Peltophorum 139
Pemba 112
Penelopides 19, 46, 65, 68, 73, 79, 191, **197,** 202, 207, Plates 9, 14
  affinis (Mindanao Tarictic Hornbill) 6, 12, 21, 38, 47, 49, 64, 67, **204–6,** 222, Plates 9, 14
  exarhatus (Sulawesi; Tarictic Hornbill) 6, 12, 15, 18, 21, 38, 47, 49, 64, 67, **198–9,** 216–17, Plates 9, 14
  manillae (Luzon Tarictic Hornbill) 6, 12, 21, 38, 47, 49, 64, 67, 79, 197, **202–4,** Plates 9, 14
  mindorensis (Mindoro Tarictic Hornbill) 6, 12, 21, 38, 47, 49, 64, 67, 197, **207–8,** Plates 9, 14
  panini (Visayan Tarictic Hornbill) 6, 12, 15, 18, 21, 38, 47, 49, 64, 67, 79, 197, **200–2,** 223, Plates 9, 14
Penguin, Emperor 13
Peninsula, Arabian 109
  Malaysia 106, 166, 173, 180, 186, 194, 210, 219
pergriseus, Ocyceros birostris **157**
Pericrotus flammeus 174
Persea 181
Petersia 251
Pharomachrus mocinno 13
Phasmatidae 103, 181
Pheasant, Great Argus 194
Philippines 14, 72–3, 79, 171, 176, 191, 197, 200, 203, 206, 208, 222, 224
Philopteridae 66–8; *see also* parasites, Mallophaga
Phoeniculus damarensis 13
Phoeniculidae, *see* Upupiformes
phylogenetics, *see* systematics
Phymateus leprosus 113
Phytolacca 254
Piciformes 5, 13
Pigeon, African Green 259
  New Guinea Crowned 13
Piper 103, 167, 181, 236
Piptadenia 113
Piptadeniastrum 244, 251
Pithecophaga jeffreyi 14
plate tectonics 72–3; *see also* geography, biogeography
play, *see* behaviour
plicatus, Aceros, see Aceros
Pliocene 72
Ploceus 259
  bicolor 113
  cucullatus 251, 262
plumage, description xiv, 89; *see also* species accounts
  development 56
  moult 7, 34–5, 49, 52–3, 89
Podocarpus 98, 103

Polemaetus bellicosus 94, 136, 139
Polyalthia 103, 164, 167, 174, 181, 217, 236, 244, 251, 258, 262
Polyboroides typus 33, 127, 136, 139, 256
Polydesmidae 168, 181
Polyscias 247, 259
Pometia 227
population biology 8, 58–60, 75–6; *see also* species accounts
predation, animal 5, 32–3; *see also* species accounts; mortality
preen gland, *see* oil gland
Premna 135, 254
productus, Bucerocophagus 66, 68
progression 25–6, 89–90
proportions 14–19
Protoxerus stangeri 151
Pseudocadia 246
Pseudospondias 244, 254
Psidium 258
Psittaciformes 13
Psittacula krameri 159
Psychidae 103, 168
Pterocarpus 115, 246
Pterospermum 169, 217, 237
Ptilolaemus 65, 100–1, 104
puff-adder 59, 99, 136
puffbird, *see* Galbuliformes
pulchrirostris, Tockus camurus **148**
Pycanthus 118, 244, 251, 254, 262
Pycnonotidae 168, 181, 236

Quelea, Red-billed 134
Quelea quelea 134
Quetzal, Resplendent 13

Ramphastidae, *see* toucan
range xiv–xv; *see also* species accounts
Raphia 98, 262, 265
Raphus cucullatus 13
Rauvolfia 247, 254, 258
Raven, Eurasian 13
Reduviidae 168
regurgitation 45
relationships 8, 63–71; *see also* systematics
rhinoceros, Buceros, see Buceros
rhinoceros, Buceros rhinoceros (Malayan Great Rhinoceros Hornbill) 11, 21, **184,** Plate 8
Rhinoceros Avis 8
Rhinoplax 65, 177
Rhinopomastidae, *see* Upupiformes
Rhoicissus 113
Rhynchaceros 64, 66, **110,** 121, Plate 4
Rhynchophoridae 244
Rhyticeros 64, 65, 68, 208–9, 212, **224,** Plate 11
Ricinodendron 262

Rio Muni 118, 146, 148, 150, 244, 250, 261
robusta, Chapinia 67, 68
roller, *see* Coraciiformes
roost, *see also* species accounts
  communal 24, 43, 44
  site 24, 31–2, 43
Rourea 167, 236
Royena 113
ruficollis, Aceros plicatus **225,** 226
rufirostris, Tockus erythrorhynchus (South African Red-billed Hornbill) 11, 20, 79, **131–4,** 136, Plates 5, 13
rugosus, Aceros corrugatus **218**
Rutaceae 108
Rwanda 95, 112, 125, 148, 253

Sabah 106, 167, 173, 186, 194, 210, 219, 234
Sagittarius serpentarius 139
Sahul Shelf 73
salt gland 39
samarensis, Penelopides affinis (Samar Tarictic Hornbill) 12, 21, **205–6,** Plates 9, 14
sanfordi, Penelopides exarhatus (South Sulawesi Tarictic Hornbill) 12, 21, **198,** Plate 9
Sanft, Kurt 9
Santiria 187
Sapindaceae 108
Sapindus 108
Sapium 254, 258
Sarawak 106, 167, 173, 186, 194, 210, 219, 234
Saudi Arabia 125
Scarabidae 168, 181
Schleichera 164
Schotia 98, 113, 134, 247
Scincidae 103
Scissirostrum dubium 217
Sciurus tennuis 195
Sclerocarya 98, 126, 134, 139
Scoliidae 181
Scorodophloeus 244
Scotopelia peli 98
Scutia 113
sea-level 72
sealing, *see* behaviour, sealing
secondary area 17–18
  length 17–18
Secretarybird 139
seed-dispersal 44, 45, 77
seed predation 44
Semien Mountains 93
semifasciatus, Tockus fasciatus (Upper Guinea Pied Hornbill) 11, 20, **117–19,** Plates 4, 13
semigaleatus, Buceros hydrocorax (Great Samar Hornbill) 12, 21, 189, **190–1,** Plate 8
Senegal 93, 118, 125, 133, 244, 264

Sersalisia 258
shape 14–19
sharpii, Ceratogymna fistulator (Sharpe's Piping Hornbill) 12, 21, **242–4**, Plates 12, 13
Shorea 108, 187–8
Shrike, Crimson-breasted 134
Sideroxylon 113, 229
Sierra Leone 93, 118, 146, 148, 150, 244, 250, 264
Singapore 167, 182, 186, 197, 206, 219
size 10, 14–19
Sloanea 167, 237
Solomon Islands 5, 73, 209, 225–6
Somalia 93, 112, 120, 125, 133, 141, 143
somaliensis, Tockus flavirostris **140**
South Africa 9, 95, 112, 125, 133, 138, 245, 258
South Carolina 164
Southeast Asia 9, 41, 72–3, 75–8, 80
Sparrowhawk, Black 139, 244
species accounts xi–xiii, **87–265**
Sphaerotheridae 168, 181
Spizaetus 41, 107
  cirrhatus 220
  lanceolatus 217
  nipalensis 14
Sri Lanka 155–6, 162
status xiv–xv; *see also* species accounts
Staudia 244, 262
Stephanoaetus coronatus 14, 119, 256, 260, 263
Sterculia 98, 103, 108, 169, 187, 229, 254
Sterculiaceae 108
stomach lining 54
Strigiformes 13
Strigops habroptilus 13
Strombosia 103, 167, 181, 236, 244, 251, 262
Strombosiopsis 244, 251, 262
Struthio camelus 13
Struthioniformes 13
Strychnos 162–4, 247, 258
studbook 80
suahelicus, Tockus alboterminatus **111**
subcylindricus, Ceratogymna, see Ceratogymna
subcylindricus, Ceratogymna subcylindricus (Upper Guinea Grey-cheeked Hornbill) 12, 20, **252**, Plate 12
subgenus 64–6, 69; *see also Aceros; Baryrhynchodes; Berenicornis; Bycanistes; Ceratogymna; incertae sedis; Lophoceros; Rhynchaceros; Rhyticeros; Tockus*
subnigra, Penelopides manillae (Polillo Tarictic Hornbill) 12, 21, **202**, Plate 9

subquadratus, Ceratogymna subcylindricus (Lower Guinea Grey-cheeked Hornbill) 12, 21, **252–5**, Plate 12
subruficollis, Aceros, see Aceros
Sudan 93, 112, 118, 120, 125, 133, 141, 146, 148, 244, 250, 253, 258, 261
Sulawesi 73, 197–9, 215–18
Sulu Island 73, 175–6
  Sea 73
Sumatra 9, 72, 100, 106, 231
Sumba Island 73, 78, 239–40
sun bathing, *see* bathing
Sunda Shelf 14, 72–3
superspecies 91, 110, 127, 160
Swan, Trumpeter 13
swidden, *see* agriculture
Swietiana 163–4
sylvestris, Buceros rhinoceros (Javan Great Rhinoceros Hornbill) 12, 21, **185–6**, Plate 8
Symplocus 103, 167, 181
syndactyl 89
systematics 61–74, 89
  cladistics 61, 62, 64–5, 69–71
  phylogenetics 61, 62
Syzigium 164, 217, 247, 258

tail length xiv, 17–19, 90; *see also* species accounts
Tanzania 9, 95, 112, 123, 125, 133, 141, 143, 245, 253, 258
tarsus length xiv, 17–19, 90; *see also* species accounts
taurus, Buceronirmus 67, 68
taxonomy, *see* systematics
Teclea 254
Tenebrionidae 113
Terathopius ecaudatus 136, 139
Terminalia 98, 139, 162, 229, 265
Terpsiphone viridis 113
territory 46–8, 49–50
Tertiary 73
Tetrameles 103, 169, 182, 229
Tetrorchidium 118
Tettigoniidae 103
Texas 94
Thailand 9, 40, 78–9, 102, 105, 106, 166, 173, 180, 194, 210, 213, 219, 231, 233
Thevetia 158
thompsoni, Buceronirmus 67, 68
ticaensis, Penelopides panini (Ticao Tarictic Hornbill) 12, 21, **200–1**, Plate 9
ticehursti, Aceros undulatus 230, 233
tickelli, Anorrhinus, see Anorrhinus
Timor Sea 78
Tinamiformes 13
Tinamou, Solitary 13
Tinamus solitarius 13

Tockus 5, 10, 13, 16, 19, 25, 33, 34, 36, 41, 46, 59, 65, 68–70, 72, 78–9, 83, **109**, 152–3, Plates 4, 5, 6, 13
  albocristatus (Long-tailed Hornbill) 4, 6, 11, 15, 17, 20, 38, 47, 64, 66, 70, **149–52**, Plates 6, 13
  alboterminatus (African Crowned Hornbill) 6, 9, 11, 15, 17, 20, 30, 32, 37, 38, 40, 47, 59, 64, 66, **110–14**, 115, 118–20, 125, 247, Plates 4, 13
  bradfieldi (Bradfield's Hornbill) 6, 11, 15, 17, 20, 30, 38, 48, 64, 66, 72, 110, 112, **114–16**, 125, Plates 4, 13
  camurus (Dwarf Red-billed Hornbill) 4, 6, 11, 15, 17, 20, 38, 48, 49, 64, 66, 146, **147–9**, Plates 6, 13
  deckeni (Von der Decken's Hornbill) 4, 6, 11, 15, 17, 20, 31, 38, 48, 64, 66, 69, 79, 127, 133, 140–1, **142–4**, Plates 5, 13.
  erythrorhynchus (African Red-billed Hornbill) 6, 8, 11, 15, 17, 20, 31, 37, 38, 39, 40, 48, 54, 64, 66, 69, 72–3, 79, 113, 127–8, **131–6**, 137–8, 140–1, Plates 5, 13
  fasciatus (African Pied Hornbill) 6, 11, 15, 17, 20, 30, 38, 47, 64, 66, 110, 112, **116–19**, 146, 243, Plates 4, 13
  flavirostris (Eastern Yellow-billed Hornbill) 6, 11, 20, 31, 38, 39, 40, 48, 64, 66, 72, 79, 127, 133, 136, **140–2**, Plates 5, 13
  hartlaubi (Dwarf Black Hornbill) 6, 11, 13, 15, 17, 20, 38, 48, 64, 66, **145–7**, 148, Plates 6, 13
  hemprichii (Hemprich's Hornbill) 6, 11, 15, 17, 20, 30, 38, 47, 64, 66, 72, 110, **119–21**, Plates 4, 13
  leucomelas (Southern Yellow-billed Hornbill) 6, 11, 15, 17, 20, 31, 37, 38, 40, 42, 48, 53, 64, 66, 69, 72, 79, 127, 128, 133, **136–40**, 140–2, Plates 5, 13
  monteiri (Monteiro's Hornbill) 6, 11, 15, 17, 20, 31, 37, 38, 47, 54, 64, 66, 69, 72, 127, **128–31**, 133, 137, 138, Plates 1, 5, 13
  nasutus (African Grey Hornbill) 6, 11, 15, 17, 20, 25, 30, 37,

Tockus cont.
  38, 40, 48, 53, 58, 64, 66, 69, 72, 79, 112, 115, 119, 122–3, **123–7**, 128, 138–9, Plates 4, 13
  pallidirostris (Pale-billed Hornbill) 6, 11, 15, 17, 20, 30, 38, 47, 64, 66, 69, 72, 112, **121–3**, 125, Plates 4, 13
  subgenus 64, 66, 69, **127**, Plate 5
Todidae, *see* Coraciiformes
tody, *see* Coraciiformes
Togo 93, 118, 133, 146, 148, 150, 244, 250, 253, 261, 264
topography, hornbill xvi
toucan 5, 37
trabeculus, Buceronirmus 67, 68
Trachyphonus erythrocephalus 126
tragopan 8
traylori, Chapinia 67, 68
Treron calva 259
triage 74
Tricalysia 118, 244
Trichilia 41, 45, 113, 121, 138, 151, 244, 246, 251, 254, 262
Trichoscypha 244, 251
Tricosanthes 167
Trochiliformes 13
trogon, *see* Trogoniformes
Trogonidae, *see* Trogoniformes
Trogoniformes 7, 13, 63
Tropicranus 65, 70
Turaco, Great Blue 13
Turkey 13

Turniciformes 13
Turnix ocellata 13
Tyto alba 98

Uapaca 244, 258
Uganda 9, 42, 45, 91, 93, 112, 118, 120, 125, 133, 141, 143, 146, 148, 150, 244, 250, 253, 261
Ulmaceae 183, 237
undulatus, Aceros, see Aceros
Upper Guinea, *see* Guinea, forests
Upupidae, *see* Upupiformes
Upupiformes 7, 13, 55, 63, 89
Urostigma 187, 195, 203
Uvaria 103, 167
Uvariopsis 254

Vangueria 254
Varingia 226
Vaspris 113
vegetation, *see* geography
vertebrae, *see* anatomy, neck
Vespidae 168, 181
Vietnam 102, 166, 180, 210, 214
vigil, Buceros, see Buceros
Viperidae 103
virola 41
Viscum 244
Vitex 217
Voacanga 254
voice xiv; *see also* species accounts; communication, vocal

waldeni, Aceros, see Aceros
walking, *see* progression

Wallacean Region, *see* geography, Wallacean
waniti, Chapinia 66, 68
Washington 168
watsoni, Bucerocophorus 67, 68
weight, *see* mass
wenzeli, Chapinia 67, 68
West Africa, *see* Africa
wing area 14–19
  length xiv, 10–21; *see also* species accounts
  loading 14–16
wood-hoopoe, *see* Upupiformes
Wood-hoopoe, Violet 13
Woodpecker, Crimson-backed 156
  Great Slaty 103, 169, 175
  Imperial 13

Xylia 105
Xylopia 118, 244, 251, 262

Yemen 125

Zaïre 93, 95, 112, 118, 123, 125, 146, 148, 150, 244–5, 250, 253, 261
Zambia 95, 112, 115, 123, 125, 133, 138, 245, 258
Zanzibar 95, 112
Zimbabwe 95, 112, 115, 125, 133, 138, 245, 258
zonatus, Buceronirmus 67, 68
zoogeography, *see* geography